ISBN 978-1-332-30291-8
PIBN 10311448

1 MONTH OF
FREE
READING

at
www.ForgottenBooks.com

By purchasing this book you are eligible for one month membership to ForgottenBooks.com, giving you unlimited access to our entire collection of over 700,000 titles via our web site and mobile apps.

To claim your free month visit:
www.forgottenbooks.com/free311448

English
Français
Deutsche
Italiano
Español
Português

www.forgottenbooks.com

Mythology Photography **Fiction**
Fishing Christianity **Art** Cooking
Essays Buddhism Freemasonry
Medicine **Biology** Music **Ancient**
Egypt Evolution Carpentry Physics
Dance Geology **Mathematics** Fitness
Shakespeare **Folklore** Yoga Marketing
Confidence Immortality Biographies
Poetry **Psychology** Witchcraft
Electronics Chemistry History **Law**
Accounting **Philosophy** Anthropology
Alchemy Drama Quantum Mechanics
Atheism Sexual Health **Ancient History**
Entrepreneurship Languages Sport
Paleontology Needlework Islam
Metaphysics Investment Archaeology
Parenting Statistics Criminology
Motivational

JOURNAL

OF THE

ASIATIC SOCIETY OF BENGAL.

VOL. LXIX.

PART II. (NATURAL HISTORY, &c.).

(NOS. I TO IV.—1900.)

EDITED BY THE

NATURAL HISTORY SECRETARY.

"It will flourish, if naturalists, chemists, antiquaries, philologers, and men of science in different parts of *Asia*, will commit their observations to writing, and send them to the Asiatic Society at Calcutta. It will languish, if such communications shall be long intermitted; and it will die away, if they shall entirely cease." SIR WM. JONES.

CALCUTTA:

PUBLISHED BY THE ASIATIC SOCIETY,

57, PARK STREET.

1901.

BAPTIST MISSION PRESS, CALCUTTA.

Dates of Issue. Part II, 1900.

No.   I.—Containing pp. 1–88, was issued on 9th July, 1900.

„   II.—Containing pp. 89–278, and Plate I (of 1898) was issued on
       26th October, 1900.

„   III.—Containing pp. 279–456, was issued on 22nd November, 1900.

„   IV.—Containing pp. 457–489, was issued on 22nd January, 1901.

LIST OF CONTRIBUTORS.

JOURNAL

OF THE

ASIATIC SOCIETY OF BENGAL.

⸺•◉•⸺

Vol. LXIX. Part II.—NATURAL SCIENCE.

No. I.—1900.

I.—*Materials for a Flora of the Malayan Peninsula.*—*By* SIR GEORGE
KING, K.C.I.E., M.B., LL.D., F.R.S., &c., *late Superintendent of the
Royal Botanic Garden, Calcutta.*

No. 11.

After about two years of unavoidable delay, I am now able to offer
to the Society a further contribution towards the completion of these
Materials. The paper now submitted gives an account of the natural
order which, in the sequence hitherto followed, falls to be numbered
XLVII. An account of the family *Myrtaceæ* which ought, according
to that sequence, to have immediately preceded this one of *Melastomaceæ*
is now in preparation, and will I hope before long be ready for presen-
tation to the Society. The present paper is not entirely my own
work, the account of the genus *Sonerila* having been most kindly
prepared for me by Dr. O. Stapf, First Assistant in the Royal Herbarium,
Kew, whose contributions to the Botany of Borneo, published by the
Linnean Society of London, have already marked him as an authority
on the Malayan Flora.

Order XLVII. MELASTOMACEÆ.

Herbs or shrubs, more rarely trees or climbers. *Leaves* opposite
or rarely whorled, generally petioled, entire or nearly so, often
palmately 3–5–7-nerved from near the base to the apex (mostly pinnate-
veined in *Memecylon*); stipules 0. *Flowers* spiked, panicled or cymose,

J. II. 1

rarely solitary or fascicled, regular, hermaphrodite. *Calyx-tube* united by vertical walls to the ovary, rarely nearly free ; limb usually 4–5-(sometimes 3- or 6-)lobed, or truncate, rarely falling off in a cap. *Petals* as many as the calyx-lobes, contorted in bud, inserted on the margin of the calyx-limb. *Stamens* as many as or more than (frequently twice as many as) the petals, inserted with them ; alternate stamens often shorter, sometimes rudimentary, filaments bent inwards in the bud ; anthers opening at the summit by one or two pores, rarely by slits down the face ; connective often appendaged near the base by bristles, tubercles or a spur. *Ovary* 4–5- (rarely 3- or 6-)celled (in *Memecylon* 1-celled) ; style simple, filiform, rarely short ; ovules very many (except in *Memecylon*) ; placentas axile, parietal or free central. *Fruit* included in the calyx-tube, capsular or berried, breaking up irregularly or by slits through the top of its cells. *Seeds* minute, very many (in *Memecylon* one only) ; albumen 0 ; cotyledons short (or in some of the *Memecyleæ* long, thin and convolute).—DISTRIB. Species about 2700, tropical, with a few subtropical ; mostly in America, but many in south-east Asia, and a few in Africa and Polynesia.

SUBORDER I. MELASTOMEÆ. *Ovary* 3–6-celled. *Ovules* very many, on placentas radiating from the axis. *Seeds* very many. *Anthers* opening by a single terminal pore (rarely by 2).

Tribe I. OSBECKIEÆ. *Seeds* curved through half a circle, minutely punctate.
Stamens all nearly alike. Fruit a berry   ...   ...   1. OTANTHERA.
Stamens very unequal   ...   .   2. MELASTOMA.

Tribe II. OXYSPOREÆ. *Seeds* straight, oblong or cuneate, raphe often ex-current. *Ovary* with the vertex usually free, conical. *Petals* more than 3. *In-florescence* not scorpioid. *Fruit* capsular.

* *Inflorescence terminal (see also* Ochthocharis).

Panicles broad, often coloured   ...   .   3. OXYSPORA.
Panicles narrow. Flowers minute   ...   ...   4. ALLOMORPHIA.

** *Inflorescence axillary, stamens equal, flowers minute.*

Stamens 4 ...   .   5. BLASTUS.
Stamens 10   ...   .   6. OCHTHOCHARIS.
Stamens 8 ...   .   7. ANERINCLEISTUS.

Tribe III. SONERILEÆ. *Seeds* straight, oblong or cuneate, often angular, raphe sometimes excurrent. *Ovary* flattened or depressed at the vertex. *Fruit* capsular.

Petals 3. Inflorescence scorpioid   ...   ...   8. SONERILA.
Petals 4. Flowers in a long-peduncled head   ...   9. PHYLLAGATHIS.

Tribe IV. MEDINILLEÆ. *Seeds* straight, cuneate or obovate, often angular. *Connective* often appendaged near the base. *Fruit* baccate.

* *Longer stamens having two long bristles attached to the base of the connective in front.*

Cymes lateral   ...   . 10. MARUMIA.
Panicles terminal   ...   ... 11. DISSOCHÆTA.

** *Longer stamens having the connective at base variously appendaged but not with two long bristles in front (equal and with two short bristles in one species.)*

Four vertical plates on the ovary	12.	ANPLECTRUM.

*** *Stamens equal or nearly so.*

Connective not at all or very shortly produced at the base, but having 2 tubercles in front and a spur behind ...	13.	MEDINILLA.
Anthers not produced at the base; connective with tufted beard at the base behind but not spurred	14.	POGONANTHERA.
Connective not produced at the base and with no appendages in front, but minutely spurred behind, not bearded	15.	PACHYCENTRIA.

SUBORDER II. ASTRONIEÆ. *Ovary* 4–5-celled; *ovules very many on parietal nearly basal placentas. Seeds very many. Anthers short, opening by slits. Fruit baccate.*

Calyx-tube smooth. Panicles terminal	16.	ASTRONIA.
Calyx-tube verrucose. Cymes small ...	17.	PTERNANDRA.

SUBORDER III. MEMECYLEÆ. *Ovary* 1-celled; *ovules about 9 on a basal short free-central placenta. Stamens equal; anthers short, opening by pores or short slits. Berry* 1-seeded.—*Leaves usually pinnate-nerved.*

Stamens 8	18.	MEMECYLON.

1. OTANTHERA, Blume.

Slender shrubs. *Leaves* membranous, entire, 3–5-nerved. *Panicles* terminal or sub-terminal, lax, cymose, few-flowered. *Calyx* bearing on its ovoid tube simple or tufted bristles, rarely glabrous, the lobes 5 or 6, equal to or shorter than the tube, deciduous. *Petals* 5–6, obovate, the apex rounded or acute. *Stamens* subequal, 10–12; anthers opening by a terminal pore, not beaked, the connective not produced at the base, or slightly biauriculate in front. *Ovary* half-inferior, with 5 or 6 bristles at its apex, 5–6-celled; style filiform, simple; ovules numerous on axile placentas. *Fruit* 5–6-celled, baccate. *Seeds* small, curved, punctate. DISTRIB. Species 7 or 8; Malaya, Burma, Philippines, N. Australia.

Calyx-tube bearing simple bristles	1.	O. celebica.
Calyx-tube quite glabrous	2.	O. nicobarensis.

1. OTANTHERA CELEBICA, Blume, Mus. Bot. Lugd. Bat. I, 56. *Leaves* narrowly elliptic, shortly acuminate: upper surface sparsely strigose, the nerves glabrous, lower usually glabrous, the main nerves (and sometimes the minor also) with sub-adpressed bristles : length 2·25–3·5 in., breadth ·5–1·5 in., petioles ·2–·4 in. long. *Calyx-tube* with simple acuminate bristles often bulbous at the base, the teeth setose. Naud., Ann. Sc. Nat. Ser. 3, XIII, 353; Cogn. in DC. Mon. Phan. VII, 342.

ANDAMAN ISLANDS; common, *King's Collectors.* DISTRIB. Celebes.

2. OTANTHERA NICOBARENSIS, Teysm. et Binn. Pl. Nov. Hort. Bogor. 29. *Leaves* lanceolate or oblong-ovate, acuminate, upper surface sparsely strigose, the nerves glabrous; lower glabrous except the strigose nerves; length 3 to 4·5 in., breadth 1·25 to 2 in., petioles ·5 to 1 in. *Calyx-tube* quite glabrous; the teeth sub-ciliate. C. B. Clarke in Hook. fil. Fl. Br. Ind. II, 522; Kurz, Journ. As. Soc. Bengal, 1876, pt. II, 131; Cogn. in DC. Mon. Phan. VII, 342.

NICOBAR ISLANDS; *Novara Expedition.*

This has been collected only by the botanists of the Austrian expedition. It is the only species of the genus with a glabrous calyx.

2. MELASTOMA, Linn.

Scaly, strigose or villous shrubs. *Leaves* petioled, oblong or lanceolate, entire, 3–7-nerved. *Flowers* terminal, solitary, clustered or panicled, rose or purple, 5- (rarely 6–7-) merous. *Calyx-tube* ovoid or campanulate, with simple (rarely with penicillate) hairs, lobes deciduous. *Petals* equal in number to the calyx-lobes. *Stamens* twice as many as the petals, very unequal, alternate longer ones with purple anthers having the connective long-produced at the base and terminating in two lobes, the shorter ones having yellow anthers, the connective not produced but with two tubercles in front. *Ovary* ovoid, more or less united to the calyx-tube, 5- (rarely 6–7-) celled, apex bearing bristles; style filiform, simple; ovules very numerous, placentas axile. *Fruit* berried, succulent or coriaceous, enveloped in the calyx-tube, bursting irregularly. *Seeds* minute, very many, curved through half a circle, minutely punctate.—DISTRIB. Species 40; Asia, North Australia and Polynesia.

Calyx-tube densely covered with long, flexuose, spreading, shining, coloured bristly hairs ·25 to ·5 in. long; young branches, petioles, and pedicels with stiff spreading bristles; flowers 2 to 3 in. in diam.   ...   1. *M. decemfidum.*

Calyx-tube, young branches, petioles and pedicels with adpressed, lanceolate acuminate, serrulate scales: cymes condensed, flowers 1·25 in. across   ...   ...   2. *M. imbricatum.*

Calyx-tube densely clothed with linear acuminate, entire or serrate scales, those of the young branches, petioles and pedicels shorter and broader; corymbs with deciduous (often large) bracts; flowers 2 to 3 in. across (only 1·25 in var. *perakensis*) ...   .   3. *M. malabathricum.*

1. MELASTOMA DECEMFIDUM, Roxb. Hort. Beng. 90; Fl. Ind. II, 406. A shrub, 5 to 10 feet high; young branches at the nodes and near the tips, petioles and pedicels with numerous stiff, spreading bristles. *Leaves* narrowly-lanceolate or oblong-lanceolate, acuminate, 3- to 5-nerved; the upper surface smooth except for a few scattered

adpressed hairs, the lower glabrous, glandular-punctate ; length 2·5 to 5 in., breadth ·6 to 1·8 in. ; petiole ·25 to ·5 in. long. *Flowers* solitary or two or three together, bright purple, 2 to 3 in. in diam., the bracts few, short, acuminate ; pedicels under ·5 in. *Calyx-tube* densely covered with long, flexuose, shining, coloured, bristly hairs ; ·25 to ·5 in. long, the teeth rather shorter than the tube, ovate-lanceolate, acuminate, glabrous inside, deciduous. *Fruit* ·75 in. in diam., truncate at the apex, ovoid to ovoid-globular, ·5 in. in diam. Jack in Trans. Linn. Soc. XIV, (1822) 6 ; DC. Prodr. III, 146 ; Naud., Ann. Sc. Nat. Ser. 3, XIII, 282 ; Bl. Mus. Bot. I, 55 ; Cogn. in DC. Mon. Phan. VII, 345. *M. sanguineum,* Sims in Bot. Mag. t. 2241 ; DC. Prodr. III, 145 ; Don in Mem. Wern. Soc. IV, 289 ; Miq. Fl. Ind. Bat. I, pt. I, 504 ; Naud. l.c. 281 ; Triana in Trans. Linn. Soc. 60 ; C. B. Clarke in Hook. fil. Fl. Br. Ind. II, 524 ; Hemsl. in Jour. Linn. Soc. XXIII, 300. *M. malabathrica,* Sims in Bot. Mag. (not of Linn) 529. *M. Gaudichaudianum,* Naud. l.c. 278. *M. macrocarpum,* Naud. l.c. 281 (not of Don). *M. porphyreum,* Bipp. et Bl. iu Flora 1831, II, 487. *M. pedicellatum,* Naud. l.c. 280 ; Cogn. in DC. Mon. Phan. VII, 346.

PENANG ; *Wallich* 4042 ! *King, Curtis* 683. MALACCA ; *Griffith* (K.D.) 2245/1 ! *Maingay* 773, 774 ! *Ridley* 3228. PAHANG ; *Ridley* 2667. KEDAH ; *Ridley* 5211. PERAK ; *Scortechini* 2123 ! *King's Collector* 1540, 1853, 8754. DISTRIB. Burma, China, Hongkong, Tonkin.

VAR. *mollis,* Clarke in Hook. fil. Fl. Br. Ind. II, 524. Young branches very hispid throughout their whole length ; leaves with numerous sub-adpressed and spreading stiff hairs on both surfaces, especially on the nerves. Vidal Syn. Pl. Filip. t. 51 ; fig. D. *M. molle,* Wall. Cat. 4046 ; Triana in Linn. Trans. XXVIII, 60 ; Cogn. in DC. Mon. Phan. VII, 346. *M. crinitum,* Naud. l.c. 524. *M. malabathricum* Blanco, Fl. Filip. Ed. III, tab. 152 (not of Linn.).

SINGAPORE ; *Wallich.* DISTRIB. Luzon ; *Cuming* 853.

2. MELASTOMA IMBRICATUM, Wall. Cat. 4047. A spreading shrub, 5 to 15 feet high ; young branches (especially at the nodes and tips), petioles, pedicels and calyx covered with adpressed, lanceolate acuminate, serrulate scales. *Leaves* elliptic-oblong to ovate-oblong, rather abruptly narrowed at the base, the apex acute or acuminate ; main nerves 5 (the two lateral slender) ; both surfaces strigose, the main nerves on the lower with adpressed scales ; length 3·5 to 8·5 in., breadth 1·5 to 4·5 in., petiole ·35 to 1·35 in. *Cymes* solitary, terminal, condensed, 1·5 to 2 in. in diam., enclosed in bud by deciduous, lanceolate bracts and bracteoles. *Flowers* 7 to 15, 1·25 in. in diam.; the pedicels ·15 in. long. *Calyx-tube* urceolate, the teeth shorter than the tube, lanceolate-acuminate, deciduous. *Petals* oblong. *Anthers* short. *Fruit* ovoid-globose or sub-globose, truncate and shortly toothed at the mouth,

pulpy, ·8 in. in diam. when ripe.   Triana in Trans. Linn. Soc. XXVIII,
60 ; C. B. Clarke in Hook. fil. Fl. Br. Ind. II, 524 ; Cogn. in DC. Mon.
Phan. VII, 355.   *M. obovatum* var. *oblongum*, Bl. ex Triana l.c. 60.

PERAK ; *Scortechini* 162 ! *King's Collector* 444, 6023, 8696, 10946 !
Wray 2980.   PENANG ; *Wallich, Curtis.* DISTRIB. Sumatra, *Forbes* 2072 !
Brit. India (Khasia and Assam) ; Tonquin.

Easily recognised by its condensed cymes and rather small flowers.

3.   MELASTOMA MALABATHRICUM, Linn. Sp. Pl. 559.   A spreading
shrub, 3 to 6 feet high ; young branches, petioles and pedicels densely
clothed with rather short, acute to acuminate, often serrulate scales.
Leaves ovate-lanceolate, ovate-oblong or elliptic, the apex acute or shortly
acuminate, petioles short, 3- to 5-nerved (the marginal pair when
present slender) ; both surfaces strigose, the hairs sparser on the upper
and pale ; on the lower the hairs more numerous (especially on the
minor nerves) and darker, the main nerves clothed with broad-based
acuminate scales : length 2 to 5·5 in., breadth ·75 to 2·5 in. ; petioles
·2 to ·5 in. long.   *Corymbs* terminal, few-flowered, enveloped in bud by
large, deciduous, ovate-cordate bracts ; *flowers* 2 to 3 in. across, the
pedicels ·2 to ·4 in. long.   *Calyx*; the tube cylindric-campanulate,
densely clothed externally with linear acuminate, entire or serrate, pale
scales, almost glabrous within ; the teeth shorter than the tube (rarely
equal to it), acute or abruptly acuminate, the apex deciduous, scaly only
near the midribs.   *Flowers* 2 to 3 in. across.   *Petals* purple.   *Fruit* sub-
globular, truncate, pulpy, ·25 in. in diam. when dry.   DC. Prodr. III,
145 ; Roxb. Hort. Beng. 33 ; Fl. Ind. II, 405 ; Wall. Cat. 4040 ; Bl. Bijdr.
1076; Bot. Reg. t. 672; W. and A. Prodr. 324; Wight Ill. t. 95 ; ◀
Dalz. and Gibs., Bomb. Fl. 92 ; Naud. in Ann. Sc. Nat. Ser. 3, XIII, 285 ;
Thwaites Enum. 106 (*a* and *β*); Benth. Fl. Aust. III, 293 ; C. B.
Clarke in Hook. fil. Fl. Br. Ind. II, 523 ; Kurz, For. Fl. I, 503, not of
Miq. Fl. Ind. Bat. I, pt. I, 507 ; Naud. in Ann. Sc. Nat. Ser. 3, XIII,
273 ; Cogn. in DC. Mon. Phan. VII, 349.   *M. affine*, D. Don in Mem.
Wern. Soc. IV, 288 ; DC. Prodr. III, 145.   *M. obvolutum*, Jack in Trans.
Linn. Soc. XIV, 3 ; Cogn. in DC. Mon. Phan. VII, 348.   *M. articulatum*,
M. heterostegium, M. novæ-hollandiæ and *M. sechellarum*, Naud. in Ann.
Sc. Nat. Ser. 3, XIII, 285, 286 and 290.   *M. velutinum*, Seem. Fl. Vit.
90.   *M. Banksii*, Cunn. ex Triana.   *Trembleya rhinanthera*, Griff. Not.
IV, 677·

In all the provinces.   DISTRIB.   British India and Malayan
Archipelago, W. China, Seychelle Islands, N. Caledonia, N. Australia.

A widely distributed species varying in reality very little in localities widely
separated. The differences have however been taken as the bases of many bad and
doubtful species. In his *Flora Australiensis*, Mr. Bentham remarks (and apparently
with justice) that the whole twenty-four species described by Naudin in Ann. Sc.

Nat. Ser. 3, XIII, pp. 283 to 293; should be reduced here. It is not without reluctance that I give four varieties as follows:—

VAR. 1. *polyantha*, Benth. Fl. Aust. III, 292. Bracts of inflorescence small and very early deciduous or altogether absent; teeth of calyx usually short; leaves not exceeding 2·5 in. in length. *M. polyanthum*, Blume in Flora for 1831, 480; Mus. Bot. I, 52, t. 6; Naudin in Ann. Sc. Nat. Ser. 3, XIII, 287; Miq. Fl. Ind. Bat. I, pt. I, 502; Triana in Linn. Trans. XXVII, 59; C. B. Clarke in Hook. fil. Fl. Br. Ind. III, 523. *M. brachyodon*, Naud. l.c. 292; Miq. Fl. Ind. Bat. I, 570, t. 8, fig. A. *M. malabathricum*, Desr. in Lam. Encyc. Bot. IV, 36; Ill. Gen. tab. 361, f. 1; Jack in Linn. Trans. XIV, 4, fig. 1, *a* to *g*; Poir. Dict. IV, 37; Bl. Bijdr. 1070. *M. erecta*, Jack l.c. 5; DC. Prodr. III, 145. *M. tidorense*, Bl. in Flora 1831, p. 482, Miq. l.c. 514. *M. Royenii*, Bl. l.c. 483. *M. tondanense*, Bl. Mus. Bot. I, 54. *M. Hombronianum*, *M. oliganthum* and *M. microphyllum*, Naud. l.c. 278, 292 and 293.

ANDAMANS; *King's Collector.* SINGAPORE; *Anderson, King.* MALACCA; *King.* PENANG; *Curtis, King.* DISTRIB. Burma, N. China, Sumatra, Java and other islands in the Malay Archipelago, Luzon, N. Australia.

VAR. 2. *adpressa*, C. B. Clarke in Hook. fil. Fl. Br. Ind. II, 523. Leaves narrowly oblong-lanceolate, smaller than in the typical plant and with harsher pubescence. *M. adpressum*, Benth. in Wall. Cat. 4081, Naud. l.c. XIII, 27; Cogn. in DC. Mon. Phan. VII, 349. *M. anophanthum*, Naud. l.c. 277.

MALACCA; *Maingay* 771 to 773 in Herb., Kew. PENANG; *Wallich, King, Curtis.* PROVINCE WELLESLEY; *Curtis.*

This is scarely worthy of separation as a variety. It is slightly smaller and more strigose than the type and differs from VAR. *polyanthum*, so far as I can see, only by the large size of the bracts of the inflorescence.

VAR. 3. *normalis*, King. Hairs of both surfaces very numerous, those of the upper sub-adpressed, of the lower sub-spreading, softer (almost silky); calyx-teeth long, adpressed hairy within. *M. normale*, Don Prodr. Fl. Nep. 220; DC. Prodr. III, 145; Naud. in Ann. Sc. Nat. Ser. 3, XIII, 289; Kurz, For. Flora I, 504; C. B. Clarke in Hook. fil. Fl. Br. Ind. II, 524; Triana in DC. Mon. Phan. VIII, 352. *M. Wallichii*, DC. l.c. 146. *M. napalense*, Lodd. Bot. Cab. t. 707. *M. pelagicum*, Naud. l.c. 279. *M. longifolium*, Naud. l.c. 293.

In most of the provinces.

VAR. 4. *perakensis*, King. Leaves more or less broadly elliptic, 5- to 7-nerved, the upper surfaces strigose (sometimes nearly glabrous in old specimens), the lower softly and rather densely pubescent; branches, petioles and pedicels with long, spreading, rather soft hairs; calyx-tube densely clothed with long, flexuose, soft, often ciliate setae, the lobes large, adpressed strigose on both surfaces : flowers occasionally only 1·25 in. in diam.

PERAK ; *Ridley* 2935 ! *Curtis* 1298 ! *Wray* 1733, 1883 ! *King's Collector* 2173, 2091, 8463 ! *Scortechini* 780. SINGAPORE ; *Hullet* 5728. SELANGORE ; *Ridley* 1996. DISTRIB. Java ; *Forbes* 1142a.

This has broader leaves more softly hairy than VAR. *normalis.* The bristles of the calyx are much longer than in any other form of *M. malabathricum* and approach in number, length and density those of *M. sanguineum,* Don. A form of this from Perak, with the calyx-hairs shorter than the type, connects it with *M. imbricatum,* Wall.

3. OXYSPORA, DC.

Large spreading shrubs with drooping branches terminated by large, lax, almost naked panicles of rose-purple flowers. *Leaves* opposite, long-petioled, large, 5- to 7-nerved, ovate, acuminate. *Panicle* long, lax, sometimes narrow, the branches decussate, the flowers on the branchlets not glomerulate, bracts very small. *Calyx-tube* ovate, cylindric or funnel-shaped, its teeth 4, short, triangular. *Stamens* 8, four large with elongated anthers and four small, or all equal, opening by a single apical pore ; the base produced and bilobed, the connective with or without an appendage. *Ovary* inferior, 4-celled, its apex glabrous ; style simple, elongate ; ovules numerous ; the placentas axile, radiating. *Capsule* dry, elongate, double fusiform, with 8 ribs. *Seeds* numerous, falcate ; the raphe lateral, produced at the apex into a point in front. DISTRIB. Seven species, Indian and Malayan.

Anthers dissimilar ; petioles not winged	...	... 1.	*O. stellulata.*
Anthers similar :—			
Petioles not winged		... 2.	*O. acutangula.*
„ winged	...	... 3.	*O. Curtisii.*

NOTE.

The genera *Allomorphia* and *Oxyspora* were so difficult of separation even before the discovery of the new species herein described (*viz., A. alata,* Scort., *O. acutangula* and *O. Curtisii*) that Baillon (*Hist. des Plantes* VII, 48) united them. In Baillon's time *Oxyspora* was distinguished mainly by having four of its eight stamens much larger than and differently coloured from the other four. The three older species (all British Indian) *O. paniculata, O. vagans,* and *O. cernua* and the new Malayan one *O. stellulata* have this character, which would form an excellent head-mark for the genus if it did not break down. The character, however, does break down, for in the two Malayan plants here published as *O. acutangula* and *O. Curtisii* the eight anthers are all equal, although in all other respects these plants have the facies of the older species of *Oxyspora.* I have referred these to *Oxyspora* as preferable to the alternative course of putting them into *Allomorphia,* and I have therefore, in order to admit them, modified the generic character of *Oxyspora* as regards anthers. *Oxyspora,* as here defined, thus depends for its separation as a genus on its open paniculate inflorescence and long double fusiform boldly-ridged capsules, while *Allomorphia* is characterised by shortly-branched panicles, on the ultimate branchlets of which the flowers are clustered in pseudo-glomeruli, while the capsules are not much longer than broad and are often urn-shaped.

1. OXYSPORA STELLULATA, King, n. sp.   A shrub, 15 to 20 feet high ; young branches, petioles, inflorescence and calyx-tube covered with pale, minute, stellate-hairy scales. *Leaves* somewhat unequal, ovate, slightly cordate at the rounded base, the apex shortly apiculate, 7-nerved (the middle nerve very strong) ; both surfaces free from hairs, but (especially the upper) with numerous minute hairy scales most numerous on the nerves, the transverse veins on the lower very distinct and straight ; length 5 to 9 in. ; breadth 2·25 to 4·5 in. ; petiole 1·25 to 4 in. *Panicle* solitary, terminal, longer than the leaves ; the branches in pairs, divaricate ; flowers in umbels of 4 to 6 on the ultimate branchlets. *Calyx-tube* funnel-shaped, sub-tetragonous, the mouth with four small triangular teeth, narrowed at the base into the short pedicel. *Petals* 4, orbicular-ovate, blunt, glabrous. *Stamens* 8, very unequal, four linear, purple and twice as long as the other 4 short, yellow. *Disc* of 4 incurved plates. *Ovary* 4-winged. *Capsule* clavate, opening by 4 broad truncate valves ; seeds minute, shortly beaked.

PERAK ; *Scortechini* 249 *in part ! King's Collector* 418, 2851 ! *Wray* 1224. DISTRIB. Sumatra, *Forbes* 3034.

2. OXYSPORA ACUTANGULA, King, n. sp.   A bush ; young branches boldly 4-angled, glabrous. *Leaves* elliptic-oblong, somewhat narrowed to the rounded base, the apex acute, 5-nerved ; upper surface glabrous, the lower minutely lepidote-pubescent on the nerves and veins ; length 4 to 6 in., breadth 1·75 to 2·5 in., petiole ·6 to ·75 in., minutely lepidote. *Panicle* solitary, terminal, usually somewhat shorter than the leaves, broadly pyramidal, lepidote-puberulous, many-flowered. *Flowers* (including the stamens) ·5 in. long. *Calyx-tube* funnel-shaped, minutely rufous-stellate lepidote, the mouth with 4 small triangular teeth. *Petals* 4, glabrous, broadly ovate, blunt. *Stamens* 7 or 8, equal, lanceolate, acuminate, slightly curved, the lobes at the base short, rounded, appendages none, but a grooved, narrow process on the back of the connective in the lower half. *Capsule* ovoid, narrowed to the equally long pedicel, 8-ridged, glabrous, the mouth with an everted rim, ·25 in. long.

PERAK ; *Wray* 329.

3. OXYSPORA CURTISII, King. A shrub ; branchlets bluntly 4-angled not winged, puberulous. *Leaves* ovate-acuminate, the base rounded, not passing into the petiole, 7-nerved, upper surface sparsely strigose, minutely lepidote, the lower glabrous, the edges glandular-serrulate ; length 5 to 7 in., breadth 3 to 4 in., petioles 1·5 to 2·5, broadly winged, the wing much expanded at the base and joining that of the opposite leaf. *Panicle* solitary, terminal, spreading, longer than the leaves, with

J. II. 2

numerous 4-angled branches, bracteate, everywhere minutely lepidote; the larger branches bearing a pair of very unequal bracts near the base exactly like the stem-leaves but smaller. *Flowers* on slender pedicels as long as the calyx-tube. *Calyx-tube* cylindric, 8-ribbed, tapering to base and apex; the mouth expanded and truncate but with 4 minute teeth. *Petals* 4, as long as the scaberulous filaments. *Stamens* 8, much exserted; the anthers longer than the filaments and longer than the calyx-tube, linear, acuminate, very slightly lobed at the base and inappendiculate. *Capsules* fusiform, ·2 in. long, much expanded at the mouth.

PERAK; *Curtis* 1300.

The only specimen of this which I have seen is in Mr. Curtis's Herbarium.

The species resembles *Allomorphia alata*, Scort. in its eight equal stamens, in the shape of its leaves and in its winged petioles; but differs in its more elongated capsules (which are fusiform instead of globose), in its larger laxer panicles and in its conspicuously bristle-serrate leaves. The two species just described form very marked connecting links between the genera *Allomorphia* and *Oxyspora.*

4. ALLOMORPHIA, Blume.

Shrubs, tall or short. *Leaves* opposite, long-petioled, large, lanceolate, ovate or orbicular, nerved, glabrous or nearly so. *Panicles* terminal, compound, with small flowers in clustered whorls. *Calyx-tube* funnel-shaped or campanulate, limb of 3 or 4 very short lobes. *Petals* 3 or 4, rose or white, small. *Stamens* 6 or 8, nearly equal; anthers attenuated at the top with one pore, cells long-produced and diverging at their bases; connective without appendage. *Ovary* 3- (rarely 4-) celled, enclosed by but nearly free from the calyx-tube; style filiform, simple; ovules very many, placentas axile, 2-fid. *Capsule* small, dry, not much longer than broad, usually urn-shaped, ribbed, opening at the top by 3–4 valves. *Seeds* very many, narrowly obtrapezoidal. DISTRIB. Species 5: whereof 4 are from the Malay Peninsula and its attached islands; 1 from Canton.

Petioles not winged :—
   Calyx-tube with 3 or 4 teeth; stamens unequal; anthers sagittate
     at base; capsule ribbed   ...   . 1. *A. exigua.*
   Calyx-tube without teeth; stamens equal; anthers only slightly
     lobed at the base; capsules not ribbed   ...   ... 2. *A. Wrayi.*
Petioles winged :—
   Branches of the panicle winged; capsule sub-globular   ... 3. *A. alata.*

1. ALLOMORPHIA EXIGUA, Blume in Flora 1831, II, 523. A shrub, 2 to 10 feet high; branches slender, subangular, puberulous or glabrous. *Leaves* elliptic-ovate to ovate-lanceolate, acuminate, those of the same pair often unequal in size; the base rounded or narrowed, not cordate;

upper surface very sparsely strigose (glabrous in old leaves) ; the lower glabrous, pale, the transverse nerves prominent; length 4·5 to 11 in.; breadth 1·5 to 5 in. ; petiole ·75 to 4 in. *Panicle* usually solitary, terminal, often longer than the leaves, lax, minutely rusty-pubescent ; bracts lanceolate or oblong, deciduous ; the *flowers* ·1 in. long, shortly pedicellate, in stalked umbels on the ultimate branchlets or in subsessile fascicles.   *Calyx* rusty-puberulous, with 3 or 4 short broad teeth. *Stamens* 6 or 8, alternately long and short, sagittate at the base and with a small linear appendage behind. *Petals* 3 or 4, rosy. *Capsule* less than ·1 in. long, boldly 6-ribbed.   Naud. in Ann. Sc. Nat. Ser. 3, XV, 310 ; Triana in Trans. Linn. Soc. XXVIII, 74 ; C. B. Clarke in Hook. fil. Fl. Br. Ind. II, 527 ; Cogn. in DC. Mon. Phan. VII, 464. *Melastoma exigua,* Jack in Trans. Linn. Soc. XIV, 10, tab. 1, fig. 2 ; DC. Prod. III, 149. *M. impuber,* Roxb. Fl. Ind. II, 405 ; Wall. Cat. 4048.   •

MALACCA ; *Griffith* 2263 & 4 (K.D.) ; *Maingay* 776 (K.D.); *Harvey.* PENANG ; *Wallich* 4048 ; *Curtis* 399 ; *Griffith* ; *King.* PERAK ; *Scortechini* 227, 383 ; *Wray* 160 ; *King's Collector* 450, 2302, 3106.   DISTRIB. Sumatra ; *Forbes* 3062.

There is some variety in this plant as regards size and inflorescence. A slender form, which never exceeds 2 or 3 feet in height and which has smaller leaves than the type, seems worthy of separation as a variety. It appears to have been so recognised by Wallich who, in distributing his Herbarium, distinguished it by the letter *a.* A less distinctly marked form is one in which the flowers are grouped on the branches of the inflorescence in dense almost sessile fascicles.

VAR. *minor,* King. Leaves narrowly elliptic, tapering much to base and apex, 2 to 4·5 in. long ; inflorescence slender, few-flowered, very lax : height only 2 to 3 feet.

PENANG ; *Wallich,* Cat. 4048a ; *Curtis* 73. PERAK ; *Scortechini* 1702 ; *Wray* 161, 3414 ; *King's Collector* 2302.

2. ALLOMORPHIA WRAYI, King, n. sp. A shrub, 2 to 4 feet high ; branches and petioles with flexuose, spreading, ferruginous hairs. *Leaves* 7-nerved, broadly ovate to rotund-ovate, the apex shortly and abruptly acuminate, the base narrowed, the edges obscurely and minutely bristle-toothed : upper surface very sparsely strigose or glabrous, often with minute, brown scales on the chief nerves ; lower glabrous ; length 5 to 9 in., breadth 3·5 to 5·5 in.; petiole 2·25 to 2·75 in. *Panicles* solitary, axillary, slightly longer than the petioles but much shorter than the leaves, glabrous or rusty puberulous towards the extremities : the branches short, spreading, rather condensed when young, few flowered.   *Calyx-tube* widely campanulate, minutely rusty-puberulous or almost glabrous, without ribs ; the mouth truncate, slighly waved but not toothed. *Petals* 4. *Stamens* 8, equal, the anthers lobed but not sagittate at the base, the back with a slight supra-basal appendage.

Capsule glabrous, sub-globular, slightly contracted below the sub-membranous truncate mouth, under ·1 in. in diam.

PERAK ; *Wray* 2483 ; *King's Collector* 2061, 2380, 2773 ; *Scortechini* 50, 425. PENANG ; *Curtis* 2008.

A species allied to *A. exigua,* Bl. but with pubescent branches and petioles, shorter and more contracted panicles, a wider, toothless calyx-tube, sub-globular capsules and equal anthers only slightly lobed at the base. In Mr. Wray's field-note on his specimens he remarks that the flowers are white and the leaves dark shining green above and crimson beneath.

3. ALLOMORPHIA ALATA, Scortechini Mss. A glabrous shrub, 3 to 6 feet high ; the branchlets broadly winged. *Leaves* ovate or elliptic-ovate, the apex shortly acuminate ; the base narrowed into the broadly-winged petiole, 5-nerved ; both surfaces glabrous, the edges shortly bristle-toothed ; length 5 to 10 in., breadth 2·5 to 4·5 in., petiole ·75 to 4 in. *Panicle* solitary, terminal, often nearly as long as the leaves, lax ; its branches diverging, 4-angled and 4-winged, very minutely rusty lepidote-puberulous. *Flowers* clustered in small umbels near the ends of the branches, the bases of the umbels sometimes with a ring of bracteoles and minute imperfect flowers. *Calyx-tube* cylindric-ovoid, the limb expanded and with 4 small, persistent teeth. *Petals* 4, twice as long as the calyx-teeth but shorter than the filaments. *Stamens* 8, equal, exserted, much longer than the calyx-tube. *Anthers* linear-acuminate, longer than the scaberulous filaments, very slightly lobed at the base and almost inappendiculate. *Capsules* sub-globular, 8-ribbed, glabrous, the mouth truncate, diam. ·1 in.

PERAK ; *Scortechini* 236 ; *Wray* 1327 ; *Curtis ; King's Collector* 572, 2047.

The leaves of the same pair differ, often considerably, in size.

5. BLASTUS, Lour.

Shrubs. *Leaves* membranous, petiolate, ovate- or oblong-lanceolate, acuminate, entire or sinuate-serrate, 3- to 5-nerved. *Flowers* small 4-merous, ebracteolate. *Calyx-tube* oblong-campanulate or shortly oblong ; the limb not expanded, truncate, minutely 4-lobed. *Petals* 4, ovate, obtuse, glabrous, convolute into a cone before expansion. *Stamens* 4, equal, the filaments thin : anther incurved subulate, opening by a single apical pore ; basal lobes divaricate, the connective inappendiculate. *Ovary* adhering to the calyx, 4-celled ; style filiform ; stigma punctiform. *Capsule* obovoid or sub-globular, slightly 4-grooved ; dehisching slowly by 4 valves. Seeds minute, numerous, irregularly recurved, reniform. DISTRIB. 3 species in Malaya, China, Cochin China and India.

BLASTUS COGNIAUXII, Stapf in Hook. Ic. Pl. t. 2311. A shrub, 6 to 10 feet high; young branches slender, as thick as a crow-quill, scaly, glabrous or puberulous. *Leaves* equal, oblong-lanceolate, narrowed at the base, the apex acuminate, 5-nerved (the marginal pair faint), the edges sub-entire; upper surface glabrous, with a few scattered, small glands; lower glandular-punctate, minutely furfuraceous on the nerves; length 4 to 7 in.; breadth 1 to 2·25 in., petiole ·3 to ·75 in. *Panicles* axillary or terminal, slender, much shorter than the leaves, glandular-scaly, the few divaricate branchlets bearing the flowers in dense terminal glomeruli of 6 to 9. *Calyx* shortly tubular, scaly, the teeth minute. *Anthers* narrowly ovate with many small yellow glands near the base. *Capsule* subglobular, truncate, ·1 in. in diam., subglabrous; seeds linear. *Ochthocaris parviflora*, Cogn. in DC. Mon. Phan. VII, 421.

PERAK; common. DISTRIB. Borneo.

This is closely allied to *B. cochinchinensis*, Lour., but differs notably in its paniculate inflorescence.

6. OCHTHOCHARIS, Blume.

Small, erect, glabrous shrubs, branches round or obtusely 4-angled. *Leaves* opposite, petioled, oblong or lanceolate, 3–7-nerved, minutely denticulate-serrulate. *Flowers* minute, in axillary clustered cymes, rarely in axillary lax cymes or in lax terminal cymose panicles. *Calyx-tube* obovoid, smooth; teeth 5, small, persistent. *Petals* 5. *Stamens* 8 or 10, equal; anthers oblong, obtuse at the top, opening with one pore, at the base shortly produced or not; connective with or without an appendage. *Ovary* inferior, 4- or 5-celled, glabrous at the apex; style simple, filiform; ovules very many, placentas axile. *Capsule* globose, 5-valved, enclosed by the membranous calyx-tube. *Seeds* very many, irregularly club-shaped. DISTRIB. Species 5 or 6, extending from Singapore to Borneo.

Connective of anthers with no appendage at the base be-
    hind —:
  Flowers in a terminal panicle, 5-merous   ...   ...   ·1. *O. paniculata.*
    ,,   fascicles, 4-merous   ...   . 2. *O. borneensis.*
Connective of anthers with an appendage at the base be-
    hind :—
  Erect; flowers 5-merous; leaves 3-nerved   ...   ... 3. *O. javanica.*
  Decumbent; flowers 4-merous; 5- to 7-nerved...   ... 4. *O. decumbens.*

1. OCHTHOCHARIS PANICULATA, Korth. in Verh. Nat. Gesch. Bot. 247 t. 64. A small shrub with slender quadrangular branches, rusty pubescent towards the tips. *Leaves* membraneous, elliptic-lanceolate, tapering to each end, the edges minutely bristle-serrate, 5- to 7-nerved (the marginal pair faint); upper surface glabrous; the lower somewhat pale,

scaly-puberulous on the nerves and prominent transverse veins; length 2·5 to 5 in., breadth 1·25 to 2 in., petiole ·35 to 8·5 in. *Panicle* cymose, terminal, spreading, less than half as long as the leaves; branches opposite, pedicels unequal, bracteolate at the base, not quite so long as the flowers. *Petals* 5, ovate-acuminate. *Stamens* 10, anthers oblong, blunt, not produced at the base and with the connective inappendiculate. *Capsule* depressed-globular, glabrous, with 5 shallow grooves, ·2 in. in diam. Blume Mus. Bot. I, 40; Naud. in Ann. Sc. Nat. ser. 3, XV, 307, with fig.; Miq. Fl. Ind. Bat. I, pt. I, 556; Triana in Trans. Linn. Soc. XXVIII, 74. C. B. Clarke in Hook. fil. Fl. Br. Ind. Il, 528; Cogn. in DC. Mon. Phan. VII, 480. *Melastoma oxyphyllum*, Benth. in Wall. Cat. 4083.

SINGAPORE; *Wallich.* DISTRIB. Borneo.

2. OCHTHOCHARIS BORNEENSIS, Blume Mus. Bot. Lugd. Bat. I, 40. A shrub, 3 or 4 feet high. *Leaves* membranous, elliptic-lanceolate, 5-nerved (the lateral pair small), acuminate, the base cuneate, the edges minutely dentate, upper surface glabrous, the lower pale, minutely and furfuraceously stellate-hairy on the main and rather prominent tranverse nerves; length 3 to 4·25 in.; breadth 1·25 to 1·75 in.; petioles sparsely stellate-hairy, unequal, ·3 to 1·8 in. long. *Flowers* in fascicles of 4 to 7, from tubercles on the stem below the leaves; the buds conical, ·1 in. long; pedicels slender, ebracteolate, ·25 to ·3 in. long. *Petals* 4, broadly ovate, narrowly acuminate. *Stamens* 8; the anthers oblong, blunt, neither produced nor appendiculate at the base. *Capsule* depressed-globular, glabrous, faintly 5-grooved, ·2 in. in diam. Naud. in Ann. Sc. Nat. Ser. 3, XV, 307; Cogn. in DC. Mon. Phan. VII, 480. *O. buruensis*, Teysm. and Binn. in Nat. Tijdschr. Ned. Ind. XXV, 426; Miq. in Ann. Bot. Lugd. Bat. I, 216.

SINGAPORE; *Ridley* 6221. DISTRIB. Borneo, Molluccas.

3. OCHTHOCHARIS JAVANICA, Blume in Flora 1831, 523. A shrub, 2 or 3 feet high. *Leaves* subcoriaceous, lanceolate or oblong-lanceolate, acute, the base cuneate, 3-nerved; both surfaces glabrous and with the nerves indistinct; the lower rather pale when dry, the edges remotely bristle-serrate; length 2 to 3·5 in.; breadth ·6 to 1 in.; petiole ·2 to ·5 in. *Cymes* or fascicles about as long as the petioles, few-flowered; pedicels slender, minutely bracteolate at the base, ·15 to ·45 in. long. *Flower*-buds ·15 in. long, much pointed. *Petals* 5, broadly ovate, abruptly acuminate. *Anthers* 5, twice as long as the filaments, minutely spurred at the base behind. *Capsule* subglobular, with 5 shallow grooves, glabrous, ·15 to ·2 in. in diam.; the placentas persistent, woody, rough on their outer surfaces. Naud. in Ann. Sc. Nat. Ser. 3, XV, 307; Miq. Fl. Ind. Bat. I, pt. I, 556; Kurz, For. Fl. I; 507. Triana Melast. 74, tab. VI, fig. 67; C. B. Clarke in Hook. fil. Fl. Br. Ind. II, 528;

Cogn. in DC. Mon. Phan. VII, 480.   *Melastoma ? littoreum,* Wall. Cat. 4087.

A sea-shore plant, SINGAPORE, JOHORE, MALACCA, PERAK. DISTRIB. Burma, Borneo, Java.

4. OCHTHOCHARIS DECUMBENS, King, n. sp. A creeping or decumbent shrub, 3 to 5 feet long; stems as thick as a swan's quill, rounded, often rooting at the nodes. *Leaves* thinly coriaceous, elliptic-lanceolate, shortly acuminate, rounded or cuneate at the base, the edges obscurely bristle-serrate, 5–7-nerved; upper surface glabrous; the lower pale and with a few minute, scattered scales; length 5 to 7·5 in.; breadth 2 to 3 in.; petioles unequal, stout, sparsely scaly, ·5 to 1·25 in. long. *Flowers* in dense, axillary cymes shorter than the petioles, the pedicels short, bracteolate. *Calyx-tube* glabrous, faintly ribbed, the teeth 4, small. *Petals* 4, broadly ovate, acuminate. *Stamens* 8; the anthers curved, tapering to the truncate apex, slightly produced at the base and with a minute tubercle behind. *Capsule* depressed-globose, glabrous, faintly 4-grooved, ·18 in. in diam.

PERAK; *King's Collector,* 2833, 10425.

7. ANERINCLEISTUS, Korth.

Shrubs; branches often round. *Leaves* petioled, ovate or lanceolate, entire, 3–7-nerved. *Flowers* small. *Calyx-tube* campanulate or funnel-shaped; lobes 4, usually very small. *Petals* 4, minute, glabrous. *Stamens* 8, equal; anthers attenuate at the top, opening by one pore, scarcely produced at the base; connective with or without a short spur. *Ovary* nearly free, 4-celled; style filiform, simple; ovules many. *Capsule* splitting by 4 large valves at the summit. *Seeds* exceedingly minute, cuneate-obovoid. DISTRIB. Species 9 or 10; Malaya and Burma.

Inflorescence umbellate; leaves small, oblong- or elliptic-lanceolate :—
    Nearly glabrous in all its parts ...   . 1. *A. macranthus.*
    More or less pilose :—
        Calyx-tube mealy tomentose, the teeth narrow, glan-
        dular ciliate on the edges   ...   . 2. *A. Scortechinii.*
        Calyx-tube not tomentose but with many long spreading
        gland-tipped hairs, the teeth minute and without hairs 3. *A. Curtisii.*
Inflorescence a large panicle; leaves large, ovate :—
    Panicle densely tomentose, leaves 7-nerved ...   ... 4. *A. floribundus.*
    Panicle clothed with short, flat, adpressed, scale-like hairs,
    leaves 5-nerved   ...   . 5. *A. sublepidotus.*

1. ANERINCLEISTUS MACRANTHUS, King, n. sp. A small shrub; young branches with pale lenticels, glabrous except for a few adpressed hairs at the apices. *Leaves* nearly equal, elliptic-lanceolate, much narrowed to the base, the apex shortly acuminate; both surfaces

glabrous, the lower pale when dry; length 2·5 to 4 in., breadth ·8 to 1·5 in., petiole ·4 to ·8 in., strigose. *Umbels* axillary or terminal, usually solitary, 4– 5-flowered, on slender, sparsely strigose peduncles shorter than the leaves. *Flowers* (including the stamens) ·5 in. long. *Calyx-tube* infundibuliform, sparsely adpressed-pilose, nearly as long as the slender strigose pedicel; the mouth with 4 narrowly triangular acuminate, erect, persistent teeth nearly as long as the tube. *Petals* 4, ovate, acuminate, glabrous, slightly longer than the calyx-teeth. *Stamens* 8, slightly unequal, the alternate 4 shorter but of the same shape as the longer 4, the base in all emarginate, scarcely lobed but with a very short process behind. Scales of disc 4, their apices broadly truncate and slightly toothed. *Capsule* globose-ovoid, ·25 to ·3 in. long.

PERAK; at elevations of 2000 to 4000 feet, *Scortechini*; *Wray* 297, 1621.

A species near to *A. hirsutus*, Korth., but differing in being glabrous and in having larger flowers which, according to Mr. Wray, are pink. The umbels of this, when in bud, are enclosed in oblong, deciduous, sparsely strigose bracts.

2. ANERINCLEISTUS SCORTECHINII, King, n. sp. A slender shrub, 3 to 4 feet high; young branches, petioles and inflorescence with many spreading and sub-adpressed, gland-tipped hairs. *Leaves* very unequal (one of each pair very small), oblong, slightly narrowed to the rounded base, the apex acuminate, the edges ciliate; main nerves 3 to 5, the minor nerves transverse, rather faint; both surfaces usually glabrous, the upper sometimes sparsely strigose, the main nerves on both and the transverse on the lower surfaces bristle-hairy; length of the larger leaf of the pair 2·5 to 4 in., breadth ·8 to 1·4 in.; petiole ·3 to ·4 in ; the smaller leaf of the pair from one-fourth to one-half as large and sub-sessile. *Inflorescence* axillary, solitary on a slender peduncle about as long as the smaller leaf, umbellately cymose, 6- to 10-flowered. *Calyx-tube* globose-campanulate, minutely mealy tomentose, about as long as the pedicel; teeth 4, narrow, reflexed, each ending in a bristle and the margins bearing 6 or 8 long straight hairs with small glandular apices. *Petals* 4, ovate, glabrous, shorter than the calyx-teeth, each with a slender, gland-tipped, reflexed hair at the apex. *Anthers* 8, equal, acute, very slightly lobed at the base and inappendiculate; *style* thick, straight; stigma small. *Ovary* crowned by obtuse, sometimes crenulate scales.

PERAK; *Scortechini* 51, 450; *Curtis*.

This closely resembles *A. Curtisii*, Stapf, but is distinguished at once by the calyx which has its tube covered with minute mealy pubescence while the teeth are long, reflexed, with long glandular hairs on the edges, whereas in *A. Curtisii* the calyx-tube bears many long glandular hairs and the teeth are triangular, minute and hairless. The leaves of *A. Curtisii* moreover are strigose between the nerves on both surfaces.

3. ANERINCLEISTUS CURTISII, Stapf in Kew. Bull. for 1892, p. 196. A small shrub, like *A. Scortechinii,* the young branches, petioles, and inflorescence spreading slightly and not gland-tipped; *leaves* as in *A. Scortechinii* but strigose on both surfaces. *Calyx-tube* with numerous subulate, gland-tipped, spreading hairs; the teeth minute, triangular and without hairs. *Anthers* blunt.

PENANG; *Curtis* 412.

4. ANERINCLEISTUS FLORIBUNDUS, King, n. sp. A shrub, about 15 feet high : young branches, petioles and panicles densely and shortly tawny-tomentose. *Leaves* unequal, ovate, shortly acuminate, 7-nerved at the rounded base, (the main nerve giving off a pair about 1 in. from the base); upper surface with a few short scattered bristles, the main nerves densely bristly-pubescent; lower surface with short coarse hairs especially on the nerves and veins; length 5 to 9 in.; breadth 4 to 5·5 in.; petiole 1 to 2 in., one leaf of each pair smaller than the other. *Panicle* terminal, solitary, longer than the leaves, much branched, the branches unequal and in pseudo-whorls, many-flowered, the ultimate branchlets few-flowered, cymose. *Calyx-tube* funnel-shaped, tapering into and longer than the pedicel, adpressed-pilose : the mouth truncate and with 4 small broadly triangular teeth and 4 alternating tufts of inwardly-directed hairs. *Petals* 4, shorter than the calyx-tube, broadly triangular or occasionally quadrate, apiculate, glabrous. *Stamens* 8, equal, or 2 smaller and sometimes suppressed; *anthers* curved, with two deep broad lobes at the base but no appendage. *Capsule* broadly obovoid, truncate, tapering at the base; seeds minute, broadly linear.

PERAK; *Scortechini* 249 in part; *Ridley* 5342; *Curtis* 1299.

5. ANERINCLEISTUS SUBLEPIDOTUS, King, n. sp. A shrub, 10 to 15 feet high; young branches, petioles and inflorescence densely clothed with short, flat, adpressed, scale-like pale hairs. *Leaves* somewhat unequal, ovate, shortly acuminate, on long petioles, 5-nerved, entire; upper surface glabrous except for a few minute scale-like hairs, the main nerves hairy like the petiole; lower surface much reticulate, minutely lepidote-hairy, the middle nerve hairy like the petiole; length 4 to 8 in., breadth 2 to 4 in.; petiole 1 to 1·5 in. *Panicle* terminal, solitary, erect in flower, pendent in fruit, usually longer than the leaves, few-branched; the ultimate branches unequal, short, few-flowered, some of them in pseudo-whorls, pedicels less than ·1 in. long. *Calyx-tube* ·15 in. long, (when dry) reddish, clothed with scattered, adpressed scale-like hairs, tubular-campanulate; the teeth 4, short, acute. *Petals* 4, ovate, acuminate, glabrous, shorter than the calyx-tube. *Stamens* 8,

J. II. 3

subequal, all yellow, curved, deeply lobed at the base but inappendiculate. *Capsule* obovoid, much tapered to the pedicel, ·15 in. long (when dry), scabrid from the stiff, strigose, scale-like hairs. *Seed* minute subulate.

PERAK; *Scortechini* 310; *King's Collector* 8068.

NOTE.

I take the opportunity to describe here the undernoted new species from Borneo.

ANERINCLEISTUS GLOMERATUS, King, n. sp. A shrub; young branches, petioles, main nerves of leaves and inflorescence densely covered with short, coarse, adpressed, pale hairs. *Leaves* narrowly oblong, narrowed towards the base, the apex acuminate; upper surface glabrous but with a few scattered glands; lower surface with a few pale hairs of unequal length on the bold transverse veins; length 3·5 to 8 in.; breadth 1 to 2 in.; petiole ·5 to 1·5 in. *Panicle* solitary, terminal, very narrow, bearing a few very short, almost sessile, 4- to 6-flowered umbels. *Flowers* on pedicels about half the length of the calyx-tube. *Calyx-tube* short, campanulate, pubescent; the mouth wide and with 4 short, triangular, acute, spreading teeth alternating with tufts of hair. *Petals* 4, narrowly oblong, spreading. *Stamens* 8, equal; *anthers* slightly lobed at the base, inappendiculate. *Ovary* 4-ridged, 4-celled; ovules numerous. *Capsule* sub-globular.

BORNEO; Sarawak, *Hullett* 257.

A species resembling *A. anisophyllus,* Stapf, in the shape of its leaves. The inflorescence is however very different; and in its short subsessile panicle recalls to one's memory *A. Beccarianus,* from which its leaves distinguish it at once.

8. SONERILA, Roxb.

Low herbs, rarely half-shrubs. *Leaves* membranous or more or less fleshy, opposite, those of a pair similar in shape, although often very different in size, or distinctly heteromorphous and then often apparently alternate, usually more or less oblique, 3–7-nerved from the base or near the base, rarely pinnate-nerved. *Flowers* pink or white, in scorpioid simple or pseudo-umbellate cymes, 3-nerved. *Calyx* sub-cylindrical, turbinate or campanulate, 3-lobed or 3-toothed. *Petals* ovate, obovate or oblong, acute, acuminate or obtuse. *Stamens* 3, equal, rarely 6 and slightly unequal; *anthers* linear, oblong or lanceolate, obtuse, acute or (often long) acuminate, minutely 2-lobed at the base, without appendages, dehiscing with apical pores. *Ovary* attached to the calyx-tube by narrow longitudinal septa, depressed at the apex, 3-celled; *style* filiform; stigma punctate or capitellate. *Fruit* enclosed in the persistent, ultimately spongy calyx-tube and forming with it a usually more or less trigonous, subcylindrical, turbinate or hemispherical false capsule, dehiscing from the centre of the depressed top with 3 valves or 6 fine teeth. *Seeds* minute, numerous, ovoid, pyramidal or clavate, smooth or asperulous; raphe usually thick, spongy. Species over 100 throughout tropical Asia.

Leaves similar, although often very unequal in size. Stamens 3. Fruit usually smooth (see No. 12–14); valves entire, distinct, exceeding the margin of the mature calyx (§ Eu-Sonerila):—
  Erect or ascending herbs with fibrous roots and without rhizome:—
    Calyx very slender (also in fruit); stem with 2 some-what raised, commissural lines. Very scantily hairy or glabrous, often much branched and small- or narrow-leaved (except No. 1) herbs:—
      Leaves ovate, 1–1·7 in. by ·7–·9 in.   ...   ...   1. *S. epilobioides.*
      Leaves much . smaller or at least very much narrower:—
        Anthers ·12–·18 in. long:—
          Leaves oblong to elliptic-oblong, ·3–·5 by ·15–·23 in. ...   .   2. *S. calaminthifolia.*
          Leaves lanceolate, 1–1·75 in. by ·15–·23 in.   ...   3. *S. hyssopifolia.*
        Anthers ·06–·09 in. long   ...   .   4. *S. erecta.*
  Calyx more or less oblong- or ovoid-campanulate; fruit turbinate or obpyramidal. Stem terete or quadrangular:—
    Leaves more or less ovate, long acuminate, 1–2 in. by ·4–1 in. or still smaller, acutely and coarsely toothed, thin, 3–5-nerved from the very base; petioles long, very slender   ...   .   5. *S. tenuifolia.*
    Leaves usually larger, not coarsely toothed:—
      Upper side-nerves starting from above the base:—
        Leaves pinnatinerved, acute or acuminate at the base ...   .   6. *S. flaccida.*
        Leaves not pinnatinerved (rarely subpinnati-nerved in No. 7); all the side-nerves springing from near the base, rarely the uppermost from near the middle:—
          Cymes distinctly peduncled:—
            Anthers subacute, ·09–·12 in. long; leaves membranous, rounded or subcordate at the base   ...   .   7. *S. andamanensis.*
            Anthers slender, acuminate, ·2–·3 in. long:—
              Stem, petioles and inflorescence minutely tomentose; leaves thinly membranous, subcordate at the base on very long and slender petioles   ...   .   8. *S. populifolia.*
              Stem, petioles and inflorescence with long hairs; leaves acute or rounded, but not subcordate at the base:—
                Leaves membranous, more or less oblong, usually acute at the base   ...   9. *S. pallida.*

Leaves somewhat fleshy, rounded at
the base :—
   Hairs more or less spreading, often
   very long, particularly on the
   petioles and near the leaf margins;
   leaves light-brown beneath   ... 10. *S. rudis.*
   Hairs adpressed, very soft; leaves
   glaucous beneath with rufous hairs 11. *S. mollis.*
Cymes sessile, reduced to few-flowered
fascicles :—
   Unbranched or almost unbranched herbs :—
     Leaves fleshy, very dark and glabrous
     above, glaucous or pale-brown with rufous
     nerves beneath; calyx with scattered
     gland-tipped hairs   ...   ... 12. *S. albiflora.*
     Leaves membranous, more or less covered
     on both sides with rufous flexuous hairs;
     calyx densely hirsute   ...   ... 13. *S. lasiantha.*
   Suffrutescent, much branched ...   ... 14. *S. suffruticosa.*
All the nerves springing from the very base of the
leaf; adult leaves quite glabrous, broadly elliptic,
fleshy   ...   . 15. *S. elliptica.*
Herbs with short stems, springing from a creeping rhi-
zome and with usually crowded to rosulate leaves and
terminal or subterminal peduncled cymes :—
Leaves 3-7 in. by 2-4 in., 7-nerved from the very base
with conspicuous subhorizontal transverse veins :—
   Stem, petioles and peduncles very succulent, stout 16. *S. succulenta.*
   Stem, petioles and peduncles usually slender   ... 17. *S. repens.*
Leaves small; uppermost side-nerves springing from
above the base; transverse veins indistinct or 0 :—
   Leaves oblong to lanceolate-oblong, 2-3·7 in. by
   ·3-1·4 in., pinnatinerved ...   . 18. *S. muscicola.*
   Leaves much smaller, not or very indistinctly
   pinnatinerved, lateral nerves springing from below
   the middle :—
     Leaves oblong-lanceolate to lanceolate, acute at
     both ends   ...   . 19. *S. saxosa.*
     Leaves ovate to elliptic or oblong, rounded or sub-
     cordate at the base :—
       Stem 1-3 in. long; leaves broad, crowded, in
       about 3 pairs; petioles ·3-·7 in. long :—
         Leaves entire, not ciliate; ·9-1·7 in. by ·6-1·2
         in.; petals ·35 in. long; anthers ·15-·2 in. long 20. *S. congesta.*
         Leaves toothed, ciliate, ·4-·8 in. by ·4-·6 in.;
         petals ·25 in. long; anthers ·12-·15 in. long ... 21. *S. Griffithii.*
       Stem shorter; leaves narrower, more numerous,
       subrosulate; ·8-1·4 in. by ·4-·6 in., often beauti-
       fully variegated; petioles up to 1 in. long,
       very slender   ...   . 22. *S. Cyclaminella.*

Leaves of each pair similar in shape, but very unequal in
size or one quite suppressed. Stamens 6. Fruit as
in Eu-Sonerila (§ Sonerilopsis, *Miq.*)   ...   ...   23. *S. heterostemon.*
Leaves of each pair very dissimilar in shape and size, one
very small and more or less rotundate or reniform-
cordate, often deciduous. Cymes usually from the axils
of the small leaves, hence often apparently leaf-opposed.
Anthers 3, oblong, obtuse, not over ·15 in. long. Fruit
more or less turbinate or semiglobose, obscurely
trigonous, usually conspicuously·muricate ; valves not or
very slightly exceeding the margin of the mature calyx,
often not or indistinctly separating from each other at the
periphery, but each of them always splitting into 2
membranous teeth in the depressed centre of the fruit.
(§ Hexodon, *Stapf*) :—
  Cymes distinctly, densely and persistently bracteate,
  distinctly and often long peduncled :—
    Cymes simple or 2 or more, sessile on a common
    peduncle, forming a dense head :—
      Cymes simple, at length 1–1·5 in. long ; leaves
      obliquely oblanceolate to obovate-oblong, 3–7 in.
      by 1–1·5 in. :—
        Leaves fleshy, glabrous with the exception of
        the minutely strigillose nerves of the underside ...   24. *S. integrifolia.*
        Leaves membranous, with rather long and spread-
        ing hairs beneath and along the margins   ...   25. *S. bracteata.*
      Cymes usually 2 or more, sessile on a common
      peduncle, very short, gathered in a dense head ;
      leaves obliquely obovate, elliptic or oblong, 4–6 in.
      by 1·75–3 in.   ...   .   26. *S. capitata.*
    Cymes usually 2 or more on a common peduncle, each
    with a special peduncle, gathered in a loose umbel ;
    leaves obliquely elliptic, 3–5 in. by 1·75–2·75 in.   ...   27. *S. caesia.*
  Cymes ebracteate, or indistinctly or deciduously bracteate :—
    Cymes sessile or subsessile ; peduncle, if any, less
    than ·5 in., when mature ; fruit muricate :—
      Cymes dense, indistinctly bracteate, sessile :—
        Bracts linear, usually minute, or suppressed ;
        calyx pubescent ; leaves fleshy, firm when dry,
        glabrous above, shortly and adpressedly pubes-
        cent or tomentose beneath   ...   ...   28. *S. Nidularia.*
        Bracts filiform, ciliate, hidden among long spread-
        ing hairs ; calyx shaggy ; leaves membranous,
        very hairy on both sides   ...   ...   29. *S. brachyantha.*
    Cymes subsessile somewhat loose, ebracteate or
    deciduously bracteate :—
      Leaves obliquely obovate-lanceolate to oblong,
      2·5–6 in. by 1–2·5 in. :—
        Leaves somewhat fleshy, firm, when dry with
        conspicuous transverse veins ; pedicels very

slender, to ·25 in. long; calyx with short, fine, spreading hairs; fruit ·08–·1 in. long, finely muricate	30. *S. microcarpa.*
Leaves membranous with conspicuous raised transverse veins; pedicels slender, ·08–·12 in. long; calyx with long spreading hairs; fruit ·15–·18 in. long, coarsely muricate	31. *S. costulata.*
Leaves obliquely elliptic, 4–6 in. by 2·2–3·5 in.; fruit ·2 in. long	32. *S. macrophylla.*

Cymes distinctly, often long, peduncled :—   •

Cymes 2 or more gathered in an umbel, or if simple, then with a pair of (often minute) leaves at the middle :—

Calyx hairy; fruit muricate :—

Cymes usually compound, minutely bracteate, axis shaggy; leaves more or less hairy all over	33. *S. paradoxa.*
Cymes usually simple, deciduously bracteate; bracts small; axis shortly hairy; leaves glabrous (in the Peninsular specimens) with the exception of the very shortly and scantily hairy nerves and veins of the underside ...	34. *S. begoniaefolia.*
Calyx glabrous; fruit smooth	35. *S. glabriflora.*

Cymes simple, ebracteate; peduncle naked :—

Softly hairy or tomentose all over :—

Leaves oblong-elliptic or obovate, 2·2–2·8 in. by 1–1·2 in., petals ·18 in. long	36. *S. elatostemoides.*
Leaves oblong to lanceolate-oblong, 3–6 in. by 1–1·5 in.; petals ·35 in. long	37. *S. bicolor.*
Shortly tomentose on stem, petioles and peduncles and pubescent on the nerves on the underside of the leaves, otherwise glabrous; fruit smooth	38. *S. Calycula.*

1. SONERILA EPILOBIOIDES, Stapf and King. An erect, sparingly branched, almost quite glabrous herb, about 6 in. high. *Stem* with 2 prominent very minutely hairy lines or quite glabrous. *Leaves* of each pair similar in shape and equal or almost equal in size, ovate, acute or subacute, acuminate at the base, minutely toothed, thinly membranous, green, glabrous or with few scattered, soft, adpressed hairs above, 1–1·75 in. long by ·7–·9 in. broad, finely 5- (rarely 3-) nerved from the very base with a few delicate side nerves higher up; petiole ·4–·5 in. long, slender. *Cymes* terminal, peduncled, rather loosely 5–6-flowered, glabrous; peduncle ·6–·8 in. long, slender; pedicels at length up to ·2 in. long, very slender. *Calyx* very slender, obconical, ·25–·3 in. long; teeth broad, triangular, mucronulate. *Petals* elliptic-oblong, acuminate-apiculate, ·2 in. long, pink. *Anthers* subacute; ·12 in. long. *Style* filiform, ·12–·15 in. long; stigma subcapitate. Mature *fruit* unknown.

KEDAH; Santow, on the limestone islands, *Curtis* 2114!

2. SONERILA CALAMINTHIFOLIA, Stapf and King. An ascending or almost prostrate branched herb, from a few inches to 1 ft. high, with very minute, spreading hairs all round or along the 2 commissural lines of the stem and branches, or glabrescent below and with scattered, adpressed, short, straight or flexuous hairs on the upper side of the leaves. *Stem* terete below, quadrangular in the upper part, purple, like the branches slender and often flexuous. *Leaves* petioled, rarely the uppermost subsessile and then sometimes apparently whorled, those of a pair similar in shape and size, symmetrical, oblong to elliptic-oblong, subobtuse, narrowed into the petiole, acutely toothed with the teeth usually mucronate and the margin revolute, rather stoutly membranous or almost fleshy, pale-green, purple along the midrib, ·3–·5 in. by ·15–·23 in., 1-nerved or with 1 very fine side-nerve on each side from near the base; petioles ·08–·12 in. or less. *Cymes* terminal, 2–5-flowered, ultimately lax or flowers solitary; peduncles filiform, ·4–·8 in. long; pedicels very slender, about ·08 in. long. *Calyx* slender, obconical-oblong, ·15–·2 in. long; teeth triangular, broad, acute. *Petals* elliptic, cuspidate, about ·33 in. long, pink. *Anthers* shortly acuminate, ·12–·17 in. long. *Style* filiform, ·2 in. long; stigma punctiform. *Fruit* oblong, subtrigonous, smooth, ·25–·3 in. by ·08–·1 in.

PERAK; Gunong Batu Pateh, 4300 ft., *Wray* 1022 !

3. SONERILA HYSSOPIFOLIA, Stapf and King. An erect, simple or branched herb, 6–9 in. high, with two lines of short, curled hairs on the stem and branches, and with whitish, flexuous, longer hairs on the upper side of the leaves. *Stem* subterete below, quadrangular above. *Leaves* sessile or petioled, those of a pair similar in shape and size, lanceolate, subacute, cuneate at the base or the uppermost sessile and rounded at the base, symmetrical, minutely and somewhat remotely toothed with very acute or mucronate teeth, membranous, green, pale below, 1–1·75 in. by ·15–·3 in., indistinctly 3-nerved from near the base; petioles very variable in length, up to ·4 in. long, or 0. *Cymes* few-flowered, at length lax; peduncles ·8–1 in. long, very slender, pedicels very slender, ·04–·08 in. long. *Calyx* very slender, almost cylindric, ·24–·28 in. by ·04 in.; teeth triangular, mucronulate. *Petals* ellipticoblong, acute, ·27 in. long, pink. *Anthers* acuminate, ·12–·18 in. long. *Style* filiform, about ·18 in. long; stigma subcapitate. *Fruit* trigonous-cylindric, slightly obconical at the base, ·5 in. by almost ·1 in.

PERAK; Gunong Hijan; *Scortechini* 1426 !

4. SONERILA ERECTA, Jack in Malay Misc. I, 7. A copiously branched herb, 1–1·5 ft. high, more or less hairy with the exception of the inflorescence, hairs flexuous, finely pointed, those of the stem and branches short, along 2 lines, those of the leaves longer, whitish, all over both surfaces. *Stem* subterete below, quadrangular above,

branched all along, branches more or less divaricate, 1 or, in luxuriant specimens, 2 from each leaf axil; often branched again. *Leaves* sessile or petioled, often apparently whorled on the upper or, in vigorous specimens, almost on all nodes, those of a pair similar in shape and size, lanceolate or ovate, acute or subacute, contracted or subcuneate at the base, rather symmetrical, entire or minutely toothed, membranous, green or greyish when very hairy, ·4–1·2 in. by ·2–·4 in., indistinctly 3-nerved from near the base; petioles slender, very variable in length, up to ·3 in. long, or 0. *Cymes* 2–9-flowered, at length very lax; peduncles filiform, ·4–2 in. long; pedicels very slender, ·04 in. long or hardly any. *Calyx* very slender, almost cylindric, ·15–·2 in. by ·03 in., teeth triangular, short. *Petals* elliptic-oblong, cuspidate-acuminate, ·15 in. long, pink. *Anthers* ovate-lanceolate, acute or subacute, ·06–·09 in. long. *Style* filiform, ·12 in. long; stigma punctiform. *Fruit* almost cylindric, slightly trigonous, smooth, ·3–·45 in. by ·06–·08 in. Jack in Hook. Bot. Misc. II, 63; Blume in Flora (1831), 491; Benn. Pl. Jav. Rar. 217; Naud. in Ann. Sc. Nat. Ser. 3, XV, 324; Miq. Fl. Ind. Bat. I, 563; Triana in Trans. Linn. Soc. XXVIII, 75; C. B. Clarke in Hook. f. Fl. Brit. Ind. II, 530; Cogn. in DC. Monogr. VII, 492; Stapf in Ann. Bot. VI, 304.

PENANG; Government Hill, *Curtis, Porter* in Wall. Cat. 4092; *Maingay* 2214 (778, Kew Distrib.). DISTRIB. Northwards as far as Moulmein.

VAR. *flexuosa*, Stapf and King. *Stems* ascending, like the very slender branches more or less· flexuous. *Leaves* lanceolate to linear-lanceolate, acuminate at the base, ·4–·8 in. by ·12–·2 in., more or less pubescent or scaberulous from very short hairs, mainly above, margins usually finely but sharply toothed and often revolute. *Cymes* 3–1-flowered with capillary peduncles. *Fruit* ·27–·35 in. long, by ·6–·8 in.

PENANG; on rocks, 2000 ft., *Curtis* 1238! PERAK; Larut, on rocks in rich, moist soil, 300 to 600 ft., *Kunstler* 2364! *Scortechini* 91!

VAR. *discolor*, Stapf and King. *Stems* ascending or erect, branches very slender, slightly flexuous, very minutely hairy along lines or almost glabrous. *Leaves* oblong-lanceolate to ovate-lanceolate, minutely toothed, narrowly revolute on the margins, very pale beneath, 3–5-nerved near the base, nerves fine, but rather distinct. *Cymes* 4–1-flowered, rather congested, also when mature. *Anthers* ·08–·09 in. *Fruit* oblong-cylindric, ·3–·35 in. by ·08 in.

PERAK; *Scortechini* 160!

5. SONERILA TENUIFOLIA, Blume in Flora 1831, 491. An erect or ascending, branched or unbranched herb, 6–12 in. high, usually with scattered, spreading, gland-tipped hairs in the upper part of the stem, on the peduncles and pedicels, and near the mouth of the calyx and with few

adpressed, finely attenuated, pale hairs on the upperside of the blades, rarely almost quite glabrous. *Stem* slender, terete below, more or less quadrangular upwards. *Leaves* of a pair similar in shape, but unequal in size, ovate to ovate-lanceolate, long acuminate, symmetrical or slightly asymmetrical and rounded at the base, acutely and coarsely toothed, thinly membranous, dark- or pale-green, the larger 1–2 in. by ·4–1 in., distinctly 3-, rarely 4–5-nerved from the very base, very faintly pinnate-nerved higher up; petioles very slender, reaching 1·5 in. in length, purple. *Cymes* 1–6-flowered, almost pseudo-umbellate; peduncle ·5–1 in. long, like the slender pedicels purple or crimson, the latter ·25–·3 in. long. *Calyx* campanulate-oblong to ovoid-oblong, about ·15 in. long; teeth distinct, triangular. *Petals* elliptic, apiculate, ·3–·35 in. long, glabrous, rose-coloured. *Anthers* lanceolate-acuminate, acute or subobtuse, ·12–·13 in. long. *Stigma* capitate. *Fruit* trigonous, obconical, ·23–·27 in. long, smooth; valves ·23 in. broad. Benn. Pl. Jav. Rar. 211, t. 44; Naud. in Ann. Sc. Nat. ser. 3, XV, 324; Miq. Fl. Ind. Bat. I, 563. Triana in Trans. Linn. Soc. XXVIII, 76; C. B. Clarke in Hook. f. Fl. Brit. Ind. VIII, 536. Stapf in Trans. Linn. Soc. 2nd Ser. IV, 156; Cogn. in DC. Monogr. VII, 502; Stapf in Ann. Bot. VI, 301.

PERAK; *Scortechini* 312! *Kunstler* 722! *Wray* 427! on Gunong Batu, 4500 ft., *Wray* 273! 406. MALACCA; Mt. Ophir, *Maingay* 2582! DISTRIB. Sumatra, Java, Borneo.

VAR. *hirsuta*, Stapf and King. Leaves hairy on both sides; hairs copious at least above, longer, wavy or curled, less adpressed than in the type; blades often very small (0·6 in. by 0·3 in.); flowers solitary.

PERAK; *Scortechini* 790! Larut, top of Gunong Bubu, 5000–5300 ft., *Kunstler* 7406! *Wray* 3841!

6. SONERILA FLACCIDA, Stapf and King. An erect or ascending, usually branched herb, ·5–1 ft. high, with a very fine, furfuraceous, dark rusty and often scanty indumentum in the lower part and on the underside of the leaves (at least on the nerves), otherwise glabrous. *Stem* quadrangular. *Leaves* of a pair similar, equal or unequal in size, oblong or oblong-elliptic, subacute or subacuminate at both ends, or the tips obtuse, sometimes decurrent at the base, entire or nearly so, symmetrical or more or less asymmetrical, thinly membranous, dark- or light-green and often spotted with white circular or elliptic spots above, whitish green beneath, 2–4 in. by ·75–1·75 in., pinnate-nerved, distinct side-nerves usually 3 on each side, the others like the tertiary nerves very faint or quite obscure; petiole up to ·6 in. long, often very short, slender. *Cymes* terminal and axillary, short, few- to 9-flowered, peduncled, peduncle very slender, ·75–1 in. long; pedicels very slender, ·08–·15 in. long. *Calyx* slender, obconical to oblong, ·13–·15 in. long, rose-coloured, teeth triangular, very short and broad. *Petals* oblong, acute, ·15 in.

long, rose-coloured. *Anthers* acute, ·12 in. long. *Style* filiform; stigma capitate. *Fruit* trigonous, truncate-obovate, ·2 in. long, smooth; valves ·12 in. broad.

PERAK; Gunong Panti, 600–800 ft., *Kunstler* 219! *Ridley* 4184! Gunong Inas, 3500 ft., *Wray* 4066! 4067!

Certain small specimens, collected by Scortechini in Perak (272), represent only a dwarf state of *S. flaccida;* their larger leaves measure 1–1·5 in. by ·6–·8 in.

7. SONERILA ANDAMANENSIS, Stapf and King. An erect or ascending, branched or unbranched herb, 3–6 in. high, more or less hirsute; particularly on the stem and petioles, with flexuous, finely pointed hairs. *Stem* reddish-brown when dry, quadrangular. *Leaves* rather approximate, those of a pair similar in shape and size, or more or less unequal; ovate to ovate-oblong, acute or subacuminate, rounded or subcordate and often slightly asymmetrical at the base, membranous, green or purple above, purplish glaucous below, length 1·5 to 3·3 in., breadth 1 to 1·7 in., 5–8-nerved from below the middle, the lower nerves more or less opposite, the uppermost 1 or 2 usually alternate, transverse veins oblique, fine or obscure; peduncle ·5–1·5 in. long. *Cymes* few- to many-flowered, much contracted, peduncles solitary and terminal, or 2–4 from the top and the uppermost leaf-axils, 1–2 in. long; pedicels slender, up to ·1 in. long, like the flowers with scanty and sometimes minutely gland-tipped hairs. *Calyx* very slender, obconical, up to ·2 in. long; teeth short, broad, triangular. *Petals* elliptic, acuminate, ·25 in. long, rose-coloured. *Anthers* ovate-lanceolate, subacute, ·09–·12 in. long. *Style* filiform, ·2–·25 in. long; stigma capitate. *Fruit* oblong with a cuneate base, ·22–·27 in. long, smooth; valves scarcely ·1 in. broad.

ANDAMANS; Mount Harriet near Port Blair, on rocks, *King's Collector* 48!

8. SONERILA POPULIFOLIA, Stapf and King. An erect or ascending, simple or sparingly branched herb, 6–9 in. high, more or less covered with minute hairs and with a few soft, adpressed, whitish, small bristles on the surface and the margins of the ultimately often glabrescent leaves, with the hairs of the inflorescence often minutely gland-tipped. *Stem* finely rusty-tomentose, subterete below, quadrangular above. *Leaves* of a pair similar in shape and equal or somewhat unequal in size, ovate, acute or acuminate, usually minutely cordate at the base, with the lobes often more or less unequal and close, subentire or toothed in the upper part, thinly membranous, light-green, 1·5–3 in. by 1·2–1·75 in., finely 7-nerved from near the base, upper pair ·2–·3 in. from the base; petioles very slender, 1–2 in. long, finely tomentose. *Cymes* few- to 12-flowered, much contracted and almost umbelliform, terminal; peduncle slender, ·5–1·2 in. long; pedicels slender, ·15–·22 in. long.

Calyx subcampanulate-oblong, ·18–·2 in. long; teeth broad, triangular. *Petals* elliptic, shortly acuminate, ·3–·4 in. long, deep- to blueish-pink. *Anthers* acuminate, tips sometimes very fine and curved, ·2–·3 in. long. *Style* filiform, ·35–·45 in. long; stigma minutely capitate. *Fruit* trigonous, truncate-obovoid, ·2–·25 in. long, smooth, often finely puberulous; valves ·15–·16 in. broad.

PERAK; *Scortechini* 136! 300–500 ft., *King's Coll.* 10055! Larut, dense jungle, 500–800 ft., *King's Coll.* 5791! Briah plains, *Wray* 4201! Tapah, *Curtis!*

The uppermost pair of leaves is often much reduced, resembling a pair of bracts. One of the leaves of the preceding pair is sometimes suppressed, whilst the peduncle and the petiole of the other leaf are so turned that the latter seems to form the continuation of the axis; hence the former appears to spring from a long petiole. This is chiefly the case with the inflorescences which terminate branches.

9. SONERILA PALLIDA, Stapf and King. An ascending, branched or unbranched herb, 6–12 in. high, hirsute all over, but chiefly on the stems and petioles, hairs pale reddish when dry, those of the inflorescence short, stiff and spreading. *Stem* decumbent at the base, rooting in the lower part, quadrangular. *Leaves* of a pair similar in shape, but usually rather unequal in size, oblong to ovate-oblong, acuminate, symmetrical or more or less asymmetrical and acute (rarely obtuse) at the base, minutely denticulate, membranous, light green, the larger 1·5–4 in. by ·8–1·8 in., finely but distinctly 5–7-nerved from near the base, the upper pair ·4–·6 in. from the base, petioles up to ·6–·75 in. long. *Cymes* few- to 8-flowered, short, on apparently terminal peduncles; peduncles slender, 1–2 in. long; pedicels ·08–·15 in. long, slender. *Calyx* slender, trigonous, obconical-campanulate, ·15–·2 in. long; teeth distinct, triangular. *Petals* elliptic-oblong, apiculate, ·45–·5 in. long, with a line of short, stiff, spreading hairs on the back, pale pink. *Anthers* acuminate, slender, ·23–·24 in. long. *Style* filiform, stigma punctiform. *Fruit* trigonous, obconical, ·25–·3 in. long, sparingly muricate, valves ·15 in. broad.

PERAK; Gunong Inas, 5000 ft., *Wray* 4100! MALACCA; Bujong, *Curtis* 3155! SELANGORE; Bukit Hitam, 2500–3500 ft., *Kelsall!* *Ridley* 7320!

10. SONERILA RUDIS, Stapf and King. A semidecumbent, sparingly branched or unbranched herb, about 1 ft. high, densely clothed with short, or often very long, fine and spreading, curved or curled hairs on the stem and petioles, with somewhat coarse, more or less adpressed hairs on both sides of the leaves and gland-tipped, spreading hairs on the peduncles, pedicels, calyx and the midrib of the petals, hairs reddish when dry. *Stem* often rooting in the lower part, terete or subquadrangular

in the upper part. *Leaves* of a pair similar, subequal or rather different in size, ovate to oblong or elliptic, acute or subacuminate, rounded at the base, entire, somewhat fleshy, green above, pale beneath, distinctly 7-nerved from near the base (upper pair ·2–·25 in. above the base) ; petioles ·4–1·2 in. long. *Cymes* 2–6-flowered, umbelliform, terminal, peduncled; peduncle slender, ·5–1 in. long ; pedicels ·08–·1 in. long. *Calyx* rather slender, subcampanulate, ·12–·15 in. long ; teeth short, broad, triangular. *Petals* elliptic to obovoid, obtuse or subacute, ·5–·6 in. by ·35–·4 in., pink, with a line of gland-tipped hairs on the back. *Anthers* acuminate, slender, ·23–·27 in. long. *Style* filiform ; stigma punctiform. *Fruit* trigonous, shortly obconical, about ·27 in. long, muricate, on stout muricate pedicels ; valves ·2 in. broad.

PERAK; *Scortechini !* Tumbung Parbat, *Scortechini* 422 ! Gunong Batu Pateh, 4500 ft., *Wray* 260. MALACCA ; Bujong, *Curtis* 3297 !

11. SONERILA MOLLIS, Stapf and King. An ascending, sparingly branched or unbranched herb, about 1 ft. high, densely and adpressedly tomentose along stem and petioles, and on the underside of the leaves along the nerves, and besides almost cobwebby on both sides of the young leaves ; all the hairs soft and reddish when dry. *Stem* often rooting in the lower part, terete or subquadrangular in the upper part. *Leaves* of a pair similar, but differing more or less in size, elliptic to ovate-elliptic, shortly and acutely acuminate, rounded at the base, entire, somewhat fleshy, very dark green and quite glabrous above when adult, pale and glabrescent beneath between the nerves, the larger 2–3 in. by 1–2 in., distinctly 5–7-nerved from near the base (upper pair of side nerves ·25–·27 in., distant from the base) ; petioles ·4—1 in. long. *Cymes* 2–4-flowered, umbelliform or flowers solitary, terminal, peduncled, glabrous; peduncle about ·5 in. long, slender ; pedicels ·2–·24 in. long, very slender. *Calyx* slender, subcampanulate, ·18–·2 in. long ; teeth very short and broad, triangular. *Petals* elliptic-oblong, acute, ·4 in. long. *Anthers* acuminate, ·2–·22 in. long. *Style* filiform ; stigma punctiform. *Fruit* trigonous, shortly obconical, ·24 in. long, smooth ; valves ·2–·24 in. long.

PERAK ; *Wray, Scortechini !* Summit of Gunong Batu Pateh, 6700 ft., *Wray* 375 !

12. SONERILA ALBIFLORA, Stapf and King. An ascending or suberect, more or less branched herb, 9–12 in. high, densely and adpressedly hirsute along stem and petioles and more sparingly on the underside of the leaves, and with gland-tipped spreading hairs on pedicels, calyx and midrib of petals. *Stem* rather slender, terete or subquadrangular in the upper part. *Leaves* of a pair similar and rather equal in size, lanceolate to ovate- or obovate-lanceolate, acute or subacuminate at both ends, entire or almost so, fleshy, very dark green (almost black when

dry) and glabrous above, pale and adpressedly hairy beneath (at least on the nerves), 1–3 in. by ·5–1·25 in., distinctly 3–5-nerved from near the base; petioles ·2–·4 in. long. *Flowers* axillary and terminal, solitary or paired; pedicels ·1–·2 in. long, slender. *Calyx* subcampanulate, ovoid, ·1 in. long; teeth distinct, broadly triangular. *Petals* oblong, apiculate, ·2 in. long, white, with a line of gland-tipped hairs beneath. *Anthers* oblong, obtuse, ·08–·1 in. long. *Style* slightly and gradually thickened upwards; stigma punctiform. *Fruit* trigonous, shortly obconical, ·2 in. long, very scantily muriculate; valves ·2 in. broad.

PERAK ; *Scortechini* 1886! Gunong Kledang, 1000 ft., *Curtis* 3293 ! *Ridley* 9691 ! *Goldham* ! Kinta in dense jungle, 3500–4000 ft., *King's Collector* 7169 !

13. SONERILA LASIANTHA, Stapf and King. An erect herb, 4–6 in. high, hirsute all over with flexuous, finely pointed, rufous hairs. *Stem* terete, with the hairs more or less adpressed. *Leaves* of a pair similar in shape, very unequal in size, obliquely lanceolate or subovate, acute, attenuated at the base, membranous, green above, pale beneath, the larger 1·5–3 in. by ·6–·8 in., with 2–3 side-nerves in the broader and 1 in the narrow half, the uppermost ·75–1 in. above the base ; petiole slender, ·3–·6 in. long. *Fascicles* few-flowered, terminal and axillary, subsessile ; pedicels rather stout, ·1–·15 in. long. *Calyx* obconical, densely hirsute, ·15–·18 in. long. *Petals* oblong, cuspidate-acuminate, ·12 in. long. *Anthers* oblong, subacute, ·06 in. long. *Style* filiform, rather stout ; stigma punctiform. *Fruit* broad, obconical, muricate, to ·25 in. long ; valves ·18–·2 in. broad.

PERAK; Gunong Bubu, *Wray* 3863! (in part).

The specimen which we have here in view is so different in habit and in the size of the comparatively long peduncled leaves from the others bearing the same number in Wray's collection, but described under *S. suffruticosa*, that we believe ourselves justified in considering it for the present as a distinct species.

14. SONERILA SUFFRUTICOSA, Stapf and King. An erect, repeatedly branched half-shrub, over 1 ft. high, shaggy all over from coarsely adpressed, crimson (reddish, when dry) hairs, or glabrescent at length at the base. *Stem* terete, woody below, hollow. *Leaves* mainly crowded near the tips of the branches, those of a pair similar in shape, but rather unequal in size, oblong to ovoid-oblong, acute, more or less asymmetrical or almost symmetrical and acute at the base, obscurely serrate or toothed, thickly membranous, dark green above, paler beneath, the larger 1–1·4 in. by ·4–·6 in., 3–5-nerved from near the base; petioles ·25 in. long to very short. *Flowers* unknown (petals white according to Wray). *Fruits* axillary, solitary or in pairs on stout short pedicels, obconical, ·2 in. long, strigose from tubercle-based hairs, or muricate from their persistent bases; valves ·12 in. broad.

PERAK; Larut, Gunong Bubu, 5000 ft., *Wray* 3863! (in part).

15. SONERILA ELLIPTICA, Stapf and King. An erect or ascending, usually unbranched herb, 6–9 in. high, with a very fine, furfuraceous, dark-rusty indumentum in the lower parts and on the young leaves, glabrous or soon glabrescent higher up. *Stem* somewhat stout and succulent, terete. *Leaves* of a pair similar in shape, slightly unequal or equal in size, broadly elliptic, rarely ovate or almost orbicular, very obtuse, usually symmetrical and rounded or subcordate at the base, minutely and inconspicuously toothed, thick, fleshy, dark-green, often mottled with white along the nerves above, waxy yellowish-green beneath, 1–2·5 in. by ·75–2 in., distinctly 5-nerved from the very base, upper nerves usually quite indistinct; petioles 1–2 in. long. *Cymes* many-flowered, dense, axis at length up to ·75 in. long; peduncle slender, 1–2·5 in. long, pedicels at length up to ·2 in. long. *Calyx* obconical, trigonous, ·12 in. long, glabrous, teeth distinct, broad, triangular. *Petals* oblong, apiculate, ·2–·23 long, glabrous, pinkish white. *Anthers* oblong, obtuse, scarcely ·1 in. long. *Stigma* punctiform. *Fruit* trigonous, obconical, ·15 in. long, smooth; valves ·12 in. broad.

PERAK; Kinta, on limestone rocks, 500–800 ft., *Kunstler* 7037! 7225! Sungie Siput, *Curtis* 3156!

16. SONERILA SUCCULENTA, Stapf and King. A succulent, erect herb, quite glabrous with the exception of a very few gland-tipped hairs on the calyx. *Stem* stout, very short to 3 in. long, very fleshy. *Leaves* few, crowded, of a pair equal, symmetrical or almost so, long-petioled, elliptic to ovate-elliptic, rather long and acutely acuminate, rounded at the base or very slightly subcordate, entire, very thinly membranous when dry, 6–7 in. by 3–4 in., 7-nerved from the base, with lax, subhorizontal transverse nerves, petioles succulent, 2–4 in. long. *Cymes* terminal and axillary, 2–5 on a long common peduncle, subebracteate, rather few-flowered, very short and dense; common peduncle stout, 4–7 in. long; special peduncles 1 to over 3 in. long, bracts very minute, subulate, the lower soon deciduous; pedicels hardly any. *Calyx* oblong-campanulate, up to ·25 in. long; teeth short, triangular. *Petals* oblong, cuspidate-acuminate, ·18 in. long. *Anthers* long-acuminate, incurved, over ·25 in. long. *Style* ·35 in. long; stigma subcapitate. *Fruit* subtrigonous, obconical, smooth, up to ·25 in. long; valves over ·15 in. broad.

PERAK; Maxwell's Hill, 3000 ft., *Scortechini* 279!

17. SONERILA REPENS, Stapf and King. A herb with a long creeping rhizome and a very short succulent stem bearing 2–3 usually much approximated pairs of leaves, with few, whitish, more or less adpressed, papilliform hairs on both sides of the leaves and with very few, minute, gland-tipped hairs on the stems, petioles and inflorescences, or glabrous with the exception of the leaves. *Leaves* of a pair similar in shape,

equal or, more usually, very unequal in size, ovate, rarely oblong, acutely acuminate, cordate, rounded or rarely subacute at the base, rather symmetrical, entire or slighty wavy and denticulate, thinly membranous, dark- or pale-green, the larger 3–7 in. by 2–4·5 in., distinctly 7-nerved from the very base, with somewhat distant transverse veins; petioles 1–2·5 in. long, slender or stout, fleshy. *Cymes* long-peduncled, often many-flowered, solitary or usually 2–4 on a common subterminal peduncle from the leaf axils or close to the top; peduncle 2–5 in. long, first slender, at length rather stout, pedicels slender, ·1–·12 in. long. *Calyx* slender, obconical-campanulate, ·2 to ·23 in. long; teeth triangular. *Petals* elliptic, acute, ·2–·25 in. long, white or greenish white. *Anthers* slender, acuminate, ·23–·27 in. long. *Style* filiform; stigma punctiform. *Fruit* trigonous, obconical with straight sides, ·22–·3 in. long, smooth; valves ·15–·16 in. long.

PERAK; 2000–4000 ft., common, *Curtis* 2015! *Scortechini* 1911! Maxwell's Hill, *Scortechini* 18/*a*! *Ridley!* Larut, on rocks in dense jungle, 2000 ft., *Kunstler* 2005! in open jungle on hill sides, 500–800 ft., *King's Collector* 5152! Kinta, *Curtis!* MALACCA; Bujong, 3000 ft., *Curtis!*

18. SONERILA MUSCICOLA, Stapf and King. A flaccid, ascending, unbranched herb, 4–6 in. high, with a creeping rhizome, with pale, fine, curved or curled hairs in the upper part of the stem and the leaves, and with scanty, gland-tipped hairs on the pedicels, calyx and on the back of the petals. *Stem* slender, weak, quadrangular. *Leaves* of a pair similar in shape and size, oblong to lanceolate-oblong, subacute or subacuminate at both ends, symmetrical or almost so, minutely toothed or almost entire, thinly membranous, pale-green, 2–3·7 in. by ·8–1·4 in., pinnate-nerved, nerves 3–4 on each side, fine, very oblique; petiole ·3–·8 in. long. *Cymes* few-flowered, terminal, peduncled, umbelliform, peduncles very slender, 1–1·5 in. long; pedicels very slender, ·08–·12 in. long. *Calyx* slender, obconical-campanulate, ·2 in. long; teeth triangular, broad. *Petals* elliptic, apiculate, ·35–·4 in. long, pink, with a few gland-tipped hairs along the middle nerve beneath. *Anthers* very slender, acuminate, tips curved, ·22–·24 in. long. *Style* filiform; stigma punctiform. *Fruit* trigonous, truncate-obovoid, ·2 in. long, smooth; valves ·12 in. broad.

KEDAH; Gunong Raya, on mossy trees, *Curtis* 2573!

19. SONERILA SAXOSA, Stapf and King. An erect, delicate herb, 2–4 in. high, scantily hairy with the exception of the glabrous flowers, hairs pale, flexuous with long, fine tips; with a slender, creeping rhizome. *Stem* very slender, quadrangular. *Leaves* in 3–4 pairs (of which the upper are rather close), those of a pair similar in shape and size, oblong-lanceolate to lanceolate, acute at both ends, rather symmetrical,

finely toothed in the upper part or almost entire, membranous, green above, purplish beneath, ·8–1·7 in. by ·3–·6 in., 5–7-nerved from below the middle; petiole ·2–·3 in. long. *Cymes* 4–7-flowered, much contracted, terminal; peduncles very slender, ·5–1·5 in. long; pedicels very slender, ·15 in. long, glabrous. *Calyx* very slender, linear-subcampanulate, ·15 in. long, teeth broad, triangular. *Petals* elliptic, acuminate, cuspidate, ·3–·35 in. long, pink. *Anthers* acuminate, ·19 in. long. *Style* filiform, ·3 in. long; stigma punctiform. *Fruit* obconical, truncate, sides almost straight, passing into the thickened pedicel, ·25 in. long; valves ·12–·15 in. broad.

PENANG; Government Hill, 2500 ft., on rocks in damp shady ravines; *Curtis!*

20. SONERILA CONGESTA, Stapf and King. An erect or suberect, rather delicate herb, 3–3·5 in. high, quite glabrous with the exception of an extremely scanty, furfuraceous, dark-rusty indumentum in the lower part; with a slender, creeping rhizome. *Stem* 1–1·5 in. long (exclusive of the peduncle), quadrangular. *Leaves* in about 3 crowded pairs, those of a pair similar in shape and size, broad, ovate to elliptic, obtuse or subobtuse, rounded or obscurely cordate at the base, rather symmetrical, subentire, thinly membranous, green, ·9–1·7 in. by ·6–1·2 in., 5–7-nerved from near the base, upper pair ·4–·6 in. from the base; petiole slender, ·6–·7 in. long. *Cymes* 4–9-flowered, contracted; peduncles slender, 1·2 in. long; pedicels slender, ·12–·15 in. long. *Calyx* slender, subcampanulate-oblong, ·15 in. long; teeth triangular, short. *Petals* elliptic-oblong, cuspidate-acuminate, ·35 in. long, pink. *Anthers* acuminate, ·15–·2 in. long. *Style* filiform, ·25–·3 in. long; stigma punctiform. *Fruit* (semimature) obovoid-oblong, ·15 in. long.

KEDAH; Gunong Chinchang, *Curtis* 2572!

21. SONERILA GRIFFITHII, C. B. Clarke in Hook. f. Fl. Brit. Ind. II, 539. An ascending, delicate herb, quite glabrous with the exception of the margins and sometimes the upper surface of the leaves; with a rather stout rhizome. *Stems* very slender, 2–3 in. long (exclusive of the panicle), rooting from the lower, soon leafless nodes, quadrangular. *Leaves* in 3–4, often crowded pairs, those of a pair similar in shape and size, broad, ovate to rotundate-elliptic, obtuse or subobtuse, obscurely cordate, rarely subobluse at the base, rather symmetrical, toothed and ciliate on the margin, membranous, green, ·4–·8 in. by ·4–·6 in., 5-nerved from near the base, upper pair ·1 in. from the base; petioles slender, ·3–·5 in. long. *Cymes* 2–3-flowered, much contracted, peduncles slender, 1–2·2 in. long; pedicels very slender, ·1 in. long. *Calyx* slender, subcampanulate-oblong, ·12 in. long; teeth triangular-ovate. *Petals* elliptic-oblong, cuspidate-acuminate, scarcely ·25 in. long, pink. *Anthers* acuminate, ·12–·15 in. long. *Style* filiform, not quite ·25 in. long; stigma

punctiform. *Fruit* truncate, obovoid-oblong, cuneate at the base when quite ripe, ·19–·23 in. long, obtusely trigonous. Cogn. in DC. Mon. VII, 513; Stapf in Ann. Bot. VI, 308.

MALACCA; Mt. Ophir, on dripping places, *Griffith* 2300! *Maingay* 2583! *Lobb* 182.

22. SONERILA ÇYCLAMINELLA, Stapf and King. A rather delicate, perfectly glabrous, almost acaulescent herb, with a creeping rhizome. *Stem* usually extremely short, quadrangular. *Leaves* in about 4 pairs, almost crowded into a rosette, those of a pair similar in shape and size, ovate to oblong-obtuse or subobtuse at both ends or subcordate at the base, rather symmetrical, undulate-crenulate or almost entire, membranous, light or dark brownish green above with silvery bands along the midrib and often also along the side, nerves more or less rich violet underneath, ·8–1·4 in. by ·4–·6, distinctly although finely 5–7-nerved below the middle, often with 1 or 2 delicate side-nerves higher up; petioles slender, ·4–1 in. long. *Cymes* 2–5-flowered, very much contracted; peduncles slender, 2–5 in. long, pedicels slender, ·08–·1 in. long. *Calyx* slender, subcampanulate-oblong, ·12–·15 in. long; teeth triangular, short. *Petals* elliptic, cuspidate-acuminate, scarcely ·25 in. long, light pink. *Anthers* acuminate, ·12 in. long. *Style* filiform, ·25 in. long; stigma minutely subcapitate. *Fruit* truncate, obovoid, obtusely trigonous, ·12–·16 in. long.

PERAK; on rocky hilltops, 800–1000 ft., *King's Collector* 10745! 10746! 10744 (in part)!

VAR. *canescens*, Stapf and King. ' Leaves more or less covered above with long, flexuous, white hairs; some of Ridley's specimens have leaves up to 3 in. by 1·2 in.

PERAK; with the type; *King's Collector* 10744 (in part)! SELANGORE; Bukit Kinta, 3000 ft., on rocks, *Ridley* 7318!

23. SONERILA HETEROSTEMONA, Naud. in Ann. Sc. Nat. Ser. 3, XV, 326, t. XVIII, fig. 4. An erect or ascending, often branched herb, ·5–2 ft. high, rarely quite dwarf, quite glabrous apart from an extremely fine, furfuraceous, rusty indumentum in the younger parts and, occasionally, a few scattered, short, whitish hairs on the upper surface of the leaves. *Stem* somewhat stout, quadrangular. *Leaves* of a pair similar in shape, but usually very unequal or one arrested at a very early stage or quite suppressed, rarely both more or less equal, usually conspicuously asymmetrical, obliquely ovate, subacute or shortly acuminate, rounded or shortly narrowed at the base, minutely toothed, membranous, metallic green, often spotted above, purplish beneath on the nerves, 1·5–4·5 in. by 1–2·3 in., 5–6-nerved from the very base with fine, lax, more or less horizontal transverse veins; petioles very unequal in length, ·5–2 in. long. *Cymes* axillary and terminal, much contracted

and compact, very many-flowered, distinctly bracteate, at length 1–2.5 in. long, peduncle ·6–1·2 in. long ; bracts spathulate or obovate, very obtuse, up to ·2 in. long, persistent ; pedicels hardly any. *Calyx* short, obconical-oblong, ·15–·16 in. long ; teeth obscure, very obtuse. *Petals* elliptic, obtuse, pink, ·15 in. long. *Stamens* 6, 3 slightly curved, purple, ·15 in. long, 3 straight or almost so, yellow, ·12–·15 in. long. *Fruit* subsessile, turbinate, ·19–·23 in. long ; valves ·19 in. broad. Miq. Fl. Ind. Bat. I, 565 ; Triana in Trans. Linn. Soc. XXVIII, 77 ; C. B. Clarke in Hook. f. Fl. Brit. Ind. II, 540. *S. obliqua,* Cogn. in DC. Monogr. VII, 515 ; and Stapf in Ann. Bot. VI, 310 (in part), not of Korth.

PERAK ; *Scortechini !* Ipoh, *Curtis* 3158 ! Changkal Serdang, *Wray* 783 ! Larut, *Scortechini* 54/a ! Goping, *Kunstler* 787 ! Tapa, *Wray* 1308. MALACCA ; *Griffith* 2302 ! 2294 ! *Maingay* 1223 ! (782, Kew Distr. partly), *Cuming* 2349 ! *Lobb* 183 ! in dense forest between Jassing and Ayer Bombon, *Maingay* 1425 ! (782, Kew Distrib. partly). MALACCA ; Batang, *Holmberg* 876 ! Ulu Gujah, *Harvey !* (dwarf specimens). SINGAPORE ; *Maingay* 3098 ! (782, Kew Distrib partly) ; Bukit Timah, *Hullet* 893 ! PAHANG ; Tahan, *Ridley !* (dwarf specimens). DISTRIB. Sumatra to Borneo.

24. SONERILA INTEGRIFOLIA, Stapf in Ann. of Bot. VI, 312. An erect or ascending, simple or branched herb, ·5–1·3 ft. high, rufously strigose on the stem, the petioles, the nerves on the underside of the leaves and the inflorescence, including the calyx, but exclusive of the bracts, hairs of the leaves very tightly adpressed, like those of the inflorescence very short. *Stem* rather robust, often swollen at the nodes, subflexuous, almost woody below. *Leaves* very dissimilar, the larger of a pair asymmetrical, rarely symmetrical, usually obliquely oblong-lanceolate to obovate-oblong, distinctly (sometimes long) acuminate, minutely cordate or acute at the base, entire, sometimes with slightly wavy margins, somewhat fleshy, soft, quite glabrous above, dark- or yellowish-green, 3–5 in. by 1–1·5 in., sub-5-nerved from near the base with the lowermost pair of nerves faint, and the uppermost (in the narrow half) ·2–1·2 in. above the base, with fine oblique transverse veins ; small leaves minute, ovate to rotundate, often cordate, sessile or shortly petioled. *Cymes* terminal and apparently leaf-opposed, peduncled, bracteate, few- to many-flowered, very dense, up to 1 in. long, peduncles very short to ·5 in. long ; rhachis often flexuous when long ; bracts oblong to linear-lanceolate, fleshy, up to ·12 in. long, often much smaller, sometimes extremely numerous and crowded ; pedicels very short or 0. *Calyx* oblong-campanulate, nearly ·1 in. long ; teeth triangular, up to ·04 in. long, acute. *Petals* oblong, acute, ·15 in. long, white to pink. *Anthers* oblong, obtuse, almost ·1 in. long. *Style* ·2 in,

long ; stigma punctiform. *Fruit* semiglobose-turbinate, ·12–·18 in. long and wide, bullate-muricate.

PERAK; Larut, 200–800 ft., *Kunstler* 1917! 2791! Changkal Serdang, *Wray* 755! Blanda Mobok, *Wray* 3954! Maxwell's Hill, *Scortechini* 16a! Hermitage, *Curtis* 1302! SELANGOR; Dusun Tua, *Ridley* 7334! Kwala Tampan Caves, *Ridley* 306!

VAR. *acuminatissima*, Stapf and King. Leaves mostly very long and finely acuminate, on the whole narrower and less asymmetric than in the type, margins often slightly wavy to remotely serrulate, not rarely with a row of white spots close to them. *Petals* white.

PERAK ; Larut, 1800–4000 ft., in dense old jungle, *Kunstler* 2004! 2161!

25. SONERILA BRACTEATA, Stapf and King. An erect or ascending, unbranched or very scantily branched herb, ·5–1·5 ft. high, softly and densely hirsute to tomentose from rufous, flexuous or curved, more or less spreading, fine hairs in all parts with the exception of the upper side of the leaves which is glabrous apart from scattered, adpressed, pale bristles. *Stem* rather stout below with swollen nodes, subflexuous, leafy part 2 to over 6 in. long. *Leaves* very dissimilar, the larger of a pair shortly petioled, somewhat asymmetrical, oblanceolate, long and finely acuminate, unequally cordate at the base with a small rounded lobe on the outer, and a still smaller or obscure lobe on the inner side, entire, ciliate along the margin, membranous, light-green, 3–7 in. by 1–1·7 in., 5-nerved from near the base (the uppermost nerve ·5–1·5 in. above the base), with oblique, transverse veins; petioles ·2 to ·4 in. long; small leaves reniform, very minute or up to ·3 in. in diam. *Cymes* terminal and axillary, long-peduncled, very dense, subcapitate at first, at length to 1·5 in. long, multibracteate, many-flowered; peduncle rather slender, up to 2 in. long; bracts linear, membranous, ciliate, up to ·12 in. long ; pedicels very short. *Calyx* shortly oblong-campanulate, ·07–·08 in. long; teeth lanceolate-triangular, about ·04 in. long. *Petals* oblong, cuspidate-acuminate, white, ·08 in. long, with a line of gland-tipped hairs on the back. *Anthers* short, oblong, obtuse, ·06–·07 in. long. *Style* ·15 in. long; stigma punctiform. *Fruit* shortly turbinate, ·15 in. long and wide, densely muricate.

PERAK; Larut, in dense old jungle, 3200–3500 ft., *Kunstler* 2133! Maxwell's Hill, *Scortechini* 12!

26. SONERILA CAPITATA, Stapf and King. An ascending, unbranched or scantily branched herb, 3–12 in. high, rufously strigillose on the stem, the petioles and the nerves on the underside of the leaves, and also in the cymes, and with few or very few scattered, short hairs on the upperside of the leaves. *Stem* prostrate at the base, stout, succulent, swollen at the nodes, leafy part up to 7 in. long. *Leaves* very dissimilar,

the larger of a pair petioled, asymmetrical, obliquely obovate, elliptic or oblong, abruptly contracted into a narrow acumen, unequally cordate at the base with a small rounded lobe (to ·2 in. long) on the broader side and gradually narrowed on the inner side, entire, membranous, light-green, 4–6 in. by 1·75–3 in., 5–8-nerved from near the base, with 2–5 nerves in the broad, and 2 in the narrow half, with oblique, on both sides distinctly raised, transverse veins ; petiole stout, ·5–1·5 in. long, small leaves orbicular-reniform, acute, cordate, sessile, up to ·4 in. in diam.   *Cymes* terminal and axillary, long-peduncled, capitate, very dense, bracteate, few- to many-flowered ; peduncles rather slender, glabrescent in the upper part, bracts numerous, linear, up to ·25 in. long ; pedicels slender, ·07–·09 in. long.   *Calyx* oblong, densely shaggy from short hairs, thickened below, ·07–·09 in., teeth narrow, triangular, ·04 in. long.   *Petals* oblong, acuminate, white or pinkish, ·12 in. long, with a line of short, thick hairs on the back.   *Anthers* short, oblong, obtuse, ·1 in. long.   *Style* filiform, ·1 in. long, stigma punctiform.   *Fruit* semiglobose, densely muricate, ·15 in. long and wide.

PERAK ; *Scortechini* 1886 ! Gunong Batu Pateh, in dense jungle, 3000–4000 ft.   *Kunstler* 8075 ! 4500 ft.   *Wray* 222 !

27.   SONERILA CAESIA, Stapf and King.   An ascending or creeping, low herb, densely hairy on the stem, petioles and the nerves and veins on the underside of the leaves, less so in the inflorescence and · with few or no hairs on the upperside of the leaves, hairs rufous, fine, straight and adpressed, particularly on the nerves, or more or less spreading on the petioles, coarse on the rhachis of the cyme and at the base of the umbels. *Stem* rather stout, prostrate below, leafy-part rarely more than ·5 in. long. *Leaves* very dissimilar, the larger of a pair long-petioled, asymmetrical, obliquely elliptic, subacuminate or subobtuse, unequally cordate at the base with a large rounded lobe (·4–·6 in. long) on the outer, and a minute or quite obscure lobe on the inner side, entire or subentire, sometimes ciliate along the margin, somewhat fleshy, blue-green above, pale, green beneath with reddish nerves and veins, 3–5 in. by 1·75–2·75 in., 6–7-nerved from near the base, 3–4 nerves in the outer (larger), 2 nerves in the inner (narrow) half, with subhorizontal or oblique transverse veins ; petiole 1–3·5 in. long ; small leaves minute, ovate-cordate, shortly petioled, or suppressed.   *Cymes* terminal and from the upper leaf-axils, usually 2–3 in peduncled umbels with small bracts at the base, bracteate, few- to very-many-flowered, very dense ; common peduncle slender, 1–2·5 in. long ; special peduncles ·25–1 in. long ; pedicels slender, ·08–·1 long ; bracts oblong, obtuse, as long as or shorter than the pedicels, glabrous, persistent.   *Calyx* obconical-campanulate, ·12 in. long, scabrid, crimson ; teeth triangular, acute, distinct. *Petals* oblong, cuspidate-acuminate, ·22 in. long, pale pink.   *Anthers*

short, oblong, obtuse, ·1 in. long.   *Style* ·25 in. long ; stigma punctiform. *Fruit* semiglobose, ·15 in. long and wide, tubercled, tubercles rounded.

PERAK ; Gunong Batu Pateh, 3,400 ft., *Wray* 1035 ! UPPER PERAK ; 300 ft , *Wray* 3442 ! 3553 !

28. SONERILA NIDULARIA, Stapf and King. An ascending, simple, rarely furcate herb, 3–8 in. high, densely rusty-tomentose on the stem, petioles and the nerves (rarely also between the nerves) on the underside of the leaves, pubescent in the inflorescence (including the calyx), other-wise glabrous ; hairs fine, flexuous, short to very short and more or less adpressed or, in the upper part of the stem, sometimes longer and more or less spreading. *Stem* prostrate at the base, stout, straight or flexuous, rooting at the base, leafy part 1–5 in. long. *Leaves* very dissimilar, the larger of a pair shortly petioled, asymmetrical, obliquely oblong to obovate-oblong, subacuminate or subobtuse, unequally cordate at the base with a larger, rounded lobe (·12–·2 in. long) on the outer and a similar, but much smaller lobe on the inner side, entire, fleshy, rather firm, dark-green above, sometimes with a row of large white spots on each side of the midrib, 3–4·5 in. by 1–2 in., 5-nerved from near the base with oblique transverse veins, the outer nerve of the inner (narrow) side marginal and often indistinct ; petiole stout, ·15–·6 in. long ; small leaves sessile, reniform or orbicular, cordate, ·2 in. or less in diam. *Cymes* terminal and axillary, sessile, minutely or obscurely bracteate, few- to many-flowered, much contracted ; pedicels very short at first, ultimately up to ·3 in. long, and stout. *Calyx* campanulate-oblong, ·12–·15 in. long ; teeth triangular, acuminate, up to ·06 in. long. *Petals* obovate-elliptic, cuspidate, almost ·25 in. long. *Anthers* short, oblong, obtuse, ·14 in. long. *Style* ·25 in. long, stigma punctiform. *Fruit* shortly turbinate, subtrigonous, muricate, ·25 in. long and wide.

PERAK ; *Scortechini* 650 ! Larut, 1000–2000 ft., *Kunstler* 2345 ! on hills in open jungle, *King's Coll.* 5764 ! Gunong Haram (?), *Scortechini* 655 ! Waterloo, common, *Curtis !*

29. SONERILA BRACHYANTHA, Stapf and King. An ascending simple or scantily branched herb, 3–8 in. high, softly hirsute or shaggy all over, hairs dense and more or less spreading on the stem, the petioles and all parts of the inflorescence (inclusive of the calyx), looser on both sides of the blades, reddish, rather long and flexuous. *Stem* rather stout, subflexuous, leafy part 1–3 in. long, branches, if any, spreading, resembling the main stem. *Leaves* very dissimilar, the larger of a pair petioled or subsessile, more or less asymmetrical, obliquely ovate-lanceo-late to oblong-lanceolate, acuminate, unequally cordate at the base with a rounded lobe (·15–·25 in. long) on the outer and a minute lobe on the inner side, entire, membranous, dark brownish-green above,· reddish or

deep-red or violet beneath, 2–4 in. by 1–1·5 in., 5-nerved from near the base, with 3 side-nerves in the broad and 2 side-nerves in the narrow half, uppermost side-nerves sometimes ·4–·5 in. above the base, with indistinct oblique transverse veins ; petioles usually short or very short, rarely up to ·5 in. long ; small leaves ovate-cordate, acute or reniform, ·15 in. long and broad, on slender, short petioles. *Cymes* terminal and axillary, solitary or in fascicles of 2–3, subsessile, indistinctly bracteate, few- to many-flowered, very dense ; peduncles very short, slender, with 2 petioled small leaflets at the base ; rhachis shaggy, bracts finely filiform, ciliate, hidden among the hairs of the rhachis ; pedicels very slender, about ·1 in. long. *Calyx* campanulate-oblong, ·12–·15 in. long, shaggy ; teeth triangular-lanceolate, ·07 in. long. *Petals* oblong, subacute, suberect, over ·25 in. long, pinkish white or pink, with a line of hairs on the back. *Anthers* short, oblong, obtuse, ·1 in. long. *Style* over ·25 in. long ; stigma punctiform. *Fruit* semiglobose-turbinate, muricate, ·2 in. long and wide.

PERAK ; *Scortechini* 1873 ! 1875 ! Goping, in dense jungle, *Kunstler* 434 ! 440 ! Larut, in dense jungle, 500–800 ft., *King's Collector* 5752 ! MALACCA ; Kinta Gunong, 1000–1500 ft., on rocky places, *King's Collector* 7179 ! Gunong Inas, 5000 ft., *Wray* 4088.

30. SONERILA MICROCARPA, Stapf and King. An ascending herb, 3–6 in. high, rusty-tomentose on the stem, petioles and the underside of the leaves, more coarsely hairy in the inflorescence (including the calyces) from short, somewhat stiff and spreading, or soft and more adpressed (underside of the leaves) hairs, and besides with scattered, longer and stouter, flexuous hairs on the upper side of the leaves. *Stem* long, prostrate at the base, somewhat stout, straight or subflexuous, leafy part 2–4 in. long. *Leaves* very dissimilar, the larger of a pair shortly petioled, asymmetrical, obliquely obovate-lanceolate or oblanceolate, abruptly acuminate, unequally cordate at the base with a larger rounded lobe (·1–·2 in. long) on the outer, and a similar, but very minute or obscure lobe on the inner side, subentire or entire or obtusely serrulate, fleshy, rather firm, dark-green above, sometimes with numerous small white spots, 2·5–4 in. by 1–1·5 in., 4–5-nerved from near the base with oblique transverse veins, petiole ·15–·4 in. long ; the small leaves sessile, reniform or orbicular, cordate, ·2 in diam. *Cymes* terminal and from the upper axils, subsessile or shortly peduncled, ebracteate, few- to many-flowered, rather lax ; peduncle very slender, if any, up to ·5 in. long ; pedicels filiform, up to ·25 in. long. *Calyx* campanulate-ovoid, ·12 in. by ·08 in. ; teeth triangular. *Petals* oblong, acute, almost ·25 in. long, like the calyx pink. *Anthers* short, oblong, obtuse, ·08–·1 in. long. *Style* ·25 in. long ; stigma punctiform. *Fruit* pale pink, subtrigonous, turbinate, minutely muricate, ·08–·11 in. long, ·15 in. broad.

PERAK ; *Scortechini !* Upper Perak, 300 ft., *Wray* 3445 ! 3446 ! 3621 !

31. SONERILA COSTULATA, Stapf and King. An ascending, unbranched herb, a few inches high, densely hirsute or tomentose on the stem, the petioles and more or less also in the inflorescence, including the calyx ; adpressedly strigillose on the nerves and veins on the underside of the leaves, and with scattered, often very few, stouter hairs on the upper side of the leaves, hairs rufous, those of the stem, petioles and inflorescence flexuous, more or less spreading. *Stem* prostrate below, subflexuous, leafy part rarely more than 1 in. long. *Leaves* crowded, the larger of a pair petioled, more or less asymmetrical or the upper sometimes almost symmetrical, obliquely (if asymmetrical) obovate-oblong or oblong, subacuminate, unequally cordate at the base with a rounded lobe (·2 in. long) on the larger and a minute lobe on the narrower half, entire, ciliolate along the margin, membranous, dark green, 3–6 in. by 1·3–2·5 in., 6- or rarely 7-nerved from near the base with 3 (rarely 4) nerves in the broad and 2 in the narrow half, with oblique, conspicuously prominent, transverse veins on both sides ; petiole stout, ·3–·5 in. long ; small leaves ovate-cordate to reniform, minute or up to ·33 in. long, on short petioles. *Cymes* terminal and in the upper axils, solitary or 2 on a common very short peduncle, ebracteate, few- or many-flowered, contracted ; rhachis very slender ; peduncles very short ; pedicels slender, ·08–·12 in. long. *Flowers* unknown. *Fruit* semiglobose-turbinate, muricate, ·15–·18 in. long and wide.

PERAK (?) ; foot of Gunong Panti, *Kunstler* 220 !

Rather closely allied to *S. Beccariana,* Cogn. ; but this has on the whole narrower, more acuminate leaves and much larger fruits.

32. SONERILA MACROPHYLLA, Stapf and King. An ascending simple herb, 3–5 in. high, softly hirsute or shaggy all over ; hairs dense and more or less spreading on the stem, the petioles and all parts of the inflorescence (inclusive of the calyx), looser on both sides of the leaves, pale reddish, rather long and flexuous. *Stem* prostrate below. *Leaves* very dissimilar, the larger of a pair petioled, asymmetrical, obliquely elliptic, acuminate, unequally cordate at the base with a large, rounded lobe (·4 in. long) on the outer and a much smaller lobe on the inner side, entire or subentire, membranous, on both sides light-brown when dry, 4–6 in. by 2·2–3 in., about 7-nerved from near the base, with 4 nerves in the broad, 2 in the narrow half and with usually indistinct, fine, oblique, transverse veins, uppermost side-nerve 1–1·25 in. above the base ; petiole stout, ·5–1·2 in. long ; small leaves ovate-cordate to reniform, up to ·25 in. long, on short, slender petioles. *Cymes* terminal and in the upper axils, solitary or 2 or a common short peduncle, ebracteate, rather many-flowered, apparently very dense owing to the long interwoven

hairs; rachis slender; common peduncle very short to ·7 in. long, slender, with a pair of spathulate-lanceolate, petioled leaflets at the point of branching; special peduncles very short; pedicels very slender, up to ·15 in. long. *Calyx* campanulate-oblong, about ·14 in. long, very shaggy; teeth triangular-lanceolate, up to ·06 in. long. *Petals* oblong, acute, suberect, ·3–·35 in. long; stigma punctiform. *Fruit* semiglobose-turbinate, muricate, about ·2 in. long and wide.

PERAK; *Scortechini!*

VAR. *laxipilosa*, Stapf and King. All parts loosely hairy with the hairs as in the type. *Leaves* up to 6 in. by 3·5 in., rather thinner. Common and special peduncles short or up to 3 in. long (together).

PERAK; Ipoh, Kinta, *Curtis* 3154! Pulau Butong, *Curtis*!

33. SONERILA PARADOXA, Naud. in Ann. Sc. Nat. Ser. 3, XV, 321. A low, creeping herb, softly hirsute or shaggy all over; hairs dense and more or less spreading on the stem, petioles and all parts of the inflorescence (including the calyx), looser on both sides of the blades, reddish, rather long and flexuous. *Stem* creeping, slender to rather stout, rooting, the leaf-bearing, terminal part rising rarely more than ·5 in. above the ground. *Leaves* crowded, very dissimilar, the larger of a pair petioled, asymmetrical, obliquely oblong or elliptic, shortly acuminate, unequally cordate at the base, with a large rounded lobe (·25–·5 in. long) on the outer and a similar but much smaller lobe on the inner side, entire or subentire, membranous, soft, light-green, 3–6 in. by 1·2–2·5 in., 6–8 nerved from near the base (3–5 nerves in the broader half), with oblique curved transverse veins; petiole ·4–2 in. long or the uppermost very short; small leaves rotundate-ovate or reniform, cordate, ·08–·4 in. in diam., on very slender petioles (·08–·6 in. long). *Cymes* terminal and from the upper axils, solitary with a pair of small petioled leaflets at the middle of the peduncle, or in umbels of 2–4, ebracteate or inconspicuously bracteate, few- to many-flowered, dense; common peduncle slender, usually 1–2 in. long; special peduncles much shorter; rhachis very shaggy; bracts linear to filiform, ciliate, short, usually hidden among the hairs of the rhachis or suppressed; pedicels ·08–·1 in. long, very slender. *Calyx* campanulate-oblong, about ·15 in. long, shaggy; teeth short, triangular. *Petals* oblong, acute, suberect, over ·25 in. long, white. *Anthers* oblong, obtuse, ·1 in. long. *Style* over ·3 in. long; stigma punctiform. *Fruit* semiglobose-turbinate, ·12–·15 in. long, ·18 in. wide, muricate-tuberculate. *S. moluccana*, Jack. Misc. I, 8; Wall. Cat. 4089; Benn. Pl. Jav. Rar. 215, (p.p.); Blume, Mus. I, 10 (p.p.); Miq. Fl. Ind. Bat. I, 562 (p.p.); C. B. Clarke in Hook. f. Fl. Brit. Ind. II, 537 (p.p.); Triana in Trans. Linn. Soc. XXVIII, 77; Cogn. in DC. Monogr. VII, 508 (p.p.); Stapf in Ann. Bot. VI, 311, 312 (p.p.); and Roxb. Flor. Ind. I, 178?

PENANG; *Wallich* Cat. 4089! *Griffith* 2298! *Maingay* 780 (Kew Distrib.) in shady, damp places, 1500–3000 ft., *Stoliczka, Hullet* 196! *King's Coll.* 1284! Pulloh Bahang, *Curtis* 411! SINGAPORE (?) ; *Lobb* 325!

Roxburgh says of his *S. moluccana*, "Habitat in insulis Moluccanis." His description is extremely short and insufficient, and there does not seem to have been a specimen in his herbarium nor was it figured by him. It is very improbable that the plant he described was identical with the Penang plant, if he received it really from the Moluccas, as the distribution of most species of the section *Hexadon* is very local, and no specimens, referrable to *S. paradoxa*, have been discovered, so far, east of the Malay Peninsula. On the other hand, it is possible that Roxburgh meant *S. malaccana* instead of "*S. moluccana*" and *insulis malaccanis* for "*ins. moluccanis*," as the editors of his Flora Indica put it. There is at least nothing in his description which would contradict the assumption that his brief diagnosis was drawn up from the Penang plant. In view of this uncertainty we have preferred to follow Naudin and to consider Roxburgh's *S. moluccana* as a "species dubia" and adopt Naudin's name for the Penang plant.

34. SONERILA BEGONIAEFOLIA, Blume in Flora (1831), 490. An ascending, usually unbranched herb, 2–6 in. high, moderately hairy with the exception of the often glabrous upper side of the leaves; hairs of the stem, petioles and the inflorescence rufous, flexuous, more or less spreading, of the leaves confined to the nerves and veins of the underside, often scanty, very short. *Stem* rather slender, rooting below. *Leaves* very dissimilar, the larger of each pair petioled, more or less asymmetrical, obliquely elliptic, subacuminate, unequally cordate at the base with a rounded lobe (·2–·3 in. long) on the outer and a much smaller on the inner side, entire or more or less obtusely serrulate, ciliolate, membranous, dark-green above, pale brown (when dry) beneath, 3–4 in. by 1·7–2·3 in., 6- sub-7-nerved from near the base (with 3–4 nerves in the broader half), with usually very conspicuous subhorizontal transverse veins; petiole ·4–1·2 in. long; small leaves ovate to rotundate, acute, cordate, very small, distinctly petioled. *Cymes* terminal and axillary, peduncled, dense, at length up to ·8 in. long, deciduously bracteate; peduncle slender, up to 1·5 in. long; bracts linear-oblong, ciliolate, up to ·1 in. long, deciduous; pedicels ·07–·1 in. long. *Calyx* campanulate-oblong, teeth broad, triangular. *Petals* ovate, acute. *Anthers* short, oblong, obtuse. *Fruit* shortly turbinate, ·18–·22 in. long and wide, muricate-tuberculate, tubercles rather coarse, acute, mostly passing into short fine bristles. Korth. in Verh. Nat. Gesch. Bot. 248, t. 54; Naudin in Ann. Sc. Nat. Ser. 3, XV, 322; Triana in Trans. Linn. Soc. XXVIII, (1873), 77. *S. moluccana*, Benn. Pl. Jav., Rar. 215; Miq. Fl. Ind. Bat. I, 562; C. B. Clarke in Fl. Brit. Ind. I, 562; Cogn. in DC. Monogr. VII, 508; Stapf in Ann. Bot. VI, 312 (all references under *S. moluccana*, p.p.).

SINGAPORE; Bukit Tunat, *Ridley* 2005! Chanchukang, *Ridley* 422! Bukit Mandu, *Ridley* 2005/*a !* JOHORE; Gunong Panti, *Ridley* 4199! DISTRIB. Sumatra, Java, South Borneo.

There being no flowers with the specimens enumerated, they have been described from Korthals, l.c. The Sumatra specimens have leaves which are more or less hairy or bristly on the upper side and represent Blume's VAR. *pilosiuscula* of *S. begoniaefolia* (Blume, Mus. I. 11) or *S. moluccana* VAR. *pilosiuscula* Stapf, l.c. A specimen from South Borneo, collected by Motley, is almost glabrous on the upperside of the leaves.

35. SONERILA GLABRIFLORA, Stapf and King. A creeping or ascending herb, a few inches high, with a long creeping rhizome, hairy on the stem, peduncles, petioles and the nerves on the underside of the leaves, otherwise glabrous; hairs reddish, straight, adpressed. *Stem* rather stout, slightly swollen at the nodes, frequently rooting, leafy part 1-4 in. long. *Leaves* very dissimilar, the larger of each pair petioled, asymmetric, obliquely elliptic, obtuse or subacute, unequally cordate at the base with a large, rounded lobe on the outer and small or obscure one on the inner side, entire or obscurely and remotely toothed, fleshy, soft, dark glossy-green above, pale with purple nerves beneath, 3-6 in. by 2-3·5 in., with 3-5 lateral nerves in the outer and 2-3 in the inner half near the base and with rather lax subhorizontal transverse veins; petiole rather stout, 1-2·5 in. long, the small leaves sessile, reniform-cordate, ·2–·25 in. in diam. *Cymes* terminal and axillary, usually 2- or 3-nate on a common peduncle, ebracteate, many-flowered, rather dense, glabrous, when ripe up to 1·3 in. long, common peduncle 1-2 in., special peduncles ·5-1·2 in. long, slender; pedicels very slender, ·1-·12 in. long. *Calyx* campanulate-ovoid, ·12-·15 in. by ·06 in.; teeth very broad and short. *Petals* elliptic, acute, almost ·25 in. long, white or tinged with pink. *Anthers* linear-oblong, obtuse, ·12 in. long. *Style* ·15 in. long; stigma punctiform. *Fruit* subtrigonous, hemispherical, quite smooth, ·15-·16 in. long and wide; valves in the depressed centre of the capsule 6, delicately membranous, fragile.

PERAK; Larut, in wet jungles, up to 100 ft., *Kunstler* 1955! 2128!

36. SONERILA ELATOSTEMOIDES, Stapf and King. An erect, ascending or creeping, unbranched or scantily branched herb, up to 6 in. high, softly tomentose in all parts, hairs rusty coloured, short, spreading and very dense on the stem and petioles, somewhat laxer in the inflorescence inclusive of the calyx, pale and longer on the leaves. *Stem* slender, subflexuous. *Leaves* very dissimilar, the larger of each pair petioled, asymmetrical, obliquely oblong-elliptic or obovate, acuminate, unequally cordate at the base with the outer lobe broad, rounded, up to ·2 in. long and the inner similar but much smaller, entire, membranous, green

above, whitish green below, 2·2-2·8 in. by 1-1·2 in., 6-7-nerved from near the base (with 3-4 lateral nerves in the broad and 2 in the narrow half) with oblique transverse veins, nerves and veins not very distinct; petiole ·18-·4 in. long; small leaves reniform-cordate, subsessile, ·15 in. in diam. *Cymes* terminal and axillary, long peduncled, few-flowered, not very dense, ebracteate; peduncle slender, ·6-1 in. long; pedicels slender, ·02 in. long. *Calyx* oblong-campanulate, ·08 in. long, teeth short. *Petals* oblong, acute, ·18 in. long, with a line of hairs on the back. *Anthers* short, oblong, obtuse, ·09 in. long. *Style* ·2 in. long; stigma punctiform. *Fruit* semiglobose, muricate, ·12 in. long and wide.

PERAK; Gunong Bubu, *Wray* 3825!

37. SONERILA BICOLOR, Stapf and King. An ascending or suberect, nearly always unbranched herb, 3-9 in. high, softly hairy all over, hairs pale, straight, fine, spreading except on the upper side of the leaves, very dense in the upper part of the stem, on the petioles and in the inflorescence inclusive of the calyx. *Stem* rather slender, subflexuous, slightly swollen at the nodes. *Leaves* very dissimilar, the larger of each pair more or less asymmetrical, shortly petioled, obliquely oblong to lanceolate-oblong, distinctly acuminate, unequally cordate at the base with a rounded lobe (to ·4 in. long) on the outer, and a minute or obscure lobe on the inner side, ciliate along the margin, membranous, soft, dark green or magenta-red above with a light green band along the midrib, purple beneath, 3-6 in. by 1-1·5 in., finely 5-nerved from near the base (uppermost nerve from ·4-·6 in. above the base, with faint oblique transverse veins; petiole ·2-·6 in. long; small leaves ovate to rotundate, cordate, shortly petioled, very small to ·4 in. long. *Cymes* terminal and axillary, peduncled, loosely few- to 10-flowered, ebracteate; peduncle slender, ·4-1·2 in. long; pedicels ·1-·2 in. long, slender, also when mature. *Calyx* oblong-subcampanulate, ·12 in. long; teeth triangular, short. *Petals* elliptic-oblong, acute, ·35 in. long, pink. *Anthers* linear-oblong, subobtuse, ·12 in. long. *Style* ·35 in. long; stigma punctiform. *Fruit* semiglobose, ·12-·15 in. long and wide, muricate, opening with 6 thinly membranous teeth in the depressed centre or with 3, 2-toothed valves.

PERAK; Ulu Salama, 500 ft., *Wray* 4159! Larut, in dense jungle, 500-800 ft., *King's Coll.* 5794! Tapa, *Baldwin*!

38. SONERILA CALYCULA, Stapf and King. An ascending or creeping herb, about 6 in. long, rusty tomentose on the stem, petioles and peduncles, and pubescent on the nerves on the underside of the leaves, otherwise glabrous; hairs short, curled or flexuous, loosely adpressed. *Stem* rather stout, straight, rooting in the lower part, leafy part about 4 in.

long. *Leaves* very dissimilar, the larger of each pair petioled, symmetrical, or almost so, lanceolate, gradually tapering towards both ends, entire, somewhat fleshy, pale-brown on both sides when dry, 3·5–5·5 in. by ·8–1·25 in., 5-nerved from the base (the outer nerves faint, submarginal) with faint oblique transverse veins; petiole ·35–·4 in. long; small leaves reniform, cordate, very minute, ·05 in. in diam., or suppressed. *Cymes* terminal and from the upper axils, peduncled, few-flowered; peduncle filiform, ·4–·5 in. long; pedicels fine, not spongy when mature, ·1–·12 in. long. *Flower* unknown. *Fruit* semiglobose, smooth, ·08–·1 in. by ·12 in., crowned by the ultimately deciduous calyx-margin the teeth of which are very broadly triangular and cuspidate.

PAHANG; Tahan River, *Ridley* 2237!

9. PHYLLAGATHIS, Blume.

Herbaceous small shrubs with very short stems. *Leaves* opposite (or the terminal leaf solitary), large, petioled, orbicular, or sub-orbicular, 7–9-nerved. *Flowers* in a peduncled dense head, purple. *Calyx-tube* campanulate, glabrous or with long bristles near the top, teeth 4 (rarely 3), acute, long-setose. *Petals* 4 (rarely 3), ovate, acute, glabrous. *Stamens* 8 (rarely 6), equal; anthers elongate, scarcely produced at the base, connective without appendage. *Ovary* adnate to the bottom of the calyx-tube, 4- (rarely 3-) celled, glabrous at the apex; style filiform; ovules very numerous, placentas large axile. *Capsule* broadly funnel-shaped, opening by 4 valves at the top. *Seeds* ellipsoid, somewhat obovoid, with glandular, hardly raised dots; raphe slightly excurrent along one side of the seed its whole length.—DISTRIB. Species 5; all Malayan.

Leaves oblanceolate	1. *P. tuberculata.*
Leaves more or less orbicular; peduncles of inflorescence many inches long and as long as or longer than the long petioles:—	
Flowers in narrow, shortly branched panicles ...	2. *P. Griffithii.*
Flowers in terminal solitary ebracteate umbels:—	
Peduncles glabrous	3. *P. Scortechinii.*
Peduncles hispid	4. *P. hispida.*
Flowers in bracteate umbels with peduncles 1 or 2 inches long	5. *P. rotundifolia.*

1. PHYLLAGATHIS TUBERCULATA, King, n. sp. *Stem* short, woody, erect, covered with small warts. *Leaves* two or three, membranous, oblanceolate, shortly acuminate, very gradually narrowed to the short hispid petiole, 7- to 9-nerved; upper surface glabrous except for a few

scattered stout bristles; lower surface coarsely strigose on the nerves, otherwise glabrous, the reticulations fine; length 10 to 16 in., breadth 3 to 6 in., petiole 0 to 1·5 in. *Peduncles* much shorter than the leaves, solitary or several from one axil, sparsely glandular-hairy, bearing at their apices a solitary, lax, involucrate, compound umbel; involucres oblong, obtuse, glabrous, ·5 in. in length or shorter, those of the umbellules smaller; pedicels shorter than the calyx-tube. *Flowers* nearly ·75 in. long. *Calyx-tube* rather widely cylindric, bearing (especially towards its base) numerous flat, shortly stalked, fleshy discoid glands; the teeth 4, short, shallow, wide, each crowned by a stalked gland. *Petals* longer than the calyx, oblong, much acuminate (forming a narrow cone in bud). *Stamens* 8; the filaments about one-fourth as long as the linear acuminate anthers. *Ovary* short, broad, truncate, deeply grooved and almost winged; style long, flattened, smooth. *Capsule* ·3 in. in diam.

PERAK; *Scortechini* 1872. *King's Collector* 7233.

At once recognisable by the curiously glandular calyx.

2. PHYLLAGATHIS GRIFFITHII, King. A shrub with a very short stem and usually only a single pair of unequal leaves. *Leaves* coriaceous, sub-rotund, 7-nerved, apex obtuse, the base cordate, edges quite entire; both surfaces glabrous, the lower lepidote and slightly puberulous on the nerves; length 5 to 8 in., breadth 3·5 to 6·5 in., petiole 4 to 8 in. *Panicle* axillary, nearly as long as the leaves, bearing in its upper half a few very short few-flowered branches, the whole lepidote especially towards the apex. *Flowers* corymbose, on scaberulous pedicels, as long as the narrowly campanulate glandular-hairy calyx-tube: mouth of calyx truncate, slightly expanded, and minutely 4-toothed. *Petals* 4, quadrate. *Stamens* much exserted: *anthers* not so long as the glabrous filaments, narrow acuminate, much sagittate at the base, inappendiculate. *Allomorphia Griffithii*, Hook. MSS. Fl. Br. Ind. II, 527. Triana in Trans. Linn. Soc. XXVIII, 74, t. VI, fig. 66c; Cogn. in DC. Mon. Phan. VII, 467.

MALACCA: *Griffith* (K.D.) 2264/1. *Maingay* (K.D.) 775; *Hullett.* PERAK; *Scortechini* 170; *King's Collector* 694. SELANGORE; *Curtis* 2333! *Ridley* 7317. PENANG; *Stolickza.*

I have removed this from *Allomorphia* to *Phyllagathis* of which it has the flowers and habit.

3. PHYLLAGATHIS SCORTECHINII, King, n. sp. *Stem* woody below, shortly creeping above and subterete. *Leaves* on very long petioles, coriaceous, reniform-rotund, cordate, the apex minutely apiculate, stoutly 9-nerved; the transverse veins bold, glabrous on the upper surface, glandular-puberulous on the lower; length 5 to 10 in.; breadth

5 to 9 in. ; petiole 4 to 9 in., stout. *Peduncles* solitary, longer and more slender than the petioles, bearing at the apex a single dense, ebracteate umbel. *Flowers* nearly ·5 in. long to the apices of the stamens, their pedicels ·25 in. long. *Calyx-tube* narrowly. campanulate, not ribbed, glabrous, very minutely lepidote ; the teeth 4, broad, rounded, shallow. *Petals* rotund-ovate, blunt, short. *Stamens* 8; the filaments about half as long as the linear acute anthers. *Ovary* grooved, short, broad ; style as long as the anthers, cylindric, glabrous. *Capsule* about ·2 in. in diam., truncate, the valves broad, truncate.

PERAK; *Scortechini* 269 ; *King's Collector* 4287. SELANGORE ; *Ridley* 7317.

This species is closely allied to *P. Griffithii* but differs notably in its solitary terminal umbels.

4. PHYLLAGATHIS HISPIDA, King, n. sp. *Stem* very short, woody, hispid. *Leaves* 2 or 3, sub-coriaceous, on long petioles, broadly ovate, cordate at the base, the apex abruptly and shortly acuminate, the edges with minute, sharp, sometimes unequal teeth, 7- to 11-nerved, the transverse nerves strong: upper surface glabrous, the lower uniformly covered with minute, scurfy pubescence, the main nerves with a few long, spreading, stout bristles; petiole densely shaggy near the base, more sparsely hispid upwards: length 5 to 10 in., breadth 3 to 7 in. ; petioles varying from 4 to 12 in. *Peduncle* axillary, shorter than the leaves and more slender than the petioles, sparsely hispid ; the umbel few-flowered, ebracteate. *Flowers* ·75 in. long ; their pedicels slender, hispid, longer than the calyx. *Calyx-tube* narrowly cylindric or cylindric-campanulate, nearly glabrous, or sparsely hispid ; the mouth with 4 broad, shallow, blunt teeth. *Petals* broadly ovate, acute. *Stamens* 8 ; *anthers* not much longer than the filaments, broadly lobed and cordate at the base, and the connective with a short, narrowly cylindric basal protuberance behind. *Ovary* very short ; *style* filiform. *Capsule* ·2 in. across.

PERAK ; *Scortechini !* *Wray* 1021, 1602, 3519. . PAHANG ; *Ridley* 2236.

5. PHYLLAGATHIS ROTUNDIFOLIA, Blume in Flora, 1831, 507. *Stem* creeping, obtusely 4-angled. *Leaves* unequal in the pairs, rotund or rotund-ovate and shortly apiculate ; both surfaces, but especially the lower, minutely lepidote ; main nerves 7 to 9, curved, radiating from the base, the transverse nerves bold, curved ; petioles unequal, from ·85 to 3·5 in. long, their interior surfaces covered with coarse black bristles; length 2 to 8 in.; breadth 1·5 to 6 in. *Inflorescence* 4- to 30-flowered ; the involucres orbicular-ovate acuminate. *Calyx-tube* ribbed, minutely lepidote ; the teeth triangular, broad at the base but with

elongate narrow apices bearing 2 or 3 bristles; filaments from one-fourth to one-half of the length of the linear stamens. *Ovary* 4-angled, the apex with a truncate cartilaginous rim. Korth. in Verh. Nat. Gesch. Bot. 252, t. 57; Naud. in Ann. Sc. Nat. Ser. 3, XXV, 332; Bot. Mag. t. 5282; Miq. Fl. Ind. Bat. I, 559; Triana Melast., tab. VI, fig. 73; C. B. Clarke in Hook. fil. Fl. Br. Ind. II, 541; Cogn. in DC. Mon. Phan. VII, 518. *Melastoma rotundifolia,* Jack in Trans. Linn. Soc. XIV, 11; DC. Prodr. III, 149.

MALACCA; *Griffith, Maingay.* SELANGORE; *Ridley* 7327. PERAK; *Scortechini, King's Collector, Wray.* DISTRIB. Burma, Sumatra, Java; common.

Rather variable as to size of leaves and as to the number of bristles on their petioles. The teeth of the calyx also vary as to the length of the acuminate apex and as to the number of bristles.

10. MARUMIA, Blume.

Twining shrubs; branches cylindric, thickened at the nodes. *Leaves* opposite, short-petioled, coriaceous, cordate at the base, 3-nerved from the base besides two submarginal nerves, entire, stellate-tomentose beneath. *Cymes* axillary; flowers 3–5, large, pedicelled, purple or white. *Calyx-tube* narrowly campanulate, tomentose (and often bristly or stellate-hairy); lobes 4, deep, persistent. *Petals* 4, obovate. *Stamens* 8, unequal; anthers elongate, opening by a single pore; connective of the longer anthers carrying in front two long bristles and behind often one or two spurs or several twisted bristles. *Ovary* at the base (or half its height) adnate to the calyx, 4-celled, densely hairy at the apex; style filiform; ovules numerous, placentas axile. *Berry* ellipsoid, crowned by the calyx-limb. *Seeds* numerous, oblong-ellipsoid, with glandular scarcely raised dots, raphe slightly excurrent along the whole length of one side.—DISTRIB. Species 10, Malaya, Borneo and the Philippines.

Calyx-tube without bristles ... . 1. *M. nemorosa.*
Calyx-tube very bristly :—
    Teeth of limb of calyx half as long as the tube,
    oblong, acute: upper surface of leaves not reti-
    culate ... ... 2. *M. rhodocarpa.*
    Teeth of limb of calyx one-fourth as long as the tube
    or less, broadly triangular, blunt; upper surface of
    leaves much reticulate ... ... 3. *M. reticulata.*

1. MARUMIA NEMOROSA, Blume in Flora XIV, (1831), 505. A straggling climber; all parts except the upper surfaces of the leaves, the petals, stamens and ovaries covered with dense rufous or pale stellate

tomentum, without bristles; the branches thickened and annulate at the nodes. *Leaves* sub-coriaceous, elliptic or oblong-lanceolate, slightly cordate at the base, the apex shortly acuminate or acute, 5-nerved (the lateral pair of nerves faint); upper surface glabrous; length 3 to 6 in.; breadth 1·35 to 2·75 in.; petiole ·2 to ·4 in. long. *Flowers* about 1·75 in. long (including the stamens), solitary or in pedunculate cymes of three from the axils of the leaves. *Calyx-tube* more or less narrowly campanulate, somewhat constricted below the limb; limb with 4 deep, triangular teeth. *Petals* broadly ovate, blunt, longer than the calyx-tube, rose-coloured. *Stamens* 8, unequal; the anthers of all linear, curved; the longer with two narrow, curved filaments at the base in front, and several smaller behind; the four smaller with two smaller, equal filaments at the base in front only. *Fruit* succulent, oblong-ovoid, constricted below the permanent calyx-teeth, sometimes sub-tuberculate and always stellate-tomentose, about 1 in. long. Blume Mus. Bot. I, 33; Naud. in Ann. Sc. Nat. Ser. 3, XV, 279; Miq. Fl. Ind. Bat. I, pt. I, 533; Triana in Trans. Linn. Soc. XXVIII, 82; C. B. Clarke in Hook. fil. Fl. Br. Ind. II, 542; Cogn. in DC. Mon. Phan. VII, 549. *M. affinis*, Korth. in Verh. Nat. Gesch. Bot. 241, t. 60; Miq. l.c. 533. *Melastoma nemorosum*, Jack in Trans. Linn. Soc. XIV, 8; DC. Prodr. III, 149; Wall. Cat. 4043.

In all the provinces except the Nicobar and Andaman Islands. DISTRIB. Sumatra, Borneo.

2. MARUMIA RHODOCARPA, Cogn. in DC. Mon. Phan. VII, 550. A powerful climber; young branches, petioles, lower surfaces of leaves and calyx densely clothed with minute, pale, stellate tomentum intermixed with numerous stout, spreading, brown bristles, the nodes somewhat swollen and with transverse lines. *Leaves* sub-coriaceous, oblong, narrowed to the rounded, minutely cordate base, the apex shortly acuminate, 5-nerved (the lateral pair slender); upper surface glabrous; length 3·5 to 5 in.; breadth 1 to 2 in.; petiole ·15 to ·2 in. *Flowers* (including the stamens) nearly 1·5 in. long, in axillary, pedunculate cymes of three. *Calyx-tube* longer than the glabrous pedicel, narrowly campanulate, constricted below the limb; limb with 4 oblong, acute teeth half as long as the tube and like the latter bearing many long, curved bristles. *Petals* white, obovate. *Stamens* 8, unequal, all linear and acuminate: the four larger with two long, filiform appendages in front and several smaller behind: the four smaller with about 4 to 6 appendages. *Fruit* shortly ovoid, crowned by the large calyx-teeth, bristly, ·4 in. in diam. *Melastoma rhodocarpum*, Wall. Cat., 4045. *Marumia echinata*, Naud. Ann. Sc. Nat. Ser. 3, XV, 280; Miq. Fl. Ind. Bat. I, pt. I, 534. *M. zeylanica*, Triana

(not of Blume) in Linn. Trans. XXVIII, 82, tab. VII, fig. 88b; C. B. Clarke in Fl. Br. Ind. II, 542.

SINGAPORE; *Wallich; Anderson* 68, 69; *Hullett* 125; *Ridley* 258; *King's Collector* 278.   MALACCA; *Maingay* (K.D.) 785; *Cuming; Griffith* (K.D.) 2270.

VAR. *sub-glabrata,* Cogn. l.c. 550.   Leaves glabrous beneath except the nerves. *M. zeylanica,* C. B. Clarke (not of Blume), VAR. *sub-glabrata,* Hook. fil. Fl. Br. Ind. II, 542.

SINGAPORE ; *Anderson* 64.

3. MARUMIA RETICULATA, Blume Mus. Bot. I, 34.   Scandent; young branches, petioles, under surfaces of leaves and calyx densely clothed with rusty stellate, more or less deciduous hairs, the young branches, petioles and especially the calyx with stout spreading bristles intermixed; the nodes swollen and annulate. *Leaves* subcoriaceous, ovate-lanceolate, 5-nerved (the lateral pair slender), minutely cordate at the base, the apex shortly acuminate; upper surface deeply reticulate, glabrous; length 2 to 3·5 in., breadth ·8 to 1·5 in., petiole ·1 to ·2 in. *Flowers* 1·25 in. long (including the anthers), axillary, either solitary on pedicels as long as themselves or in pedunculate cymes of three; the middle flower being nearly sessile, the lateral pair on short pedicels. *Calyx-tube* densely hispid externally, the teeth 4, short, broadly triangular and blunt, tomentose on the inner surface. *Petals* ovate, sub-acute, glabrous. *Stamens* 8, the anthers somewhat unequal in length but all linear, curved and with two long appendages at the base in front. *Fruit* ovoid, (unripe) ·5 in. in diam., crowned by the calyx-teeth. Miq. Fl. Ind. Bat. I, pt. I, p. 535; C. B. Clarke in Hook. fil. Fl. Br. Ind. II, 542; Cogn. in DC. Mon. Phan. VII, 551. *M. stellulata,* Korth. (not of Blume) Ver. Nat. Gesch. Bot. 243. *M. oligantha,* Naud. in Ann. Sc. Nat. Ser. 3, XV, 281 ; Miq. l.c. 534.

MALACCA; *Griffith* (K.D.) 2269; *Maingay* 784. DISTRIB. Sumatra, Java.

11. DISSOCHÆTA, Blume.

Shrubs, usually twiners. *Leaves* opposite, petioled or nearly sessile, elliptic or oblong-lanceolate, rounded at the base, 5-nerved from the base (the two submarginal nerves slender), entire. *Flowers* in terminal, sometimes leafy panicles, purple or white, bracts large or small. *Calyx-tube* campanulate-cylindric or funnel-shaped, densely stellate-tomentose, pubescent or glabrous; limb obscurely 4-lobed or entirely truncate, more rarely distinctly 4-toothed, persistent. *Petals* 4. *Stamens* 8, unequal, 4 shorter sometimes wanting, connective of the 4 longer with 2 long bristles in front at the base. *Ovary* adnate to the calyx, 4-celled, apex glabrous or densely hairy; style filiform; ovules very many, placentas axile. *Berry* ovoid or elliptic, crowned by the calyx-limb; *Seeds* elipsoid, flattened on the side of the raphe.—DISTRIB. Species 25; throughout Malaya to the Philippines.

Stamens 8 (four of them sometimes imperfect) :—
  Young branches, under surfaces of leaves and panicles
  covered with persistent stellate-tomentum :—
    The stellate-tomentum not mixed with bristles;
    the nodes with transverse interpetioler lines :—
      Flowers 1·75 in. long   ...   .   1. *D. annulata.*
      Flowers ·75 in. long   ...   .   2. *D. punctulata.*
    The stellate-tomentum mixed with spreading
    persistent bristles; flowers ·75 in. long   ...   3. *D. hirsuta.*
  Young branches, under surfaces of leaves and panicles
  covered at first with stellate-tomentum which ulti-
  mately more or less disappears :—
    Bracts of panicle obovate, persistent ...   ...   4. *D. bracteata.*
    Bracts of panicle linear, deciduous   ...   ...   5. *D. pallida.*
  Young branches, under surfaces of leaves and panicles
  covered with minute simple scales not stellate   ...   6. *D. gracilis.*
Stamens 4 :—
  Panicles ebracteate, or bracts, if any, caducous :—
    Flowers ·3 in. long; fruit ·15 to ·2 in. in diam.;
    mouth of calyx-tube 4-toothed; petals broadly
    oblong, blunt; stamens blunt, not appendiculate
    at the apex   ...   .   7. *D. celebica.*
    Flowers ·5 in. long; fruit ·2 to ·25 in. in diam.;
    mouth of calyx-tube truncate not toothed;
    petals ovate, acute; stamens narrowed to the
    appendiculate apex   ...   .   8. *D. intermedia.*
  Panicles with persistent, oblong bracts :—
    Young branches at first rusty stellate-hairy,
    finally glabrous; branches of panicle divari-
    cating, lax   ...   .   9. *D. anomala.*
    Young branches with persistent, pale, adpressed,
    stellate hairs; branches of panicle short,
    condensed   ...   .   10. *D. Scortechinii.*

  1. DISSOCHÆTA ANNULATA, Hook. fil. ex Triana in Trans. Linn. Soc. XXVIII, 83. A strong climber; young branches, petioles, under surfaces of leaves and inflorescence covered with dense, rusty, deciduous stellate tomentum, the nodes thickened and slightly annulate. *Leaves* coriaceous, ovate-oblong, cordate at the base, the apices shortly and rather abruptly acuminate, upper surface at first sparsely stellate-hairy, ultimately glabrous, the nerves and reticulations bold; length 2·5 to 3·5 in.; breadth 1 to 1·8 in.; petiole ·2 to ·5 in. *Panicle* solitary ter-minal, much longer than the leaves, lax, the branches divaricating,

cymose, 1- to 3-flowered; bracts small, caducous. *Flowers* 1·5 in. long (including the stamens), pedicels much shorter than themselves. *Calyx-tube* narrowly campanulate or funnel-shaped, the mouth widened and with 4 blunt, triangular teeth, densely stellate-hairy outside. *Petals* obovate-oblong, blunt, reflexed. *Stamens* 8; curved, elongate-linear, acuminate, with two bristle-like appendages at the base in front, laciniate behind. *Fruit* ellipsoid, succulent, crowned by the enlarged teeth of the calyx-tube. C. B. Clarke in Hook. fil. Fl. Br. Ind. II, 543; Cogn. in DC. Mon. Phan. VII, 557. *Melastoma bracteatum*, Wall. Cat. 4044 (in part).

PENANG; *Wallich* 4044; *Griffith* (K.D.) 2268; *Maingay* (K.D.) 788; *Curtis* 740. MALACCA; *Wallich.* SINGAPORE; *Hullet* 213; *Ridley* 5187. JOHORE; *King's Collector* 224. PERAK; *Scortechini* 235. DISTRIB. Borneo.

2. DISSOCHÆTA PUNCTULATA, Hook. fil. ex Triana in Linn. Trans. XXVIII, 83. Young branches thickened but not annulate, clothed like the petioles, under surfaces of the leaves and the inflorescence with minute, rusty scales. *Leaves* coriaceous, oblong-ovate, the base rounded and not cordate, the apex acute or very shortly acuminate; 5-nerved, the marginal nerves very slender; upper surface glabrous; length 3 to 4·5 in.; breadth ·75 to 1·75 in.; petiole ·25 to ·4 in. *Panicles* solitary, terminal, several times longer than the leaves, the branches and branchlets short and the latter cymosely few-flowered, bracts small, linear, deciduous. *Flowers* ·75 in. long (including the stamens), on pedicels shorter than themselves. *Calyx-tube* infundibuliform, the mouth with a broad, everted, wavy, obscurely toothed edge, deciduously scaly. *Petals* 4, ovate, sub-acute, glabrous. *Anthers* 8, equal, curved, cylindric, attenuated to the 1-pored apex, the base with two long, geniculate, upward-curving, narrow, flattish appendages. *Fruit* succulent, urceolate, slightly warted, sub-glabrous, crowned by the slightly enlarged mouth of the calyx, ·4 in. long. C. B. Clarke in Hook. fil. Fl. Br. Ind. II, 543; Cogn. in DC. Mon. Phan. VII, 555.

MALACCA; *Griffith* (K.D.) 2291! *Maingay* 789. SINGAPORE; *Ridley* 3918, 4803. SELANGORE; *Ridley* 2015. JOHORE; *Ridley* 3246, 2106. PENANG; *Walker, etc.*

3. DISSOCHÆTA HIRSUTA, Hook. fil. ex Triana Trans. Linn. Soc. XXVIII, 83. A strong creeper with slender branches only slightly thickened at the nodes; all parts except the upper surfaces of the leaves and the petals densely clothed with deciduous, stellate hairs mixed with long, stiff, spreading, curved hairs. *Leaves* membranous, lanceolate or ovate-lanceolate, 5-nerved (the lateral pair slender), the base rounded and slightly cordate, the apex shortly acuminate; upper surface

glabrous, length 3 to 4·5 in.; breadth 1 to 1·75 in.; petiole ·15 to ·3 in. *Panicle* solitary, terminal, thickened at the nodes, broadly pyramidal, much branched. *Flowers* ·75 in. long. *Calyx-tube* longer than the pedicel, cylindric-tubular, the mouth not everted but with four linear elongate teeth. *Petals* 4, broadly lanceolate, blunt. *Stamens* 8' equal, all perfect, elongate and narrow, the apices much prolonged into a rather thin appendage, the base with two delicate, filiform appendages. *Fruit* ovoid-globose, crowned by the persistent calyx-limb, densely setose-lepidote, ·4 in. in diam. Cogn. in DC. Mon. Phan. VII, 556.

JOHORE; at the base of Gunong Panti; *King's Collector* 197; *Ridley* 4185. DISTRIB. Borneo.

4. · DISSOCHÆTA BRACTEATA, Blume in Flora, 1831, 495' Young branches with a transverse ridge at the nodes, petioles and panicles more or less densely clothed with sub-deciduous, rusty, stellate hairs, and the under surfaces of the leaves sparsely so. *Leaves* membranous, 5-nerved (the marginal pair faint), ovate-lanceolate, the base rounded and often sub-cordate, the apex acute or shortly acuminate; upper surface glabrous, the lower sparsely stellate-hairy; length 3 to 4·5 in.; breadth 1·25 in.; petiole ·2 to ·3 in. *Panicles* much bracteate, both axillary and terminal, the former shorter, the latter longer, than the leaves; the branches few, divaricate. *Flowers* 1·2 in. long (including the stamens), in cymes at the ends of the branchlets, each subtended by, and while in bud enveloped in, a membranous, obovate, blunt, stellate-pubescent, more or less permanent bract about ·5 in. long. *Calyx-tube* twice as long as the pedicels, cylindric-campanulate, the mouth slightly widened and minutely 4-toothed. *Petals* broadly elliptic, blunt, glabrous. *Stamens* 8, much curved, long, linear, the base with two long flattened, linear, upward-pointing appendages. *Fruit* ellipsoid, crowned by the rim of the calyx-tube, ·6 in. long. Miq. Fl. Ind. Bat. I, pt. I, 529; Triana in Trans. Linn. Soc. XXVIII, 84; C. B. Clarke in Hook. fil. Fl. Br. Ind. II, 543; Cogn. in DC. Mon. Phan. VII, 598. *D. bracteosa*, Naud. in. Ann. Sc. Nat. Ser. 3, XV, 76; Miq. Fl. Ind. Bat. I, pt. I, 527. *Melastoma bracteatum*, Jack in Trans. Linn. Soc. XIV, 9; Wall. Cat. 4044, partly.

PENANG; *Wallich*, Cat. 4044; *Curtis* 2298. . MALACCA; *Maingay* 791. DISTRIB. Borneo; *Haviland.*

5. DISSOCHÆTA PALLIDA, Blume in Flora, 1831, 500. A shrubby creeper, 20 to 50 feet long; young branches thickened and with inter-petiolar ridges at the nodes, sparsely covered with minute stellate-hairy scales like the petioles, under surfaces of the leaves and the panicles. *Leaves* sub-coriaceous, 5-nerved (the lateral pair faint), elliptic-ovate to ovate, the base rounded and sub-cordate, the apex shortly acuminate

or acute; upper surface glabrous; length 2·5 to 5 in.; breadth 1·2 to 2·2 in.; petiole ·3 to ·5 in. *Panicles* both axillary and terminal, the former shorter than the leaves or slightly exceeding them, the latter longer, all rather lax, the branchlets divaricating, cymose, and with a few short, linear, deciduous bractlets. *Flowers* nearly 1 in. long (including the stamens). *Calyx-tube* oblong-campanulate or funnel-shaped, densely lepidote-stellate; the mouth expanded and obscurely 4-toothed. *Petals* obovate-oblong, blunt, glabrous. *Stamens* 8, usually equal (four sometimes shorter or obsolete); basal processes long, linear, sub-erect. *Fruit* cylindric-campanulate, sub-glabrous, crowned by the slightly enlarged limb of the calyx, ·25 in. long. Blume, Mus. Bot. I, 36 (excl. syn. Korth.); Naudin in Ann. Sc. Nat. Ser. 3, XV, 69, tab. 4 fig.; Miq. Fl. Ind. Bat. I, pt. I, 528; Triana in Trans. Linn. Soc. XXVIII, 83, tab. VII, fig. 89*b*; C. B. Clarke in Hook. fil. Fl. Br. Ind. II, 544; Cogn. in DC. Mon. Phan. VII, 557. *Melastoma pallida,* Jack in Trans. Linn. Soc. XIV, 12; DC. Prodr. III, 150; Wall. Cat. 4049. *Dissochæta ovalifolia* and *D. superba,* Naud. l.c. 76 and 77. *D. astrotricha,* Miq. l.c. Suppl. 318.

In all the provinces except the Nicobar and Andaman Islands; common.

The four stamens which are usually shorter than the other are sometimes obsolete.

6. DISSOCHÆTA. GRACILIS, Blume in Flora, 1831, 498. A straggling or scandent shrub; young branches slender, bluntly 4-angled, the nodes swollen and transversely ridged, minutely scaly like the petioles, leaves and panicles. *Leaves* 5-nerved (the lateral pair slender), broadly lanceolate or oblong-lanceolate, rounded at the base and shortly acuminate at the apex; length 3 to 4·5 in.; breadth 1·5 to 1·8 in.; petiole ·2 to ·4 in. *Panicles* axillary and terminal, the former shorter than, and the latter longer than the leaves, slender, spreading, lax, many-flowered; bracts few, narrowly oblong, caducous. *Flowers* ·35 in. long, on filiform pedicels longer than themselves. *Calyx-tube* narrowly campanulate, minutely stellate-pubescent, the mouth glabrous, sub-truncate, everted, and obscurely 4-toothed. *Petals* 4, broadly ovate or sub-orbicular, blunt. *Stamens* 8; four large, perfect, rather short, with a broad truncate 2-pored apex; the four imperfect small, narrow; all with two erect, filiform flat basal appendages. *Fruit* sub-globular, crowned by the narrow limb, almost glabrous, ·1 to ·15 in. in diam. Korthals Verb. Nat. Gesch. Bot. 237; Naud. in Ann. Sc. Nat. Ser. 3, XV, 75; Miq. Fl. Ind. Bat. I, pt. I, 526; Triana in Trans. Linn. Soc. XXVIII, 83, tab. VII, fig. 89*c*; C. B. Clarke in Hook. fil. Fl. Br. Ind. II, 544; Cogn. in DC. Mon. Phan. VII, 559. *Melastoma gracile,* Jack

in Trans. Linn. Soc. XIV, 14; DC. Prodr. III, 149. *M. fallax,* Wall.
Cat. 4080. ? *M. glauca,* Griff. Ic. Pl. As. 637. *M. vacillans,* var. *pallens,*
Blume, Bijdr. 1074.

In all the provinces except the Nicobar and Andaman Islands.
DISTRIB. Java, Borneo.

7. DISSOCHÆTA CELEBICA, Blume, Mus. Bot. I, 36. A slender
creeper; young branches, petioles and under surfaces of the leaves, also
the panicles, densely clothed with rusty, scurfy, stellate tomentum.
Leaves 3-nerved, membranous, lanceolate or oblong-lanceolate, the base
rounded, the apex shortly acuminate; upper surface glabrous except the
stellate-pubescent midrib; length 2·5 to 5 in.; breadth 1 to 2 in.; petiole
·2 to ·25 in. *Panicles* lateral and terminal, the former slightly longer and
the lateral several times longer than the leaves, ebracteate, the branches
divaricating, the ultimate branches cymose, 3-flowered. *Flowers* ·3 in.
long (including the stamens). *Calyx-tube* oblong-campanulate, slightly
widened and 4-toothed at the mouth, somewhat longer than the pedicel.
Petals 4, broadly oblong, blunt. *Stamens* 4; anthers equal, short, blunt
and with no apical appendage, opening by 2 apical pores. *Fruit*
sub-globular, crowned by the narrow limb of the calyx, faintly 8-ribbed,
sparsely pubescent or sub-glabrous, ·15 to ·2 in. in diam. C. B. Clarke
in Hook. fil. Fl. Br. Ind. II, 544; Cogn. in DC. Mon. Phan. VII, 561;
Miq. Fl. Ind. Bat. I, pt. I, 530; Triana in Trans. Linn. Soc. XXVIII, 83.
D. microcarpa, Naud. in Ann. Sc. Nat. Ser. 3, XV, 72; Miq. l.c. 523.
D. bancana, Miq. l.c. 529. *Melastoma fallax,* Wall. Cat. 4050; ? Jack
in Trans. Linn. Soc. XIV, 13. *M. rubiginosum,* Wall. Cat. 4052, partly.

In all the provinces except the Nicobar and Andaman Islands;
common. DISTRIB. Bangka, Celebes, Borneo.

VAR. *contracta,* King. *Panicle* solitary, terminal, short, condensed,
not longer than the leaves.

PERAK; *King's Collector* 2911.

8. DISSOCHÆTA INTERMEDIA, Blume in Flora, 1831, 493. A somewhat slender creeper, resembling *D. celebica* in its other parts, but
with larger flowers (·5 in. long), pointed in bud; larger fruit (·2 to ·25 in.
in diam.); calyx-tube with a truncate, toothless mouth; ovate, acute
petals, and longer stamens narrowed to and appendiculate at the apex.
Blume, Mus. Bot. I, 35, tab. V; Naud. in Ann. Sc. Nat. Ser. 3, XV, 72;
Miq. Fl. Ind. Bat. I, pt. I, 524; Triana in Trans. Linn. Soc. XXVIII,
83; tab. VII, fig. 89 f.; C. B. Clarke in Hook. fil. Fl. Br. Ind. II, 544;
Cogn. in DC. Mon. Phan. VII, 562. *Melastoma rubiginosum,* Wall. Cat.
4052 (in part).

MALACCA; *Griffith* (K.D.) 2287; *Helfer* (K.D.) 2286. PENANG and
SINGAPORE; *Wallich.* PERAK; *Scortechini.* DISTRIB. Java, Borneo.

9. DISSOCHÆTA ANOMALA, King, n. sp.   A creeper 15 to 20 feet long; young branches slender, slightly thickened at the nodes, at first sparsely rusty stellate-hairy, afterwards glabrous and . sub-glaucous. *Leaves* ovate-oblong, the base broadly rounded, the apex shortly, abruptly and bluntly acuminate, 5-nerved (the two lateral nerves faint); upper surface glabrous; the lower sparsely stellate-hairy on the midrib and nerves, otherwise glabrous; length 3·5 to 6·5 in.; breadth 1·75 to 2·75 in.; petiole sparsely hispid, ·2 in. long. *Panicles* axillary and terminal, the former half as long and the latter twice as long as the leaves, stellate-pubescent especially at the thickened nodes; the branches divaricate, trichotomous, lax, bracteate at the divisions; the bracts ·3 in. long, oblong, blunt, involute, stellate-tomentose, their edges ciliate. *Flowers* ·3 in. long, in cymes of three. *Calyx-tube* shorter than the pedicel, narrowly campanulate, densely stellate-tomentose; the mouth without teeth, truncate, not everted. *Petals* glabrous, ovate, acuminate, forming a pointed bud. *Stamens* 4, equal and all perfect, subsessile, broad, blunt, the lateral basal appendages filamentous. *Fruit* broadly campanulate, crowned by the narrow calyx-limb, glabrous, ·15 in. in diam.

PERAK; *King's Collector* 2258, 10468.

This plant forms a collecting link between the genera *Anplectrum* and *Dissochæta* as they are defined in this work. It agrees in externals with *Anplectrum pallens*, and has the 4 stamens of that genus, but their anthers have the elongate basal processes so well developed in *Dissochæta* and not at all represented in *Anplectrum*.

10. DISSOCHÆTA SCORTECHINII, King, n. sp.   Scandent; young branches slender, the nodes swollen and transversely ridged, thinly clothed with minute, pale, stellate hairs. *Leaves* ovate-lanceolate, slightly cordate at the broad base, the apex with a short, blunt point; 5-nerved (the lateral pair small); upper surface glabrous; lower rusty in colour and bearing sparse, white, stellate hairs longer than those on the stem; length 2·5 to 3·75 in.; breadth 1·5 to 2 in.; petiole densely rusty stellate-tomentose, ·1 in. long. *Panicle* solitary, narrow, terminal, shorter than the leaves, densely rusty stellate-tomentose, the branches short and few-flowered, bearing many oblong, blunt or spathulate, pale, 3-nerved almost glabrous bracts longer than the flower-bud, the lower ones much larger. *Flowers* ·3 in. long, their pedicels short. *Calyx-tube* narrowly campanulate, at first densely but afterwards sparsely stellate-tomentose; the mouth truncate, without teeth, waved but not everted, glabrous. *Petals* 4, glabrous, orbicular-ovate, acuminate, forming a pointed bud. *Stamens* 4, all equal and perfect, short, broadly ovate, the base with two long, erect, filiform appendages, the broad apex with a small, pale,

subacute appendage. *Fruit* unknown. *D. intermedia*, Scort. MSS. (not of Blume), in Herb. prop.

PERAK; *Scortechini* 23, 34. PENANG; *Curtis* 1301.

12. ANPLECTRUM, A. Gray.

Twining shrubs. *Leaves* subcoriaceous, or rarely coriaceous, opposite, short-petioled, entire, oblong, narrowed upwards, 3–5-nerved from the base. *Flowers* white, in terminal panicles sometimes leafy at the base. *Calyx-tube* funnel-shaped or ovoid, limb obscurely 4-lobed or truncate. *Petals* 4. *Stamens* 4 perfect, rarely 8, anthers attenuated upwards, opening by one pore, connective at base shortly appendaged or subnude, never with two long bristles in front, rarely with a long appendage and two small erect bristles; imperfect stamens 4, 2, or 0. *Ovary* 4-celled, free at the apex, with 4 vertical ridges; style simple; ovules many, placentas axile. *Berry* ovoid or globose, crowned with the calyx-limb. *Seeds* very many, small, falcate, obovoid; raphe long, lateral. DISTRIB. Species about 18, in Malaya and the Philippines.

Stamens 4 perfect :—
    Nodes of the young branches and of the lower part of the panicle with conspicuous, stellately lepidote, bristly annuli ... ... 1. *A. lepidoto-setosum.*
    Nodes of the young branches and of the panicle with a small smooth annulus or faint transverse ridge :—
        Calyx-tube funnel-shaped ; stamens 4, all perfect ; young branches stellate-hairy ... 2. *A. glaucum.*
        Calyx-tube ovoid to globular-ovoid ; stamens 8, the anthers of 4 of them narrow and imperfect ; young branches glabrous or nearly so ... ... 3. *A. pallens.*
    Nodes not annulate ; stamens 8, the anthers of 4 of them narrow and imperfect ; young branches and panicles densely stellate-scaly ... ... 4. *A. divaricatum.*
    Stamens 8 perfect ... ... 5. *A. anomalum.*

1. ANPLECTRUM LEPIDOTO-SETOSUM, King, n. sp. Young branches slender, conspicuously annulate at the nodes, sparsely clothed with coarse, spreading hairs with thickened points, the very youngest also with deciduous, stellate hairs. *Leaves* oblong-lanceolate, 5-nerved ; the base rounded and minutely cordate, the apex shortly caudate-acuminate ; both surfaces glabrous except for a few coarse hairs near the petiole ; the lower shining and pale when dry ; length 4·5 to 6 in. ; breadth 1·1 to 1·8 in. ; petiole very short, attached to the cup-shaped, densely rufous stellate-tomentose node and like it with scattered

bristle-hairs. *Panicle* terminal, solitary, shorter than the leaves, everywhere densely rusty-tomentose with long bristles intermixed, annulate at the bases of the short, spreading, few-flowered branches. *Calyx-tube* cylindric; the mouth undulate-truncate, obscurely toothed, its outer surface stellate-lepidote with a few long bristles near the mouth. *Petals* 4, ovate, glabrous. *Stamens* 8 (4 large and 4 small) ; the large broad, and with a short, grooved ridge on the back near the base, blunt; the 4 small linear, very acuminate. *Capsule* ovoid-globular, truncate, nearly glabrous, ·15 in. in diam.

PERAK ; *Scortechini* 2106.

2. ANPLECTRUM GLAUCUM, Triana in Trans. Linn. Soc. XXVIII, 84 (*excluding much of the synonymy*). Scandent, to 20 or 30 feet ; young branches 4-grooved, stellate-hairy like the petioles and main nerves of the leaves and the inflorescence, the nodes inconspicuously annulate or transversely ridged. *Leaves* oblong-lanceolate, rounded or slightly narrowed to the often slightly cordate base, the apex shortly acuminate or acute, boldly 3-nerved ; both surfaces glabrous except for some scattered, stellate hairs ; length 3 to 5 in. ; breadth 1 to 2 in. ; petiole ·5 to ·25 in. *Panicle* large, terminal, solitary, several times larger than the leaves, pyramidal, its branches divaricate, many-flowered. *Flowers* drooping, ·4 or ·5 in. long (including the stamens). *Calyx-tube* funnel-shaped, sub-glabrous ; the mouth truncate, not toothed, everted with age. *Petals* ovate-lanceolate with truncate bases, the apex acuminate. *Stamens* 4 ; *anthers* much curved, all perfect, their bases not lobed but with a corrugated membranous process in front. *Ovary* prominently 4-winged. *Fruit* truncate, sub-globular, glabrous, ·2 in. long (when dry). C. B. Clarke in Hook. fil. Fl. Br. Ind. II, 545 ; Cogn. in DC. Mon. Phan. VII, 566. *A. cyanocarpum,* Kurz in Journ., As. Soc., 1877, pt. 2, p. 78 (not of Triana). *Melastoma glauca,* Jack in Trans. Linn. Soc. XIV, 15 ; DC. Prodr., 151. *M. cernuum,* Wall. Cat. 4055 (not of Roxb.). *Osbeckia tetrandra,* Roxb. Fl. Ind. II, 224. *Dissochæta glauca,* Blume in Flora, 1831, p. 501. *D. spoliata,* Naud. in Ann. Sc. Nat. Ser. 3, XV, 69, t. 4, fig. 1.

In all the provinces, common. DISTRIB. Sumatra, Java, Borneo.

3. ANPLECTRUM PALLENS, Blume, Mus. Bot. I, 38. Scandent, to 30 or 40 feet ; young branches terete, glabrous or minutely puberulous, especially near the slightly thickened and transversely ridged nodes, round. *Leaves* oblong, the base rounded, the apex abruptly, bluntly and shortly sub-caudate-acuminate, 5-nerved (the marginal pair slender), the edges (when dry) slightly recurved ; upper surface glabrous, the lower minutely and scantily stellate-puberulous ; length 1·5 to 4·5 in., breadth ·5 to 1·5 in.; petiole ·1 to ·2 in. *Panicles* axillary (about as

long as the leaves) and terminal (much longer than the leaves), slender, spreading, lax, rather few-flowered, with a small, blunt, oblong, deciduous bract under each branch, finely rufous stellate-pubescent. *Flowers* about ·4 in. long (including the stamens). *Calyx-tube* ovoid to globular-ovoid, densely rufous-puberulous when young, glabrous when old, the mouth slightly expanded and with very small teeth. *Petals* ovate, acute, glabrous. *Anthers*; the four large much curved and sub-acute; the rudimentary linear, acuminate. *Capsule* globose-obovoid, glabrous, ·15 in. in diam. Naud. in Ann. Sc. Nat. Ser. 3, XV, 303; Triana in Trans. Linn. Soc. XXVIII, 303; C. B. Clarke in Hook. fil. Fl. Br. Ind. II, 545; Cogn. in DC. Mon. Phan. VII, 564. *Melastoma petiolare*, Wall. Cat. 4053.

In all the provinces except the Andaman and Nicobar Islands; not uncommon. DISTRIB. Sumatra, Borneo.

4. ANPLECTRUM DIVARICATUM, Triana in Trans. Linn. Soc. XXVIII, 84 (in part), tab. VII, fig. 90b. Scandent, to 20 or 30 feet; young branches obscurely quadrangular, and like the petioles, nerves of the leaves on both surfaces, and the inflorescence, densely covered with tawny, stellate scales, not annulate at the nodes. *Leaves* lanceolate, acute or sub-acute, the base minutely cordate, 5-nerved, the marginal pair slender; both surfaces, but especially the lower, with glandular-punctate scales between the nerves and veins; length 2 to 3 in.; breadth ·6 to 1·1 in.; petiole ·15 to ·25 in. *Panicle* solitary, terminal, pyramidal, several times as long as the leaves, the branchlets divaricate, each with two ovate-lanceolate, ciliate, furfuraceous, deciduous bracts at its base and three flowers in a cyme at the apex. *Flowers* ·4 in. long, on short pedicels. *Petals* oblong, acute. *Calyx-tube* narrowly obovoid, campanulate, truncate, densely furfuraceous stellate-tomentose. *Anthers* 8; the 4 large thick and much curved, obtuse; the smaller narrow. *Capsule* globose-obovoid, with a narrow, everted rim, length ·15 to ·2 in. C. B. Clarke in Hook. fil. Fl. Br. Ind. II, 546; Cogn. in DC. Mon. Phan. VII, 567. *Melastoma divaricatum*, Willd. Spec. Pl. II, 596; DC. Prodr. III, 150. *M. polyanthum*, Benth. in Wall. Cat. 4051. *Disso-chæta divaricata* and *D. pepericarpa*, Naud. Ann. Sc. Nat. Ser. 3, XV, 70 and 71. *D. anceps*, Naud. l.c. 70. *D. palembanica*, Miq. Fl. Ind. Bat. Suppl. 317.

MALACCA; *Griffith* (K.D.) 2288/1: *Maingay* (K.D.) 794: *Harvey*, PENANG; *Wallich* 4051. PERAK; *King's Collector* 369. DISTRIB. Java, Borneo, Sumatra.

5. ANPLECTRUM ANOMALUM, King and Stapf, n. sp. A woody creeper, 20 to 100 feet long; young branches as thick as a wheat-straw, terete, covered with stellate, rusty scurf. *Leaves* coriaceous, obovate or

oblong, blunt, much narrowed to the base, 3-nerved, glabrous on the upper, rusty stellate-hairy on the lower surface like the petioles and inflorescences; length 1 to 1·5 in.; breadth ·6 to ·9 in.; petiole ·2 in. *Panicles* terminal, sometimes leafy, 3 to 6 in. long, lax, the branches in pairs, divaricate, many-flowered, minutely bracteolate at the divisions. *Flowers* ·35 in. long, the pedicels somewhat longer. *Petals* oblong, obtuse, waxy, reflexed, pale greenish-white. *Stamens* 8, equal; anthers inflexed in aestivation, lanceolate-subulate, the base of the lobes produced into an elongated halbert-shaped process with two erect subulate processes at its broad upper end. *Fruit* (not quite ripe) ovoid-globose, greenish-yellow, ·2 in. in diam.

PERAK; *King's Collector* 5779, 10357.

This plant differs from *Anplectrum*, as the genus has hitherto been limited, in having 8 anthers, each of which has a very much produced halbert-shaped basal process, from which two erect hair-like appendages originate at the upper or broad end. The plant agrees better with *Anplectrum* than with any other *Melastomaceous* genus, but it might possibly be better treated as the basis of a new one.

13. MEDINILLA, Gaud.

Branching shrubs, erect or scandent. *Leaves* opposite or whorled, rarely alternate, entire, often fleshy, mostly glabrous, usually longitudinally 3–9-nerved. *Flowers* in terminal panicles or lateral cymes, white or rose, with or without bracts, 4- or 5-, rarely 6-merous. *Calyx-tube* ovoid or cylindric, limb truncate or obscurely toothed. *Stamens* twice as many as the petals, equal or nearly equal (rarely unequal); anthers opening at the top by one pore; connective not (or very shortly) produced at the base but having two tubercles in front and a spur behind. *Ovary* inferior, 4–6-celled, usually glabrous at the apex; style filiform; ovules very many, placentas axile. *Berry* crowned by the limb of the calyx. *Seeds* very many, ovoid or subfalcate, raphe often thickened and excurrent.—DISTRIB. Species about 100; mainly in Malaya, East Bengal and Ceylon; a few in the Fiji Archipelago and in the East African islands.

Flowers 4-merous:—
  Leaves alternate ...   . **1.** *M. scandens.*
  Leaves in whorls (large) ...   . **2.** *M. speciosa.*
  Leaves opposite :—
    Flowers in terminal panicles, anthers dissimilar ... **3.** *M. heteranthera.*
    Flowers in lateral panicles, anthers similar :—
      Anther-cells with tubercles at their bases in front,
      and a short spur from the connective behind :—
        Flowers 1 in. long ...   . **4.** *M. venusta.*

Flowers under ·5 in. long :—
    Leaves petiolate ... ... 5. *M. Hasseltii.*
    Leaves sessile ... ... 6. *M. Scortechinii.*
    Anther-cells with tubercles at the base in front,
    but no spur from the connective behind ... 7. *M. Maingayi.*
Flowers 5-merous :—
  Leaves in whorls :—
    Leaves elliptic or obovate, blunt, 1 to 2·75 in. long,
    fruit ·15 in. in diam. ... ... 8. *M. Clarkei*
    Leaves oblanceolate or narrowly obovate, acuminate,
    3 to 5 in. long ; fruit ·4 in. in diam. ... ... 9. *M. crassinervia.*
    Leaves elliptic-rotund, blunt, 2·5 to 5·5 in. long;
    fruit ·35 in. in diam. ... ... 10. *M. perakensis.*

1. MEDINILLA SCANDENS, King, n. sp.   A climber, 15 to 30 feet long, rooting and adhering to trees ; the stems rough, as thick as a swan's quill. *Leaves* alternate, glabrous, long-petioled, subcoriaceous, elliptic or ovate-oblong, shortly acuminate, the base cuneate ; nerves 5 to 7, mostly from the midrib above its base ; length 4·5 to 9 in. ; breadth 2·5 to 6 in. ; petioles 1·5 to 7 in. *Flowers* ·4 in. long, in dense fascicles in the axils of fallen leaves, on rusty-puberulous, minutely bracteolate pedicels. *Calyx-tube* narrowly campanulate ; the limb very slightly expanded, truncate, very obscurely toothed. *Petals* 4, ovate-oblong, acute. *Stamens* 8 ; *anthers* narrowly elliptic, with a long apical 1-pored beak ; and at the base a short, broad, blunt process from the connective behind, the lobes of the anthers slightly produced in front and minutely tuberculate.

PERAK ; *Scortechini* 86 and 150 ; *King's Collector* 1814.

This resembles *M. alternifolia,* Blume, but has a much shorter spur from the connective at the base of the anthers. It has also larger leaves and more numerous flowers in the fascicles.

2. MEDINILLA SPECIOSA, Blume in Flora, 1831, p. 515.   A glabrous shrub or small tree, not epiphytal ; young branches as thick as the little finger, 3- or 4-angled ; the bark shining, pale when dry, bearing at the nodes numerous stout, subulate bristles ·5 in. long. *Leaves* large, subcoriaceous, in whorls of 3 (rarely of 4) or in pairs, sessile, or very shortly petiolate, oblanceolate or obovate-oblong, sometimes elliptic, acute, the base cuneate, nerves 7 to 9 mostly from the midrib above its base, all except the lowest pair bold, the veins slender ; length 6 to 12 in. ; breadth 2·75 to 5 in. *Panicles* lateral and terminal, 4 to 8 or even 14 in. long, on peduncles equally long, many-flowered ; the branches with a whorl of small reflexed bracts at their bases ; whorled, spreading, minutely bracteolate at the divisions. *Calyx-tube* cupular, slightly constricted below the narrow, minutely 4-toothed limb. *Petals* 4, ovate-

acute. *Fruit* ovoid, ·3 in. in diam.   Bot. Mag. t. 4321 ; Morren in Ann.
Soc. Hort. Gand. V, 281 ; Naud. in Ann. Sc. Nat. Ser. 3, XV, 291 ; Miq.
Fl. Ind. Bat. I, pt. I, p. 540 ; Triana in Linn. Trans. XXVIII, 87 ;
C. B. Clarke in Hook. fil. Fl. Br. Ind. II, 549. *Melastoma eximium,*
Blume Bijdr. (not of Jack).   *Melastoma speciosum,* Reinw. ex Blume,
in Flora, 1831, 516.

 MALACCA ; *Maingay* (Kew Distrib.) 798. PENANG ; *Hullett* 203 ;
Curtis 874 ; *King's Collector* 1595. PERAK ; *Wray* 3218 ; *King's Collector*
2652. DISTRIB. Moluccas, Java, Sumatra.

 3. MEDINILLA HETERANTHERA, King, n. sp. Epipytal and terrestrial; branches slender, smooth, reddish when fresh, drying dark
purplish-brown. *Leaves* of the pairs somewhat unequal, ovate-lanceolate
to lanceolate, acuminate, the base slightly narrowed, glabrous, 3-nerved ;
length 2·5 to 5·25 in.; breadth 1 to 1·8 in.; petioles ·3 to ·8 in. *Panicles*
terminal on the branches, half as long as the leaves, corymbosely
cymose, 5- or 6-flowered, minutely bracteolate. *Flowers* ·75 in. long,
their pedicels much shorter. *Calyx-tube* cylindric, the mouth wide and
with 4 broad, shallow teeth. *Petals* 4, ovate-lanceolate, shortly acuminate. *Stamens* 8 ; *anthers* unequal, the larger four twice as long as
the shorter four, all curved, much acuminate and with two tubercles
at the base in front, the shorter 4 with a short spur on the connective
behind, the larger with no spur. *Fruit* globular-ovoid, crowned by the
wide calyx-limb, ·35 in. in diam.

 PERAK ; *Scortechini* 341 ; *King's Collector,* 3291, 3644, 6304, 6904 ;
Wray 397 ; at elevations of from 3000 to 4500 feet.

 This resembles *M. Horsfieldii,* Miq.,—a species from Java and Borneo—which
however has 5-merous flowers, obovate petals and leaves of thinner texture.
According to Scortechini the petals are waxy white tinged with red and the stamens
are yellow.

 VAR. *latifolia.* *Leaves* broadly elliptic, shortly acuminate, the base
cuneate, 2·5 to 3·75 in. long and 1·5 to 2 in. broad.

 PERAK ; *King's Collector* 8017 ; *Wray* 268.

 4. MEDINILLA VENUSTA, King, n. sp. Epiphytal, 2 to 4 feet long,
glabrous; branches stout, with large, scattered tubercles, glabrous.
Leaves large, opposite, thinly coriaceous, sessile, elliptic, shortly acuminate, narrowed to the base, boldly 3-nerved above the base with often
a faint, small, basal, marginal pair ; length 6 to 9 in.; breadth 2·75 to
4 in. *Cymes* much shorter than the leaves, laxly umbellate, axillary
or from the axils of fallen leaves, about 6-flowered ; the pedicels
slender, bibracteolate at the base. *Flowers* nearly 1 in. long. *Calyx-
tube* campanulate, ·35 in. long, with a narrow, obscurely 4-toothed limb.

Petals 4, broadly ovate, acute.· *Stamens* 8; *anthers* equal, linear-lanceo-late, much acuminate, curved, the tubercles at the front of the base as long as the posterior basal spur from the connective. *Fruit* unknown.

PERAK; *King's Collector* 2390.

5. MEDINILLA HASSELTII, Blume in Flora, 1831, p. 513. Epiphytal on trees, 3 or 4 feet high; branches slender, terete, pale, more or less prominently warted (the warts black). *Leaves* opposite, coriaceous, oblong-lanceolate, acuminate, narrowed to the rounded base, 3-nerved, with sometimes a faint, additional lateral pair; length 4·5 to 5·5 in.; breadth 1·25 to 1·75 in; petioles ·15 to ·35 in. *Cymes* axillary or from the axils of fallen leaves, less than half as long as the leaves, (more than half as long in VAR. *Griffithii*), broader than long; the branches divari-cate, 8–12-flowered, minutely bracteolate. *Flowers* ·35 in. long. *Calyx-tube* campanulate-cylindric, somewhat constricted below the minutely 4-toothed mouth. *Petals* 4, obovate-oblong. *Stamens* 8, equal; the anthers linear-oblong, somewhat curved, the base with two short, black, conical protuberances in front and a similar one behind. *Fruit* globular, truncate, ·2 to ·25 in. in diam. Miq. Fl. Ind. I, pt. I, 542; C. B. Clarke in Hook. fil. Fl. Br. Ind. II, 547; Cogn. in DC. Mon. Phan. VII, 586. *Melastoma laurifolium* in Wall. Cat. 4084 (not of Blume)·. *Medinilla crassifolia,* Triana in Trans. Linn. Soc. XXVIII, 86 (in part).

MALACCA; *Griffith* (Kew Distrib.) 2282; *Maingay* (Kew Distrib.) 797, *Wallich* 4084. PERAK; very common. SINGAPORE; *Anderson.* SUNGEI UJONG; *Ridley* 2205, SELANGORE; *Curtis* 2334; *Ridley* 286. PANGKORE; *Curtis* 1642.

DISTRIB. Java, Sumatra.

VAR. *Griffithii,* C. B. Clarke in Hook. fil. Fl. Br. Ind. II, 547. *Cymes* much branched, more than half as long as the leaves, many-flowered.

MALACCA; *Griffith* (Kew Distrib.) 2282.

6. MEDINILLA SCORTECHINII, King, n. sp. Epiphytal, 3 or 4 feet high; stems as thick as a goose-quill, the bark brown (when dry), sparsely verrucellate. *Leaves* opposite, coriaceous, glabrous, sessile, stem-clasping, oblong-ovate to oblong, shortly acuminate, the base sub-cordate, 3-nerved, with occasionally a faint pair at the margin; length 4·5 to 6·5 in.; breadth 1·5 to 2·75 in. *Panicles* cymose, axillary, rather shorter than the leaves, very lax, spreading; the branches slender, minutely bracteolate at the divarications; the branchlets compressed, sometimes 2-winged. *Flowers* nearly ·4 in. long, on pedicels as long as themselves. *Calyx-tube* campanulate, the mouth not much

expanded, obscurely 4-toothed. *Petals* 4, oblong, acute, reflexed. *Stamens* 8; anthers curved, linear-lanceolate, with a short 1-pored apical process; the base with a short, sharp, downward-pointing spur from the connective; the bases of the anther-cells each with a linear, curved, small tubercle as long as the spur, and like it dark in colour when dry. *Fruit* ovoid-globular, ·2 in. in diam.

PERAK; *Scortechini* 307, 478, 622; *Curtis* 1297; *Wray* 391, 1739; *King's Collector* 4188.

The nearest ally of this is *M. javanensis*, Bl.

7. MEDINILLA MAINGAYI, C. B. Clarke in Hook. fil. Fl. Br. Ind. II, 549. A small epiphyte with slender, terete, pale, smooth branches. *Leaves* opposite, thinly coriaceous, obovate, with rounded apices and cuneate bases, obscurely 3-nerved, the lower surface rather paler than the upper when dry; length ·65 to 1·25 in.; breadth ·5 to ·75 in.; petiole ·05 to ·2 in. *Cymes* much shorter than the leaves, with short, divaricate, broadly bracteolate branches, few-flowered. *Flowers* ·25 in. long, their pedicels shorter. *Calyx-tube* narrowly campanulate, with a slightly expanded, minutely 4-toothed limb. *Petals* 4, lanceolate, sparsely strigose outside. *Stamens* 8; anthers lanceolate, without protuberences at the base in front, but with a short spur behind. *Fruit* unknown. Cogn. in. DC. Mon. Phan. VII, 586.

MALACCA; *Maingay* (Kew Distrib.) 806, 807. PERAK; *Wray* 3781. SINGAPORE; *Ridley* 1652, 2018. PAHANG; *Ridley* 2663.

8. MEDINILLA CLARKEI, King, n. sp. A small epiphyte; young branches with dark, rough, tubercled bark. *Leaves* in whorls of three or four, broadly elliptic to obovate, blunt or subacute, the base cuneate; the upper surface rugulose and green when dry, the lower pale-brown; length 1 to· 2·75 in., breadth ·75 to 1·8 in.; petiole ·4 to ·8 in., puberulous. *Cymes* about as long as the leaves or slightly longer, from the axils of fallen leaves, on slender pedicels from ·5 to ·75 in. long, the branches whorled, spreading, minutely bracteolate at the divisions; pedicels slender. *Flowers* 20 to 30, ·3 in. in length. *Calyx-tube* cupular, the mouth truncate and usually obscurely toothed, sometimes distinctly 5-toothed. *Petals* 5, broadly ovate to rotund, blunt. *Stamens* 10; anthers linear-lanceolate; the base with two small tubercles in front, and a small spur behind. *Fruit* globular-truncate, ·15 in. in diam. *M. rosea*, C. B. Clarke in Hook. fil. Fl. Br. Ind. II, 547 (*not of Gaudichaud*).

MALACCA; *Griffith* (Kew Distrib.) 2282; *Maingay* (Kew Distrib.) 796; *Stoliczka* in Herb. Calc. PERAK; *Scortechini* 243; *Wray* 206, 412, 3831, 4084; *King's Collector* 7333; at elevations of from 3000 to 5000 feet.

M. rosea, Gaud., to which this plant has been referred in the Flora of British India, is a tetramerous species from the Marianne Islands. It has, morever, larger flowers than this and larger, more acute leaves. This plant varies in the size of its leaves: specimens collected at the highest elevations having the largest leaves. The structure of the flowers is, however, uniform.

9. MEDINILLA CRASSINERVIA, Blume in Flora, 1831, 510. Branches with pale bark, the older terete, the youngest striate when dry. *Leaves* in whorls of 3, coriaceous, broadly oblanceolate, or narrowly obovate, shortly and abruptly acuminate, much narrowed to the base, 3-nerved from a little above the base, occasionally with two short lateral faint nerves from the very base; length 3 to 5 in.; breadth 1·5 to 2 in.; petiole ·35 to ·75 in. *Flowers* in short fascicles on the stem below the leaves, in few-flowered pedunculate cymes, mixed with a few solitary, on pedicels ·5 in. long. *Calyx-tube* ovoid-campanulate, the mouth truncate and almost entire. *Stamens* 10, subequal; the basal anterior processes broad and about as long as the filiform posterior spur. *Fruit* globose with a cylindric truncate mouth, ·4 in. across: Blume Rumphia I, 15; Miq. Fl. Ind. Bat. I, pt. I, 545; Cogn. in DC. Mon. Phan. VII, 574. *M. macrocarpa,.* Clarke (not of Blume) in Hook. fil. Fl. Br. Ind. II, 547.

SINGAPORE; *Ridley* 1637. PENANG; *Curtis* 2225. PERAK; *Wray* 1821. MALACCA; *Maingay* (Kew Distrib.) 799. DISTRIB. Borneo.

True *M. macrocarpa,* Bl., is represented in the Kew Herbarium by a single specimen collected by Blume in the Moluccas. The flowers on it have, as described by the author of the species, an irregularly toothed calyx-limb. The plant now described differs in having an almost entire truncate limb, and I follow Cogniaux in referring it to *M. crassinervia,* Bl. In the Flora of British India it is, however, referred to *M. macrocarpa,* Bl.

10. MEDINILLA PERAKENSIS, King, n. sp. Epiphytal; branches terete, glabrous, tubercled. *Leaves* in whorls of 3 or 4, coriaceous, elliptic-rotund, blunt, the base rounded and narrowly cordate, glabrous; 5-nerved, the lateral pair of nerves faint; length 2·5 to 5·5 in.; breadth 1·75 to 3·75 in.; petioles ·6 to 1·2 in. *Panicles* cymose, on rather long peduncles from the axils of fallen leaves, shorter than the leaves, lax, 12- to 20-flowered; branches spreading, whorled, 2-3-chotomous. *Flowers* ·5 in. long, their pedicels ·35 in. *Calyx-tube* cupular; the limb but little expanded, cut into 5 shallow, broad teeth. *Petals* 5, oblong. *Anthers* 10, curved, with 2 yellow tubercles at the base in front and a short spur behind from the connective. *Fruit* ·35 in. in diam.; the seeds oblong, obtuse, with an excurrent tail, the testa pitted.

PERAK; *Scortechini* 410; *Wray.*

Collected only by the late Father Scortechini and Mr. Wray. According to the field-note of the former, the petals and anthers are white and the fruit blueishcarnation.   The nearest ally of this is *M. montana,* Cogn.—a New Guinea species—which has however, longer flowers, a wider calyx-tube, more slender branches which are moreover smooth, and narrower leaves narrowed at the base.

14. POGONANTHERA, Blume.

Shrubs; branches round, minutely·scaly. *Leaves* opposite, petioled, oblong or ovate, entire, glabrous, 3-nerved. *Flowers* small, pulverulent, in small, terminal panicles having opposite, cymose branches. *Calyx-tube* narrowly campanulate, subquadrangular; limb 4-toothed. *Petals* 4, oblong-lanceolate. *Stamens* 8, equal; anthers oblong, acute, opening by a terminal pore, not produced at the base; connective at the base bearded behind with a tuft of hairs, not spurred. *Ovary* half-inferior, 4-celled, with a tuft of hairs at the apex; style filiform; ovules very many, placentas axile. *Berry* small, globose, 4-celled, crowned with the calyx-limb. *Seeds* very many, obovoid-oblong, smooth. DISTRIB. Species 2; Malayan.

POGONANTHERA PULVERULENTA, Blume in Flora, 1831, 521.   An epiphytic shrub, all parts (but especially the calyx-tube) bearing pale yellow scales; the stems sparsely lenticellate. *Leaves* rather fleshy, ovate-oblong to oblong-lanceolate, acute or shortly acuminate, the base more or less narrowed and bituberculate, 3–5-nerved; the margins obscurely crenate and slightly reflexed; upper surface glabrous, the lower paler and minutely pulverulent; length 3 to 6·5 in.; breadth 1·75 to 3 in.; petiole ·25 to ·6 in., stout. *Panicles* terminal, 2 to 3 in. long and equally broad, cymose, the branches spreading. *Calyx-tube* sub-cylindric, constricted below the expanded 4-toothed mouth, very scaly. *Petals* 4, oblong, densely scaly externally. *Stamens* 8; anthers lanceolate, the connective with a tuft of hairs at the base behind. *Fruit* pisiform, ·15 in. in diam. Korth. Verh. Nat. Gesch. Bot. t. 65; Griff. Notul. IV. 678; Miq. Fl. Ind. Bat. I, pt. I, 553; Triana in Trans. Linn. Soc. XXVIII, 89; C. B. Clarke in Hook. fil. Fl. Br. Ind. II, 550; Cogn. in DC. Mon. Phan. VII, 610. *P. reflexa,* Blume in Flora, 1831, 521; Mus. Bot. Lugd. Bat. I, 24; Naud. Ann. Sc. Nat. Ser. 3, XV, 303, tab. 15, fig. 1; Triana l.c. 89; Beccari Malesia, II, 241, tab. LIX, 4–5. *P. squamulata,* Korth. (ex Blume) Mus. Bot. I, 24. *Melastoma reflexa,* Reinw. ined. (ex Blume in Flora, 1831, 521). *M. rubicunda,* Jack in Trans. Linn. Soc. XIV, 19; Wall. Cat. 4086. *M. pulverulenta,* Jack in Trans. Linn. Soc. XIV, 19; DC. Prodr. III, 149; Blume in Bijdr., 1072.

SINGAPORE; PERAK; MALACCA; PENANG; common. DISTRIB. Java, Sumatra, Borneo.

I have followed Mr. Clarke in reducing *P. reflexa*, Bl. here, as I can find no tangible character to separate it. *P. reflexa* is said to have white tumid petals not toothed on the margin, while typical *P. pulverulenta* is described as having red petals with a single tooth on each margin. The union of the two species was suggested by Naudin.

15. PACHYCENTRIA, Blume.

Glabrous, often scandent shrubs, with cylindric or obscurely angled, pulverulent branches. *Leaves* somewhat fleshy, oblong or ovate-lanceolate, entire or obscurely crenulate. *Flowers* small, rose-coloured, corymbose, the pedicels 2-bracteolate, 4-merous. *Calyx-tube* ovoid or turbinate, the part beyond the ovary angular, constricted below the obscurely 4-toothed mouth. *Petals* ovate or oblong, subacute or acuminate. *Stamens* 8, equal; anthers linear-oblong or subulate, rostrate at the apex and minutely 1-pored; the connective not produced at the base, inappendiculate in front but minutely spurred at the back. *Ovary* adherent beyond its middle, 4-celled, its apex free, conic, angled; style filiform, the stigma obtuse or capitate. *Berry* globose, crowned by the limb of the calyx. *Seeds* dimidiately obovoid, the raphe lateral. DISTRIB. About 12 species, all Malayan.

PACHYCENTRIA TUBERCULATA, Korth. Ver. Nat. Gesch. Bot. 246, t. 63. Epiphytic; branches as thick as a swan's quill, glabrous below but with rusty scurf near the apices, the bark pale; the roots bearing woody tubercles. *Leaves* somewhat fleshy, narrowly elliptic-oblong, subacute, narrowed at the base, the edges entire, 3-nerved; length 2·5–4·5 in.; breadth ·9–1·25 in.; petioles ·15–·2 in. *Panicles* terminal or axillary, pedunculate; the branches spreading, cymose, 2–2·5 in. long and as wide. *Flowers* ·25 in. long, the pedicels shorter. *Calyx-tube* campanulate, the mouth truncate, obscurely 4-lobed. *Petals* 4, lanceolate. *Stamens* 8, equal, shortly spurred at the base behind. *Fruit* globular, glabrous, ·15 in. in diam. Blume, Mus. Bot. Lugd. Bat. I, 23; Miq. Fl. Ind. Bat. I, pt. I, 552; Triana in Linn. Trans. XXVII, 89, tab. VII, fig. 95a.

PERAK; *King's Collector* 1707, 10569; *Wray* 3422; *Scortechini* 260, 550, 1961. SINGAPORE; *Anderson* 55. PENANG; *Curtis* 347; *Hullett* 158. DISTRIB. Borneo, Burma (Tenasserim, *Griffith*).

16. ASTRONIA, Blume.

Shrubs with opposite, petioled, ovate or oblong, entire, 3-nerved leaves. *Flowers* in terminal panicles, small, white or purple. *Calyx-tube* campanulate; limb irregularly truncate or 3–8-lobed. *Petals* 4–5. *Stamens* 8–10–12, equal; filaments short, broad; anthers short, obtuse, opening by slits down the front, connective spurred at the base or unappendaged. *Ovary* inferior, 2-5-celled, glabrous at the apex; style short, stigma capitellate; ovules numerous, placentas axile, nearly basal. *Capsule* finally breaking up irregularly. *Seeds* very many, linear, raphe excurrent.—DISTRIB. Species 24; in Malaya and the Pacific Islands.

ASTRONIA SMILACIFOLIA, Triana in Trans. Linn. Soc. XXVIII, 152. Young shoots, petioles, under surfaces of the young leaves and inflorescence rufous-lepidote. *Leaves* oblong, tapering to each end, the transverse nerves stout and distant; length 3·5–5·5 in., breadth 1·5–2·5 in.; petiole ·75–1 in. *Panicle* usually terminal, condensed, 1–2 in. in diam.; branches numerous, short, the pedicels shorter than the globular, minutely 5-toothed calyx-tube. *Petals* reflexed, obovate. *Fruit* subglobular, truncate at the apex, ·25 in. in diam. C. B. Clarke in Hook. fil. Fl. Br. Ind. II, 550 ; Cogn. in DC. Mon. Phan. VII, 1094. *Melastoma smilacifolia*, Wall. Cat. 4057.

PENANG; *Wallich, Curtis.* MALACCA; *Maingay* (K.D.) 808. PERAK; *Scortechini* 683; *Wray* 2813.

VAR. *lepidophylla*, Scort. MSS. Arboreous; inflorescence, leaves on the under surface and petioles (when young) densely covered with deciduous scales.

PERAK; *Scortechini* 1875; *King's Collector* 7270, 2027.

This variety is described by Scortechini and Kunstler as a tree 50–80 feet high, the typical form never being more than a large bush.

17. PTERNANDRA, Jack.

Large shrubs or trees, glabrous or minutely pubescent. *Leaves* sub-coriaceous or coriaceous, opposite, short-petioled or sub-sessile, entire, 3-5-nerved. *Flowers* solitary and axillary on long ·peduncles, or in axillary or teminal, often very short, and clustered cymes ; the pedicels often 2-bracteolate. *Calyx-tube* campanulate or hemispheric, tesselate, verrucose, or covered with more or less adpressed, often puberulous scales ; the mouth truncate, often 4-toothed. *Petals* 4, ovate or oblong, blueish or white. *Stamens* 8, equal in length, but the anthers of some of them often imperfect, perfect anthers broad, blunt, shortly spurred behind but never in front, dehiscing by slits, the filaments stout, often geniculate. *Ovary* inferior, 4-celled : the apex glabrous, depressed or flat ; *style* filiform ; *stigma* clavate ; *ovules* numerous, placentas sub-basal. *Berry* subglobose or ovoid, truncate or surmounted by the calyx-teeth, scaly or smooth. *Seeds* cuneate-ovoid, or obovoid, angular.—DISTRIB. Species about 12; in Malaya and the Philippines.

Calyx-tube tesselate outside, not covered with distinct
scales ...   . 1. *P. cœrulescens.*
Calyx-tube covered with large distinct scales :—
    Young branches and under surfaces of leaves
    rusty-pubescent; calyx-tube ·2-·3 in. long, covered
    with more or less spreading scales; teeth of calyx
    spreading, elongate, acuminate   ...   ... 2. *P. echinata.*
    Young branches and under surfaces of leaves
    glabrous; calyx-tube ·15 in. long, its scales
    adpressed; teeth of calyx broadly triangular, blunt,
    reflexed   ...   . 3. *P. Griffithii.*

1. PTERNANDRA CŒRULESCENS, Jack in Mal. Misc. II, 61. A tree;
young branches cylindric with deciduous, dark-brown, glabrous bark and
slightly thickened nodes with obscure transverse ridges. *Leaves*
chartaceous or sub-coriaceous, broadly ovate, ovate-lanceolate, ovate-
oblong or elliptic, much narrowed at the base, the apex shortly
acuminate, 3–5-nerved; both surfaces glabrous; length 2·5–5 in. (10
in. in var. 2); breadth 1·25-2·5 in., (to 5 in. in var. 2) petiole ·1-·2 in.
Flowers in short, axillary, pedunculate cymes (often several from one
axil), or in terminal cymes, shorter than the leaves. *Calyx-tube*
cylindric-campanulate, ·15 in. long, tesselate; the mouth truncate but
with 4 small, erect, triangular teeth. *Petals* thick, ovate, reflexed after
expansion. *Stamens* 8, equal in length; the filaments short, geniculate;
perfect anthers 4 or 5, broadly ovate, blunt, shortly spurred behind, the
remaining 3 or 4 imperfect, as long as but much narrower than the
perfect. *Fruit* turbinate or sub-hemispheric, truncate, nearly smooth,
·15-·3 in. in diam. Wall. Cat. 4077; Triana in Trans. Linn. Soc.
XXVIII, 153; Kurz, For. Fl. I, 509 and in Journ. As. Soc. 1877, pt. II,
79; C. B. Clarke in Hook. fil. Fl. Br. Ind. II, 551; Cogn. in DC. Mon.
Phan. VII, 1103. *Ewyckia cyanea,* Blume Rumph. I. 24, t. 8; Miq.
Fl. Ind. Bat. I, pt. I, 568; Triana l.c. *E. Jackiana,* Walp. Rep. V. 724.
Apteuxis trinervis, Griff. Notul. IV, 672.—*Nov. Gen.* Roxb. Fl. Ind. II, 225.

In all the provinces except the Andaman Islands.

A common and variable plant of which four forms seem worthy of separation as
varieties. These, however, pass into each other by numerous connecting specimens.
One variety (*Jackiana*) differs from the typical plant in having few-flowered almost
sessile cymes; a second (*capitellata*) has sub-sessile cymes and much larger leaves
and the third (*paniculata*) is probably only an example of fasciation.

VAR. 1. *Jackiana,* Clarke in Fl. Br. Ind. II, 551. *Flowers* in very
short, few-flowered, almost sessile, axillary cymes. *Leaves* as in the
typical form but with slightly longer petioles.

In all the provinces except the Andaman and Nicobar Islands,
equally abundant with the typical form.

VAR. 2. *capitellata*, King. *Leaves* thicker in texture than in the typical form (sub-coriaceous), broadly elliptic to sub-orbicular, with 3 very strong nerves and a fainter marginal pair; length 4·5–10 in.; breadth 2·75–5 in., petiole ·2–·3 in. *Flowers* in dense, very shortly-stalked, axillary glomeruli composed of numerous 3-flowered cymes very much shorter than the leaves. *Pternandra capitata*, Jack in. Mal. Misc. II, addenda prefixed to the paper p. 3; Wall. Cat. 4079; W. and A. Prodr. 325; Triana in Trans. Linn. Soc. III, 153; Kurz, For. Fl. I, 509 and in Journ. As. Soc. 1877, pt. II, 79; C. B. Clarke in Hook. fil. Fl. Br. Ind. II, 551; Cogn. in DC. Mon. Phan. VII, 1103. *Ewyckia capitellata*, Walp. Rep. V, 724; Miq. Fl. Ind. Bat. I, pt. I, 568. *E. medinilliformis*, Naud. in Ann. Sc. Nat. Ser. 3, XVIII, 261.

SINGAPORE; *Wallich* 4079. PENANG; *Curtis* 67; *King.* PERAK; *Scortechini* 43, 1043; *Wray* 1971. MALACCA; *Maingay* 802 (K D.); *Helfer* (K.D.) 2279.

VAR. 3. *paniculata*, King. *Flowers* in large, lax, terminal, much branched, few-flowered, leafy and bracteolate panicles. *Leaves* of the stem 2–6 in. long and from ·9–2·75 in. broad, those of the panicle from ·75–2 in. long and ·15–·8 in. broad. *P. paniculata*, Benth. in Wall. Cat. 4080; C. B. Clarke in Hook. fil. Fl. Br. Ind. II, ·551; Cogn. in DC. Mon. Phan. VII, 1104. *Ewyckia latifolia*, Blume Mus. Bot. I, 6. *E. cyanea*, var. *latifolia*, Korth. ex Miq. Fl. Ind. Bat. I, pt. I, 568. *E. paniculata*, Miq. l.c. Suppl. 321. *Pternadra latifolia*, Triana in Linn. Trans. XXVIII, 153.

PENANG; *Wallich* 4080; *Curtis* 2768. MALACCA; *Griffith* (K.D.) 2273. PERAK; *Scortechini* 248, 1303; *Wray* 92. DISTRIB.; Borneo, Bangka.

2. PTERNANDRA ECHINATA, Jack, Mal. Mis. II, n. 9 and add. prop. 3. A small tree; young branches quadrangular, thickened and with transverse ridges at the nodes, minutely rusty-pubescent. *Leaves* sub-coriaceous, boldly 3-nerved, lanceolate or oblong-lanceolate, 3-nerved, narrowed to the base, shortly acuminate at the apex; upper surface glabrous; the lower puberulous or sub-glabrous, minutely reticulate; length 2·5–4·5 in.; breadth ·75–1·75 in.; petiole ·05–·2 in. *Flowers* axillary and solitary on pedicels longer than themselves, or in threes in terminal pedunculate cymes, the pedicels of both sets of flowers with one or more pairs of curved, linear-oblong bracteoles. *Calyx-tube* widely campanulate, ·2–·3 in. long, closely covered with triangular rusty-pubescent scales, those nearest the mouth longest, most acute, and most persistent; the mouth truncate and with 4 narrow, acuminate teeth. *Petals* broadly ovate-quadrate, abruptly and shortly acute, the edges undulate, blue. *Anthers* broadly ovate, on thick short filaments.

Fruit sub-hemispheric, truncate, sub-echinate, ·35 in. in diam. Wall. Cat. 4078. *Kibessia echinata*, Cogn. in DC. Mon. Phan. VII, 1108. *Kibessia simplex*, Korth. Verh. Nat. Gesch. Bot. 253; Blume, Mūs. Bot. I, 9; Triana in Trans. Linn. Soc. XXVIII, 152; C. B. Clarke in Hook. fil. Fl. Br. Ind. II, 552. *Kibessia cupularis*, Dcne in Deless. Ic. Sel. V, t. 5; Ann. Sc. Nat. Sér. 3, XV, 317. *K. acuminata*, Dcne in Ann. Sc. Nat. Ser. 3, V, 316; Triana in Trans. Linn. Soc. XXVIII, 153.

MALACCA and SINGAPORE; not uncommon; many collectors.

I cannot see how *K. acuminata*, Dcne, is to be distinguished as a species and I reduce it here without any hesitation.

VAR. *pubescens*, King. ·Bases of leaves somewhat rounded and sub-cordate; young branches, under surfaces of leaves and panicles with much minute rusty pubescence. *P. echinata*, Jack, Wall. Cat. 4078a. *Kibessia pubescens*, Dcne in Ann. Sc. Nat. Ser. 3, V, 318; Triana in Linn. Trans. XXVIII, 152; C. B. Clarke in Hook. fil. Fl. Br. Ind. II, 552; Cogn. in DC. Mon. Phan. VII, 1108.

PENANG, MALACCA, PERAK.

I cannot see what claim this has to specific rank. To me it appears to be a variety and not a very distinct one of *P. echinata*, Jack. Wallich did not even regard it as a variety and issued it as true *P. echinata*. This form, in the three provinces where it occurs, appears to be very common.

3. PTERANDRA GRIFFITHII, King, n. sp. A small tree; young branches cylindric, very little thickened at the nodes, glabrous. *Leaves* thinly coriaceous, elliptic to ovate-oblong, narrowed at the non-cordate (cordate in var.) base, the apex very shortly acuminate or acute, 3-nerved; both surfaces glabrous, shining; length 2·25–4 in.; breadth 1–2 in.; petiole ·15–·2 in. *Flowers* in 2–3- rarely 5–7-flowered, axillary, bracteolate cymes shorter than the leaves, rarely in crowded, terminal cymes; bracteoles ovate, acute, minute. *Calyx-tube* widely campanulate, ·15 in. long, covered with adpressed, triangular, puberulous scales; the mouth with 4 large, blunt, triangular teeth. *Petals* orbicular-ovate, undulate, abruptly and shortly apiculate-spreading, not calyptrate. *Stamens* 8, equal; the anthers short, thick, about as long as the filaments, gibbous at the base behind, inserted at an obtuse angle on the filaments. *Fruit* globular-ovoid, truncate at the mouth, covered by the persistent scales, under ·2 in. in diam.

MALACCA; *Griffith* (K.D.) 2272/1; PENANG; *Curtis* 953.

Griffith's specimens of this (2272/1) have been referred by M. Cogniaux (DC. Mon. Phan. VII, 1110) to *Rectomitra tuberculata* Bl., but comparison in the Kew Herbarium with two authentic specimens of that plant collected in Sumatra and issued from the Leiden Herbarium shew that this differs from Blume's plant. Specimens of this were originally collected by Griffith in 1845, and as none had been gathered until Curtis's in 1886, the species is presumably a rare one.

VAR. *cordata*, King.   Leaves with cordate bases.
PENANG; *Curtis* 453.   PERAK; *Wray* 1994.

18. MEMECYLON, Linn.

Shrubs or *trees*, glabrous. *Leaves* opposite, short-petioled or sessile, coriaceous or sub-coriaceous, orbicular, ovate or lanceolate, entire, pinnate-nerved or rarely 3-nerved. *Flowers* usually in small, axillary, rarely terminal, simple or panicled cymes or umbels. *Calyx-tube* campanulate, glabrous; limb dilated, truncate or shortly 4-lobed. *Petals* 4, blue or white, rarely reddish. *Stamens* 8, equal, filaments long; anthers short, opening by slits in front, connective ending in a horn behind. *Ovary* inferior, 1-celled; apex glabrous, surmounted by a convex or depressed disc with 8 radiating grooves; style filiform, simple; ovules 6–12, whorled on a free-central placenta. *Berry* globose or ellipsoid, crowned with the calyx-margin, 1-seeded. *Seed* large, cotyledons convolute.—DISTRIB. Species about 130; numerous in South-East Asia and its islands; a few extending into Polynesia and Australia, several in tropical Africa.

Leaves boldly 3-nerved from base to apex   ...   ...   1. *M. oligoneuron.*
Leaves with pinnate nervation, sessile or subsessile :—
    Main nerves of leaves distinctly visible when dry, inter-
    arching but not forming (except in No. 3) a bold intra-
    marginal nerve; leaves thinly coriaceous or membranous,
    small, not exceeding 4 in. in length :—
        Young branches boldly 4-angled or winged; inflor-
        escence very shortly stalked (the stalk not manifest) :—
            Young branches 4-winged :—
                Cymes solitary, 3- or 4-flowered; leaves narrowly
                lanceolate, ·6 to 1 in. broad   ...   ...   2. *M. epiphyticum.*
                Cymes solitary, 8- to 10-flowered; leaves ovate- or
                oblong-lanceolate, 1·35 to 1·85 in. broad   ...   3. *M. fruticosum.*
                Cymes several from the same axil, compoundly
                umbellate, 1·5 to 2·5 in. long, many-flowered,
                pubescent; leaves elliptic much narrowed to each
                end ...   .   4. *M. pubescens.*
            Young branches 4-angled, never winged and some-
            times sub-terete :—
                Cymes 3- to 5-flowered : leaves lanceolate, 1 to 2
                in. broad   ...   .   5. *M. dichotomum.*
            Young branches terete; inflorescence with a manifest
            peduncle   ...   .   6. *M. Kunstleri.*

Main nerves of leaves distinct when dry, prominent on
the lower surface and anastomosing with a bold intra-
marginal line, coriaceous, more than 4 in. long :—
  Inflorescence manifestly pedunculate :—
    Peduncles several in each axil, many-branched ; fruit
    ellipsoid   ...   .   7. *M. caloneuron.*
    Peduncles solitary, few-branched ; fruit globular ... 8. *M. Hullettii.*
  Inflorescence sessile or on a very short peduncle :—
    Flowers large, the mouth of the calyx ·2 in. in diam. :—
      Arboreous; leaves sub-acute; flowers in fascicles
      of 12 to 20, their pedicels ·15 in. long, stout; calyx
      truncate, not toothed   ...   .   9. *M. Maingayi.*
      Shrubby; leaves acuminate; cymes 3- or 4-flowered ;
      pedicels ·4 in. long, slender; calyx-limb 4-toothed 10. *M. Kurzii.*
    Flowers small ; mouth of the calyx under ·2 in. in diam.:—
      Leaves slightly narrowed or rounded at the base,
      rarely minutely sub-cordate :—
        Young branches not winged below the nodes ;
        main nerves of leaves 18 to 20 pairs ; fruit ·7 in.
        in diam.   ...   .   11. *M. heteropleurum.*
        Young branches with 4 short wings below each
        node (sometimes obscure) ; main nerves 12 to 14
        pairs ; fruit ·35 in. in diam.   ...   ...   12. *M. costatum.*
      Leaves distinctly cordate at the base and quite
      sessile, amplexicaul   ...   .   13. *M. amplexicaule.*
Main nerves of leaves indistinct on both surfaces when
dry; leaves coriaceous or thinly so :—
  Leaves with broad cordate bases, sessile, amplexicaul:—
    Branches terete, fruit large, globular   ...   ... 14. *M. microstomum.*
    Branches 4-angled; fruit ellipsoid   ...   ... 15. *M. coeruleum.*
  Leaves much narrowed at the base, never cordate,
  petiolate :—
    Inflorescence in axillary glomeruli or in very shortly-
    peduncled (not manifest) cymes :—
      Mouth of calyx entire in the expanded flower :—
        Flowers in fascicles, their pedicels slender;
        leaves often 4 in. long ; leaves brown underneath
        when dry   ...   .   16. *M. campanulatum.*
        Flowers in short umbellate sub-sessile cymes ;
        young branches bi-sulcate : fruit globular; leaves
        pale yellowish underneath when dry   ... 17. *M. minutiflorum.*
      Mouth of calyx 4-toothed :—
        Teeth of calyx long, sharp, its fundus narrowed ;
        flower buds narrowly conical; cymes many-
        flowered; fruit globular, ·2 in. in diam.; leaves
        shortly acuminate, 1·5 to 2·5 in. long   ... 18. *M. myrsinoides.*
        Teeth of calyx short, acute, its fundus narrow ;
        cymes few-flowered, fruit globular, ·3 in. in
        diam.; leaves very acuminate, 2 to 2·5 in. long... 19. *M. laevigatum.*

Teeth of calyx short, acute, its fundus rounded;
fruit globular, ·25 in. in diam.; flower pedicels
with acicular bracteoles; leaves much acuminate,
2·8 to 5·5 in. long ... ... · 20. *M. cinereum.*
Inflorescence manifestly pedunculate :—
Fruit ellipsoid :—
Young branches terete; calyx-limb truncate;
leaves elliptic-oblong or elliptic, 2·5 to 4·75 in.
long ... ... 21. *M. oleaefolium.*
Fruit globular :—
Branches 4-angled :—
Leaves rhomboid or elliptic-rhomboid; inflorescence under ·5 in. long; calyx saucer-shaped
with wide, minutely 4-toothed mouth; fruit ·2
in. in diam. ... ... 22. *M. pauciflorum.*
Leaves oblong to elliptic, much tapered to each
end; inflorescences 1 in. or more in length,
several in each axil; calyx with narrow
fundus, the mouth wide (·1 in.) and obscurely
4-toothed; fruit ·5 in. in diam. ... ... 23. *M. elegans.*
Branches terete :—
Mouth of calyx with 4 broad, shallow teeth;
flowers 4 to 6 in a compound umbel; fruit ·25
in. in diam.; leaves caudate-acuminate · ... 24. *M. acuminatum.*
Mouth of calyx truncate or with 4 obscure
teeth :—
Cymes many-flowered, on peduncles not
longer than the leaf-petioles; calyx with
wide, obscurely toothed mouth and narrow,
cup-shaped tube; young branches not bi-sulcate ... ... 25. *M. garcinioides.*
Cymes or peduncles very slightly if at all
longer than the petioles; calyx not toothed;
young branches deeply bi-sulcate under the
nodes ... ... 26. *M. andamanicum.*
Cymes or peduncles several times longer
than the leaf-petioles :—
Peduncles solitary; leaves thinly coriaceous ... ... 27. *M. intermedium.*
Peduncles several from the same axil;
leaves coriaceous ... ... 28. *M. edule.*

1. MEMECYLON OLIGONEURON, Blume, Mus. Bot. I, 354. A small
tree or shrub; young branches slender, terete, their bark pale-brown.
Leaves thinly coriaceous, brown below, greenish brown above when dry,
oblong to ovate- or elliptic-oblong, shortly and obtusely acuminate,
boldly 3-nerved from the cuneate base, transverse nerves invisible;
length 2-4 in.; breadth 1·3-1·65; petiole ·05-·15 in. *Flowers* small

('05 in. long), on pedicels about ·1 in. long, densely crowded in clusters on small tubercles in the axils of leaves or of fallen leaves. *Calyx-tube* cupular, but little contracted at the base, the mouth wide with four broad shallow teeth. *Fruit* unknown. Miq. Fl. Ind. Bat. I, pt. 1, 574; Cogn. in DC. Mon. Phan. VII, 1132. *M. trinerve,* Hassk. Cat. Hort. Bog. 259 (not of DC.). *Myrtus oligoneura,* Korth. ex Blume l.c. 354.

PERAK; *Scortechini* 1309; *King's Collector* 2513, 10280. PENANG; *Curtis* 1065, 1446, 2220, 10920., DISTRIB. Java, Borneo.

A species easily recognised by its 3-nerved leaves.

2. MEMECYLON EPIPHYTICUM, King, n. sp.   An epiphytic shrub; branches rather stout, strongly angled and with short ear-like projections just below the nodes. *Leaves* thinly coriaceous, narrowly oblong-lanceolate, acute or acuminate, somewhat narrowed to the rounded sub-cordate base; main nerves 9 or 10 pairs, indistinct on the lower surface, invisible on the upper; length 1·75–3 in.; breadth ·6–1 in.; petiole ·05 in. *Cymes* in pairs, axillary, 3–4-flowered, on slender pedicels ·1–·15 in. long, bracteate at the apex; pedicels half as long as the peduncle. *Calyx-tube* cupular, rounded at the base; the mouth deep and wide (·05 in. across), undulate, truncate. *Fruit* globular, smooth, ·25 in. in diam.

PERAK; on trees, *King's Collector* 5184; *Wray* 2727.

A species allied to *M. dichotomum,* Clarke, but with smaller leaves, more boldly angled branches, smaller, less numerous flowers, and cymes on more slender peduncles.

3. MEMECYLON FRUTICOSUM, King, n. sp.   A shrub, 6–8 feet high; young branches boldly 4-winged especially near the slightly thickened nodes, the bark pale-brown. *Leaves* ovate-lanceolate or oblong-lanceolate, chartaceous, shortly acuminate, slightly narrowed to the rounded base; main nerves 7–9 pairs, interarching ·15 in. from the margin, somewhat conspicuous on the lower but indistinct on the upper surface; length 3–4 in.; breadth 1·35–1·85 in.; petiole ·05 in. *Cymes* usually in pairs, axillary, on short peduncles, 8–10-flowered. *Flowers* on pedicels with acute bracteoles at their bases. *Calyx-tube* shortly campanulate, tapering much to the base (obconic), the mouth ·075 in. wide, with 4 shallow obscure teeth, or truncate; the buds not very conical. *Fruit* globose-ovoid, constricted below the thick persistent calyx-limb, ·35 in. long and ·25 in. in diam. (unripe).

PERAK; *King's Collector* 2971, 3265, 3425.

Approaching *M. dichotomum* and *M. sub-dichotomum* but with differently shaped fruit.

4. MEMECYLON PUBESCENS, King.   A tree, 30–70 feet high; young branches somewhat slender, pale-brown, 4-angled. *Leaves* coriaceous,

elliptic, shortly and abruptly acuminate, the base much narrowed; yellowish green on the lower surface and olivaceous on the upper when dry; main nerves 7–10 pairs, quite distinct on the lower surface, less so on the upper, curved and interarching ·1 in. from the margin. *Cymes* 1–3 from the axils of leaves, often unequal, proliferously umbellate, from 1·5–2·5 in. long, always pedunculate, the peduncle and all its branches 4-angled, bracteolate at the divisions, sparsely and deciduously rusty-pubescent. *Flowers* densely clustered at the apices of the thickened secondary peduncles, pedicellate; the pedicels with numerous sharply acuminate bracteoles at their bases. *Calyx-tube* campanulate, much narrowed at the base, the mouth rather more than ·05 in. wide, truncate but with 4 minute, acicular teeth. *Fruit* globular, ·15 in. in diam. (unripe). *M. grande*, Retz., var. *pubescens*, Clarke in Hook. fil. Fl. Br. Ind. II, 558; Cogn. in DC. Mon. Phan. VII, 1153.

MALACCA; *Griffith* (Kew Distrib.) 2336. PERAK; *King's Collector* 6089, 10760. SINGAPORE; *Ridley* 10390.

Ripe fruit of this is unknown.

5. MEMECYLON DICHOTOMUM, C. B. Clarke in Herb. Kew. A slender shrub, 6–8 feet high; young branches slender, acutely 4-angled (even 4-winged) below the slightly thickened nodes; the bark pale-brown. *Leaves* thinly coriaceous, almost sessile, lanceolate to ovate-lanceolate, much acuminate, often caudate; the base rounded or slightly narrowed; main nerves 6–8 pairs, curved, interarching rather far from the margin, often indistinct; length 2·5–4 in.; breadth 1–2 in.; petiole very short (under ·05 in.). *Cymes* 2–5-flowered, solitary, axillary and terminal; peduncles very short, 4-angled; pedicels with two ovate, acute bracteoles at their apices embracing the calyx. *Calyx-tube* campanulate, tapering to the base (obconical); minutely glandular outside when dry, the mouth with 4 broad, shallow lobes when young, truncate and almost entire when old; buds rather large, conical. *Fruit* globular, crowned by the narrow calyx-limb, smooth when ripe, about ·5 in. in diam. *M. elegans*, var. *dichotoma*, C. B. Clarke in Hook. fil. Fl. Br. Ind. II, 554; Cogn. in DC. Mon. Phan. VII, 1138.

MALACCA; *Griffith* (Kew Distrib.) 2324; *Maingay* (K.D.) 818, 820. PERAK; *Wray* 2989; *King's Collector* 3239, 5036, 5297, 10783. PAHANG; *Ridley* 2609.

I restore for this species the MS. name originally given to it by Mr. C. B. Clarke in the Kew Herbarium. Mr. Clarke subsequently reduced it, as a variety, to *M. elegans*, Kurz, of which there were, at the time he made the reduction, no good specimens. Now that there are excellent examples of *M. elegans*, it is clear that *M. dichotomum* is not near that species.

There are specimens in the Calcutta Herbarium of what appear to be other species allied to this. But the material of all is imperfect and I describe none of them.

6. MEMECYLON KUNSTLERI, King, n. sp. A tree, 40–60 feet high; young branches slender, terete, very pale-grey. *Leaves* chartaceous, drying brown (palest on the lower surface), elliptic-oblong, bluntly acuminate; the base rounded and often minutely cordate; main nerves 7 or 8 pairs, ascending, faint on the lower and almost invisible on the upper surface when dry; length 2·75–4·5 in.; breadth 1–2 in.; petiole ·05–·1 in. *Peduncles* from the axils of fallen leaves or axillary, bracteolate, ·35–·6 in. long, umbellately panicled, bracteolate at the divisions, 4-angled like the pedicels; ultimate umbels 4–6-flowered, on the thickened ends of the secondary peduncles, pedicels bracteolate at the base. *Calyx-tube* cup-shaped, shallow, the mouth obscurely 4-toothed. Young fruit ellipsoid, crowned by the thick, shallow, obscurely 4-toothed limb of the calyx, ·3 in. long, and ·15 in. in diam.

PERAK; *King's Collector* (Kunstler) 8195, 10419.

This is known only by Mr. Kunstler's two suites of specimens. One of these sets bears no fruit; the other no flowers. The leaves on the former are rather smaller than those on the second but the venation is the same and I assume that they belong to one species. The terete branchlets, associated as they are with an inflorescence which is 4-angled in all its branches, even down to the pedicels and the ellipsoid fruit, distinguish the plant.

7. MEMECYLON CALONEURON, Miq. Fl. Ind. Bat. Suppl. 321. A tree; branchlets and leaves as in *M. costatum*, Miq., but the latter with fewer nerves. *Flowers* in axillary, pedunculate, many-branched cymes, 1–2 in. long, the flowers in dense glomeruli on the thickened apices of the ultimate branchlets; all the peduncles boldly 4-angled or winged; fruit ellipsoid, ·35 in. long (including the small persistent calyx-limb). *M. costatum*, Miq., var. *ellipsoidea*, Blume Mus. Bot. I, 361; Cogn. in DC. Mon. Phan. VII, 1136.

MALACCA; *Maingay* (Kew Distrib.) 813. PERAK; *Wray* 3235; *King's Collector* 6945, 8505. DISTRIB. Java; Sumatra, *Forbes* 2696; Borneo.

The inflorescence and fruit are so different from those of *M. costatum*, Miq., that I have followed Miquel in treating this as a species. Miquel did not however, recognise that his *M. caloneuron* really covers Blume's variety *ellipsoidea* of his own species *M. costatum*.

8. MEMECYLON HULLETTII, King, n. sp. Young branches slightly ridged near the nodes, otherwise terete, the bark pale-brown. *Leaves* chartaceous, ovate-oblong, gradually narrowed to the acuminate apex; the base broad, abruptly rounded, slightly cordate; main nerves about 15 pairs, thin but distinct on the lower surface, horizontal; length 6–8 in.; breadth 2·25–3 in.; petiole under ·1 in. *Peduncle* solitary, axillary, 1·5–2·5 in. long, slender, bearing at its apex a single or compound

few-flowered umbel; the flower-pedicels longer than the calyx and, like the peduncles of the secondary umbels, rugulose. *Calyx-tube* campanulate, narrowed to the base; the mouth truncate, with 4 very obscure shallow teeth. *Fruit* globular, crowned by the rather large calyx-limb, ·25 in. in diam. (not quite ripe).

JOHORE; on Gunong Pulai, *Hullett* and *King* 253; *Lake* and *Kelsall* 4073.

A very well-marked species near *M. amplexicaule,* Roxb., at once distinguished by its elongately acuminate, broad-based leaves and long-peduncled umbels.

9. MEMECYLON MAINGAYI, Clarke in Hook. fil. Fl. Br. Ind. II, 557. A tree, 20–40 feet high; branches stout, terete, somewhat thickened at the nodes, the bark pale-brown when dry. *Leaves* coriaceous, nearly sessile, elliptic-oblong, sub-acute, slightly narrowed to the rounded, sometimes slightly cordate base; in length 6·5–9 in.; breadth 3–4·25 in.; main nerves 12–15 pairs, not very prominent, interarching inside the margin. *Flowers* large for the genus, in few-flowered (12–20) fascicles from the axils of the leaves or of fallen leaves; peduncles and pedicels about ·15 in. long, bracteolate. *Calyx-tube* widely cupular, truncate, toothless, ·2 in. in diam. when dry. *Petals* obtuse in bud. *Fruit* unknown. Cogn. in DC. Mon. Phan. VII, 1139.

MALACCA; *Maingay* (Herb. prop.) 1422. PERAK; *King's Collector* 4726.

An arboreal species with large flowers and thick branches, allied to *M. amplexicaule* but well distinct.

10. MEMECYLON KURZII, King. A glabrous shrub; young branches terete, swollen under the nodes, the bark pale when dry. *Leaves* thinly coriaceous, sub-sessile, ovate-oblong, shortly acuminate, slightly narrowed to the rounded base; main nerves 15–20 pairs, rather straight, interarching ·25 in. from the margin, faint; length 8·5–10 in.; breadth 3·25–4·5 in.; petiole about ·1 in. long, stout. *Flowers* large, on slender bi-bracteolate pedicels ·4 in. long; the cymes 3- or 4-flowered, from the axils of fallen leaves, solitary or several together; peduncle short (only ·15 in. long). *Calyx-tube* campanulate, ·2 in. long; the mouth ·2 in. wide, wavy and with 4 broad teeth. *Fruit* ellipsoid, somewhat curved, ·75 in. in length (including the persistent limb of the calyx) and ·4 in. in diam. *M. subtrinervium,* Miq., VAR. *grandiflora,* Kurz in Journ. As. Soc. Beng. 1876, pt. II, 131; C. B. Clarke in Hook. fil. Fl. Br. Ind. II, 565; Cogn. in DC. Mon. Phan. VII, 1143.

NICOBAR ISLANDS; *Kurz, King's Collector* 509.

The leaves of this when dry are pale-brown on the upper and pale-olivaceous on the lower surface. They are different in shape from those of *M. subtrinervium,* Miq., of which Kurz makes this a variety. The flowers of the latter are moreover small and in slender pedunculate cymes.

11. MEMECYLON HETEROPLEURUM, Blume, Mus. Bot. Lugd. Bat. I, 362. ˙A tree, 30 and 40 feet high; young branches rather slender, terete, the bark pale-brown or cinereous when dry. *Leaves* thinly coriaceous, broadly oblong-lanceolate or elliptic-oblong, shortly and rather bluntly acuminate, the base rounded or narrowed, not cordate, shortly petiolate, pale-brown with sometimes a touch of green on both surfaces when dry; main nerves 18–20 pairs, straight, interaching close to the margin, thin but very distinct on the lower surface when dry; length 5–7 in. rarely 10 in.; breadth 2–3·5 in.; petiole only ·1 in. *Flowers* pointed in bud, in dense cymes from the axils of the leaves or from those of the old leaves; the peduncle ·25 in. long; pedicels shorter, bi-bracteolate. *Calyx-tube* widely cupular, narrowed to the base, truncate, ·1–·15 in. in diam. when dry. *Fruit* globular, ·5 in. in diam. Miq. Fl. Ind. Bat. I, pt. I, 579; C. B. Clarke in Hook. fil. Fl. Br. Ind. II, 557; Cogn. in DC. Mon. Phan. VII, 1140.

MALACCA; *Griffith* 2337 (Kew Distrib.); *Maingay* (K.D.) 816. PENANG; *Curtis* 814. SINGAPORE and SELANGORE; *Ridley.* PERAK; *King's Collector, Wray, Scortechini :* many Nos.; common. DISTRIB. Sumatra and Borneo.

VAR. *olivacea,* King. Leaves rounded at the base, large, 10–14 in. long and 3·5–4·25 in. broad, with a strong olivaceous tint when dry : fruit not seen.

PERAK; *King's Collector* 500, 2778, 10872; *Wray* 1310.

This variety differs (as far as it is represented by dried specimens) from typical *M. heteropleurum,* Bl., only in the size of its leaves and their colour when dried; fruit of it is unknown, the flowers and shape of leaves are exactly those of the type.

12. MEMECYLON COSTATUM, Miq. in Verh. Ned. Inst. 1850, p. 29. A tree, 30–60 feet high; young branches terete, but with 4 short wings below the nodes. *Leaves* thinly coriaceous (drying pale-brown with a tinge of yellowish-green), oblong- or oblong-lanceolate, sometimes oblong-ovate, shortly acuminate, the base rounded or slightly narrowed, not cordate, penni-nerved; the main nerves 12–14 pairs, stout, curved, anastomosing at ·25 in. from the margin with a bold lateral nerve; length 4·5–7·5 in.; breadth 1·75–3 in.; petiole very short, stout. *Flowers* crowded in axillary glomeruli, 1 in. or less in diameter; their pedicels short (·1–·05 in.), the bracteoles minute, triangular. *Calyx* cup-shaped, truncate, slightly narrowed at the base. *Petals* ·2 in. in diam. *Fruit* globose, ·35 in. in diam. Miq. Fl. Ind. Bat. I, pt. I, 573; Triana in Linn. Trans. XXVIII, 157; Blume, Mus. Bot. I, 360; C. B. Clarke in Hook fil. Fl. Br. Ind. II, 558; Cogn. in DC. Mon. Phan. VIII, 1136. *M. grande,* Bl. Bijdr. 1095 (not of Retz.).

PERAK; *King's Collector* 10785. DISTRIB.; Java, Sumatra (*Forbes* 1442).

13. MEMECYLON AMPLEXICAULE, Roxb. Fl. Ind. II, 260. A shrub, 8–12 feet high ; branches rather slender, terete between, but 4-angled and sometimes 4-winged below the nodes. *Leaves* (tinged with greenish-yellow when dry) sessile or nearly so, often semi-amplexicaule, ovate-oblong or ovate-lanceolate, sub-acute or shortly and bluntly acuminate, broadest a little above the cordate base, penni-nerved; the main nerves 9–12 pairs, not prominent, interaching inside the margin ; length 3·5–6 in.; breadth 1–2·5 in. *Flowers* ·2 in. long, crowded in dense, axillary glomeruli 1 in. or less in diameter; their pedicels very short (lengthened to ·25 in. in fruit) and with minute bracteoles. *Calyx* campanulate, truncate, much narrowed to the base. *Petals* sub-rotund, ·2 in. in diam. *Fruit* globose, ·3 in. in diam. Wight Ic. 279. Naud. in Ann. Sc. Nat. Ser. 3, XVIII, 277; Miq. Fl. Ind. Bat. I, pt. I, 580; C. B. Clarke in Hook. fil. Fl. Br. Ind. II, 559 (in part); Cogn. in DC. Mon. Phan. VII, 1139 (in part). *M. depressum*, Benth. in Wall. Cat. 4101 (in part); Triana in Linn. Trans. XXVIII, 158 (in part). *M. cordatum*, Wall. Cat. 4100 (in part). *M. coerulum*, Triana in Linn. Trans. XXVIII, 158 (in part).

In all the Provinces except the Andaman and Nicobar Islands ; common.

The petals of this are white tinged with pink. The plant described by Roxburgh under the name *M. amplexicaule* is a Malayan one, as he distinctly states. The species from the South of India which has, in most of the synonyms above quoted, been treated as identical with this is, in my opinion, quite distinct. It has smaller and proportionately broader leaves, and the flowers, which are smaller and more numerous, are in fascicles from the axils of fallen leaves. This is allied to *M. costatum*, and like it, this has the stems often 4-winged below the nodes ; the leaves are also sessile or nearly so, but they differ from those of *M. costatum* in invariably being cordate at the base.

14. MEMECYLON MICROSTOMUM, Clarke in Hook. fil. Fl. Br. Ind. II, 557. A tree, 40–70 feet high ; branches terete, rather slender, dark greyish-brown when dry. *Leaves* very coriaceous, sessile and almost amplexicaul, oblong or narrowly elliptic, sub-acute or obtuse, the base rounded and slightly cordate, very opaque, the nerves very indistinct ; length 3·25–4·5 in.; breadth 1·3–2 in. *Flowers* numerous, small, less than ·1 in long (excluding the exserted stamens), crowded in dense axillary glomeruli, pedicels filiform. *Calyx-tube* infundibuliform, constricted in its lower third, the mouth wide truncate. *Petals* pale yellowish-green. *Fruit* large (·6 in. in diam.), globular, the persistent calyx-limb small. Cogn. in DC. Mon. Phan. VII, 1147.

MALACCA ; *Maingay* (Kew Distrib.) 821 ; PERAK ; *Wray* 1137 ; *King's Collector* 10588. SINGAPORE ; *Ridley* 2033. PENANG ; *Curtis* 766.

The leaves of this, when dry, are olivaceous on the upper and pale-brown on the lower surface. The species resembles *M. amplexicaule* but differs notably in its large globular fruit.

15. MEMECYLON COERULEUM, Jack. in Mal. Misc. I, 26. A shrub, 5–15 feet high; branchlets often 4-angled near the apices, otherwise terete, slender, the bark pale-brown when dry. *Leaves* sessile, coriaceous, opaque, oblong or ovate-oblong, obtuse or sub-acute, broadest a little above the rounded, cordate base, the midrib distinct but the main nerves faint and the reticulations obsolete; length 2·5–4·75 in.; breadth 1–2·5 in. *Flowers* rather numerous, in dense, axillary, condensed glomerulate cymes, the peduncle ·25 in. long, the pedicels shorter than the flowers, each with two broad, acute bracteoles. *Calyx-tube* short, widely campanulate, narrowed to the base, the mouth wide truncate. *Petals* conical in bud. *Fruit* narrowly ellipsoid, ·4 in. long and ·25 in. in diam. (including the deep, persistent calyx-limb). Miq. Fl. Ind. Bat. I, pt. I, 580; Triana in Linn. Trans. XXVIII, 158 (excl. syn. *M. amplexicaule*, Roxb.); Kurz, For. Flora B. Burma I, 511; C. B. Clarke in Hook. fil. Fl. Br. Ind. I, 559; Cogn. in DC. Mon. Phan. VII, 1163. *M. grande*, Smith in Rees' Cyc. XXIII (not of Retz). *M. cordatum*, Wall. Cat. 4100 (partly); Griff. Not. IV, 673. *M. manillanum*, Naud. in Ann. Sc. Nat., Ser. 3, XVIII, 276; Miq. l.c. 576. *M. lutescens*, Presl. Epim. Bot. 208 (not of Naud.).

In all the provinces; not uncommon. DISTRIB. Philippines.

16. MEMECYLON CAMPANULATUM, Clarke in Hook. fil. Fl. Br. Ind. II, 563. Young branches rather slender, terete, their bark pale-brown. *Leaves* coriaceous, elliptic, sometimes with a short blunt apical point, the base always much and abruptly narrowed, nerves invisible; length 3–4·5 in.; breadth 1·25–2·2 in.; petiole ·1–·15 in. *Flowers* on slender pedicels, ·1–·15 in. long, bracteolate at the base and crowded in dense fascicles in the axils of the leaves or of the fallen leaves, the buds of the petals shortly conical. *Calyx-tube* campanulate, blunt at the base and somewhat contracted below the wide truncate limb. *Fruit* unknown. Cogn. in DC. Mon. Phan. VII, 1162.

MALACCA; *Griffith* (Kew Distrib. 2325).

In its leaves this much resembles *M. oleafolium*, Bl., but the flowers of that species are in lax, few-flowered, pedunculate umbels, whereas the flowers of this are in dense, epedunculate fascicles.

17. MEMECYLON MINUTIFLORUM, Miq. Fl. Ind. Bat. Suppl., 323. A tree, 30–70 feet high; young branches slender, with a broad, angularly-margined groove on each side; the bark pale, smooth. *Leaves* thinly coriaceous, drying yellowish-green beneath, narrowly elliptic, cordate-acuminate, the base much narrowed; main nerves very indistinct.

Inflorescence twice as long as the petioles, many-flowered; the peduncles often two or three from the same axil, each bearing several 2–4-flowered umbels; pedicels stout, bracteolate at the base. *Calyx-tube* cup-shaped, not tapered to the base, the mouth expanded, truncate and ·05 in. wide, glandular-hairy when young like the bluntly conical petal-bud and the pedicels. *Fruit* crowned by the minute calyx-limb, depressed globular, smooth, ·35 in· in diam. and ·3 in. deep. Cogn. in DC. Mon. Phan. VII, 1169. *M. acuminatum*, Sm., VAR. *flavescens*, Clarke in Hook. fil, Fl. Br. Ind. II, 562; Cogn. in DC. Mon. Phan. VII, 1152.

MALACCA; *Griffith* (Kew Distrib.) 2325/2. PENANG; *Curtis* 815. PERAK; *King's Collector* 5027, 6105, 6265, 8724. DISTRIB. Sumatra.

Triana considers this a distinct species and I think he is right. *Fruit* however is wanting to complete our knowledge of the form. The Perak specimens agree perfectly with the type sheet named *M. minutiflorum*, Miq., in Herb. Calcutta. The species is not, as was suggested by Kurz, identical with *M. lilacinum*, Zoll. and Moritzi.

18. MEMECYLON MYRSINOIDES, Blume, Mus. Bot. I, 356. A tree, 30–40 feet high (rarely a shrub); young branches terete, slender, with pale-grey bark. *Leaves* thinly coriaceous, drying brown (palest on the lower surface), narrowly elliptic or ovate-lanceolate, the apex very acuminate, much narrowed to the cuneate base; main nerves invisible on both surfaces; length 1·5–2·5 in.; breadth ·9–1·25 in.; petiole ·15–·25 in. *Flowers* numerous, in very short-peduncled cymes, densely clustered together in the same axil; pedicels about the length of the calyx, bracteolate at the base. *Calyx-tube* campanulate, much narrowed to the base, the mouth less than ·05 in. wide, with 4 long (for the genus) acute teeth; *petals* in bud forming a long narrow cone, acuminate. *Fruit* globular, the size of a grain of black pepper. Miq. Fl. Ind. Bat. I, pt. I, 577; Triana in Linn. Trans. XXVIII, 158 (excl. syn.); Cogn. in DC. Mon. Phan. VII, 1160; excl. syn. *M. lilacinum*. *M. capitellatum*, Blume, Bijdr. 1091 (not of Linn.).

PENANG; *Curtis* 2219. JOHORE; *Ridley* 2026. PERAK; *Wray* 2258; *King's Collector* 1851, 3517, 5923, 8828. DISTRIB.; Sumatra; *Forbes* 2953; Java; Bangka.

VAR. *lilacina*, King. Young branches with two deep, sharply-margined grooves; leaves broadly elliptic, yellowish on the under surface when dry, cymes not crowded (only two in an axil). *M. lilacinum*, Zoll. & Mor. Syst. Verzeich., 9; Naud. in Ann. Sc. Nat. Ser. 3, XVIII, 281; Miq. Fl. Ind. Bat. I, pt. I, 575.

PENANG; *King's Collector* 1457; *Curtis* 100. SINGAPORE; *Ridley* 6218. PERAK; *King's Collector* 10442. DISTRIB.; Java, *Zollinger* 178.

J. II. 11

19. MEMECYLON LAEVIGATUM, Blume, Mus. Bot. Lugd. Bat. I, 358.
A small tree; young branches very slender, terete, the bark pale.
Leaves thinly coriaceous, broadly ovate or elliptic, more or less rostrate-
acuminate, the base cuneate; main nerves obscure; length 2–2·5 in.;
breadth 1–1·75 in., petiole ·1–·15 in., opaque, when dry dull dark-
brown, the lower surface slightly paler than the upper. *Cymes* mostly
from the nodes of fallen leaves, small, few-flowered, the peduncle very
short (·1 in. long), pedicels also very short. *Flowers* small (less than ·1
in. long), their buds pointed; *calyx-tube* campanulate, much tapered to
the base, the mouth with 4 acute, small teeth. *Fruit* globular, ·3 in.
in diam., smooth. Miq. Fl. Ind. Bat. I, pt. I, 576; Triana in
Linn. Trans. XXVIII, 157; C. B. Clarke in Hook. fil. Fl. Br. Ind. II,
561 (excl. VARS.); Kurz, For. Flora. I, 513; Cogn. in DC. Mon. Phan.
VII, 1159. *M. Myrilli*, Blume, Mus. Bot., 357; Miq. l.c. 578.
M. pachyderma, Wall. Cat. 4104. *M. Vosmaerianum*, Scheff. in Flora,
1870, 249.

MALACCA; *Ridley* 1767. SINGAPORE; *Ridley* 1815, 1906, 2026, 4805.
PERAK; *Scortechini* 81; *Wray* 2091; *King's Collector* 3768. SELANGORE;
Ridley 2024. DISTRIB.; Burma, *Helfer* 2328; *Wallich* 4104; Bangka,
Java, Borneo.

20. MEMECYLON CINEREUM, King, n. sp. A shrub; young branches
rather slender, terete, sulcate on two sides, the bark dark-cinereous
when dry. *Leaves* coriaceous, drying very dark cinereous-brown on
the upper surface, somewhat paler on the lower, lanceolate or ovate-
lanceolate, much acuminate, the base rounded but more often cuneate;
main nerves 8–10 pairs, invisible on both surfaces or nearly so,
length 2 8–5·5 in.; breadth 1–2·2 in.; petiole ·15–3 in. *Peduncles*
axillary or from the leafless nodes, not much longer than the petioles,
glomerulate, many-flowered; pedicels short, stout, with small acicular
bracteoles at the base. *Calyx-tube* cupular, with a rounded base; the
mouth expanded, ·1 in. wide, undulate and with 4 acute, triangular
teeth. *Fruit* globular, the persistent calyx-limb small, ·25 in. in diam.,
smooth.

PERAK; *Scortechini* 394, 2035; *King's Collector* 3143, 10758.

21. MEMECYLON OLEAEFOLIUM, Blume, Mus. Bot. I, 359. A tree,
30–60 feet high; young branches rather slender, terete, smooth, the
bark very pale. *Leaves* coriaceous, elliptic-oblong or elliptic, the apex
obtusely acuminate, the base much narrowed, when dry of a pale olivaceous-brown colour on both surfaces, the upper the darker; main
nerves 8–10 pairs, obscure; length 2·5–4·75 in.; breadth 1–2·25 in.;
petioles ·15 to ·3 in. *Peduncles* 1–3 in one leaf-axil, several times
longer than the petiole (elongating in fruit), bearing at the apex

numerous, crowded, 2-3-flowered umbellules with a semi-circular bract
at the bases of their short, stout, 4-angled peduncles. *Flowers* with
conical buds, less than ·1 in. in diam., on slender pedicels longer than
themselves, bracteolate at their bases. *Calyx* hemispheric; the mouth
truncate, entire. *Fruit* ovoid-elliptic, crowned by the short calyx-limb,
·4 long and ·25 in. in diam. Miq. Fl. Ind. Bat. I, pt. I, 579 (excl. syn.) ;
Cogn. in DC. Mon. Phan. VII, 1150. *M. Horsfieldii,* Miq. Fl. Ind. Bat.
I, pt. I, 572. *M. grande,* Retz, VAR. *Horsfieldii,* Clarke in Hook. fil. Fl.
Br. Ind. II, 558; Cogn. in DC. Mon. Phan. VII, 1153 (excl. syn.
M. celastrinum, Kurz from both). *M. lampongum,* Miq. Fl. Ind. Bat.
Suppl. 321.

MALACCA ; *Maingay* (Kew Distrib.) 811. SINGAPORE ; *Ridley* 6414.
PERAK ; *Scortechini* 2069 ; *King's Collector* 426, 5187, 4420, 4439, 8571.
DISTRIB. Bangka; *Horsfield ;* Sumatra ; *Forbes* 3213.

This has been treated by Messrs. Clarke and Cogniaux as a variety of *M. grande*
of Retz, a species originally described by its author from specimens sent to him by
Koenig, who collected in Southern India. Retz's description is very short and, as
Mr. Clarke points out, would suit several species. The species of *Memecylon* have
not, as a rule, a wide distribution, and very few indeed of them are common to
S. India or Ceylon and to the Malay Peninsula. I think it, therefore, in the absence
of his type specimen, advisable to consider Retz's name as properly belonging to
the Ceylon plant represented by Thwaites's C.P. 3442. Both Messrs. Clarke and
Cogniaux treat as belonging to typical *M. grande,* Retz, the Singapore plant issued by
Wallich as No. 4472 of his Catalogue under the name *M. laxiflorum.* This plant is
now represented only by fruiting specimens which do not, in my opinion agree with
any other *Memecylon* in Herb. Kew. The inflorescence in Wallich's specimens is
2·5 in. long, pedunculate, and laxly compound-umbellate. When flowers shall be
forthcoming it will probably be found necessary to let the species *M. laxiflorum*
stand good.

Thwaites's C.P. which I assume, in the absence of a type specimen, to be equal
to the type of *M. grande,* Retz, does not in my opinion resemble the four forms
which the two distinguished botanists just mentioned agree in treating as varieties
of it, sufficiently closely to warrant such treatment of the latter. I would venture
to dispose of them as follows :—

VAR. Horsfieldii = M. oleaefolium, *Bl.*   VAR. khasiana = M. celastrinum, *Kurz.*
VAR. pubescens = M. pubescens, *King.*   VAR. merguica = M. merguica, *King.*

M. Cogniaux has inadvertently described the fruit of *M. oleaefolium* as globose,
whereas in his original description of it Blume writes " *fructibus ellipsoideis.*"

22. MEMECYLON PAUCIFLORUM, Blume, Mus. Bot. I, 356. A small
tree ; young branches 4-angled, slender, pale-brown. *Leaves* coriaceous,
rhomboid or elliptic-rhomboid, drying brown, the lower surface paler,
the apex blunt and often retuse, the base acute or subacute; nerves 6
or 7 pairs, invisible or very faint; length 1-1·5 in.; breadth ·35-1 in. ;
petiole under ·1 in. *Cymes* umbellate, axillary, on slender peduncles ·1-·2
'in. long ; flowers 7-10, small, on slender pedicels bracteolate at the base

and about 15 in. long. *Calyx-tube* shortly campanulate, or saucer-shaped, with a large, wide, sharply and minutely 4-toothed mouth. *Petals* acuminate. *Stamens* and style much exserted. *Fruit* depressed-globular, smooth, crowned by the toothed calyx, ·2 in. in diam. Miq. Fl. Ind. Bat. I, pt. I, 578; Kurz, For. Flora Burma I, 514; C. B. Clarke in Hook. fil. Fl. Br. Ind. II, 555; Cogn. in DC. Mon. Phan. VII, 1169. *M. capitellatum*, Spanoghe in Linnaea, XV, 203 (not of Linn.). *M. umbellatum*, Benth. Fl. Austral III, 293 (non Burm.). *M. australe*, Muell. ex Triana in Linn. Trans. XXVIII, 159.

ANDAMAN ISLANDS; very common. DISTRIB. Burma (*Helfer* 2332); Chittagong; Australia; Timor.

The Penang specimens have narrower, less rhomboid leaves than those from the Andamans.

23. MEMECYLON ELEGANS, Kurz in Journ. As. Soc. Beng. 1872, pt. II, 307. A glabrous shrub; young branches slender, boldly 4-angled, sometimes winged, the bark pale. *Leaves* coriaceous, pale yellowish, the upper surface tinged with green when dry, oblong to elliptic, much acuminate, the base very cuneate; main nerves invisible or very indistinct; length 3·5–5·5 in.; breadth 1·4–2 in.; petiole ·15–·3. *Flowers* ·15 in. long, their pedicels longer, (·2 in.), slender, angled. *Cymes* axillary, several together, pedunculate, simply or trichotomously umbellulate; peduncles ·3–·75 in. long, 4-angled. *Calyx-tube* somewhat large for the genus, cup-shaped, narrowed to the base, ·1 in. wide at the undulate, obscurely 4-lobed mouth. *Petals* blue, broadly ovate, acuminate. *Fruit* globular, smooth, ·5 in. in diam. Kurz, For. Flor. Burma I, 514; C. B. Clarke in Hook. fil. Fl. Br. Ind. II, 554; Cogn. in DC. Mon. Phan. VII, 1138.

ANDAMAN ISLANDS; very common.

VAR. *minor*, King. *Cymes* usually solitary, the pedicel slender, short; fruit only ·2 in. in diam. (? ripe); leaves 2–3 in. long.

ANDAMANS; *King's Collectors.*

Smaller than the typical form in all its parts. The flower buds also differ somewhat from those of the typical form.

24. MEMECYLON ACUMINATUM, Smith in Rees Cyclop. XXIII, 4. A tree, 30–50 feet high; young branches slender, terete, the bark brown, smooth. *Leaves* thinly coriaceous, drying pale olivaceous-brown, the surfaces concolourous, ovate to ovate-lanceolate, caudate-acuminate, the base cuneate; main nerves invisible; length 1·5–2·25 in.; breadth ·8–1·4 in.; petiole ·1–·15 in. *Cymes* solitary or in pairs, axillary, umbellate, on peduncles several times longer than the petioles. *Flowers* 6–8 in a compound umbel; pedicels bracteolate at the base, slender,

twice as long as the flowers. *Calyx-tube* cup-shaped, the fundus broad, slightly constricted below the thick, short, undulate, broadly 4-toothed limb. *Petals* conical in bud. *Fruit* globular, somewhat depressed, smooth, crowned by the narrow calyx-limb, ·25 in. in diam. Triana in Trans. Linn. Soc. XXVIII, 158; DC. Prodr. III, 6; Clarke in Hook. fil. Fl. Br. Ind. II, 562; Cogn. in DC. Mon. Phan. VII, 1152 (excl. VAR. *flavescens*).

MALACCA; *Griffith* (Kew Distrib.) 2325; *Maingay* 810; *Derry* 1041; *Ridley* 3297, 3298, 4574. JOHORE; *Ridley* 4656. PERAK; *King's Collector* 3458, 6754.

25. MEMECYLON GARCINIOIDES, Blume, Mus. Bot. I, 358 (excl. VAR. B). A tree, 20–40 feet high; young branches terete, slender, pale-brown. *Leaves* thinly coriaceous, oblong-ovate or elliptic, abruptly and rather obtusely acuminate, the base cuneate, drying pale-brown tinged with olive, the under surface the palest; main nerves invisible; length 3–5·5 in., breadth 1·2–2 in.; petiole ·05–·1 in. *Cymes* axillary and in the axils of old leaves, umbellate, many-flowered, on short peduncles (·2 in. long, longer in fruit); pedicels slender, bracteolate at the base, ·1–·15 in. long. Flower-buds acute. *Calyx-tube* small and cup-shaped, the mouth very wide (nearly ·1 in.), truncate, but with four minute, acute teeth. *Fruit* globular, smooth, pale when dry, ·2 in. in diam. Cogn. in DC. Mon. Phan. VII, 1152.

MALACCA; *Derry* 1240. *Maingay* (Kew Distrib.) 817. PERAK; *Scortechini* 2033; *Wray* 2961, 3203; *King's Collector* 1984, 2938, 7123, 10034; SINGAPORE; *Ridley* 8118. SELANGORE; *Ridley* 7333. DISTRIB. Sumatra, *Blume, Forbes* 2970, 3108; *Borneo, Beccari* 536.

26. MEMECYLON ANDAMANICUM, King, n. sp. A shrub; young branches slender with faint grooves below the nodes, the bark pale-brown. *Leaves* chartaceous, brown on the upper and greenish-yellow on the lower surface when dry, oblong-lanceolate, gradually and bluntly acuminate, the base cuneate; main nerves 10–12 pairs, interarching near the edge, sub-horizontal; length 2·25–3 in.; breadth ·75–1 in.; petiole ·25–·3 in. *Peduncles* unequal, ·2–·4 in. long, in pairs in the axils of leaves or of fallen leaves, bearing at their apices several 3–5-flowered umbels, bracteolate at the divisions, flower-pedicels as long as the calyx, minutely bracteolate at the base. *Calyx-tube* campanulate, tapered below, the mouth truncate, nearly ·2 in. wide. Bud of petals conical. *Fruit* depressed-globular, crowned by the small calyx-limb, yellowish, ·2 in. in diam.

ANDAMAN ISLANDS; *King's Collectors,* 357, 452. NICOBAR ISLANDS.

A species with leaves somewhat like those of *M. garcinioides,* Bl., but narrower. In its inflorescence it resembles *M. acuminatum,* Sm., but the peduncles are longer than in that species. The inflorescence also resembles that of *M. intermedium*, Bl., but when young it is covered with a yellow waxy coat; the pedicels and peduncles are moreover much shorter than in *M. intermedium*. The leaves resemble those of the latter species in shape but are of a thinner texture so that the nerves are visible though faint.

27. MEMECYLON INTERMEDIUM, Blume, Mus. Bot. I, 358. A tree, 20–40 feet high; young branches slender, terete, pale cinereous. *Leaves* thinly coriaceous, broadly ovate, shortly and bluntly acuminate, the base cuneate, greenish above and brown beneath when dry; main nerves invisible or nearly so; length 2·75–3·5 in.; breadth 1·25–2 in.; petiole ·25–·35 in. *Cymes* large, crowded, in the axils of leaves or of fallen leaves, usually in pairs, on peduncles several times longer than the petioles, compoundly umbellate; pedicels slender, bracteolate at the base, ·1 in. long. *Calyx-tube* cup-shaped, with a wide, truncate, edentate or minutely toothed limb. *Fruit* not seen (globose *fide* Cogniaux). Triana in Linn. Trans. XXVIII, 157; C. B. Clarke in Hook. fil. Fl. Br. Ind. II, 561; Cogn. in DC. Mon. Phan. VII, 1158. *M. umbellatum*, Blume, Bijdr. 1094 (not of Burm.) Naud. in Ann. Sc. Nat. Ser. 3, XVIII, 273; Miq. Fl. Ind. Bat. I, pt. I, 575. *M. garcinioides*, Bl., VAR. *elongatum*, Blume, Mus. Bot. I, 358.

PERAK; *Scortechini* 1036. DISTRIB. Sumatra; Java.

This resembles *M. garcinioides*, Bl., very closely, but differs in inflorescence, the cymes of this being larger, on longer peduncles.

28. MEMECYLON EDULE, Roxb., Corom. Plants I, t. 82. A shrub or small tree; young branches terete, pale when dry. *Leaves* coriaceous, drying brown, the lower surface paler, both often with an olivaceous tinge, elliptic or ovate, the apex sub-acute or shortly and bluntly acuminate, the base usually cuneate but sometimes rounded; main nerves 5–8 pairs, very inconspicuous, ascending; length 2–4 in.; breadth ·85–2·25 in.; petiole ·1–·35 in. *Peduncles* several together, unequal in length, longer than the petioles, axillary, umbellately cymose, many-flowered; pedicels longer than the calyx. *Calyx-tube* cupular, narrowed to the base, the limb truncate, sometimes obscurely 4-toothed. *Fruit* globular, crowned by the small calyx-limb, ·25 in. in diam.

Only two of the numerous varieties of this species occur in our region. These are as follows:—

VAR. 1. *typica*. Leaves usually under 3 in. long, dull, tinged with yellow when dry, acute or obtuse. *M. edule*, Roxb. Fl. Ind. II, 260; DC. Prodr. III, 6; Wall. Cat. 4107; Dalz. & Gibs. Bomb. Fl. 93; Kurz, For. Fl. I, 512. *M. edule*, VAR. *a*, Thwaites Enum. 111. *M. umbellatum*, Burm. Fl. Zeyl. t. 31. *M. tinctorium*, Kœn. ex W. & A. Prodr. 319; Wight Ill. t. 31. *M. globiferum*, Wall. Cat. 4108. *M. pyrifolium*, Naud. in Ann. Sc. Nat. Ser. 3, XVIII, 277.

SINGAPORE ; *Ridley* 4084, 6054. MALACCA ; *Griffith* (Kew Distrib.) 2327 ; *Maingay* (K.D.) 812 ; *Derry* 1028. KEDAH ; *Ridley* 2627, *Curtis* 2627. DISTRIB. India, Ceylon.

VAR. 2. *ovata*, C. B. Clarke in Hook. fil. Fl. Br. Ind. II, 563. Leaves large, often 4–4·5 in. long, acute or acuminate at ·the apex, the base rounded or cuneate, shining when dry ; fruit black when ripe and somewhat succulent. *M. ovatum*, Sm. ex Kurz, For. Fl. I, 512. *M. edule*, VAR. γ, Thwaites Enum. 110. *M. umbellatum*, Hb. Heyne in Wall. Cat. 4109. *M. tinctorium*, VAR. β, W. & A. Prodr. 319. *M. prasinum*, Naud. in Ann. Sc. Nat. Ser. 3, XVIII, 275. *M. grande*, Wall. Cat. 4103, partly. *M. lucidum* and *M. pyrifolium*, Presl. Epim. Bot. 209, 210.

ANDAMAN ISLANDS ; not common. NARCONDAM and GREAT COCO ISLANDS ; *Prain.* PERAK ; *King's Collector* 4175 ; *Scortechini* 917. PENANG ; *Curtis* 723. SINGAPORE ; *Ridley* 6532. DISTRIB. India, Malayan Archipelago.

DOUBTFUL SPECIES.

M. amabile, Bedd. VAR. *malaccensis*, Clarke in Fl. Br. Ind. II, 555. This is founded by its author on the very imperfect material afforded by Maingay's specimens (Kew Distrib. 819).

M. laxiflorum, Wall. Cat. ; see note under *M. oleaefolium*, Blume.

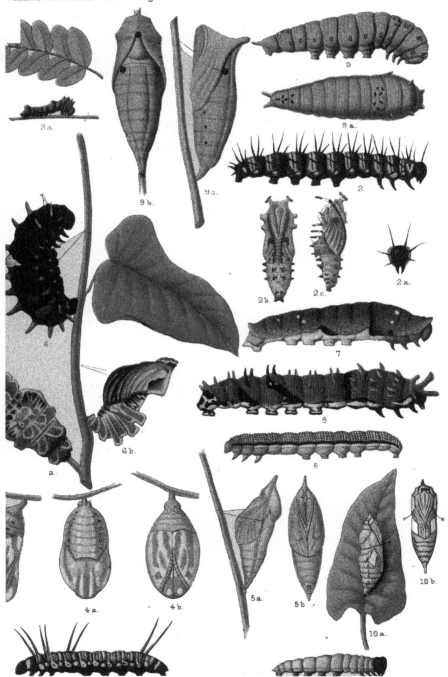

3 a.

9 b.

9 c.

3

9 a.

2

2 a.

6

6 b.

a.

2 b.

2 c.

7

8

5

4 a.

4 b.

5 a.

5 b.

10 a.

10 b.

JOURNAL

OF THE

ASIATIC SOCIETY OF BENGAL.

---●◉●---

Vol. LXIX, Part II.—NATURAL SCIENCE.

No. II.—1900.

II.—*Note on four Mammals from the neighbourhood of Darjeeling.*—*By* W. P. MASSON. *Communicated by the Natural History Secretary.*

[Received 14th January; Read 4th April, 1900.]

Mr. G. C. Dudgeon, F.E.S., having recently published in the Proceedings of the Asiatic Society of Bengal, p. 111 (1899), a paper entitled "Mammalia not hitherto recorded from the Darjeeling District and Sikhim," I venture to lay my experience gained by having collected largely during a period of nearly twenty years in that region before the members of the Society, as regards the animals referred to by Mr. Dudgeon.

Ursus malayanus, Raffles. With regard to this animal in the year 1883 I wrote to "*The Field*" under the nom-de-plume of "Pulteney" a note entitled "Bear Shooting in Darjeeling":—"Jerdon, in his 'Mammalia of India'—and I see Mr. Sterndale follows him—has only one species of black bear found on the Himalayas (*Ursus tibetanus*).* Now I know two distinct kinds, *U. tibetanus* and a smaller species. *U. tibetanus* seldom climbs trees, but the smaller species always does.

---

* Mr. W. T. Blanford in 'The Fauna of British India—Mammalia,' p. 197, n. 93, uses the name *U. torquatus,* Wagner, for this species in preference to *U. thibetanus,* Cuvier, the older name, for the reason that this bear is not found in Tibet itself, and is therefore misleading. [*Ed.*]

J. II. 12

It is entirely a very glossy black, with the exception of a very narrow white mark on the chest, sending up a branch on each side in front of the shoulder; the nose is buff-coloured or white. I have shot many of both species, and have now in my possession skins of both. Most of the villagers about the hills could have told Mr. Sterndale that there were two kinds of bears, one called 'bhooe bhaloo,' or ground bear, and another 'rook bhaloo,' or tree bear. I have always found the bhooe bhaloo or ground bear (*Ursus tibetanus*) very much more numerous than the rook bhaloo or tree bear." In the same note I mentioned that on the 23rd of October, 1883, a friend, my brother and myself went to Birch Hill Park about 9 P.M. one moonlight night, and we saw three bears in a small oak eating the acorns; and on the following night counted no less than five in a single tree, all of them were rook bhaloos. These numbers are unusual, as a rule not more than a couple of bears are seen in one tree. I may remark that I have since found that *U. tibetanus* in the Darjeeling District does climb trees, as I have shot them in oaks when the acorns are ripe in October. Also that the smaller species, *U. malayanus*, principally affects oaks and chestnuts, in which they form rude nests by breaking off the smaller branches and piling them into a heap amongst the larger branches; and that the examples of *U. malayanus* I have shot were of the normal form described by Mr. Blanford in which the crescentic white patch on the chest does not have the apex prolonged into a white streak on the abdomen. The claws are short. The hillmen dread the rook bhaloo very much more than the larger species, and they all agree that if disturbed it attacks at once.

Atherura macrura, Linnæus. I have had the Asiatic Brush-tailed Porcupine from below Ging, from near the Rumam river, and from the Rohtak Valley—all in Sikhim. One of my collectors brought me a fine specimen of the species from beyond Sundukpho on the Nepalese frontier, over 11,000 feet elevation.

Nemorhædus bubalina, Hodgson. I have shot the Himalayan Goat-antelope or Serow on some rocky steep hills close to Sundukpho at about 11,000 feet elevation, have seen them shot at Senchal at about 8,000 feet, have again got them on some rocky steep ground below Soom and Singtom at about 4,000 feet, and again on some very rocky ground on the Sikhim side of the Rumam river at about the same elevation.

Cemas goral, Hardwicke. I have shot the Goral near Philot and near Tongloo, at about 12,000 feet elevation, again on the landslip between Soom and Singtom, and numbers are to be found on some very rocky and precipitous ground on the banks of the Rumam River.

III.—*On a new method of treating the properties of the circle and analogous matters.*—By PROMOTHONATH DUTT, M.A., B.L. *Communicated by the Natural History Secretary.*

[Received 12th March ; Read 4th April, 1900.]

According to Euclid the circle is defined as a plane figure, which is such that the length of any straight line drawn from a certain point within the circle to the boundary is constant.

A circle may also be defined as the locus of a point which moves so that the ratio of its distances from two fixed points is constant. This proposition has been proved as prop. 4 of the Theorems and Examples on Bk. VI in Hall and Stevens's edition of Euclid, page 361. There the proposition has been given in the following words : "Given the base of a triangle and the ratio of the other two sides, to find the locus of the vertex." The proof shows that the locus is a circle. I propose to take this property of the circle as my starting point, and to deduce other properties from it. I shall first of all proceed to show how the centre of the circle can be found from the definition adopted.

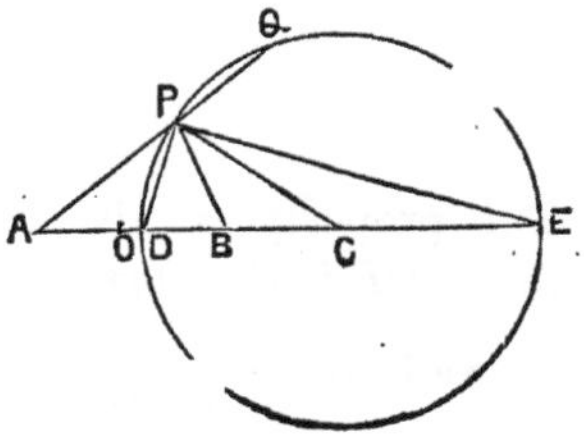

Let A, B be two given points, and PDE be the circle, so that whatever the position of P, the ratio of AP to BP is constant.

Fig. 1.

Then $\dfrac{AP}{BP} = \dfrac{AD}{BD}$

$\therefore \angle APD = \angle BPD$ (prop. 3, Euc. Bk. VI).

Also $\dfrac{AP}{BP} = \dfrac{AE}{BE}$

$\therefore \angle BPE = \angle QPE$ (prop. A. Euc. Bk. VI).

$\therefore \angle EPD$ is a right angle.

Take O as the middle point of DE.

By a well-known rider (Ex. 2, on prop. 32, Bk. I, Hall and Stevens page 100).

We have $OP = OD = OE$.

$\therefore O$ is the centre of the circle.

According to the definition adopted, it will be found that AB is divided harmonically at D and E (Hall and Stevens's Geometry, Example I, Bk. VI, page 360). It will appear from Example III, that the straight line through B drawn at right-angles to the diameter DE is the polar of A with respect to the circle. Example II shows that if O be the middle point of AB, $OD.OE = OB^2$. Example I at page 233 (Hall and Stevens) shews that the rectangle AC, BC is equal to the square on the radius.

Let us write the property in the form $r_1 = mr_2$, where $r_1 = AP$, $r_2 = BP$. Describe two circles with the fixed points as centres, and radii equal to a, b so that $a = mb$, then the equation of the circle can be reduced to the form $r_1 - a = m\ (r_2 - b)$, which means geometrically that the distance of any point on the circle $r_1 = mr_2$ from the circumferences of the circles described with the fixed points as centres are in the fixed ratio m.   (1).

The form $r_1 - a = m\ (r_2 - b)$ shows that the circle passes through the intersections of $r_1 = a$, and $r_2 = b$, and it is evident, therefore, that the three circles $r_1 = a$, $r_2 = b$, and $r_1 = mr_2$ will co-intersect if $a = mb$.

The proposition may be enunciated geometrically as follows :—

Let circles be described with the fixed points as centres, so that their radii are in the ratio m. The circle which represents the locus $r_1 = mr_2$ passes through the intersections of these circles.   (2).

If PT be the tangent at P we

Fig. 2.

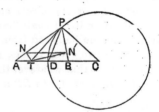

have
$$\frac{\cos APT}{\cos BPT} = \frac{\dfrac{dr_1}{ds}}{\dfrac{dr_2}{ds}}$$

$$= \frac{dr_1}{dr_2}$$

$$= m.$$

This property can also be deduced without the use of the differential calculus. CP is the normal at P.

∴ we have $\dfrac{\sin CPA}{\sin CPB} = \dfrac{\cos APT}{\cos BPT}$.

But $\dfrac{\sin CPA}{\sin OPB} = \dfrac{\dfrac{\sin CPA}{\sin PCA}}{\dfrac{\sin OPB}{\sin POB}} = \dfrac{\dfrac{CA}{AP}}{\dfrac{CB}{BP}}$.

But from similar triangles we can prove that $\dfrac{CA}{CB} = m^2$ (*vide* the figure of prop. 3, Hall and Stevens, page 361).

Also $\dfrac{AP}{BP} = m$

∴ $\dfrac{\cos APT}{\cos BPT} = m$.   (3).

Let N, N' be the feet of the perpendiculars from T on AP and BP. Then $PN = PT \cos APT$, $PN' = PT \cos BPT$.

∴ $PN = m.\ PN'$.

Also $AP = m.\ BP$.

∴ $AN = m.\ BN'$.

But the circle described on AT as diameter cuts AP at N and that on BT cuts BP at N'.

The proposition may be enunciated geometrically thus :—

If the tangent at any point P to a circle meets the line joining the fixed points A, B in T and on AT, BT are described circles cutting AP, BP in N, N' respectively, then the ratio of AN to BN' is the same as that of AP to BP.   (4).

As an alternative and purely geometrical proof of prop. (4) the following is given.*

Join NN' and PD.

We have $\angle CPB = \angle CPD - \angle BPD = \angle CDP - \angle APD = \angle PAD$.

∴ $\angle ATN$ being the complement of $\angle PAD$ is equal to $\angle BPT$, the complement of $\angle CPB$.

But P, N, T, N' lie on a circle.

∴ $\angle N'PT = \angle TNN'$.

∴ $\angle ATN = \angle TNN'$.

∴ AB is parallel to NN'.

∴ $\dfrac{AN}{BN'} = \dfrac{AP}{BP}$.

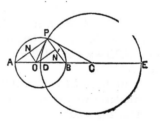

Fig. 3.

Let the circle described on AB as diameter cut the given circle in P. Then if O be the middle point of AB, we have $OD.OE = OB^2 = OP^2$.

But $OD.OE = CO^2 - CD^2$

$$= CO^2 - CP^2.$$

∴ $CO^2 = CP^2 + OP^2$.

By Euclid I, 48, $\angle OPC$ is a right-angle.

Therefore, OP is the tangent at P to the given circle, and it is the normal to the circle described on AB as diameter. Therefore, the circle described on the line joining the fixed points as diameter cuts the given circle at right angles.   (5).

Also as proved before, $PN = m. PN'$.

∴ $ON' = m. ON$ as $PNON'$ is a rectangle.

Therefore, if the circle described on the line joining the fixed points cuts the given circle at P, and O be the point, where the tangent at P meets AB, then the distances of AP, BP from O are in their inverse ratio.   (6).

---

* For this proof I am indebted to my nephew, Babu Benodebihari Dutt of the Sanskrit College, who has given me much assistance in the composition of this paper. P.D.

The property follows also from the consideration of the equality of the areas of the triangles APO and BPO.

Let AP cut the circle again in P', then

Fig. 4.

$$\frac{AP}{BP} = \frac{AP'}{BP'}$$

$$\therefore \frac{AP}{AP'} = \frac{BP}{BP'}$$

or $\dfrac{AP}{PP'} = \dfrac{BP}{BP'-BP} = \dfrac{BP}{P'Q}$ where $BQ =$

    BP.

$\therefore PQ$ is $\parallel AB$.   (7).

Let PB produced meet the circle in R

then $\dfrac{AP}{PB} = \dfrac{AR}{BR}$

$\therefore \dfrac{AP}{AR} = \dfrac{BP}{BR}$

$\therefore \angle PAB = \angle BAR.$ (8).

Similarly it can be shewn that if $P'B$ meet the circle again in R' then

$$\angle P'AB = \angle BAR'.$$

$\therefore A, R', R$ lie on the same straight line.   (9).

Again, in the two triangles $APB, AR'B$ we have $\angle PAB = \angle R'AB$ also $\angle APB = \pi - \angle P'PR = \pi - \angle P'R'R = \angle AR'B$, and the side AB is common

$\therefore$ we have $BP = BR', AP = AR'.$

Similarly we can show that $BP' = BR$ and $AP' = AR.$ (10).

Also $BQ = BP.$

$\therefore BP'^2 - BR'^2 = P'R' . P'Q.$ (11).

Since $BP = BQ = BR'$ we can show that the angle QPR' is a right-angle.   (12).

If $R'Q'$ be drawn parallel to AB, similarly $\angle QQ'R'$ is a right-angle, and as PQ, $R'Q'$ are each parallel to AB, $QPR'Q'$ is a rectangle.   (13).

And it is evident that P, Q, Q', R' lie on a circle, the centre of which is B.   (14).

The results of propositions (7) to (14) may be summed up as follows :—

If a straight line be drawn through the fixed point A, cutting the circle in P and P' and PB, $P'B$ meet the circle again in R and R', and PQ and $R'Q'$ are drawn parallel to AB, then A, R', R lie on one straight line. A circle with centre B passes through P, Q, Q', R'. AB bisects the angle $P'AR'$ and P, R' and P', R are respectively symmetrical with respect to AB.

Now, let us consider the properties of the system of circles, obtained with the same fixed points A, B, by varying the ratio of AP to BP. We have $\dfrac{OA}{OB} = m^2$.

$$\therefore \frac{AB}{OB} = m^2 - 1 \text{ or } OB = \frac{AB}{m^2 - 1}.$$

Now, AB is constant $\therefore OB \propto \dfrac{1}{m^2 - 1}.$ (15).

If $m = 1$, $OB = \infty$, and this must be so as in this case, the locus reduces itself to a straight line, which may be taken as the case of a circle with an infinite radius.

It will appear from (5) that all the circles of the system are intersected orthogonally by the circle described on AB as diameter.

If r be the radius of the circle, $r^2 = OA \times OB = m^2 \times \left(\dfrac{AB}{m^2 - 1} \right)^2.$

$$\therefore \quad r = AB \times \frac{m}{m^2 - 1}.$$

or $AB = r \left(m - \dfrac{1}{m} \right).$ (16).

This shows that the radius remains the same if m be changed into $-\dfrac{1}{m}$.

Geometrically this means that if we describe a circle making $BP = m \cdot AP$, the radius will be the same, but the centre will be on the other side, and this also appears from the consideration of symmetry.

The relation $r_1 = mr_2$ reduces to $r = ma$ where a is fixed, and r and m vary. Writing the relation in the form $\dfrac{r}{m} = a$, we arrive at another result, which may be enunciated as follows :—

If two circles be described, one on the line joining the fixed points as diameter, and the other with centre B and radius equal to a; then if any of the system of circles drawn with the same fixed points A, B and having the relation $r_1 = mr_2$ cut the circle described round B as centre in P, and the straight line AB in D, and DE be drawn perpendicular to AB, meeting the circle described on AB as diameter in E, then the length of AP varies as the square of the co-tangent of the angle DAE.

We have $AP = BP \times \dfrac{AD}{BD}$

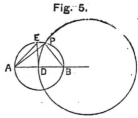

Fig. 5.

$$= BP \times \frac{AD}{DE} \times \frac{DE}{BD}$$

$$= BP \cot^2 DAE$$

$$= a \cot^2 DAE.$$

Therefore, the position of the point P is known if the magnitude of the angle BAE is given.

Also for any point on the circle described round B with a fixed radius $AP \tan^2 BAE$ is constant and equal to the radius. (17).

Now, if AP, BP vary but the ratio $\dfrac{AP}{BP}$ remains constant, the ratio $\dfrac{AD}{BD}$ will remain constant, and also the angle BAE.

Therefore, if A be fixed and any point B is taken in AB, and D taken such that $\dfrac{AD}{BD}$ is constant, the locus of the point where the perpendicular at D meets the circle described on AB as diameter is a fixed straight line through A. (18).

The preceding propositions will, I hope, show the utility of the method. Many other properties may be similarly deduced, and this paper is put forward in the hope that it will lead to the deduction of other important and useful results by others.

I beg to add that the properties alluded to before, may, I think, be utilised in some physical experiments in which two motions are produced simultaneously at two different places, but as I have not the opportunity or ability to carry them out myself, I only beg to give a brief outline of the method in the hope that others will be able to work it out.

Experimental measurement of the velocity of sound from observations in a railway train.—If a railway train travels at a constant speed, and a sound be produced at a distance at the same moment, when the train leaves the station, the sound will be heard in the train at only two points in the line of motion, formed by its intersection with the circle with the source of sound and the station as the fixed points. From the observations the ratio of the velocity of sound to that of the railway train can be calculated, as shewn below. The following proof applies to the case in which the line of motion is at right-angles to the line joining the railway station to the source of sound.

Let B be the station and A the source of sound in the figure in Ex. 3, Hall and Stevens page 361, and let H and K be the points where the sound from A is heard. Then, from the observation of the time since leaving the station, an observer can find the distance BH, if the speed of the railway train be known. AB can be found by direct measurement, and thence AH. The ratio of AH to BH gives that of the velocity of the sound to that of the train.

IV.—*Note on a method of detecting free Phosphorus.*—By P. Mukerji, B.Sc., *Professor of Chemistry, Presidency College, Calcutta. Communicated by the Natural History Secretary.*

[Received 12th March ; Read 4th April, 1900.]

The following statement is made in Roscoe's Treatise on Chemistry and in Watts' Dictionary of Chemistry: Phosphorus does not directly combine with Hydrogen.

The principle employed for the detection of free Phosphorus in the present note is the phosphorescence of the vapour of Phosphorus diluted with Hydrogen gas.

The following is the *apparatus* employed : A three-necked Woulffe's bottle ; to the middle neck of which there is attached by a cork a tube about eleven inches long and of half an inch diameter, the upper end of the tube being closed with a cork. To another neck a safety funnel with a long tube is attached, and in the third neck a short jet is inserted. The tube of the safety funnel dips into the liquid in the bottle ; the middle tube and the jet enter the bottle only for a short distance. The tube attached to the middle neck is somewhat *loosely* fitted so that traces of air may enter the bottle. The jet and the funnel are fitted tightly. The capacity of the bottle may be a litre or less.

For the above-mentioned apparatus a simpler one has been substituted in several experiments ; namely, a small flask of about 184 c.c. capacity, and fitted with a funnel with a stopcock and also with a jet.

The method of carrying out the test, and the appearance :—Zinc and dilute Sulphuric acid are introduced into the bottle, and the gas issuing from the jet is observed. If in a *dark* room, there is no glow perceived, then the materials employed are free from uncombined Phosphorus. When the bottle is so hot through the chemical action between the zinc and acid that it can scarcely be touched, *i.e.*, when the temperature of the liquid is about 60° to 70° C., and that of the gas in the upper part of the bottle is 45° to 50° C., the cork is removed from the top of the tube attached to the middle neck and the substance suspected to be

Phosphorus or to contain this in a free state is introduced through the tube into the bottle, and then the cork is quickly inserted. If an organic mixture suspected to contain free Phosphorus has to be introduced, it may be poured in through the thistle funnel, or better still through the neck to which the jet is attached; the jet is quickly put back into the neck after introduction of the mixture. Soon after the introduction of Phosphorus (free) the gas issuing from the jet is found to glow in a dark room, a sheaf of light emanating from the jet. The liquid inside the bottle glows, and luminous flashes are also seen now and then. If now the cork at the top of the middle tube be removed, the light will sink down through the jet, and the gas escaping at the top of the middle tube will glow. On replacing the cork the glow will reappear at the jet. By alternately removing and replacing this cork, the glow may be made *to move downwards and upwards* through the jet *at the option of the operator.* [If the second form of apparatus is used, it is only necessary to take off and replace the stopcock below the funnel, and the glow will appear alternately in the funnel and at the jet.]

Fresh quantities of dilute Sulphuric acid and Platinic chloride may be introduced if the glow shews sign of ceasing.

Delicacy of the test :—In one experiment, 7 mgrs. of Phosphorus were cut into two larger and two smaller pieces; and one of the smaller pieces (about 1·5 mgr.) was introduced into the second form of apparatus; the appearance noted above was observed. After about half an hour the zinc was washed two or three times and kept in the flask under water. About twenty-four hours later this zinc without any further addition of Phosphorus again shewed the glow at the jet and in the funnel.

In another experiment the first form of apparatus was employed, the capacity of the bottle being a little over a litre. 112 grammes of zinc, 250 c.c. of acid and water, and 30 c.c. of milk were introduced into the bottle; next 2 mgr. of Phosphorus were introduced. The jet, the middle tube, and the bottle all glowed in the way described above. It may be here stated that Fresenius's limit for Mitscherlich's test is 1·5 mgr. of Phosphorus in 150 grammes of mixture; the second experiment shews that the present method is *at least* as delicate as Mitscherlich's.

Remarks on the process :—The use of nascent Hydrogen as a reagent for the detection of Phosphorus is not new. Valentine mentions it in his Qualitative Analysis; but his method is not the same as the present method. He observes the green core of the hydrogen jet *after ignition,* and he does not use a dark room. Moreover, his apparatus is Marsh's Apparatus. His test is not a test of free Phosphorus only but indicates indifferently either Phosphorus, phosphide, phosphite or hypophosphite, since any one of these substances will give a *flame* with a green core on treatment with nascent hydrogen, *i.e.*, with zinc and dilute Sulphuric acid. It need hardly be stated that the method described in this paper is not the Blondlot-Dussard Method, since the Hydrogen jet is *not* ignited, and the appearance observed is not the green core of the flame of *burning* Hydrogen. The phenomena observed in the present method are very similar to those described by Crookes in page 489 of his *Select Methods*, Third Edition. Crookes, however, uses a somewhat complicated distilling and condensing apparatus, and employs a lamp for distilling the Phosphorus : the lamp again necessitates the use of materials for preventing its light from hindering the observation of the glow. Moreover, in his method, the glow is *not under control.* The apparatus described above is simple ; no lamp is required for heating, and consequently no precaution is necessary, as to the reflected light of the lamp being mistaken for the glow of Phosphorus. Again, the operator can by using the present method make the glow move up and down as often as he likes and with great ease. Finally, there is no risk of explosion, since hydrogen *continues* to be evolved even when air is sucked in through the jet (by opening the top of the middle tube)'; air cannot rush in to any large extent. Lastly, Phosphorus can not only be detected by this method, but it can also be estimated quantitively by passing the vapour thus evolved into silver nitrate solution.

After determining the exact method in which the present test is to be carried out, the influence of a good many substances upon the glow described above was tried ; and by comparing the results obtained with those stated in connection with Mitscherlich's method by Crookes and others it appears that the *present method is superior to Mitscherlich's.* This will be seen from what will be stated presently. It may be first of all noted that common bazaar zinc was used ; the zinc contained some Phosphorus in the combined state (and also some Arsenic). Hence in contact with dilute Sulphuric acid the zinc evolved a jet of hydrogen, which burnt with a green core on ignition, but which gave no glow without introduction of free Phosphorus. It follows that for the *detection* of Phosphorus, it is not necessary to use pure zinc. It may be suggested that steam could be substituted for nascent Hydrogen in the second form of apparatus described above. The objection to steam is that the glow is not as controlable as it is with nascent Hydrogen ; moreover there is the risk of explosion after suction of air through the jet by removing the lamp from below the flask and *then re-heating.* The risk of explosion will be distinctly greater in presence of mustard oil, phosphoretted hydrogen, &c., and in presence of Iodine, Sulphuretted

Hydrogen, Ether, &c. ; nascent Hydrogen has also a decisive superiority as will appear later on. Then again on *boiling* Phosphorus in a complex aqueous mixture (when the temperature must be higher than 100° C.) some portion of the element may get oxidised by water if by nothing else and cannot then be vaporised.

As already stated some of the substances in presence of which Phosphorus was detected by this method were such as are known to interfere with the glow in the older methods. Milk, boiled rice, flour, silica, though not included in this category were thus tried ; the glow was well observed both at the jet and at the top of the middle tube after addition of Phosphorus. Sodium or potassium hypophosphite solution freshly prepared and filtered from particles of Phosphorus gave no glow, though phosphoretted hydrogen was apparently evolved as the gas at the jet burnt with a green core upon ignition : the glow appeared soon after introduction of a minute piece of Phosphorus. Phosphoretted hydrogen therefore does not interfere with the glow in this method. Sodium phosphite also gave no glow though it evolved phosphoretted hydrogen. Nitrous fumes are said to stop the glow of phosphorus, but it was found that in this method nitrate or nitrate mixed with chloride did not prevent the glow. Mustard oil was tried : it did not interfere with the glow; the jet became blocked by a whitish oily deposit after some time, but there was the usual glow in the funnel (of the second form of apparatus). The gas did not take fire and there was no explosion. Sulphuretted hydrogen and also Iodine are stated to interfere with or stop the glow of Phosphorus. In the present method the result obtained in presence of either of these substances may be regarded as satisfactory. In one experiment 10 c.c. of a saturated solution of Iodine in potassium Iodide were introduced in three portions ; the glow was seen as usual, and the issuing gas did not colour starch paper to any appreciable degree. In another experiment about four-and-a-half grammes of powdered ferrous sulphide were introduced in two portions, and about 3 mgr. of Phosphorus were used. The glow was observed at the jet and in the funnel : next the jet was changed and the new jet also gave the glow. Ether, alcohol, and oil of turpentine were also tried. Oil of turpentine stopped the glow altogether ; but it was found that so far as the present method is concerned, the failure may be easily remedied. A piece of Phosphorus, about 2 mgr., was introduced into boiled rice ; oil of turpentine, 2·5 c.c., was next added and then a little water, and the mixture shaken. Now the turpentine was removed by decantation after adding some water ; the mixture was next quickly washed (by decantation) with Alcohol and finally with water. The mixture probably still contained traces of turpentine oil ; but on introduction into the hydrogen-evolving bottle, the glow was seen as usual. Next 5 c.c. of alcohol were introduced into the bottle, the glow continued at the jet, and was also observed at the middle tube, though it was a little less bright. As regards either it is as prejudicial to the glow as oil of turpentine : but there is this difference that the glow appears *after a time* on introduction of a fresh piece of Phosphorus.

It may be stated here that after the experiment is over, the bottle or flask used for the experiment should be filled with water before opening it in air.

It will be seen from what has been stated above that the present method will greatly facilitate the detection of Phosphorus in food and in cases of poisoning. The apparatus described is simple ; and Zinc and Sulphuric acid can be procured in any town. The glow is seen on a magnified scale ; and there is no possibility of its being mistaken, especially as no lamp is required. Moreover, many substances that hinder the glow of Phosphorus in the older methods do not affect the glow as observed by the present method.

I have lastly to state that my assistant, Haridas Saha, M.A., has helped me considerably in carrying out these experiments.

V.—*Note on the occurrence of* Rhodospiza obsoleta (*Licht.*) *in the Tochi Valley.*—*By* Capt. H. J. Walton, I.M.S.

[Received 18th May ; Read 6th June, 1900.]

Lieut. S. R. Douglas, I.M.S., has very kindly been collecting birds for me in the Tochi Valley for some months. Amongst some that I recently received from him, from Datta Khel, are two specimens of *Rhodospiza obsoleta* (Licht.).

Mr. Oates, in vol. ii, "Fauna of British India, Birds," p. 223, mentions this Desert Rose-Finch as a species likely to be found in India, but I can find no previous record of it having been procured within the limits of the empire.

There is no doubt about the identification ; I have compared the birds with those in the Indian Museum, collected by Stoliczka in Turkestan, and Mr. Finn also agrees that they are *R. obsoleta.* As my two specimens are carbolised, I am not sure of the sexes : the one that I take to be the male is a larger bird than the other, and has the lores and a very narrow frontal line black, the upper tail-coverts contrast with the sandy-brown back in being rather bright rufous, the inner secondaries are rather broadly edged with pale buff.

The other bird, which I take to be the hen, has the lores buff, and the upper tail-coverts and inner secondaries concolorous with the sandy-brown back.

The first specimen was shot on April 14th, 1900. Its measurements, taken in the flesh, are :—

Long : tot : 6·45 inches.
     al : 3·3   ,,
     caud : 2·37   ,,
     culm : 0·4   ,,
     tars : 0·625   ,,

The smaller bird was shot on April 18th, 1900, and measures

Long : tot : 6·1   inches.
     al : 3·25   ,,
     caud : 2·25   ,,
     culm : 0·43   ,,
     tars : 0·625   ,,

The colours of the soft parts were not noted.

VI.—*On the Birds collected and observed in the Southern Shan States of Upper Burma.*—*By* Col. C. T. Bingham, F.Z.S., *and* H. N. Thompson, F.Z.S. *Communicated by the Natural History Secretary.*

[Received 16th May ; Read 6th June, 1900.]

The Southern Shan States may roughly be said to be bounded on the north by the Northern Shan States which are separated from them by the Nam Tu or Myitgnè river; on the east by Chinese territory and the French possessions on the Mèkong river; on the south-east and south by the Siamese Shan States, and the semi-independant state of Karreni ; and on the west by the settled districts of Pyinmana, Kyauksè, and Meiktila.

A large portion of this territory consists of stretches of plateau land at elevations from 2500 to 5000 ft. above the level of the sea, but low-lying hot valleys and high ranges of hills with peaks rising to 8000 ft. and upward also occur. The flora and fauna of the States is consequently very diversified and well-worth thorough exploration and working out.

During last cold weather (1899-1900) accompanied by Mr. H. N. Thompson, Deputy Conservator in charge of the forests, Southern Shan

States, I made a short tour in some of the most western of the States. It was on this tour that the greater part of the material on which this paper is founded was collected, but many additions have been made by Mr. Thompson of birds collected or observed and identified by him previous to and after our joint tour.   We were fortunate in being able to spend a few days at various camps on Loi-San-Pa, a mountain about 8000 ft. in elevation* situated in the Möng Köng State.   Here the birds procured were most interesting and I think fairly representative of the Avi-fauna of the country.

Altogether something over 350 specimens representing 239 species were obtained during the tour or subsequently by Mr. Thompson, while 62 species, specimens of which were not procured, were seen, noted and identified beyond a doubt.

In the list which follows, one or two of the birds entered under already described species, vary more or less in size and colouring from their types.   These might possibly be separated as distinct.   I have however while noting the points of difference preferred to consider them merely local races of well-known species found in the Himalayas, Assam, other parts of Burma, &c.   Two species however, one procured at Kalaw 4300 ft., the other on Loi-San-Pa at 6000 ft. are in my opinion distinctly new.   One of these is a pretty little Fly-catcher belonging to the genus *Cyornis*, the other a remarkable Bulbul allied in habits to *Hypsipetes*, but differing structurally and in colour from all species of that genus, and from all known Bulbuls.   I have ventured to propose (*vide* Annals and Magazine of Natural History, London, series vii, vol. v, No. 28 (April, 1900), pp. 357–359), a new genus *Cerasophila* for the aberrant Bulbul, and to name it *C. thompsoni* after Mr. Thompson, who was the first to shoot a specimen of the bird, recognizing it at the time as probably an undescribed species.

The arrangement followed in the subjoined list is that adopted in the volumes on the Birds, Fauna of India series, by Messrs. Oates and Blanford.   An asterisk prefixed to the current number in the list shows that the species was seen and identified but not procured.

I have to express my thanks to Major A. Alcock, I.M.S., Superintendent of the Indian Museum, Calcutta, who was good enough to send a trained taxidermist to accompany my camp; and to Mr. Frank Finn, the Deputy Superintendent of the Museum, who kindly assisted me in comparing and identifying the birds obtained by me with the series of birds in the collection of the Indian Museum.

---

* The actual heights on this mountain, which has a double cap, are Loi-San-Pa 8002 ft., Loi-Sang the northern rocky peak 8129 ft, above sea-level.   (H.N.T.).

Family CORVIDÆ.

*1 (4). Corvus macrorhynchus, Wagler.
Faun. Brit. Ind., Birds, I, p. 17.

Very common throughout the more wooded portions of these States. Destroys and eats a large number of wild birds' eggs during the breeding season.

*2 (8). Corvus insolens, Hume.
Faun. Brit. Ind., Birds, I, p. 21.

Common, but only in the villages and towns, not in the forest.

*3 (10). Pica rustica, Linn.
Faun. Brit. Ind., Birds, I, p. 24.

Only once met with at Möng Löng (3000 ft.) in the Möng Köng State (C.T.B.). Common in the Eastern States of Hopong, Möng Pöng, Möngnai, &c. I have met it close to the banks of the Mekong river, as low down as 1600 ft. Breeds in March and April. (H.N.T.).

4 (12). Urocissa occipitalis, Blyth.
Faun. Brit. Ind., Birds, I, p. 26.

Common in the drier parts of the States. One or two of the specimens procured had the lower plumage strongly tinged with a wash of ochreous-red.

5 (14). Cissa chinensis, Bodd.
Faun. Brit. Ind., Birds, I, p. 28.

Decidedly rare. One specimen procured at Kalaw (4400 ft.).

6 (18). Dendrocitta himalayensis, Blyth.
Faun. Brit. Ind., Birds, I, p. 32.

Common on the plateau in the north-west corner of the Yatsauk State, elevation 3600 ft. Has a pleasant metallic call. (H.N.T.).

7 (25). Garrulus leucotis, Hume.
Faun. Brit. Ind., Birds, I, p. 39.

Very common in the States of Yatsauk and Möng Köng. Prefers dry oak forest and *Indaing*, and goes about in large parties. (H.N.T.).

Near Yatsauk I procured two out of a counted flock of over thirty. (C.T.B.).

Seems most plentiful at elevations of between 2500 and 3500 ft. One specimen was procured on Loi-San-Pa at 6000 ft.

*8 (31). Parus atriceps, Horsf.
Faun. Brit. Ind., Birds, I, p. 46.
Found along the western border of the States. (H.N.T.).

9 (32). Parus minor, Temm.
Faun. Brit. Ind., Birds, I, p. 48.
Procured wherever there were pine trees. Common at Kalaw 4400 ft., and on Loi-San-Pa up to 6000 ft. The green colour of the nape extends further down the back than in any Tenasserim specimens.

10 (41). Machlolophus spilonotus, Blyth.
Faun. Brit. Ind., Birds, I, p. 54.
Loi-San-Pa 6000 ft. and upwards. Confined to dense evergreen forests.

11 (52). Paradoxornis guttaticollis, A. David.
Faun. Brit. Ind., Birds, I, p. 62.
I put up several individuals of this species when beating the forests for barking deer on the plateau in the north-west corner of the Yatsauk State at an elevation of 3800 ft. I procured two specimens last April on the Salween-Mekong watershed at 6000 ft. (H.N.T.).

Family Crateropodidæ.

12 (64). Dryonastes chinensis, Scop.
Faun. Brit. Ind., Birds, I, p. 74.
One specimen on Loi-San-Pa 5000 ft. (C.T.B.). Common on the Mènètaung range on the west border of the Yatsauk State. Is a very noisy bird. (H.N.T.).

13 (67). Dryonastes sannio, Swinh.
Faun. Brit. Ind., Birds, I, p. 76.
Common above 4000 ft.

14 (70). Garrulax belangeri, Less.
Faun. Brit. Ind., Birds, I, p. 79.
Procured along the road at various places. Does not seem to go higher than the edge of the Myèlat plateau at 3000 ft.

15 (72). Garrulax pectoralis, Gould.
Faun. Brit. Ind., Birds, I, p. 80.
Procured two specimens near Yatsauk at 2800 ft. Noticed several
J. ii. 14

times in the low-lying hot valleys in the States. One of the specimens procured has the whole lower plumage washed with ochreous-yellow.

*16 (73). GARRULAX MONILIGER, Hodgs.
Faun. Brit. Ind., Birds, I, p. 81.
Noticed on one occasion mixed up in a flock of *G. pectoralis.*

17 (86). TROCHALOPTERUM MELANOSTIGMA, Blyth.
Faun. Brit. Ind., Birds, I, p. 92.
Loi-San-Pa between 5000 and 6000 ft. A great skulker.

18 (87). TROCHALOPTERUM PHŒNICEUM, Gould.
Faun. Brit. Ind., Birds, I, p. 93.
I procured specimens of this species on the Salween-Mekong watershed at 7000 ft. elevation. I observed it on one occasion on the slopes of Loi-San-Pa. (H.N.T.).

Specimens from Taunggyi differ, as noted below, from Mr. Blanford's description of this bird in Faun. Brit. Ind., Birds :—

No black supercilium ; the head above and nape dark slatey-brown shading into olive-brown on the back, and paling to olive-green on the rump. The tail above is dark brown with conspicuous cross-barrings of dusky black ; the outer three tail feathers on the underside suffused with crimson-brown and tipped broadly with pale orange ; under tail coverts also tipped with orange. (C.T.B.).

19 (106). ARGYA GULARIS, Blyth.
Faun. Brit. Ind., Birds, I, p. 107.
Confined to the hot low portions of the States.

20 (116). POMATORHINUS SCHISTICEPS, Hodgs.
Faun. Brit. Ind., Birds, I, p. 116.
Common from 3000 ft. and upwards. Its loud " hoot-hoot-hoot " was one of the first sounds to greet one in the morning.

21 (129a). POMATORHINUS IMBERBIS, Salvadori.
Faun. Brit. Ind., Birds, IV, *App.*, p. 479.
At 4400 ft. Taunggyi, April and May. (H.N.T.) I have procured this species in the Ruby Mines District at 6000 ft. in April. (C.T.B.). Iris blood red. (H.N.T.):

22 (139). PYCTORHIS SINENSIS, Gm.
Faun. Brit. Ind., Birds, I, p. 137.
Not uncommon, but a great skulker. One specimen procured near Taunggyi at 5000 ft.

23 (145). PELLORNEUM SUBOCHRACEUM, Swinh.
Faun. Brit. Ind., Birds, I, p. 142.
Common. Breeds in April (H.N.T.). I came across it only twice during the tour, and I entirely missed its incessant call of "pretty dear"—"pretty dear" so commonly heard in the Tenasserim forests. (C.T.B.).

24 (153). CORYTHOCICHLA BREVICAUDATA, Blyth.
Faun. Brit. Ind., Birds, I, p. 148.
Taunggyi, a single specimen. Hitherto I believe only procured on Moolayit mountain in Tenasserim.

25 (163). ALCIPPE NEPALENSIS, Hodgs.
Faun. Brit. Ind., Birds, I, p. 157.
Confined to well-wooded parts. Procured near Taunggyi at 4000 ft.

26 (176). MIXORNIS RUBRICAPILLUS, Tick.
Faun. Brit. Ind., Birds, I, p. 167.
Near Taunggyi 4000 ft. Found occasionally in well-wooded parts, especially in bamboo jungle. Has a most monotonous call, which it keeps up throughout the day in the hot weather. (H.N.T.).

27 (182). SITTIPARUS CASTANEICEPS, Hodgs.
Faun. Brit. Ind., Birds, I, p. 172.
One specimen procured on Loi-San-Pa at over 7000 ft. elevation. I have only seen this bird in dense evergreen forest at 7000 ft. and above. It sometimes clings to the stem of a tree like a Tree-Creeper (*Certhia*). (H.N.T.).

28 (188). MYIOPHONEUS EUGENII, Hume.
Faun. Brit. Ind , Birds, 1, p. 179.
This species is in my opinion barely separable from *M. temminckii*. The points of difference noted by Oates are the absence of the white tippings to the upper wing coverts, and the larger bill in *M. eugenii*. Two specimens were procured on the same stream on Loi-San-Pa at about 6000 ft., one has conspicuous white tips to the upper wing coverts, the other has not. Both however have large and massive bills. I have therefore considered them both as belonging to Hume's species

29 (203). SIBIA PICAOIDES, Hodgs.
Faun. Brit. Ind., Birds, I, p. 195.
Loi-San-Pa about 6000 ft. I have only found this bird at high

elevations. It goes about in small flocks, and has a loud clear whistling call, which however is not kept up so long or so monotonously as that of *Lioptila melanoleuca*. It extends to the east to the Mekong-Salween watershed, a few miles west of the town of Kengtung. (H.N.T.).

30 (206). LIOPTILA MELANOLEUCA, (Tick.), Blyth.
Faun. Brit. Ind., Birds, I, p. 198.

31 (207). LIOPTILA CASTANOPTERA, Salvadori.
Faun. Brit. Ind., Birds, I, p. 199.

The distribution of these two species is rather remarkable. After leaving the station at Thazi on the Burma Railway and proceeding by road to Taunggyi and Fort Stedman, the large civil and military stations in the Southern Shan States, the first *Lioptila* met with is *L. castanoptera*, which is common and the only species of the genus I found at Kalaw 4400 ft. It no doubt extends along the range of hills bordering ·the western side of the Myealat plateau to Karenni, from whence it was first sent by Signor Fea. This species so far as my observations go does not extend to the east of Kalaw, as on the range of hills including Taunggyi and extending to the south to Fort Stedman and northwards towards·the hills in the Möng Köng State including Loi-San-Pa I only found *L. melanoleuca*. At the same time it has to be noted that Oates (*loc. cit.*) records having examined a specimen of *L. castanoptera* which was obtained at Fort Stedman,* where during several days collecting I only saw and procured *L. melanoleuca*, and Mr. Thompson informs me that both species occur to the east on the Salween-Mekong watershed.

32 (208). LIOPTILA ANNECTENS, Blyth.
Faun. Brit. Ind., Birds, I, p. 199.

Common on Loi-San-Pa at 6000 ft. and upwards. A restless bird, always on the move, whistling as it twines in and out among the twigs and leaves.

33 (212). ACTINODURA RAMSAYI, Wald.
Faun. Brit. Ind., Birds, I, p. 202.

Loi-San-Pa 4000 to 7000 ft. A decidedly rare bird in these States. It has a weird cry, which it monotonously repeated for two or three hours at a time from the topmost branch of a tree. (H.N.T.).

---

* Query. Was this specimen procured in the hills to the *West* of the Fort Stedman lake? If so the Kalaw hills run out in that direction.

34 (222). SIVA SORDIDA, Hume.
Faun. Brit. Ind., Birds, I, p. 210.
Loi-San-Pa 6000 ft. and upwards. Hunts about the twigs, leaves, and flowers like a tit.

35 (228). ZOSTEROPS SIMPLEX, Swinh.
Faun. Brit. Ind., Birds, I, p. 215.
Met with at Kalaw and throughout the States at elevations of 3000 ft. and above. It is somewhat remarkable that none of the allied species of *Zosterops* were seen.

36 (236). CUTIA NEPALENSIS, Hodgs.
Faun. Brit. Ind., Birds, I, p. 222.
I observed this species once or twice on the slopes of Loi-San-Pa, but was unable to procure a specimen. I have shot it further to the east on the Salween-Mekong watershed at 6000 ft. in dense evergreen forest. (H.N.T.).

37 (238). PTERUTHIUS ÆRALATUS, Tick.
Faun. Brit. Ind., Birds, I, p. 225.
Loi-San-Pa 7000 ft. Found in heavily-wooded localities at elevations of 5000 ft. and upwards. Not common. Keeps to the tops of the high trees searching the leaves and moss-covered stems for insects. (H.N.T.).

38 (240). PTERUTHIUS INTERMEDIUS, Hume.
Faun. Brit. Ind., Birds, I, p. 227.
One ♀ procured on Loi-San-Pa at about 6000 ft.

39 (243). ÆGITHINA TIPHIA, Linn.
Faun. Brit. Ind., Birds, I, p. 230.
Common in the low hot valleys up to an altitude of 3000 ft. (H.N.T.). Particularly plentiful at Sinhè, foot of Taunggyi 2900 ft.

40 (247). CHLOROPSIS AURIFRONS, Temm.
Faun. Brit. Ind , Birds, I, p. 234.
Fairly common everywhere in the States. One specimen was procured at 5000 ft. on Loi-San-Pa.

41 (249). CHLOROPSIS HARDWICKII, Jard. and Selby.
Faun. Brit. Ind., Birds, I, p. 236.
Found above 4000 ft., and very common on Loi-San-Pa, where numbers crowded the wild cherry trees which were in full bloom.

42 (255). MELANOCHLORA SULTANEA, Hodgs.
Faun. Brit. Ind., Birds, I, p. 241.

Seems to be rare in the States. One ♂ shot out of a small party in thick forest on a stream in the Möng Pöng State, elevation about 2500 ft.

43 (257). MESIA ARGENTAURIS, Hodgs.
Faun. Brit. Ind., Birds, I, p. 244.

Loi-San-Pa at 6000 ft. and above in dense evergreen bushes.

44 (261). PSAROGLOSSA SPILOPTERA, Vigors.
Faun. Brit. Ind., Birds, I, p. 249.

Met with once, at Sinhè 2900 ft., where a small flock were busy searching the flowers on a group of trees of the *Butea frondosa.*

45 (270). HYPSIPETES CONCOLOR, Blyth.
Faun. Brit. Ind., Birds, I, p. 261.

Very common in all the well-wooded localities at an elevation of 4000 ft. and upwards.

46 (270 *bis*). CERASOPHILA THOMPSONI, Bingh.
Ann. & Mag. Nat. Hist., Lond, seventh series, vol· v (1900), p. 358.

Loi-San-Pa 6000 ft. Mr. Thompson was the first to procure a specimen of this remarkble Bulbul, close to our camp on some wild cherry trees which were at the time of our visit in full bloom. Subsequently watching by the same groups of trees we were able to secure more specimens. The trees were crowded with birds, *Lioptila melanoleuca, Chloropsis hardwickii, Arachnotheras,* and Sun-birds. Every now and then parties of *Hypsipetes concolor,* and with them parties of this species would alight, busily work over the cherry blossoms, sending them in a shower to the ground, and then fly off. *Cerasophila* was most conspicuous with its snow-white head, and in flight, voice, and habits closely resembled *Hypsipetes.*

A very dark-coloured race or species was seen by me in 1897 on the high range dividing the Trans-Salween State of Möng-Tun from that of Möng-Hsat. (H.N.T.).

It is very likely this dark-coloured bird will turn out to be *Hypsipetes leucocephalus,* Gmelin, a Chinese species. (C.T.B.).

47 (272). HEMIXUS FLAVALA, Hodgs.
Faun. Brit. Ind., Birds, I, p. 263.

Loi-San-Pa up to 7000 ft. Not common.

48 (275). HEMIXUS MACCLELLANDI, Horsf.
Faun. Brit. Ind., Birds, I, p. 265.
Loi-San-Pa 5000 to 6000 ft.   Fairly common, a very silent bird.

49 (277). ALCURUS STRIATUS, Blyth.
Faun. Brit. Ind., Birds, I, p. 266.
Loi-San-Pa above 7000 ft., rare.   The two specimens procured are
remarkably large fine individuals with richer colouring and more yellow
about them than Himalayan specimens.

50 (279). MOLPASTES BURMANICUS, Sharpe.
Faun. Brit. Ind., Birds, I, p. 269.

51 (280). MOLPASTES NIGRIPILEUS, Blyth.
Faun. Brit. Ind., Birds, I, p. 270.
Both these species are fairly common in the States, the former
affecting the low hot valleys and rarely ascending above 4000 ft.; the
latter common round Taunggyi, at 4800 ft. and above.   My belief is
that the two species interbreed at Moulmein.   I once found a nest in
April, starting the female, an undoubted *M. nigripileus*, from it.   She
flew on to a tree close by and was joined by another Bulbul, which so
far as I could make out had the breast much darker, like that of
M. burmanicus.   Both birds jerked about the tree in an excited manner,
abusing me the while.   On examining the nest I found the eggs were
hard set, and marking the spot I withdrew.   My intention was to let
the eggs hatch out and then to take and rear the young birds, and see
what species they turned out to be.   Unfortunately I found two days
later that the nest had been destroyed, probably by a cat, or squirrel.

In the Southern Shan States, specimens which seem intermediate
between the two species were procured.   In these the colour of the neck
and throat is much darker than in true *M. nigripileus*, though not by
any means so dark as in true *M. burmanicus*.

52 (287). XANTHIXUS FLAVESCENS, Blyth.
Faun. Brit. Ind., Birds, I, p. 275.
Not uncommon in the well-wooded parts of the States.

53 (288). OTOCOMPSA EMERIA, Linn.
Faun. Brit. Ind., Birds, I, p. 276.
Universally distributed.

54 (290). OTOCOMPSA FLAVIVENTRIS, Tick.
Faun. Brit. Ind., Birds, I, p. 278.
A very common bird in all the States.   Much more of a jungle bird
than the preceding species.

55 (292). SPIZIXUS CANIFRONS, Blyth.
Faun. Brit. Ind., Birds, I, p. 280.
Loi-San-Pa 6000 ft. and upwards. Notwithstanding the peculiar shape of its head and bill, in habits and voice it is a true Bulbul. Procured also by Mr. Thompson on the Salween-Mèkong´ watershed at 7000 ft., and on the Mènètaung borders of Upper Burma at 6000 ft.
I also got specimens at Bernardmyo 7000 ft. (C.T.B.).

56 (296). IOLE VIRESCENS, Blyth?
Faun. Brit. Ind., Birds, I, p. 284.
Möng Köng and Yatsauk States at about 3000 ft. elevation. Two specimens procured answer, so far as plumage goes, fairly well to Oates' description (*loc. cit.*) and to Tenasserim specimens in the Indian Museum, but the Southern Shan States' bird seems conspicuously larger, and may hereafter be separated. Measurements taken in the flesh. Length 7·8″, Wing 3 7″, Tail 3·4″, Tarsus 0·7″, Bill from gape 0·6″. Bill horny-brown, lighter at the base of the lower mandible; legs and feet fleshy-brown; iris dark grey.

57 (297 *bis*). PYCNONOTUS XANTHORRHOUS, Anderson.
Faun. Brit. Ind., Birds, I, p. 286 (*foot note*).
Loi-San-Pa 6000 ft. I procured it also on the Salween-Mèkong watershed. (H.N.T.) It was not uncommon at Kyatpyin 4500 ft. in the Ruby Mines District, in April. (C.T.B.)

58 (306). PYCNONOTUS BLANFORDI, Jerdon.
Faun. Brit. Ind., Birds, I, p. 291.
Very rare in the States. One specimen procured just on the border below Wetpyuyè 2800 ft. Lower it is common. All the specimens procured had remarkably white ear-coverts.

Family SITTIDÆ.

59 (317). SITTA NEGLECTA, Wald.
Faun. Brit. Ind., Birds, I, p. 301.
The common Burmese Nuthatch was procured at elevations from 1000 to 3000 ft. Breeds in April in holes in the large banyan trees that are so common near the Shan villages.

60 (318). SITTA NAGAENSIS, Godw.-Aust.
Faun. Brit. Ind., Birds, I, p. 302.
Met with on several occasions above 3000 ft.

*61 (319). SITTA MAGNA, Wardlaw Ramsay.
Faun. Brit. Ind., Birds, I, p. 303.

Exceedingly rare. I noticed this species both ou Loi-San-Pa and at Taunggyi. From its size it cannot be confounded with any other Nuthatch. (H.N.T.) Two specimens, a ♂ and a ♀, sent me from Taunggyi by Mr. Thompson. The ♂ is similar in plumage to the ♀, but of a darker slatey-blue above, and conspicuously larger. Both ♂ and ♀ differ from the description in the Fauna volume in having the feathers of the head from the forehead to the nape lying between the two black bands that run from the nostril backwards, tipped with white. This speckled band is very conspicuous. (C.T.B.).

62 (325). SITTA FRONTALIS, Horsf.
Faun. Brit. Ind., Birds, I, p. 307.
Common in low hot valleys.

Family DICRURIDÆ.

63 (327). DICRURUS ATER, Hermann.
Faun. Brit. Ind., Birds, I, p. 312.
Occurs throughout the States. It was often observed, but only one specimen was shot at Thammakan 4300 ft.

64 (333). DICRURUS CINERACEUS, Horsf.
Faun. Brit. Ind., Birds, I, p. 318.
This Drongo was fairly common, and was procured up to 6000 ft.

*65 (334). CHAPTIA AENEA, Vieill.
Faun. Brit. Ind., Birds, I, p. 318.
Observed just below Kalaw at 3500 ft.

66 (335). CHIBIA HOTTENTOTTA, Linn.
Faun. Brit. Ind., Birds, I, p. 320.
Common. (H.N.T.) One specimen procured on Loi-San-Pa at 6000 ft.

67 (339). BHRINGA REMIFER, Temm.
Faun. Brit. Ind., Birds, I, p. 324.
Rather uncommon. (H.N.T.). Several individuals were observed in heavy forest near Möng Löng at about 2000 ft. elevation. One was procured at Sinhè at about 3000 ft.

*68 (340). DISSEMURUS PARADISEUS, Linn.
Faun. Brit. Ind., Birds, I, p. 325.
Common in the well-wooded valleys at medium elevations. (H.N.T.)

Family CERTHIIDÆ.

69 (344).   CERTHIA DISCOLOR, Blyth

Faun. Brit. Ind., Birds, I, p. 331.

Loi-San-Pa 7000 ft. This appears to be rather rare in these States. I observed altogether about four individuals on the higher slopes of Loi-San-Pa, and have noticed it nowhere else. (H.N.T.)

Family SYLVIIDÆ.

70 (389).   MEGALURUS PALUSTRIS, Horsf.

Faun. Brit. Ind., Birds, I, p. 383.

Procured at several places from 2000 ft. to 5000 ft. Common in the elephant-grass jungles near the larger streams. (H.N.T.)

71 (404).   HERBIVOCULA SCHWARZI, Radde.

Faun. Brit. Ind., Birds, I, p. 399.

Loi-San-Pa 6000 ft. Taunggyi 5500 ft.

72 (405).   PHYLLOSCOPUS AFFINIS, Tick.

Faun. Brit. Ind., Birds, I, p. 401.

Möng Löng at about 2000 ft. Not common.

73 (414).   PHYLLOSCOPUS PULCHER, Hodgs.

Faun. Brit. Ind., Birds, I, p. 407.

Loi-San-Pa 6000 to 7000 ft. Two specimens.

74 (417).   PHYLLOSCOPUS SUPERCILIOSUS, Gmelin.

Faun. Brit. Ind., Birds, I, p. 409.

Very common during the cold weather. This bird was a perfect nuisance by its abundance, always getting shot in mistake for something else. It was procured up to 7000 ft. on Loi-San-Pa.

75 (423).   ACANTHOPNEUSTE PLUMBEITARSUS, Swinh.

Faun. Brit. Ind., Birds, I, p. 414.

Loi-San-Pa 3000 ft. Rare. Specimens with and without the double bar on the wing (*A. viridiamus ?*) were shot.

76 (468).   PRINIA BLANDFORDI, Wald.

Faun. Brit. Ind., Birds, I, p. 454.

Seen only once at Sinhè at about 3000 ft., when several specimens were procured. All have an exceedingly rufous tinge with very little green about them. The sub-terminal black patches on the tail feathers are very large and well-marked.

Family LANIIDÆ.

77 (474). LANIUS COLLURIOIDES, Lesson.
Faun. Brit. Ind., Birds, I, p. 462.
Universally distributed.

78 (475). LANIUS NIGRICEPS, Franklin.
Faun. Brit. Ind., Birds, I, p. 463.
Fairly common up to 6000 ft. Frequents localities well-provided with grass jungle. (H.N.T.)

79 (484). HEMIPUS PICATUS, Sykes.
Faun. Brit. Ind., Birds, I, p. 471.
Common in wooded localities. Particularly plentiful at Kalaw 4500 ft.

80 (488). TEPHRODORNIS PONDICERIANUS, Gmelin.
Faun. Brit. Ind., Birds, I, p. 475.
Confined to rather dry forest up to 4000 ft. (H.N.T.)

81 (490). PERICROCOTUS SPECIOSUS, Lath.
Faun. Brit. Ind., Birds, I, p. 479.
One of the commonest birds in the States.

82 (500). PERICROCOTUS PEREGRINUS, Linn.
Faun. Brit. Ind., Birds, I, p. 487.
Common in dry forests at low and medium elevations.

83 (505). CAMPOPHAGA MELANOSCHISTA, Hodgs.
Faun. Brit. Ind., Birds, I, p. 491.
Fairly common in dry forest up to 4500 ft. (H.N.T.).

*84 (510). GRAUCALUS MACII, Less.
Faun. Brit. Ind., Birds, I, p. 496.
Confined to the dry oak and *Dipterocarpus* forests up to 4000 ft. (H.N.T.).

85 (512). ARTAMUS FUSCUS, Vieill.
Faun. Brit. Ind., Birds, I, p. 498.
Very common in some places. Probably migrates locally from State to State. Wherever it is met with it is found in large flocks. Very partial to old clearings (*taungyas*) where it perches on the dead tree tops, sallying forth every now and then catching insects on the wing. (H.N.T.).

Family ORIOLIDÆ.

86 (515). ORIOLUS TENUIROSTRIS, Blyth.
Faun. Brit. Ind., Birds, I, p. 503.
I procured one specimen of this species at Payingon in the north-western corner of the Yatsauk State. (H.N.T.).

*87 (521). ORIOLUS MELANOCEPHALUS, Linn.
Faun. Brit. Ind., Birds, I, p. 506.
Both heard and seen on several occasions during the tour in the low hot valleys.

88 (522). ORIOLUS TRAILLII, Vigors.
Faun. Brit. Ind., Birds, I, p. 508.
Kalaw 4400 ft. One specimen procured. Rare ; seems to go about in parties of three and four.

Family EULABETIDÆ.

89 (524). EULABES INTERMEDIA, A. Hay.
Faun. Brit. Ind., Birds, I, p. 511.
Common in the large forests at the foot of the hill ranges in the States. (H.N.T.).

Family STURNIDÆ.

90 (538). STURNIA MALABARICA, Gmelin.
Faun. Brit. Ind., Birds, I, p. 527.
Two specimens procured below Kalaw at about 3000 ft.

91 (539). STURNIA NEMORICOLA, Jerd.
Faun. Brit. Ind., Birds, I, p. 528.
Common in the more wooded portions of the States where it replaces *S. malabarica.* (H.N.T). Not observed on Loi-San-Pa.

92 (546). GRACULIPICA NIGRICOLLIS, Payk.
Faun. Brit. Ind., Birds, I, p. 534.
Found occasionally in small parties in the plateaus bordering some of the rivers up to an altitude of 3000 ft. Seen also in April building a large conspicuous nest of grass, rags and feathers. I have lately found a nest on the top of a cactus bush near Yatsauk. (H.N.T.).
Procured in the paddy fields round Möng Pöng and Möng Löng at 2000 to 3500 ft.

*93 (549).   ACRIDOTHERES TRISTIS, Linn.
Faun. Brit. Ind., Birds, I, p. 537.

*94 (552).   ÆTHIOPSAR FUSCUS, Wagler.
Faun. Brit. Ind., Birds, I, p. 539.
Both these species were seen and identified near the lake at Fort Stedman.

95 (553).   ÆTHIOPSAR GRANDIS, Moore.
Faun. Brit. Ind., Birds, I, p. 541.

96 (554).   ÆTHIOPSAR ALBICINCTUS, Godw.-Aust. and Wald.
Faun. Brit. Ind., Birds, I, p. 541.
Both these Mynas are common throughout the States, especially round cultivation and villages. We procured them from Thamakhan to Möng Löng 3000 ft. and above.

*97 (556).   STURNOPASTOR SUPERCILIARIS, Blyth.
Faun. Brit. Ind., Birds, I, p. 543.
Fairly common.   (H.N.T.).

Family MUSCICAPIDÆ.

98 (560).   SIPHIA STROPHIATA, Hodgs.
Faun. Brit. Ind., Birds, II, p. 8.
Loi-San-Pa 6000 ft.

99 (562).   SIPHIA ALBICILLA, Pall.
Faun. Brit. Ind., Birds, II, p. 10.
Bow-ya-that 3000 ft. One specimen.

100 (569).   CYORNIS MELANOLEUCUS, Tick.
Faun. Brit. Ind., Birds, II, p. 18.
I procured two specimens of this bird in the Yatsauk State at 4000 ft., and other specimens on Loi-San-Pa 6000 ft.   (H.N.T.).

101 (570).   CYORNIS ASTIGMA, Hodgs.
Faun. Brit. Ind., Birds, II, p. 19.
One specimen procured at Yebok in the Yatsauk State at about 3000 ft.

102 (571).   CYORNIS SAPPHIRA, Tick.
Faun. Brit. Ind., Birds, II, p. 20.
Loi-San-Pa at 4500 ft.

103 (572).   CYORNIS OATESI, Salvadori.
Faun. Brit. Ind., Birds, II, p. 20.
Fairly common in the evergreen forests on the slopes of Loi-San-Pa from 4000 ft. upwards.

104 (575).   CYORNIS RUBECULOIDES, Vigors.
Faun. Brit. Ind., Birds, II, p. 23.
Met with on several occasions at 2000 ft. and above.   Common at Kalaw 4400.   Two specimens procured.

105 (577).   CYORNIS MAGNIROSTRIS, Blyth.
Faun. Brit. Ind., Birds, II, p. 26.
Lower slopes of Loi-San-Pa at between 3000 and 4000 ft.   One specimen.

106 (578 *bis*).   CYORNIS BREVIROSTRIS, Bingh.
Ann. & Mag. Nat. Hist., Lond., seventh series, vol. v, p. 359.
Kalaw 4000 ft.   A true *Cyornis* in habits and coloration, but with a remarkable bill approaching in shape that of *Chelidorhynx*.

107 (579).   STOPAROLA MELANOPS, Vigors.
Faun. Brit. Ind., Birds, II, p. 28.
Not uncommon at elevations of 3500 ft. and upwards.   (H.N.T.).
I found this species nesting in holes on the side of the cart road from Bernardmyo (6000 ft.) to Mogôk, Ruby Mines District, in April.

108 (592).   CULICICAPA CEYLONENSIS, Swains.
Faun. Brit. Ind., Birds, II, p. 38.
Very common.

109 (593).   NILTAVA GRANDIS, Blyth.
Faun. Brit. Ind., Birds, II, p. 40.
This beautiful species was fairly common on Loi-San-Pa.   It seems entirely confined to dense evergreen forests at high altitudes.

110 (594).   NILTAVA SUNDARA, Hodgs.
Faun. Brit. Ind., Birds, II, p. 41.
Not so often met with on Loi-San-Pa as the preceding species. Confined like it to altitudes above 6000 ft.

111 (599).   TERPSIPHONE AFFINIS, Hay.
Faun. Brit. Ind , Birds, II, p. 47.
One specimen was procured in the Yatsauk State at 3500 ft. (H.N.T.).

112 (601). Hypothymis azurea, Bodd.
Faun. Brit. Ind., Birds, II, p. 49.
Met with on the western borders of the States below 3000 ft

113 (603). Chelidorhynx hypoxanthum, Blyth.
Faun. Brit. Ind., Birds, II, p. 51.
Rare. One specimen, too much damaged for preservation, was procured on Loi-San-Pa at 6000 ft.

114 (605). Rhipidura albicollis, Vieill.
Faun. Brit. Ind., Birds, II, p. 53.
Common. The only Fantail met with.

Family Turdidæ.

115 (608). Pratincola caprata, Linn.
Faun. Brit. Ind., Birds, II, p. 59.
We met with this bird everywhere, apparently occurs all over the States.

116 (610). Pratincola maura, Pall.
Faun. Brit. Ind., Birds, II, p. 61.
Generally found near grass jungle sparsely provided with trees.

*117 (611). Pratincola leucura, Blyth.
Faun. Brit. Ind., Birds, II, p. 63.
Common on the shores of the Inlè lake near Fort Stedman, and from thence to Bow-ya-that along the road, frequenting the high grass skirting the lake and swampy depressions by the side of the road. (H.N.T.). Near Bow-ya-that I saw several, but failed to secure a specimen. (C.T.B.).

118 (615). Oreicola ferrea, Hodgs.
Faun. Brit. Ind., Birds, II, p. 66.
Very common.

119 (633). Henicurus immaculatus, Hodgs.
Faun. Brit. Ind., Birds, II, p. 85.
Rather rare ; confined to the beds of rocky streams. It does not ascend to any great height. (H.N.T.). Two specimens were procured at elevations between 3000 and 4000 ft. I saw it once on Loi-San-Pa at about 6000 ft. (C.T.B.).

120 (638). CHIMARRHORNIS LEUCOCEPHALUS, Vigors.
Faun. Brit. Ind., Birds, II, p. 89.
This species, so common in the valley of the Chindwin, seems rare in
the States. Two specimens were procured near streams in dense evergreen forest.

121 (641). RUTICILLA AUROREA, Pall.
Faun. Brit. Ind., Birds, II, p. 93.
The common Redstart of the central plateau.   (H.N.T.).

122 (646). RHYACORNIS FULIGINOSUS, Vigors.
Faun. Brit. Ind., Birds, II, p. 98.
Met with only on one occasion, when a specimen was secured at the
foot of Loi-San-Pa at about 2500 ft. Confined to streams with heavily-
wooded banks. (H.N.T.).

*123 (663). COPSYCHUS SAULARIS, Linn.
Faun. Brit. Ind., Birds, II, p. 116.
Extremely common at all altitudes, except in very dense jungle.

*124 (664). CITTOCINCLA MACRURA, Gmelin.
Faun. Brit. Ind., Birds, II, p. 118.
Not uncommon in the well-wooded portions of the States.   (H.N.T.).

125 (676). MERULA BOULBOUL, Lath.
Faun. Brit. Ind., Birds, II, p. 130.
One ♀ was procured by me on the plateau in the north-west corner
of the Yatsauk State at 3600 ft. Several others were seen, but I was
not fortunate enough to bag any of them. The males were very rare,
I saw only two against six or seven females met with. At that time
of the year they seem confined to the evergreen strips of jungle bordering the streams. (H.N.T.).

*126 (677). MERULA ATRIGULARIS, Temm.
Faun. Brit. Ind., Birds, II, p. 131.
One was observed by us both quite close to our camp on Loi-San-
Pa 6000 ft.

*127 (679). MERULA PROTOMOMELÆNA, Cab.
Faun. Brit. Ind., Birds, II, p. 133.
This thrush is common at Taunggyi 4700 ft. During the rainy
season it has a most delightful song that it keeps up for hours. The

song can be heard at a distance of very nearly a mile. They breed in the strip of jungle bordering the vegetable garden at Taunggyi. It also occurs on the Loi-Lem range.

128 (690). Petrophila erythrogastra, Vigors.
Faun. Brit. Ind., Birds, II, p. 143.
One specimen, Kalaw 4400 ft.

129 (693). Petrophila cyanus, Linn.
Faun. Brit. Ind., Birds, II, p. 146.
Common. Every rest-house on the road from Thazi to Taunggyi seemed to be frequented by one or more of these birds. A most silent bird. (H.N.T.).

130 (698). Oreocincla dauma, Lath.
Faun. Brit. Ind., Birds, II, p. 152.
Seems to be rare. One specimen was procured at the foot of Loi-San-Pa 2500 ft.
I have shot it on the Salween-Mèkong watershed at 6500 ft. (H.N.T.).

131 (710). Cinclus pallasi, Temm.
Faun. Brit. Ind., Birds, II, p. 164.
Rare. I have only observed this Dipper in one other locality in these States, in streams in the trans-Salween Sub-State of Mèsègun on the Siamese frontier. (H.N.T.).
Two specimens, a pair, were shot by Mr. Thompson at the foot of Loi-San-Pa 2500 ft. Both specimens when fresh had the distinctive mark of " the legs plumbeous in front."
These birds are difficult to shoot. Affecting steep, rocky, densely wooded streams, they keep flying on ahead as you approach and dodge round the corners rapidly, barely affording a glimpse of their dark bodies.

Family Ploceidæ.

132 (724). Ploceëlla javanensis, Less.
Faun. Brit. Ind., Birds, II, p. 180.
Common. We met with this bird everywhere in grass jungles near water.

133 (726). Munia atricapilla, Vieill.
Faun. Brit. Ind., Birds, II, p. 183.
Fairly common in the grass plains in the lake valley. (H.N.T.).

J. ii. 16

134 (728). UROLONCHA STRIATA, Linn.
Faun. Brit. Ind., Birds, II, p. 185.
Met with once in the Möng Pöng Sub-State at about 2500 ft.

135 (735). UROLONCHA PUNCTULATA, Linn.
Faun. Brit. Ind., Birds, II, p. 189.
One specimen was procured by me on the plateau in the north-west
corner of the Yatsauk State at 3500 ft. in March, 1900. Several flocks
were noticed at the time. (H.N.T.),

*136 (739). SPORÆGINTHUS FLAVIDIVENTRIS, Wallace.
Faun. Brit. Ind., Birds, II, p. 193.
Not uncommon round the borders of the lake at Fort Stedman
3000 ft.

Family FRINGILLIDÆ.

137 (761). CARPODACUS ERYTHRINUS, Pall.
Faun. Brit. Ind., Birds, II, p. 219.
Loi-San-Pa at about 4000 ft. Very local in its distribution.
(H.N.T.).

*138 (776). PASSER DOMESTICUS, Linn.
Faun. Brit. Ind., Birds, II, p. 236.
The Common Sparrow was seen occasionally only at Fort Stedman.

139 (779). PASSER MONTANUS, Linn.
Faun. Brit. Ind., Birds, II, p. 240.
The common house sparrow of these States. (H.N.T.). Was
observed common round most of the towns and villages.

140 (781). PASSER FLAVEOLUS, Blyth.
Faun. Brit. Ind., Birds, II, p. 242.
Rather common. Both this and the preceding species build in my
house at Taunggyi. (H.N.T.).

141 (791). EMBERIZA PUSILLA, Pall.
Faun. Brit. Ind., Birds, II, p. 254.
Met with on several occasions in small flocks at Taunggyi 4700 ft.,
and on Loi-San-Pa at 6000 ft.

142 (797). EMBERIZA AUREOLA, Pall.
Faun. Brit. Ind., Birds, II, p. 259.
Rare. (H.N.T.). One specimen was shot out of a flock near
Bow-ya-that, at about 3000 ft. elevation.

143 (801). EMBERIZA RUTILA, Pall.
Faun. Brit. Ind., Birds, II, p. 263.
Rare. I have procured this species on the Salween-Mèkong watershed at 6000 ft. close to the town of Kengtung. (H.N.T.).

144 (803). MELOPHUS MELANICTERUS, Gmelin.
Faun. Brit. Ind., Birds, II, p. 265.
This beautiful Bunting was met with plentifully, avoiding only heavy forest.

Family HIRUNDINIDÆ.

145 (809). COTILE SINENSIS, J. E. Gray.
Faun. Brit. Ind., Birds, II, p. 273.
Common in suitable localities. (H.N.T.). One specimen procured at Möng Löng at about 2500 ft. Several were observed flying in and out of burrows in a perpendicular bank on the Nam-lan stream.

146 (818). HIRUNDO SMITHII, Leach.
Faun. Brit. Ind., Birds, II, p. 280.
Common and a permanent resident on the plateau. (H.N.T.).
Numbers were seen perched on the telegraph wires along the road from Kalaw to Taunggyi.

147 (820). HIRUNDO STRIOLATA, Temm.
Faun. Brit. Ind., Birds, II, p. 281.
Common and a resident species on the central plateau, being found there and at Taunggyi at all times of the year. (H.N.T.). One specimen shot near Möng Löng.

148 (823). HIRUNDO ERYTHROPYGIA, Sykes.
Faun. Brit. Ind., Birds, II, p. 283.
One specimen was procured in the Möng Köng State about 3000 ft. elevation out of several that were hawking about. It was seen nowhere else, and apparently was migrating, for 10 days later, returning to the same camp, not a single individual was to be seen.

Family MOTACILLIDÆ.

149 (827). MOTACILLA LEUCOPSIS, Gould.
Faun. Brit. Ind., Birds, II, p. 288.
Met with commonly along the road-sides and by the streams.

*150 (839). Limonidromus indicus, Gmelin.
Faun. Brit. Ind., Birds, II, p. 300.
Rare. Confined to the low well-wooded valleys. (H.N.T.).

151 (841). Anthus maculatus, Hodgs.
Faun. Brit. Ind., Birds, II, p. 304.
This Pipit was excessively common, alike on the bare plateau and in the wooded valleys. Near Hlaingdet on the Thazi-Taunggyi road I came on a rather large green tree snake (*Tragops prasinus*) which had got hold of an individual of this species by the head. The bird was still alive and fluttering vigorously, but when released from the snake's mouth seemed unable to fly away and expired in about an hour.

152 (847). Anthus rufulus, Vieill.
Faun. Brit. Ind., Birds, II, p. 308.
Common. This and the preceding species were the only Pipits seen, neither *A. striolatus* nor *A. richardi* was met with.

Family Alaudidæ.

*153 (861). Alauda gulgula, Frankl.
Faun. Brit. Ind., Birds, II, p. 326.
Very common on the high plateaux. (H.N.T.).

154 (873). Mirafra microptera, Hume.
Faun. Brit. Ind., Birds, II, p. 336.
One specimen was procured at Kalaw 4300 ft. It is common in the plains below at Thazi.

Family Nectariniidæ.

155 (882). Æthopyga seheriæ, Tickell.
Faun. Brit. Ind., Birds, II, p. 348.
Common from the lowest altitudes up to 3600 ft. I have lately procured several specimens in the Yatsauk State. The males of this species are very pugnacious and chase each other about. A male in possession of a tree will not allow another to come near it. The males of *Æ. dabryi* and *Æ. gouldiæ* are much more tolerant of each others presence. (H.N.T.).

156 (888). Æthopyga gouldiæ, Vigors.
Faun. Brit. Ind., Birds, II, p. 352.
Loi-San-Pa at 6000 ft.; observed nowhere else. (H.N.T.).

157 (889). ÆTHOPYGA DABRYI, J. Verr.
Faun. Brit. Ind., Birds, II, p. 353.
Loi-San-Pa 6000 ft. and above. Confined to high altitudes. Common on Loi-San-Pa. (H.N.T.).

158 (890). ÆTHOPYGA SATURATA, Hodgs.
Faun. Brit. Ind., Birds, II, p. 354.
Loi-San-Pa at 6000 ft. and upwards. Very common on the flowers of the wild cherry.

159 (895). ARACHNECHTHRA ASIATICA, Lath.
Faun. Brit. Ind., Birds, II, p. 359.
Confined as a rule to the lower altitudes in the States, rarely ascending above 4000 ft. (H.N.T.). Two specimens procured, Taunggyi 4700 feet, and Kalaw 4300 feet.

160 (906). ARACHNOTHERA MAGNA, Hodgs.
Faun. Brit. Ind., Birds, II, p. 369.
Fairly common in well-wooded parts. Has a very distinct call which I have heard mimicked by *Chloropsis hardwickii.* (H.N.T.). Loi-San-Pa from the foot up to about 6000 feet.

Family DICÆIDÆ.

161 (912). DICÆUM CRUENTATUM, Linn.
Faun. Brit. Ind., Birds, II, p. 376.
Found at elevations from 1000 to 4000 feet. (H.N.T.). This pretty little species was common at most places along our route, working about bunches of creepers and flowers. It was not observed either at Taunggyi or on Loi-San-Pa above 4000 feet.

162 (914). DICÆUM CHRYSORRHŒUM, Temm.
Faun. Brit. Ind., Birds, II, p. 378.
Less common than the preceding species. (H.N.T.). One specimen procured at about 2500 ft. elevation.

163 (915). DICÆUM IGNIPECTUS, Hodgs.
Faun. Brit. Ind., Birds, II, p. 378.
Loi-San-Pa at 4000 ft. and upwards.

164 (917). DICÆUM OLIVACEUM, Wald.
Faun. Brit. Ind., Birds, II, p. 380.
Kalaw and Loi-San-Pa at 4000 feet and upwards.

165 (925). PACHYGLOSSA MELANOXANTHA, Hodgs.
Faun. Brit. Ind., Birds, II, p. 386.

This very beautiful Flower-pecker was only twice seen and procured, a ♂ in the Yatsauk State at about 2500 feet elevation, and a ♀ on Loi-San-Pa at 6000 feet. Like all the *Dicæidæ* it is a restless little bird, frequenting bunches of creepers and flowers and twisting in and out among them in incessant motion.

Family PICIDÆ.

166 (948). GECINUS STRIOLATUS, Blyth.
Faun. Brit. Ind., Birds, III, p. 20.

Seen only once, and a specimen procured in the Yatsauk State at 2500 ft. elevation.

167 (950). GECINUS OCCIPITALIS, Vigors.
Faun. Brit. Ind., Birds, III, p. 22.

Met with on several occasions, but not above 4000 feet.

168 (954). GECINUS NIGRIGENIS, Hume.
Faun. Brit. Ind., Birds, III, p. 26.

Not uncommon in the dry forests. One specimen at about 2500 ft. shot in company with a number of *Garrulus leucotis.*

169 (960). HYPOPICUS HYPERYTHRUS, Vigors.
Faun. Brit. Ind., Birds, III, p. 32.

It appears to prefer dry oak forest to other jungle. (H.N.T). Kalaw 4300 ft., Yatsauk State 2500 ft. to about 4000 ft.

170 (968). DENDROCOPUS ATRATUS, Blyth.
Faun. Brit. Ind., Birds, III, p. 40.

Loi-San-Pa at 6000 feet and above.

171 (975). IYNGIPICUS CANICAPILLUS, Blyth.
Faun. Brit. Ind., Birds, III, p. 46.

Met with in dry forest on the plateau between Yatsauk and Möng Pong at about 3000 feet. Occurs at Taunggyi 4800 ft.

172 (988). TIGA JAVANENSIS, Ljung.
Faun. Brit. Ind., Birds, III, p. 61.

Common in dry forest. This species goes about in mobs in company with *Gecinus nigrigenis,* Jays, Drongos, &c. (H.N.T.).

*173 (996). HEMILOPHUS PULVERULENTUS, Temm.
Faun. Brit. Ind., Birds, III, p. 71.

*174 (998). THRIPONAX FEDDENI, Blanford.
Faun. Brit. Ind., Birds, III, p. 73.
Both these birds were met with during our tour more than once, but no specimens were obtained.   The former affects more dense and moister forests than the latter, which is essentially a bird of the dry forests.

*175 (1003). IYNX TORQUILLA, Linn.
Faun. Brit. Ind., Birds, III, p. 78.
Common on the Taunggyi plateau in the winter.   (H.N.T.).

Family CAPITONIDÆ.

176 (1007). MEGALÆMA VIRENS, Bodd.
Faun. Brit. Ind., Birds, III, p. 86.
One specimen Loi-San-Pa at 6000 feet.   Universally distributed over the States in high forest over 3000 feet.   (H.N.T.).

177 (1009). .THEREICERYX LINEATUS, Vieill.
Faun. Brit. Ind., Birds, III, p. 88.
Common.   Occasional specimens have the sides of the neck, breast, and upper abdomen pale dirty brown without any lineation.

178 (1012). CYANOPS ASIATICA, Lath.
Faun. Brit. Ind., Birds, III, p. 92.
Was commonly met with.

179 (1018). CYANOPS RAMSAYI, Wald.
Faun. Brit. Ind., Birds, III, p. 97.
Found in dense forests at altitudes of 5800 ft. and above. (H.N.T.).
Common on Loi-San-Pa above 6000 ft.

180 (1019). XANTHOLÆMA HÆMATOCEPHALA, P. L. S. Müller.
Faun. Brit. Ind., Birds, III, p. 98.
Common in the lower hot valleys.   One specimen was procured on Loi-San-Pa above 6000 ft.

Family CORACIADÆ.

*181 (1023). CORACIAS AFFINIS, McClell.
Faun. Brit. Ind., Birds, III, p. 105.
Common.

182 (1025).   Eurystomus orientalis, Linn.
Faun. Brit. Ind., Birds, III, p. 107.
Common in heavy jungle in the northern parts of the Yatsauk
State.   (H.N.T.).

Family Meropidæ.

183 (1026).   Merops viridis, Linn.
Faun. Brit. Ind., Birds, III, p. 110.
Common.   Ascends up to 6000 ft., at which altitude one specimen
was procured on Loi-San-Pa.

184 (1030).   Melittophagus swinhoii, Hume.
Faun. Brit. Ind., Birds, III, p. 114.
Common in some of the low valleys of the States.   (H.N.T.).   One
shot below Kalaw at about 3000 ft., but too much damaged to preserve.

185 (1031).   Nyctiornis athertoni, Jard. & Selby.
Faun. Brit. Ind., Birds, III, p. 115.
Rather uncommon in the States.   (H.N.T.).   Two specimens pro-
cured between Taunggyi and Fort Stedman.   Elevation about 3500 ft.

Family Alcedinidæ.

*186 (1033).   Ceryle varia, Strickland.
Faun. Brit. Ind., Birds, III, p. 119.
Seen during the tour near several of the streams.   Common around
the lake at Fort Stedman.

*187 (1034).   Ceryle lugubris, Vigors.
Faun. Brit. Ind., Birds, III, p. 121.
Occasionally found at the head waters of some of the larger streams
in very dense dark forest.   (H.N.T.).

188 (1035).   Alcedo ispi$_D$a, Linn.
Faun. Brit. Ind., Birds, III, p. 122.
Common.

*189 (1043).   Pelargopsis gurial, Pearson.
Faun. Brit. Ind., Birds, III, p. 129.
Common in the low valleys.   (H.N.T.).

190 (1044).   Halcyon smyrnensis, Linn.
Faun. Brit. Ind., Birds, III, p. 132.
Met with commonly along our route.

191 (1045). HALCYON PILEATA, Bodd.
Faun. Brit. Ind., *Birds*, III, p. 133.
Rather rare. Found at low altitudes only. (H.N.T.).

Family BUCEROTIDÆ.

*192 (1051). DICHOCEROS BICORNIS, Linn.
Faun. Brit. Ind., *Birds*, III, p. 142.
The Great Hornbill was both seen and heard on several occasions in the Yatsauk State.

193 (1053). ANTHRACOCEROS ALBIROSTRIS, Gmelin.
Faun. Brit. Ind., *Birds*, III, p. 145.
Common. At the town of Möng Pöng 2000 ft. elevation we were encamped under a *Ficus* tree in fruit, which was all day long frequented by this species in crowds.

194 (1054). RHYTIDOCEROS UNDULATUS, Shaw.
Faun. Brit. Ind., *Birds*, III, p. 147.
I have only seen and shot this species in the evergreen forests skirting the banks of the Mèkong river in the Kengtung State. (H.N.T.).

195 (1057). ACEROS NEPALENSIS, Hodgs.
Faun. Brit. Ind., *Birds*, III, p. 149.
Rare. Confined to the heavy forest on the highest hill ranges. (H.N.T.). One specimen procured on Loi-San-Pa at about 7000 ft.

Family UPUPIDÆ.

196 (1067). UPUPA INDICA, Reich.
Faun. Brit. Ind., *Birds*, III, p. 161.
The Indian Hoopoe was fairly common in the Yatsauk State.

Family CYPSELIDÆ.

197 (1086). MACROPTERYX CORONATA, Tick.
Faun. Brit. Ind., *Birds*, III, p. 180.
Fairly common in the more wooded portions of the States.
Other Swifts and Swiftlets were seen, but none sufficiently close for identification. Among the latter a Palm Swiftlet was fairly common at Kalaw, but no specimens were secured.

Family CAPRIMULGIDÆ.

198 (1091). CAPRIMULGUS ASIATICUS, Lath.
Faun. Brit. Ind., Birds, III, p. 186.
This species was frequently put up along the road in scrub jungle. It does not seem to affect high forest.

199 (1095). CAPRIMULGUS INDICUS, Lath.
Faun. Brit. Ind., Birds, III, p. 190.
One specimen seen and procured on Loi-San-Pa at about 7000 ft.

*200 (1096). LYNCORNIS CERVINICEPS, Gould.
Faun. Brit. Ind , Birds, III, p. 192.
Common all over the States in suitable localities. (H.N.T.). Its ringing trisyllabic cry was heard once or twice on the plateau between Yatsauk and Möng Pöng.

Family TROGONIDÆ.

201 (1101). HARPACTES* ERYTHROCEPHALUS, Gould.
Faun. Brit. Ind., Birds, III, p. 200.
Common. Confined to the lower valleys. (H.N.T.). One specimen was shot at Wetpyuyè 2800 ft., but unfortunately was so damaged by the shot as to be passed preservation.

Family CUCULIDÆ.

*202 (1104). CUCULUS CANORUS, Linn.
Faun. Brit. Ind., Birds, III, p. 205.
Very common all over the States in the hot weather. (H.N.T.).

203 (1108). HIEROCOCCYX SPARVERIOIDES, Vigors.
Faun. Brit. Ind., Birds, III, p. 211.
One specimen procured at Kyauk-ku in the northern portion of the Yatsauk State about 3200 ft. in March. (H.N.T.).

204 (1113). CACOMANTIS MERULINUS, Scop.
Faun. Brit. Ind., Birds, III, p. 218.
Met with only between Fort Stedman and Sinhè. At Bow-ya-that 3500 ft. it was common, affecting the trees and bushes along the roadside.

---

* *Harpactes*, Swainson, 1837' is pre-occupied by *Harpactes*, Templeton, 1834, used for a genus of *Arachnida*.

205 (1114). PENTHOCERYX SONNERATI, Lath.
Faun. Brit. Ind., Birds, III, p. 219.
Kalaw at 5000 ft.   One specimen.

206 (1120). EUDYNAMIS HONORATA, Linn.
Faun. Brit. Ind., Birds, III, p. 228.
Very common up to altitudes of 4000 ft.   (H.N.T.).

207 (1123). RHOPODYTES TRISTIS, Less.
Faun. Brit. Ind., Birds, III, p. 232.
Rare.   (H.N.T.).   Seen and procured at Naupandet 2000 ft., and at
Wetpyuyè 2800 ft.

*208 (1130). CENTROPUS SINENSIS, Steph.
Faun. Brit. Ind., Birds, III, p. 239.
Common.   (H.N.T.).

Family PSITTACIDÆ.

209 (1142). PALÆORNIS FINSCHI, Hume.
Faun. Brit. Ind., Birds, III, p. 254.
This is the common Paraquet met with throughout the States.

210 (1150). LORICULIS VERNALIS, Sparrm.
Faun. Brit. Ind., Birds, III, p. 261.
I have shot this species in the State of Möngpan and Mawkmai.
(H.N.T.).

Family STRIGIDÆ.

211 (1152). STRIX FLAMMEA, Linn.
Faun. Brit. Ind., Birds, III, p. 264.
Found at Taunggyi.   I have taken the nests on low bushes and
rocks on swampy ground near Taunggyi.   There were three young in
each nest.   (H.N.T.).

212 (1160). SYRNIUM INDRANI, Sykes.
Faun. Brit. Ind., Birds, III, p. 275.
On the Crag at Taunggyi 4400 ft.   (H.N.T.).

Family ASIONIDÆ.

213 (1166). KETUPA JAVANENSIS, Less.
Faun. Brit. Ind., Birds, III, p. 283.
I have shot this species in trans-Salween Mawkmai.   (H.N.T.).

*214 (1170). Huhua nepalensis, Hodgs.
Faun. Brit. Ind., Birds, III, p. 287.
Occasionally found in the heavy forest at the foot of the high ranges.

215 (1180). Athene brama, Temm.
Faun. Brit. Ind., Birds, III, p. 301.
Common in the Pein Valley Mawkmai State. (H.N.T.). One specimen procured just above Nanpandet at about 2000 ft. elevation.

216 (1183). Glaucidium cuculoides, Vigors.
Faun. Brit. Ind., Birds, III, p. 305.
Wetpyuyè 2800 ft. One specimen procured.

*217 (1187). Ninox scutulata, Raffl.
Faun. Brit. Ind., Birds, III, p. 309.
Very common all over the States, except at the higher elevations. (H.N.T.).

Family Pandionidæ.

218 (1189). Pandion haliaëtus, Linn.
Faun. Brit. Ind., Birds, III, p. 314.
I have procured this bird on the lake at Fort Stedman, and also on the Loikaw river.

Family Vulturidæ.

*219 (1191). Otogyps calvus, Scop.
Faun. Brit. Ind., Birds, III, p. 318.
Common. (H.N.T.). Two specimens were seen at close quarters round the carcass of a dog near the Fort Stedman lake.

*220 (1196). Pseudogyps bengalensis, Gmelin.
Faun. Brit. Ind., Birds, III, p. 324.
Common. Generally met with during our tour.

Family Falconidæ.

221 (1202). Aquila bifasciata, J. E. Gray.
Faun. Brit. Ind., Birds, III, p. 336.
This is the commonest true Eagle that is found in the Southern Shan States. Leaves the country in May. (H.N.T.). One specimen at Bow-ya-that 3000 ft.

222 (1205). Aquila maculata, Gmelin.
Faun. Brit. Ind., Birds, III, p. 340.
I have shot this species close to the town of Mongnai. (H.N.T.).

*223 (1208). Hieraëtus pennatus, Gmelin.
Faun. Brit. Ind., Birds, III, p. 344.
Sparsely distributed all over the States, except on the highest hills in the winter.

224 (1212). Spizaëtus limnaëtus, Horsf.
Faun. Brit. Ind., Birds, III, p. 351.
Very common in the heavy jungles at the foot of the hill ranges. (H.N.T.). Two specimens obtained at about 2500 ft. elevation in the Möng Pöng Sub-State.

225 (1217). Spilornis cheela, Lath.
Faun. Brit. Ind., Birds, III, p. 357.
Common in the low valleys. I procured a young specimen with only just the faintest traces of spots on the plumage of the lower parts. (H.N.T.).

226 (1221). Butastur liventer, Temm.
Faun. Brit. Ind., Birds, III, p. 364.
Common. Confined to dry *eng* forest up to 3500 ft. I saw an individual chase a Shikra (*Astur badius*) and make it drop a lizard that it was carrying off. The Buzzard then got possession of the lizard. (H.N.T.). One specimen shot near the foot of Loi-San-Pa about 3000 ft.

*227 (1228). Haliastur indus, Bodd.
Faun. Brit. Ind., Birds, III, p. 372.
Confined to the low valleys. Often seen on the shores of the Inlè lake, Fort Stedman. (H.N.T.).

*228 (1229). Milvus govinda, Sykes.
Faun. Brit. Ind., Birds, III, p. 374.
Common. Disappears in the rains and returns at the end of October. (H.N.T.).

*229 (1230). Milvus melanotis, Temm.
Faun. Brit. Ind., Birds, III, p. 377.
Common in the well-wooded States. (H.N.T.).

230 (1232). Elanus cæruleus, Desf.
Faun. Brit. Ind., Birds, III, p. 379.
Seen once or twice in the Yatsauk State. One specimen obtained in the plains between Bow-ya-that and Hèho 4000 ft.

*231 (1236). Circus melanoleucus, Forster.
Faun. Brit. Ind., Birds, III, p. 385.
Common. Does not leave the plateau till the end of May. (H.N.T.).

*232 (1237). Circus æruginosus, Linn.
Faun. Brit. Ind., Birds, III, p. 387.
Common. Was seen more than once near the lake at Fort Stedman, and at Yatsauk.

233 (1238). Circus spilonotus, Kaup.
Faun. Brit. Ind., Birds, III, p. 388.
Rare. One specimen procured near Möng Pöng at about 2000 ft.

234 (1244). Astur badius, Gmelin.
Faun. Brit. Ind., Birds, III, p. 398.
Specimens of only the lighter Burmese form (*Astur poliopsis,* Hume) were procured between Taunggyi and Fort Stedman, and at Kalaw.

235 (1246). Lophospizias trivirgatus, Temm.
Faun. Brit. Ind., Birds, III, p. 401.
Rare. I have shot it at Taunggyi. (H.N.T.).

236 (1247). Accipiter nisus, Linn.
Faun. Brit. Ind., Birds, III, p. 402.
Rare. One specimen procured on the road between Taunggyi and Fort Stedman at about 5000 ft. elevation. The specimen, a ♂, is a remarkably large one. Length 17″, Wing 10·1″.

*237 (1255). Falco perigrinator, Sundev.
Faun. Brit. Ind., Birds, III, p. 415.
Has bred for the last three years on the steep rock known as " The Crag " at Taunggyi. (H.N.T.).

*238 (1257). Falco jugger, G. E. Gray.
Faun. Brit. Ind., Birds, III, p. 419.
Common in the low hot open valleys. (H.N.T.).
We saw but failed to secure an individual of this species which

frequented a group of trees in the great paddy plain lying between the town of Möng Löng and the foot of Loi-San-Pa in the Möng Köng State, elevation about 3000 ft.

239 (1260). FALCO SUBBUTEO, Linn.
Faun. Brit. Ind., Birds, III, p. 422.
Passes through as a migrant. (H.N.T.). One specimen, a very richly coloured ♂, was obtained near the town of Yatsauk.

240 (1261). FALCO SEVERUS, Horsf.
Faun. Brit. Ind., Birds, III, p. 423.
Found sparingly in the well-wooded States. I have obtained it at Yatsauk. (H.N.T.).

*241 (1262). ERYTHROPUS AMURENSIS, Gurney.
Faun. Brit. Ind., Birds, III, p. 424.
This species passes through the States every year on its journey back to Northern Asia. It migrates in parties of several hundred individuals, and only remains in any one place for a day or so. The date on which the birds appear in the Southern Shan States is very constant, and is about the 15th April. On that date I observed it in Mawkmai in 1897, and in 1898. I saw it in large parties in the Yatsauk State on the 18th April. Again in 1899 I met with it near Kengtung on the 16th April, while this year (1900) I noticed large numbers of them at Taunggyi on the 14th April. (H.N.T.).

242 (1265). TINNUNCULUS ALAUDARIUS, Gmelin.
Faun. Brit. Ind., Birds, III, p. 428.
Common. (H.N.T.). One specimen procured at Kalaw 4300 ft.

243 (1267). MICROHIERAX EUTOLMUS, Blyth.
Faun. Brit. Ind., Birds, III, p. 432.
Common. Not found above 4000 ft. as a rule. (H.N.T.). Two specimens procured in the Yatsauk State, one at Wetpyuyè.

Family COLUMBIDÆ.

244 (1271). CROCOPUS PHŒNICOPTERUS, Lath.
Faun. Brit. Ind., Birds, IV, p. 5.
Common in heavy hill jungle. (H.N.T.). Procured on the fruiting *Ficus* trees at Sinhè 4000 ft., and in the Yatsauk State.

*245 (1281). Treron nepalensis, Hodgs.
Faun. Brit. Ind., Birds, IV, p. 14.
Common on the high ranges.   (H.N.T.).

246 (1283). Sphenocercus sphenurus, Vigors.
Faun. Brit. Ind., Birds, IV, p. 16.
Common on the high ranges.   Found up to 6000 ft.   (H.N.T.).
A ♀ was obtained on Loi-San-Pa at about 7000 ft.

247 (1287). Ducula griseicapilla, Wald.
Faun. Brit. Ind., Birds, IV, p. 22.
Common up to all altitudes at suitable times of the year.   .(H.N.T.).
Only one specimen was procured during our tour ; and it was only seen
and heard in January on Loi-San-Pa at and above 6000 ft.

248 (1291). Chalcophaps indica, Linn.
Faun. Brit. Ind., Birds, IV, p. 26.
Confined to the low-lying hot valleys.   I once procured the skin of
a new species of this genus in the Mawkmai forests.   This bird
differed in having a *pure white* (not grey) rump.   The bill and feet
were of a horny-brown colour and not red.   Otherwise it agreed
with *C. indica*.   Unfortunately the skin was destroyed by dogs before
I could return to head-quarters.   I noticed several individuals in the
forest in which I shot the specimen noted above.

249 (1297). Dendrotreron hodgsoni, Vigors.
Faun. Brit. Ind , Birds, IV, p. 33.
" I have procured a most extraordinary bird for these hills (the
eastern slopes of the Crag at Taunggyi), *viz.*, the Spotted Wood-pigeon.
This bird according to Blanford is confined to the Himalayas and
Eastern Thibet at altitudes of from 10,000 to 13,000 ft. in the summer.
I procured it at Taunggyi last Sunday 22nd April (the hottest time of
the year here is the end of April), on the Crag at about an altitude of
5600 ft.   1 saw several others at the same time." (Extract from a letter
dated 24th April from Mr. Thompson).

250 (1304). Turtur orientalis, Lath.
Faun. Brit. Ind., Birds, IV, p. 40.
This dove was very common.   Several specimens were procured.

*251 (1309). Turtur cambayensis, Gmelin.
Faun. Brit. Ind., Birds, IV, p. 45.
Common.   (H.N.T.).   Observed during the tour near the towns of
Möng Pöng and Möng Löng.

252 (1312). MACROPYGIA TUSALIA, Hodgs.
Faun. Brit. Ind., Birds, IV, p. 49.
Decidedly rare. I lately procured a very fine specimen in the
N.-W. portion of the Yatsauk State at an altitude of 3600 ft.

Family PHASIANIDÆ.

253 (1328). GALLUS FERRUGINEUS, Blyth.
Faun. Brit. Ind., Birds, IV, p. 75.
Common. I have not found this species at altitudes above 4500 ft.
(H.N.T.).

254 (1331). PHASIANUS HUMIÆ, Hume.
Faun. Brit. Ind., Birds, IV, p. 80.
Found at Kalaw, on the hill behind Taunggyi, very common in
trans-Salween Mawkmai at 3500 ft., and I have recently put up a
magnificent male at Payingon on the plateau in the N.-W. corner of
the Yatsauk State. (H.N.T.).

255 (1340). GENNÆUS LINEATUS, Vigors.
Faun. Brit. Ind., Birds, IV, p. 92.
One ♂ out of a small party of three or four shot at Wetpyu-yè
2800 ft.

*256 (1352). BAMBUSICOLA FYTCHII, Anderson.
Faun. Brit. Ind., Birds, IV, p. 110.
Common. (H.N.T.). Met with on one occasion only during the
tour, at Taunggyi 4800 ft.

257 (1356). COTURNIX COROMANDELICA, Gmelin.
Faun. Brit. Ind., Birds, IV, p. 116.
Common at Hèho in the rains.

258 (1363). ARBORICOLA RUFIGULARIS, Blyth.
Faun. Brit. Ind., Birds, IV, p. 126.
One ♂ specimen procured on Loi-San-Pa at about 6000 ft.   It
belongs to the sub-species or variety named *A. tickelli* by Hume.

259 (1374). FRANCOLINUS CHINENSIS, Osbeck.
Faun. Brit. Ind., Birds, IV, p. 138.
Very common on the plateaux but was not seen or heard on Loi-
San-Pa. This bird has a habit of sitting on trees by the roadside, from
whence it keeps uttering its characteristic call,

Family TURNICIDÆ.

260 (1382). TURNIX PUGNAX, Temm.
Faun. Brit. Ind., Birds, IV, p. 151.
Common. (H.N.T.). One specimen was procured in the Yatsauk
State at about 2500 ft.

Family RALLIDÆ.

261 (1402). GALLINULA CHLOROPUS, Linn.
Faun. Brit. Ind., Birds, IV, p. 175.
This species was common on the lake at Fort Stedman and on the
marsh near Yatsauk.

*262 (1404). PORPHYRIO POLIOCEPHALUS, Lath.
Faun. Brit. Ind., Birds, IV, p. 178.
The same remark applies to this as to the preceding species.

*263 (1405). FULICA ATRA, Linn.
Faun. Brit. Ind., Birds, IV, p. 180.
Apparently not so common as the two preceding species.

Family GRUIDÆ.

*264 (1410). GRUS SHARPII, Blanf.
Faun. Brit. Ind., Birds, IV, p. 189.
Close to the lake of Fort Stedman a couple of parties of this species
were seen sufficiently close for identification. So far as my observation
goes this bird does not build nearly as large a nest as *G. antigone.* Its
cry too seems to me to differ somewhat.

Family GLAREOLIDÆ.

265 (1425). GLAREOLA ORIENTALIS, Leach.
Faun. Brit. Ind., Birds, IV, p. 214.
Yatsauk State, on the Honan plain. (H.N.T.).

266 (1427). GLAREOLA LACTEA, Temm.
Faun. Brit. Ind., Birds, IV, p. 216.
In the valley of the Yong-Mawkmai State. (H.N.T.).

Family PARRIDÆ.

267 (1428). METOPIDIUS INDICUS, Lath.
Faun. Brit. Ind., Birds, IV, p. 218.
Very common on all the tanks. (H.N.T.).

268 (1429). HYDROPHASIANUS CHIRURGUS, Scop.
Faun. Brit. Ind., Birds, IV, p. 219.
A few were seen and one shot near Bow-ya-that.

Family CHARADRIIDÆ.

269 (1432). SARCOGRAMMUS ATRINUCHALIS, Blyth.
Faun. Brit. Ind., Birds, IV, p. 226.
Very common. (H.N.T.). One specimen at Hèho.

.*270 (1435). HOPLOPTERUS VENTRALIS, Wagl.
Faun. Brit. Ind., Birds, IV, p. 229.
Common on the banks of the larger rivers, such as the Salween and Mèkong. (H.N.T.).

271 (1439). CHARADRIUS FULVUS, Gmelin.
Faun. Brit. Ind., Birds, IV, p. 234.
Common on the shores of the lake at Fort Stedman and at Hèho. (H.N.T.).

272 (?). ÆGIALITIS *sp.?*
A Sand Plover was procured by me near Hèho 4200 ft., but was carelessly lost by my servant who was carrying it before it was examined and identified.

273 (1460). TOTANUS HYPOLEUCUS, Linn.
Faun. Brit. Ind., Birds, IV, p. 260.
Met with at Kalaw, Fort Stedman, Bow-ya-that, &c.

274 (1461). TOTANUS GLAREOLA, Gmelin.
Faun. Brit. Ind., Birds, IV, p. 261.
Sinhè 4300 ft.
Near the lake at Fort Stedman and on the road to Bow-ya-that a few Stints of kinds were seen but not procured or identified, I am sorry to say.

275 (1482). SCOLOPAX RUSTICULA, Linn.
Faun. Brit. Ind., Birds, IV, p. 283.
Common. Several are shot every year at Taunggyi. (H.N.T.).

276 (1483). GALLINAGO NEMORICOLA, Hodgs.
Faun. Brit. Ind., Birds, IV, p. 285.
Common in the marshes to the south of Taunggyi, where several couple are shot every year. (H.N.T.).

277 (1484). GALLINAGO CŒLESTIS, Frenzel.
Faun. Brit. Ind., Birds, IV, p. 286.

278 (1485). GALLINAGO STENURA, Kuhl.
Faun. Brit. Ind., Birds, IV, p. 289.
Both species of Snipe are common. The latter is perhaps more plentiful than the former.

279 (1487). GALLINAGO GALLINULA, Linn.
Faun. Brit. Ind., Birds, IV, p. 292.
A few are shot every year on the shores of the Fort Stedman lake. It is also found at Hopong Saga and Mawkmai. (H.N.T.).

280 (1488). ROSTRATULA CAPENSIS, Linn.
Faun. Brit. Ind., Birds, IV, p. 293.
Very common. (H.N.T.).

Family LARIDÆ.

281 (1491). LARUS BRUNNEICEPHALUS, Jerdon.
Faun. Brit. Ind., Birds, IV, p. 301.
Two specimens procured on the Fort Stedman lake.

282 (1496). HYDROCHELIDON HYBRIDA, Pallas.
Faun. Brit. Ind., Birds, IV, p. 307.
Found on the Fort Stedman lake. I have not noticed it anywhere else except on the Salween river. (H.N.T.). Three specimens procured.

Family PHALACROCORACIDÆ.

*283 (1526). PHALACROCORAX CARBO, Linn.
Faun. Brit. Ind., Birds, IV, p. 340.

*284 (1528). PHALACROCORAX JAVANICUS, Horsf.
Faun. Brit. Ind., Birds, IV, p. 342.
Both species are common in the States in suitable localities. (H.N.T.).

*285 (1529). PLOTUS MELANOGASTER, Pennant.
Faun. Brit. Ind., Birds, IV, p. 344.
Very common. (H.N.T.). I passed within a few feet of one seated on a fishing stake in the lake at Fort Stedman. (C.T.B.).

Family IBIDIDÆ.

*286 (1544). PLEGADIS FALCINELLUS, Linn.

Faun. Brit. Ind., Birds, IV, p. 364.

Not uncommon. (H.N.T.). A small flock was seen but unfortunately out of shot, in the swamp near the town of Möng Löng 3000 ft.

Family CICONIIDÆ.

287 (1548). DISSURA EPISCOPUS, Bodd.

Faun. Brit. Ind , Birds, IV, p. 370.

One shot at Möng Pöng 2500 ft. Seems, in the States, to keep to the banks of streams.

*288 (1550). LEPTOPTILUS DUBIUS, Gmelin.

Faun. Brit. Ind., Birds, IV, p. 373.

Rare. I have seen it close to the Mèkong river and once at Taunggyi. (H.N.T.).

Family ARDEIDÆ.

*289 (1566). ARDEOLA BACCHUS, Bonap.

Faun. Brit. Ind., Birds, IV, p. 394.

Found throughout the States. (H.N.T.).

I believe I saw *A. grayi* too, several times, but no specimen was shot, and I cannot be certain that I identified it correctly. (C.T.B).

*290 (1568). NYCTICORAX GRISEUS, Linn.

Faun. Brit. Ind., Birds, IV, p. 397.

Common. (H.N.T.).

Family ANATIDÆ.

291 (1589). DENDROCYCNA JAVANICA, Horsf.

Faun. Brit. Ind., Birds, IV, p. 430.

Very common.

292 (1591). NETTOPUS COROMANDELIANUS, Gmelin.

Faun. Brit. Ind., Birds, IV, p. 433.

Common. Specimens of both the above were shot on the Fort Stedman lake.

293 (1593). ANAS PŒCILORHYNCHA, Forst.

Faun. Brit. Ind., Birds, IV, p. 436.

Common. This and the next species meet at the lake, Fort Stedman, and intermediate individuals (hybrids) are often shot. (H.N.T.).

294 *(1593 bis)*.   Anas zonorhyncha, Swinh.*
Common.   (H.N.T.).   Met with during our tour on the Fort Stedman lake and at Möng Pöng.

295 (1595).   Chaulelasmus streperus, Linn.
Faun. Brit. Ind., Birds, IV, p. 440.
Abundant at Mongnai and on the marshy ground bordering the Thabet Chaung (stream) in the Hopong and Waung Wun States.

296 (1600).   Dafila acuta, Linn.
Faun. Brit. Ind., Birds, IV, p. 447.
Common.   (H.N.T.).

297 (1601).   Querquedula circia, Linn.
Faun. Brit. Ind., Birds, IV, p. 449.
Common.   (H.N.T.).   This and the preceding species was not uncommon on the Fort Stedman lake in December and January.

*298 (1602).   Spatula clypeata, Linn.
Faun. Brit. Ind., Birds, IV, p. 452.
Rather rare.   Found on the Fort Stedman lake and the Mongnai tanks.   (H.N.T.).

299 (1606).   Nyroca ferruginea, Gmelin.
Faun. Brit. Ind., Birds, IV, p. 460.
I have shot this species on the small tank at Mongnai.   (H.N.T.).

Family Podicipedidæ.

300 (1615).   Podicipes cristatus, Linn.
Faun. Brit. Ind., Birds, IV, p. 473.
I once obtained a pair on the Inlè lake, Fort Stedman, in December.

*301 (1617).   Podicipes albipennis, Sharpe.
Faun. Brit. Ind., Birds, IV, p. 475.
Common.   (H.N.T.).

* This Duck is not given in the Faun. Brit. Ind., Birds, but is mentioned in part ii, p. 148, of the second part of Mr. Eugene W. Oates' "A Manual of the Game Birds of India" (1899), and is described in the Cat. Birds B. M., vol. xxvii, p. 211, n. 13 (1895).   Mr. Oates calls this species the Chinese Grey Duck, and proposes for it and for two other species the new generic name of *Polionetta*.   It has been previously recorded from Indian limits from Kentung in the Shan States in the "Asian" Newspaper (10th January, 1899), according to Mr. Oates.   [Ed.].

VII.—*On the Form of Cormorant inhabiting the Crozette Islands.*—*By F. Finn, B.A., F.Z.S., Deputy Superintendent of the Indian Museum.*

(With exhibition of specimen.)

[Received 30th May; Read 6th June, 1900.]

In the British Museum Catalogue of Birds, vol. xxvi, p. 394, Mr. W. R. Ogilvie-Grant refers the Cormorant of the Crozette Islands to *Phalacrocorax verrucosus*, with a mark of doubt. It may therefore be of interest to ornithologists to know that we have in the Indian Museum, among the specimens belonging to the Asiatic Society, the type of Blyth's *Hypoleucus melanogenis* (sic), which came from the Crozettes, and was received in 1860 from the late Mr. E. L. Layard, on behalf of the Government Museum, Cape Town, together with numerous skins of South African Mammals and Birds. As the reference, in J.A.S.B., vol. xxix, p. 101, has apparently been overlooked by the author of the Catalogue, I give it below in full:—"*Hypoleucus melanogenis*, nobis, *n.s.* Very like *H. varius* (Gm., *Ph. hypoleucos*, Gould), of Australia, but distinguished by its black cheeks and crest-feathers $1\frac{3}{4}$ in. long. Wing $10\frac{1}{2}$ in. Tail 5 in. Bill to forehead $2\frac{1}{16}$ in. Foot 4 in. From the ' Crozettes.' "

The asterisk indicates that the species was new to the Asiatic Society's collection. The skin was mounted, but has been long ago dismounted, and is now in a rather bad state. It bears part of a paper label with the name, locality, and reference to description in Blyth's handwriting. A later label repeats the information, with the addition of the donor's name.

On comparing the specimen in question with the excellent figures and descriptions of the *Hypoleucus* group of Cormorants in the volume of the British Museum Catalogue above referred to, it is seen to be obviously most closely allied to *Phalacrocorax albiventer* and *P. verrucosus*, being generally purple-black above, glossed green on wings, and white below from chin to vent, and the feet having evidently been orange, not black as in *P. varius*, with which Blyth compared it. The chin-feathering extending barely before the gape, and the black cheeks noticed by Blyth evidently ally it to the two first-named species; and it further agrees with the former of these in the black not extending down the side of the lower jaw, and in the possession of a moderate white alar bar; the smaller dimensions, however, bring it nearer to *P. verrucosus*, with which it almost exactly agrees in this respect, and to this species I should be inclined to refer it, unless it be considered distinct, in which case the Crozette Island Cormorant will stand as *Phalacrocorax melanogenys* (Blyth).

VIII.—*On Two Rare Indian Pheasants.*—*By* F. FINN, B.A., F.Z.S.,
· *Deputy Superintendent of the Indian Museum, and* LIEUTENANT H. H.
TURNER.

(With exhibition of specimens.)

[Received 30th May; Read 6th June, 1900.]

The Pheasants I am dealing with in the present brief paper were
met with in the Chin Hills during the present year by Lieutenant H. H.
Turner, who is responsible for the accompanying notes in brackets, and
who submitted his specimens of the birds to me for identification.

Phasianus humiæ.

A typical male of this species, agreeing with the Manipur form, of
which we have a specimen in the Museum, not with the nearly white-
rumped Ruby Mines race or species discriminated by me in J.A.S.B.,
vol. lxvi, pt. 2, p. 523, and since named by Mr. Oates (*Ibis*, 1898, p. 124)
Calophasis burmannicus.

. [I had left my camp, which was pitched about six miles from Fort
White, on the evening of the 6th March, to go after some Hill Partridge,
which one of my men had seen just below my camp; not seeing any
signs of them, I walked on for about a mile, and was returning along
the road (the Fort White—Kalemyo road), when glancing down the *khud*
I saw something grey disappearing in the long grass just below me. I
immediately started to go after it, when I saw what appeared to me a
light blue streak just disappearing. I immediately fired, but it was
with faint hopes that I walked up to the spot, as not only did I think
the bird had disappeared before I shot, but I had just at the moment
of shooting slipped. I was therefore very much delighted when I saw
the blue streak tumbling down the *khud* below me. I immediately
went after him and secured him; as I was descending the original
grey bird, which was evidently the female, got up and flew a short
distance. I walked her up, and my dog again put her up; unfortunately,
owing to the thick jungle, I was unable to get a shot. Walking on,
however, I put up another, whether a cock or hen I could not say, as
it was already dusk. I fired, but the bird flew away, and although I
believe it dropped, I could not find it. These birds when I saw them
were feeding amongst the dry leaves which littered the ground.

. The next evening I tried the upper side of the road and put
several (four at least) of these same birds out of some long grass on
a steep hill-side. I only managed to get one long shot, which was
not successful. I again tried the next morning, and was successful in
bagging another; my dog put it up on our right, and flying very

low through the bushes it crossed just in front of me. Unfortunately the bird was not well skinned and I had to throw this specimen away.

The specimen that I have retained is a full-grown cock; the other one was a young cock without the long tail; the plumage was otherwise identical with that of the other bird. The hill on which I obtained these specimens was between 4000 and 5000 feet high, being one of the spurs of the Chin Hills running down into the Kale valley, and the birds were close to the Fort Kalemyo—Fort White road, just about at milestone 20. The latitude is approximately 23°, and the longitude approximately 94°.]

Gennæus davisoni ?

A pair of Kalij procured by Lieutenant Turner appear to come nearest to what Mr. W. R. Ogilvie-Grant (Cat. B. M. Birds, vol. xxii, p. 304), and Dr. W. T. Blanford (Faun. Brit. Ind., Birds, vol. iv, p. 95), call by the above name. The male resembles that of *G. horsfieldi* in size, and in form of tail, and possesses equally well-marked white bands on the feathers of the lower back and rump. It differs, however, markedly in having the feathers of the upper surface behind the head, wings, and tail, *all* clearly though finely pencilled with white irregular lines, the line on the white-tipped back and rump feathers which precedes the white band at the tip being separated from it by a black space about equal in breadth to the band itself. A few of the feathers of the black under-surface are similarly but less clearly pencilled with white upon their outer webs.

It is thus intermediate between *G. horsfieldi* and *G. andersoni*[*] (with the type of which I have compared it) but far more nearly approaches the former, not having the long curved tail and crest of the latter, and with the white pencillings much reduced in breadth and less regular. There is also no tendency to plain whiteness in the middle or to blackness in the outer tail-feathers, all being equally black with white pencilling. It must also very closely resemble Mr. E. W. Oates' *G. williamsi* from the same district, ("The Game Birds of India," vol. i, p. 342) ; but that bird has the light markings of the plumage buff and not white, except for the barring on the rump. It is quite different in character from the Asiatic Society's specimens of forms intermediate between *G. horsfieldi* and *G. lineatus*, the white markings in these being blurred and indistinct.

---

[*] I have dealt with the question of the type of *G. andersoni* of Messrs. Elliot, Blanford, and Ogilvie-Grant in the *Ibis* (1899, p. 331), and photographs of this are now in the British Museum. The *G. andersoni* of Mr. Oates' work is the *G. davisoni* of the "Fauna" and the "Catalogue." [F.F.]

The female is like that of *G. horsfieldi*, but paler and duller, and with the stippling of the plumage more distinct; the black lateral tail-feathers also are pencilled with white, as in the male, but less regularly, and the pencilling tends to die out at the tips of the outer pairs.

This sex also therefore appears to closely correspond with Mr. Oates' *G. williamsi*, except again for the white pencilling, which also does not tend to disappear on the outermost feathers to the extent indicated by Mr. Oates. The facial skin in both birds has evidently been red, and the bill and feet horn-coloured, as in *G. horsfieldi.*

This form is thus distinguishable from any as yet described, and, if it should be thought worthy of a name, might be called *G. turneri*, after its discoverer; but I am very much averse to giving names to these numerous and variable forms of *Gennæus*, and hence range it provisionally under that subspecies of *G. horsfieldi* to which it bears the most resemblance. To *G. lineatus* I should say that it had no relationship, unless *G. andersoni* be indeed, as Dr. Blanford suggests, a cross between this species and *G. nycthemerus.* But without breeding together *G. nycthemerus*, *G. horsfieldi*, and *G. lineatus* in confinement and noting the result, I do not see how the status of these numerous more or less pencilled forms of the genus is to be settled. The buff pencilling in some of the species described by Mr. Oates is peculiarly difficult to account for, except on the supposition that they are all really distinct, which seems to me most improbable *à priori.*

[The two (cock and hen) Kalij were shot by me on the 15th March, 1900, on the road between Kalewa and Minza, Kalewa being an important village on the Chindwin from whence supplies are sent up to the Chin Hills; Minza is a small village 20 miles or so to the south of Kalewa, the birds being shot about 12 miles from Kalewa. When first I saw them they were in the thick bushes on the bank of a small stream, but they immediately flew into the bamboo jungle, and it was here that I shot them. I shot another the next morning close to the same spot, but my dog so spoilt the skin that I had to eat the bird instead of skinning it, and very excellent eating it was. I had already shot a hen of this species (I think there is little doubt that it *was* the same, though rather darker than the female I mentioned above) at Yasygo, a village in the Kale valley about 50 miles to the north. It was shot, however, at such close quarters that the skin was ruined.

The latitude in which the above pair were found is approximately 23°, and the longitude approximately 94° 30′; the elevation was between 600 and 800 feet.]

IX.—*Notes on the Structure and Function of the Tracheal Bulb in male Anatidæ.—By* F. FINN, B.A., F.Z.S., *Deputy Superintendent of the Indian Musuem.*

[Received 30th May ; Read 6th June, 1900.]

The peculiar bony, or bony and membranous, bulbs found at the base of the trachea in so many males of the duck tribe, are well known to be confined to individuals of that family and sex alone. But, well known as the *Anatidæ* are or should be, from the fact that so many species are shot for food or kept in captivity, thus making specimens easily available for dissection, this structure of the trachea does not appear to have been recorded in some even of the best known species, nor does its functional importance ever appear to have been fully realized. It is my purpose here to note briefly some specific peculiarities in the form of the organ which seem to have escaped the notice of naturalists, and to conclude with some observations on its presumable use.

Nettopus coromandelianus.

In writing a popular series of articles in the *Asian* newspaper in 1898-99, destined for sportsmen who wished to identify the ducks they shot, I was careful to note the structure of the trachea in each species when I was able to give the information, in hopes of interesting my readers in this point. Thus I mentioned that the male of the above species was devoid of a tracheal bulb, a fact I had ascertained from my own examination of the trachea of this bird. Dr. W. T. Blanford, in a letter to me on the subject, has informed me that this observation is new, and I am hence induced to record it here. I find, however, that Sir E. Newton (*Ibis*, 1863, p. 459), says of *Nettopus auritus*, "Unlike most other ducks, the trachea of the male in this species is of a very simple form, there being scarcely any enlargement whatever at the lower extremity." It is possible, therefore, that the absence of a tracheal bulb may characterize the genus, as appears to be the case with *Erismatura*, in which genus MacGillivray, presumably referring to *E. rubida*, states that "There is no expansion or tympanum, as in other ducks " (Orn. Biogr., vol. iv, 1838, p. 331, *fide* Beddard, Structure and Classification of Birds, p. 464, footnote). Pallas certainly states (Zoog. Ross. As., vol. ii, p. 285), that such a tympanum is absent in *E. leucocephala,* a fact which I have been able personally to verify (see J.A.S.B., vol. lxvi, pt. 2, p. 527). It would, however, be rash to assume that such deficiency of the tracheal bulb must necessarily apply to all the species of these genera, since among the members of the genus

Oidemia, O. nigra alone is exceptional in having no difference of this kind between the sexes.

Aix galericulata.

Well-known as is the Mandarin duck, I have been unable to find any reference to a description of the tracheal structure of the male, and I therefore venture to record that there exists in this sex of the species a large, rather flattened, thin-walled bony bulb of rounded outline, somewhat resembling that of the Muscovy (*Cairina moschata*).

Cairina moschata × Anas boschas.

It may be of interest to note that the male hybrid between these two domestic ducks, judging from two specimens of each species and of the hybrid examined by me, agrees with the Mallard in the size and form of the tracheal bulb, but has this rather thinner-walled, thus approaching the structure of the very thin-walled bulb found in the Muscovy drake, the very flattened and expanded form of which is characteristic.

Casarca rutila.

I have been unable to discover any mention of the form of the male trachea in this species, and so was unable to give any account of it when writing the series above-mentioned. Since then, however, I have obtained a specimen, and find the bulb present and bony-walled, but very small, only about the size of a pea, thus differing very much from the great double inflation found in *Tadorna cornuta*.

FUNCTION OF THE TRACHEAL BULB IN DRAKES.

There can be no doubt that the use of this peculiar structure is to modify the voice of the owner, as was pointed out, indeed, by Yarrell ("British Birds," article Mallard) many years ago; but the fact needs drawing attention to, as it has been doubted, for Coues (*fide* Dresser, "Birds of Europe," article American Bittern) states that "The curious cartilaginous or osseous bulbs at the lower larynx of most Ducks seem to have no influence on the voice." But nothing could be further from the truth. Yarrell (*loc. cit.*) approvingly quotes Gilbert White's observation on the sexual difference of voice in Ducks; and careful study of the living birds will show that in nearly every case where the tracheal enlargement exists in the male, he emits a correspondingly different note from the female—generally one of a weaker character, if not always. The accompanying table will make this point clear.

With a bony bulb in trachea of male.

Name.	Voice of Male.	Voice of Female.
Carina moschata.	A panting hiss.	A sharp quack.
Aix galericulata.	A snorting whistle (*weesh!*)	A quacking sneeze (*atch!*)
Chenalopex ægyptiacus.	A husky chatter.	A harsh barking quack.
Eunetta falcata.	A whistling cry like a duckling's.	A quack five times repeated.
Dafila acuta.	A faint weak quack.	A harsh single quack.

With a partly bony, partly membranous bulb in trachea of male.

Rhodonessa caryophyllacea.	A melodious double call (*wugh-ah!*)	Not known; but never gives the male note, while the drake often calls.
Nyroca baeri.	A faint low quack.	A harsh croak (*karr!*)
Nyroca africana.	,,	,,
Merganser castor.	A harsh croak (*karr!*)	A short abrupt quack.

Of course many more well-known cases could be given, but I have in the above table merely given some, which being imperfectly or not at all recorded, have attracted my special notice. *Casarca rutila* is an exception proving the rule, since in this bird the tracheal bulb of the male is, as above noted, singularly small, and there appears to be no difference between the voice of the sexes in consequence. Authors in describing the sexual difference in the voices of various *Anatidæ*, often speak as if the quack were common to both sexes, and another note peculiar to the male alone. At other times no sexual distinction in voice is mentioned, as in Hume's account of the Pintail (*Dafila acuta*) "Game Birds of India," vol. iii, p. 193). The same explanation will, I think, account for both errors, the birds having been studied from a distance when in flocks, and the cries of the separate birds thus having been confused. With regard to our two domestic ducks (*Anas boschas* and *Cairina moschata*) the drakes seem quite incapable of giving the duck's note, and *vice versâ*, and I think the same rule will be found to apply throughout. That there are certain species, like *Nettopus coromandelianus* and *Œdemia nigra*, in which the male has no tracheal bulb and yet gives a different call from the female, in no way invalidates the importance of this structure in the others; for the voices of male birds, in the *Gallinæ* and *Passeres* for instance, often differ from those of females without any corresponding difference in anatomical structure. The tracheal bulb in the *Anatidæ*, where found, merely, in my opinion, makes such difference necessary and permanent.

X.—*Note on* Calinaga, *an aberrant genus of Asiatic Butterflies.—By*
LIONEL DE NICÉVILLE, F.E.S., C.M.Z.S., &c.

[Received 5th June; Read 4th July, 1900.]

In the " Memoirs and Proceedings of the Manchester Literary and
Philosophical Society," vol. xliii, part iv, n. 11, pp. 1-23, pls. iv,
v and vi (1899), appears a paper by Mr. John Watson entitled " On
Calinaga, the Single Genus of an aberrant Sub-Family of Butterflies,"
in which he has brought together a mass of most interesting facts
mostly based on his own original observations regarding these butterflies.
This paper should be carefully studied by everyone interested in the
phylogeny of this anomalous genus.

The following brief notes will shew the very various positions in
the natural order of Butterflies in which *Calinaga* has been placed by
different writers :—

Dr. F. Moore in 1857 when first describing the genus placed it in
the subfamily *Nymphalinæ* next following the genus *Hestina*, Westwood ;
in 1865 he placed it at the end of the *Papilionidæ*, the *Pieridæ* being
retained as a distinct family ; in 1895 he replaced it in the family *Nymphalidæ*, but proposed a new subfamily for it, the *Calinaginæ*, which
will follow the subfamily *Nymphalinæ*. In 1861 Dr. C. Felder placed
it in the family *Nymphalinæ* after the genus *Penthema*, Doubleday.
Mr. W. F. Kirby in 1871 placed the genus second in the subfamily
Papilioninæ, but in 1877 he removed it to the *Nymphalinæ*; in 1894
he stated that it belonged to the *Nymphalidæ*. In 1872 Herr Gustav
Weymer noted that *Calinaga* does not belong to the *Papilioninæ*, but to
the *Pierinæ*. In 1873 Mynheer P. C. T. Snellen placed *Calinaga* in the
Nymphalinæ. M. Charles Oberthür in 1881 noted that the antennal
characters of the genus place it between a genus of *Pierinæ* (*Leuconea*)
and a genus of *Papilioninæ* (*Parnassius*). M. H. Lucas in 1884 placed
it in the same position as M. Oberthür in 1881. Dr. A. G. Butler in
1885 placed *Calinaga* in the *Nymphalinæ*, but said that he had no doubt
that its proper place is in the subfamily *Satyrinæ*. In 1886 Dr. O.
Staudinger placed the genus in the subfamily *Nymphalinæ*. In 1886,
I placed it in the *Nymphalinæ* wholly on account of the forelegs of the
imago of the male being pectoral unfitted them for use in walking. At
that date I had seen no female, the forelegs of which have since been
found to be perfect but unfit for use in walking, being very small and
pectoral as in the opposite sex. Dr. E. Schatz in 1887 placed *Calinaga*
in the " Diademen " group, which is the fourth group in his seventh
family " Nymphaliden." For *Diadema* the genus *Hypolimnas* is intended, as *Diadema* having been proposed in 1817 for a genus of *Crustacea*

cannot be used for the genus of Butterflies proposed by Boisduval in 1832. In 1888 Mr. H. J. Elwes placed the genus in the *Nymphalinæ.* In 1892 Mr. J. H. Leech placed it in the *Nymphalinæ.* In 1898 Dr. Karl Jordan placed the subfamily *Callinaginæ* [sic!] in the family *Nymphalidæ.* Lastly in 1899 Mr. John Watson noted that the genus has "A great similarity to the *Papilionidæ* (subfamilies *Papilioninæ, Parnassiinæ,* and *Pierinæ*), and to the *Nymphalidæ* (subfamily *Danainæ,* but not to the subfamily which includes the genus *Hypolymnas**)." In writing to me on February 10th, 1900, Mr. Watson says that the genus "Must be placed next to (not in) the *Danainæ.*"

Mr. Watson notes that the egg as far as he has been able to study it with very insufficient and imperfect material, *i.e.,* the broken upper portions of two eggs of *Calinaga davidis,* Oberthür, obtained from the dried body of the female, shews that *Calinaga* is near to the *Danainæ, i.e.,* to the genera *Hestia* and *Danais.* I have nothing to urge against this conclusion, in fact I agree with it so far that I think it probable that *Calinaga* is allied to the subgenus *Radena* of the genus *Danais.* The acquisition of a perfect freshly deposited egg would, I may note, settle definitely at once and for ever the position of this anomalous genus. I may note that Mr. W. Doherty has pointed out that *Radena nicobarica,* Wood-Mason and de Nicéville, has 16 vertical ribs on the egg, and that *Radena vulgaris,* Butler, has from 14 to 16 vertical ribs, and 25 cross-lines. When the egg of *Calinaga* is discovered it will be in. teresting to note if the vertical ribs and cross-lines agree with those of *Radena.* Mr. Doherty describes the eggs of the *Danainæ* as "Much higher than wide, leathery, radiate, with numerous broad flattened ribs and distinct cross-lines, reticulate over a small area at the apex." The micropyle of the eggs of butterflies is of little or no value for classi. ficatory purposes.

As regards the foreleg, which in the male is very hairy, imperfect, having but a single joint to the tarsus, while in the female it is perfect, having five tarsal joints, the terminal joint furnished with a pair of claws, a pair of paronychia, and a pulvillus, Mr. Watson notes that the female of *Calinaga* "Shews in its tarsal structure the most ancient type of leg of the whole of the *Nymphalidæ.*" I have examined the forelegs of both sexes of *Calinaga,* which I could not do in 1886 when I dealt with the genus in the second vol. of "The Butterflies of India, Burmah and Ceylon," as at that date I had seen no female, from which it is clear that the genus cannot be placed in M. Constant Bar's group *Tetrapoda,* in which both sexes have the forefeet or tarsi imperfect, the *Tetrapoda* being

---

* Recte *Hypolimnas.*

equivalent to the family *Nymphalidæ*, but that *Calinaga* should find a place in the *Heteropoda*, which has the forefeet imperfect in the male but perfect in the female, the *Heteropoda* being equivalent to the families *Lemoniidæ* and *Lycænidæ*. But to place *Calinaga* in either of those families on the structure of the forelegs only is obviously out of the question, as in every other respect the genus shews no relationship to those families. The forelegs of both sexes of *Calinaga* are highly specialised, and have lost all functional characters, being extremely small and quite unfitted for walking. This is a characteristic of the *Nymphalidæ*, but the fact that the male has a nymphalid foreleg, while the female has in structure a foreleg, which, but for its ridiculously small size unfitting it for use, would place the genus amongst the *Lemoniidæ*, *Lycænidæ*, *Papilionidæ* and *Hesperiidæ* which have six fully developed ambulatory legs in the female, removes the genus from any known family of butterflies as hitherto diagnosed, the structure of the forelegs having of recent years been largely used as a primary basis on which to divide the families.

Calinaga has a prototype in the genus *Pseudergolis*, usually also placed in the subfamily *Nymphalinæ*, but for which Dr. Karl Jordan has proposed the subfamily name *Pseudergolinæ*, the foreleg of the male of which has a single joint to the tarsus, the female having five tarsal joints, the terminal joint furnished with a pair of claws, a pulvillus, and bifid paronychia. In both sexes in *Pseudergolis* the forelegs are very distinctly longer both actually and proportionally than in *Calinaga* (in while they are ridiculously short), but are still quite unfitted for walking. *Pseudergolis* therefore is another genus that can find no place as far as I can see in any existing family of butterflies as hitherto diagnosed. Its transformations (*Pseudergolis wedah*, Kollar) are known fortunately, and have been figured by me in Journ. Bomb. Nat. Hist. Soc., vol. xi, p. 371, n. 83, pl. U, figs. 9a, 9b, *larva;* 9c, 9d, 9e, 9f, *pupa* (1898). The full grown larva reminds one somewhat of that of the genus *Apatura* (= *Potamis*, Hübner, to follow Dr. F. Moore), subgenus *Rohana*, Moore, while the pupa is like that of the genus *Athyma*. Mr. Watson says that *Pseudergolis* " In point of general resemblance and neuration is undoubtedly in the *Precis* and *Junonia* section of the *Nymphalinæ*." Mr. W. Doherty, who first discovered the abnormality in the forelegs of the female, placed it in the *Apaturidæ*, " Differing but little from its neighbours in other points [except the feet], if the feet offered really reliable characters " (Journ. A.S.B., vol. lx, pt. 2, p. 12 (1891). Mr. Doherty placed *Pseudergolis* in his *Junonia* group, which embraced the genera *Junonia*, *Precis*, *Pseudergolis* and *Rohana* (Journ. A.S.B., vol. lv, pt. 2, p. 123, n. 82 (1886).

With regard to the antennæ of *Calinaga*, Mr. Watson says that they are "Similar in general form to those of *Danais* and *Euplœa*, and in scaling are like those of the *Parassiinæ*." In my opinion the antennæ superficially most nearly resemble those of the genus *Luehdorfia* of the *Papilioninæ*, with which they agree very closely in length and shape. In *Calinaga* they are too short and stout to agree well with the *Danainæ*, and too long to agree with *Parnassius*, besides which the antennæ of the latter genus are more abruptly clubbed than in *Calinaga*, and the club is thicker. The structure of the legs and the neuration will, however, in my opinion, remove *Calinaga* entirely from the *Papilionidæ*.

With regard to the basal cell of the hindwing in *Calinaga*, which is caused by the peculiar conformation of the costal, subcostal and median nervures together with another vein called by Mr. Watson the "interno-costal nervule," that writer sums up the evidence as regards *Calinaga* that in formation it is "Similar to *Parnassius*, and dissimilar to the *Danainæ*, and still less similar to the genus *Hypolimnas*." I have nothing to add to this. The basal cell is found in the *Danainæ*, *Morphinæ*, *Pierinæ*, and *Papilioninæ*.

As regards the general facies of *Calinaga* I agree with Mr. Watson that it reminds one of the genus *Aporia* of the *Pierinæ*, still more so of some species of *Parnassius*, but in my opinion most of all to certain species of the subgenus *Radena* of the *Danainæ*, say *Radena meganira*, Godart, for the dark *Calinaga buddha*, Moore, and *Radena juventa*, Cramer, for the lighter *Calinaga davidis*, Oberthür, and *C. cercyon*, mihi. Neither of these species of *Radena*, however, occur where the *Calinagas* are found. The disposition and extent of the markings of the two genera *Radena* and *Calinaga* is very markedly similar. *Calinaga* is almost unique in the heavily hairy, ferruginous clothing of the thorax of both sexes of all the known species, which strongly contrasts with the black abdomen, and this hairiness is characteristic of *Parnassius*. The coloration feature of the thorax is also noticeable in *Parnassius citrinarius*, Motschulsky, of which *P. glacialis*, Butler, is a synonym, but in the latter species the coloured hairy clothing of the thorax is of much less extent, forming merely a collar, and is rather paler, more fulvous, in shade. I may note that the hairless, leathery abdomen of *Calinaga* reminds one of that of the genus *Danais*.

I can extend the geographical range of the genus *Calinaga* considerably beyond that given by Mr. Watson, as it is found in the Western as well as in the Eastern Himalayas, but it is a noticeable fact that as far as is at present known *C. buddha* occurs in the Western Himalayas only in the Kulu Valley, near Dalhousie, in Chumba, at Murree, and possibly near Mussoorie, there being an immense gap of nearly 600 miles before it is again met with in Sikkim in the Eastern Himalayas ; while to the east *C. davidis*, Oberthür, occurs in Central as well as in Western China, while *C. sudassana*, Melvill, is found in Upper Burma and Upper Siam.

Dr. George Watt obtained *Calinaga brahma*, Butler, which is a synonym of *C. buddha*, Moore, " Near Assam " according to Dr. Butler who described it, but Mr. J. H. Elwes has noted that the specimens were probably collected in the Naga Hills on the march from Manipur to Kohima. In Mr. Elwes' collection is a female specimen of *C. buddha* from Chumba in the Western Himalayas. *C. buddha* was originally described from N. India, presented by Colonel Buckley, who collected largely in the neighbourhood of Mussoorie in the Western Himalayas, but no specimens from the Mussoorie region have been obtained in recent years. Excepting Dr. Watt there is no European in India as far as I know who has seen a *Calinaga* alive except Mr. A. Grahame Young, who informs me that it is not very rare in the Kulu Valley, also in the Western Himalayas; that it frequents the banks of heavily wooded streams from 3,500 to under 6,000 feet elevation above the sea, never higher than that; that it is purely a forest insect, never found amongst mere brushwood but always amongst trees; and occurs generally from about 25th March to 20th May. He notes that once, and once only, he saw three altogether, in July, 1872. It is very fond of settling on wet sand or gravel, when disturbed it flies off with a strong *Papilio*-like flight, and that it is very strong on the wing. Last year (1899), Mr. Young tells me that his native collector reported that he saw over twenty specimens, but caught only five. In Kulu it is very local, Mr. Young knows of two spots only where it occurs.

In 1886 I noted that " The species [of *Calinaga*] are probably mimetic." I think so still, and that they mimic species of *Danais*, *C. buddha*, *C. davidis*, and *C. cercyon* mimicking *D. limniace*, Cramer, *C. sudassana* mimicking *D. sita*, Kollar (= *D. tytia*, Gray), and *C. lhatso*, Oberthür, possibly the same species; the mimic and its model in all cases being found in the same locality.

In conclusion I may note that the opinion I expressed in 1886 that " The proper position of the genus is amongst the *Nymphalidæ*, though a knowledge of its transformations is necessary for determining its exact position in that family " remains unaltered. Till its egg is properly known (eggs extracted from the body of a long deceased specimen are not very satisfactory, though very much better than nothing), and its larva and pupa are discovered, especially the young larva on first emergence from the egg, the position of the genus must remain uncertain. It is not a hesperid, papilionid, lycænid or lemoniidid, so must come into the family *Nymphalidæ*. Various authors have placed it in the *Danainæ, Satyrinæ, Nymphalinæ, Pierinæ*, and *Papilioninæ*. Its most probable position is I think in the subfamily *Danainæ* as Mr. Watson states, and I would place it tentatively after *Hestia* and *Ideopsis* and before *Danais*, next before the subgenus *Radena*. If, however, the male has no extrusible tufts of hairs at the end of the abdomen as have all the *Danainæ*, it cannot be placed in that subfamily. I think it probable that these tufts are absent. The structure of the forelegs of the female in correlation with that of the male removes the genera *Calinaga* and *Pseudergolis* from any hitherto known family of butterflies, if the sequence and definition of the rhopalocerous families is to be primarily based on the structure of the forelegs of the imago as has been done of recent years. But I think these two genera must come into the existing families as aberrant genera, the genus *Calinaga* amongst the *Danainæ* and the genus *Pseudergolis* amongst the *Nymphalinæ*. In all five species of *Calinaga* are known. I have given a list of them in Journ. A.S.B., vol. lxvi, pt. 2, p. 551 (1897).

XI.—*Notes on birds collected in Kumaon.*—By CAPTAIN
H. J. WALTON, I.M.S.

[Received 6th June ; Read 4th July, 1900.]

The birds enumerated in the following list were collected by me in British Garhwal and the Almora district of Kumaon during the months of April—July, 1899.

The nature of the duty on which I was engaged rendered it necessary for me to visit nearly the whole of British Garhwal, and most of the district lying east of Almora almost to the boundary of Nepal. Unfortunately my camps had to be selected without any reference to their merits as collecting grounds, and, indeed, as I was travelling nearly every day, most of the birds were shot actually from the roads.

Garhwal is a large district, extending far up into the Himalayas, bounded on the north by Thibet, on the west by Native Garhwal, and on the south and east by Kumaon. Although there are some very high peaks and mountain ranges, as a rule the valleys up which the roads run are rather low-lying, averaging about 6,000 feet, but varying from about 4,000 to 12,000 feet above sea-level. The sides of the valleys are almost everywhere covered with tree jungle, but from Chamoli to Yoshimath, and again in the neighbourhood of the Mana and Niti passes, the hillsides are almost bare. In parts the jungle is very thick indeed, and as it is precisely in such places that many of the birds are met with, unless one has plenty of time at one's disposal, one is sure to overlook a great many species. The Garhwalis do not seem to take much interest in birds, and they are by no means skilful in marking down those that are shot. I lost a great many birds owing to the difficulty of finding them in thick undergrowth.

The climate of Garhwal, in summer, presents great and sudden variations. As one marches along, it is often very cold on the higher roads, and then, after a steep descent into a low-lying valley, the heat becomes very trying. This, of course, is a good thing in one respect, as the birds met with necessarily vary very much, even in the course of a single march.

At the beginning of April the weather was cold with a good deal of rain. The latter part of the month of May and the beginning of June were hot and fine. The rains began about the 10th of June, and soon became very heavy indeed. At the end of June and throughout July the birds were very silent : the jungles were wet and slippery and full of small but voracious leeches, and, consequently, not very attractive.

Nearly all the birds were breeding at the time I visited Kumaon : I regret that I had too little time at my disposal to devote any attention to nests. The choice lay between birds and eggs, and I preferred the former. All the pheasants were breeding or preparing to breed, and had retired into the depths of the jungles, consequently I did not see very many. I was shown a breeding-place of the White-bellied or Snow-pigeon (*Columba leuconota*) at the lower end of the Mana pass, but on May 20th, the date of my visit, the eggs had not been laid. I hope to receive some from the headman of the neighbouring village, later on. I also saw two breeding sites of the Alpine Swift (*Cypselus melba*) in rocky precipices above the Alaknanda river, but they were quite inaccessible.

I left Naini Tal on April 9th, and marching *viâ* Almora and Ranikhet, entered Garhwal on April 18th. I went first to Kedarnath, which is about 10,000 feet above sea-level in the north-west of the district. The roads were crowded with pilgrims on their way to the famous temple. These people come every year from all parts of India, most of them marching up the Ganges valley from Hurdwar, as soon as the fair there is over. I met pilgrims from Hyderabad (Deccan) and Quetta. The pilgrimage season lasts for about six months, and one can only hope that the spiritual benefits received are at all in proportion to the physical discomforts undergone by the pilgrims, many of them very old men and women. It was bitterly cold at Kedarnath, and snowed hard the day I was there (May 4th).

From Kedarnath I went *viâ* Ukhimath and Yoshimath to the

Mana Pass, which I reached on May 19th. By that date the weather was beginning to get much warmer, though the pass was not yet open. Thence my route was to the Niti pass. On May 31st, two Thibetans, the first of the year, had just crossed the pass, but they reported that they had been much delayed on the road by heavy snow-drifts. From the Niti pass I marched nearly due south, entering the Almora district on June 15th. I proceeded to Pithoragarh, in the east of the district, and returned to Naini Tal on July 17th.

The numbers and nomenclature adopted in the following list are those given by Messrs. Oates and Blanford in " The Fauna of British India, Birds."

1. *Corvus corax.*—I did not see any Ravens in Garhwal, but, according to native accounts, they visit the higher valleys in winter.

4. *Corvus macrorhynchus.*—The Jungle-Crow is common in all the valleys, except the very highest.

7. *Corvus splendens.*—The Indian House-Crow only occurs about the larger hill stations in Kumaon : I did not see any in Garhwal.

12. *Urocissa occipitalis.*—The Red-billed Blue Magpie is very common, up to about 6,000 feet, wherever there is tree or bush jungle.

13. *Urocissa flavirostris.*—Common at rather higher elevations than the last species.

18. *Dendrocitta himalayensis.*—Very common : seen generally in small parties of from four to six.

24. *Garrulus lanceolatus.*—Common up to about 6,000 feet : this bird feeds a great deal on the ground.

26. *Garrulus bispecularis.*—Also common : occurs up to rather higher elevations than *G. lanceolatus.*

29. *Graculus eremita.*—I did not see the Red-billed Chough below 10,000 feet. It was common near Badrinath, and the Mana pass. There were none at Kedarnath.

30. *Pyrrhocorax alpinus.*—The Yellow-billed Chough was very common, and very tame, about Malari, and at one or two other places in Garhwal. The birds flew about the camp in a very unconcerned way. They have a much softer note than the Red-billed Chough.

31. *Parus atriceps.*—Common, at moderate altitudes.

34. *Parus monticola.*—Very numerous.

35. *Ægithaliscus erythrocephalus.*—Very common.

42. *Machlolophus xanthogenys.*—I only saw the Yellow-cheeked Tit once—near Naini Tal. A small party were feeding in some low trees on the banks of a small stream. I shot one specimen, but although I was constantly on the look-out for more, I did not see the bird again. I fancy, therefore, that it must be rather locally distributed in Kumaon.

44. *Lophophanes melanolophus.*—The Crested Black Tit is common throughout the district.

69. *Garrulax leucolophus.*—Tolerably common in the lower hills.

76. *Garrulax albigularis.*—The White-throated Laughing-Thrush is very common. One of its notes is a regular hiss—like a snake rather than a bird. It is an inquisitive bird, and as a rule not at all shy. My Scottish terrier was an object of great interest to this and other kinds of Laughing-Thrushes. Though they never actually mobbed it, as Drongos and Mynas often do, they often followed the dog for some distance through the jungle. I saw about twenty birds of this species following a marten (*Mustela flavigula*) in the same way.

82. *Trochalopterum erythrocephalum.*—This is such a skulker that it appears to be much rarer than it probably is. The stomach of one I shot was full of small snails.

90. *Trochalopterum variegatum.*—Fairly common. It has the usual habits of the Laughing-Thrushes; is very noisy, and goes about in small troops among scattered bushes and thin jungle.

99. *Trochalopterum lineatum.*—Very common indeed. It has a tri-syllabic call, preceded by a low trill: the latter is not heard, unless one is very close to the bird.

125. *Pomatorhinus ruficollis.*—Another inveterate skulker in thick undergrowth: I never saw more than three birds together.

174. *Stachyrhidopsis pyrrhops.*—In thick jungle; not often seen.

187. *Myiophoneus temmincki.*—Common everywhere near running water, from the lowest valleys to above the snows. It is a thirsty bird, and very fond of bathing.

204. *Lioptila capistrata.*—The Black-headed Sibia is very common, especially on the edge of forest. It has a very loud, shrill song, which it sings perched upon some conspicuous branch. Besides this it has a variety of very harsh notes. Out of six specimens that I collected I find that the back and scapulars of one (a young male) are almost concolorous with the lower plumage, which is much duller rufous than in the others.

221. *Siva cyanuroptera.*—I saw and shot the Blue-winged Siva a few miles from Almora, where it was common. I did not notice it elsewhere.

226. *Zosterops palpebrosa.*—I did not see this White-eye in Garhwal: it was common in parts of the Almora district.

232. *Ixulus flavicollis.*—A retiring bird, seen among thick undergrowth, in tree jungle.

249. *Chloropsis hardwickii.*—Not seen in Garhwal: locally common in the Almora district.

261. *Psaroglossa spiloptera.*—Mr. Oates separates this bird from the Starlings, and he is no doubt right. All the same its habits are very like those of Mynas, except that it is never seen on the ground. It is very noisy and flies about in flocks. It has a very harsh call-note. A young bird had brownish irides.

269. *Hypsipetes psaroides.*—Very common at moderate elevations. It has many different, mostly harsh, notes. One of them is like the " mew " of a young kitten.

283. *Molpastes intermedius.*

284. *Molpastes leucogenys.*—This and *M. intermedius* are the common Bulbuls of Kumaon : they do not occur much above 6,000 feet.

315. *Sitta himalayensis.*—Very common and generally distributed. It is a wonderful tree climber, and progresses with equal ease, vertically, upwards, or downwards.

316. *Sitta cinnamomeiventris.*—Replaces *S. himalayensis* at the lower elevations : very common at Naini Tal.

325. *Sitta frontalis.*—I obtained only one specimen in the Almora district.

328. *Dicrurus longicaudatus.*—The Ashy Drongo is very common throughout the district. The iris in young birds is dark brown.

341. *Certhia himalayana.*—Common everywhere. It has a rapidly repeated note, which is very loud for such a small bird.

343. *Certhia nepalensis.*—I obtained one specimen, a male, near Yoshimath, Garhwal, on June 6th. The testes were so large that it was probably breeding then. This Tree-Creeper does not appear to have been previously recorded west of Nepal.

382. *Franklinia gracilis.*—Franklin's Wren-Warbler was common at moderate elevations.

405. *Phylloscopus affinis.*—Tickell's Willow-Warbler was very common and breeding in Garhwal, above 8,000 feet : it has a loud song.

418. *Phylloscopus humii.*—I am uncertain whether a Willow-Warbler that I obtained should be referred to *P. superciliosus* or to *P. humii* : the coronal band is certainly quite distinct, and, therefore, according to " The Fauna Brit. Ind., Birds," vol. i, page 410, more resembles *P. superciliosus.* On the other hand, the second primary is intermediate in length between the eighth and ninth, and the bird should therefore be *P. humii* (loc. cit., p. 411). According to the geographical distribution given by Mr. Oates, it is more *likely* to be *P. humii.* Even with the help of Mr. Brooks's paper in Str. Feath., vol. vii, on the subject, the identification remains undecided in my mind.

422. *Acanthopneuste viridanus.*—Common.

428. *Acanthopneuste occipitalis.*—Another common species at rather

high elevations: evidently breeding at about 10,000 feet, at the end of May.

434. *Cryptolopha. xanthoschista.* — Hodgson's Grey-headed Fly-catcher-Warbler is common everywhere.

458. *Suya crinigera.*—Common throughout the district. It has an agreeable, but weak song; also a "chiff-chaff"-like call, repeated several times in quick succession.

475. *Lanius nigriceps.*—Common in the east of the Almora district, especially about Pithoragarh; not seen in Garhwal.

476. *Lanius erythronotus.*—Common up to about 7,000 feet.

495. *Pericrocotus brevirostris.*—This was the only Miniret I saw; it is generally distributed.

505. *Campophaga melanoschista.*—I procured specimens and saw this bird almost throughout the whole district.

518. *Oriolus kundoo.*—Common in many of the lower valleys.

538. *Sturnia malabarica.*—I only saw this bird once; near Almora.

549. *Acridotheres tristis.*—Common in the Almora district: only noticed at Chamoli, in Garhwal, where it was apparently breeding in holes in the cliffs, in company with *Cypselus melba* and *C. affinis.*

558. *Hemichelidon sibirica.*—I got some specimens near Niti, at about 10,000 feet; I did not notice it elsewhere.

568. *Cyornis superciliaris.*—Common.

579. *Stoparola melanops.*—Very common at low elevations.

592. *Culicicapa ceylonensis.*—Common.

595. *Niltava macgrigoriæ.*—The Small Niltava is met with among low thick undergrowth in the more open, low-lying valleys of Garhwal.

598. *Tersiphone paradisi.*—The "Ribbon-bird" is common in the lower parts of the district.

603. *Chelidorhynx hypoxanthum.*—Common, especially in the higher forests. It has the habit of opening its tail out fan-wise, as it hops about bushes.

608. *Pratincola caprata.*—Common.

610. *Pratincola maura.*—Common.

615. *Oreicola ferrea.*—The Dark-grey Bush-Chat is met with from the lowest valleys to nearly 11,000 feet. It was common, in May, about Niti, almost up to the snow line.

630. *Henicurus maculatus.*—Common on the lower streams and about small waterfalls. I saw a couple feeding in a flooded rice field at some distance from running water. It is rather shy and restless, and frequently perches on low bushes near the edge of a stream. The smallest brook suffices to attract this bird. By quietly approaching one of the many culverts that span small streams on the road between Naini

Tal and Almora, one was pretty sure to get a glimpse of a Forktail or two.

637. *Microcichla scouleri.*—I only saw the Little Forktail at one place, near Kedarnath, where it was very common. It has a short twittering song, which it sings in the intervals of searching under stones for food. It is remarkable to see such a frail little bird standing "knee deep" in the most rapid torrents.

638. *Chimarrhornis leucocephalus.*—Common throughout the valleys of Garhwal : it was especially numerous about Kedarnath and Badrinath, in May.

644. *Ruticilla rufiventris.*—This was the only Redstart that I saw, and that only near the Niti Pass. In the plains of India, in the cold weather, this is quite a tame, confiding bird, but I found it very wary in Garhwal, and had some difficulty in procuring a specimen. This, a male, shot on May 31st, had very large testes, and if not already breeding, would certainly have done so very soon. I had a long but fruitless search for nests of this species.

646. *Rhyacornis fuliginosus.*—Common about all the lower streams in Kumaon.

651. *Calliope pectoralis.*—I saw this bird at Mana and Niti in May. It has a long and pretty, though weak song. All that I shot were males. I noticed a plain-looking bird, perching on stones and low bushes, and rather shy : it was quite a surprise to me, on shooting one, to see the bright crimson throat.

654. *Ianthia rufilata.*—I only got one specimen, a hen, in Garhwal, in June.

663. *Copsychus saularis.*—Common at moderate elevations, throughout the whole of Kumaon.

672. *Merula albicincta.*—The White-collared Ouzel is fairly common in Garhwal. It frequents rather open forest. Both male and female have white tips, as well as white shafts, to the under tail coverts. The margin of the eyelids is bright yellow.

676. *Merula boulboul.*—A common forest bird. It is one of the best songsters I know, and its loud and varied notes are a striking refutation of the old calumny that Indian birds "don't sing." It especially frequents the tops of high trees.

677. *Merula atrigularis.*—I only saw and obtained one specimen, at Dwarahath, Almora. It was feeding on wild cherries near the Dak Bungalow.

678. *Merula unicolor.*—Tickell's Ouzel was not very common. I noted the following colours in a ♂, shot on June 26th, which, from the condition of its testes, must have been breeding—

Iris—Dark brown.

Bill—Yellow throughout.

Tarsi—Greenish-yellow; feet, more yellow; claws, yellowish-horny.

683. *Geocichla wardi.*—Rather a retiring bird; met with only in moderately open forest.

690. *Petrophila erythrogastra.*—Not nearly so common as *P. cinclorhyncha.* "Rock-thrush" seems an inappropriate name for this bird in the Himalayas. I generally saw it perched near the top of a high tree.

691. *Petrophila cinclorhyncha.*—Very common at low elevations. The call-note is a single, rather loud whistle. The Blue-headed Rock-Thrush, like the last species, is often seen on high trees; it goes about in small parties of three or four individuals.

693. *Petrophila cyanus.*—Not very common, but still seen throughout most of the district. A male, shot near Niti, at about 10,200 feet, on May 30th, had testes about the size of peas.

695. *Turdus viscivorus.*—Common at high elevations in Garhwal, even above the snow. It is very pugnacious towards its own kind. I never heard it utter other than very loud, harsh notes.

701. *Oreocincla mollissima.*—This bird is distinctly rare in Garhwal. I obtained one specimen near Yoshimath, among snow drifts at an altitude of about 13,000 feet.

704. *Zoothera monticola.*—I only met with the Large Brown Thrush in thick forest: it does not seem to be at all common. The claws of the anterior toes are white; that of the posterior dusky.

709. *Cinclus asiaticus.*—This was the only Dipper I obtained: it is very common.

727. *Uroloncha acuticauda.*—Hodgson's Munia is common in the lower valleys.

735. *Uroloncha punctulata.*—Common in the same localities as the last species.

746. *Pyrrhula erythrocephala.*—Frequents low ringall undergrowth in jungle; not at all shy.

755. *Propasser pulcherrimus.*—Very common about Mana and Niti, above 10,000 feet. It frequents low bushes, going about in considerable flocks. Although this is certainly a handsome bird, its specific name would be more appropriate to *P. rhodochrous,* which is a lovely bird in life.

758. *Propasser rhodochrous.*—Only noticed about Badrinath (10,284 feet). It was very common there in May; in the Mana and Niti districts it was replaced by the last species.

761. *Carpodacus erythrinus.*—Very common indeed in the lower valleys. Oates gives its range up to 10,000 feet, but I did not see it in Garhwal above 7,000 feet.

767. *Carduelis caniceps.*—I shot one, a ♂, out of a large flock, perched on a high fir tree, near Niti, at over 10,000 feet: this was the only occasion on which I saw this species.

771. *Metoponia pusilla.*—A common, familiar bird near Niti. The birds were sitting about, like sparrows, on the house tops of the village of Bampa, in May. Like sparrows, too, they roosted at night in large numbers on one or two particular trees, with the same noisy twittering as each batch of new-comers arrived and settled down for the night. The reproductive organs were in an advanced stage of development, and the bird must, I think, nest in the neighbourhood. The forehead and anterior portion of the crown are rather orange than crimson in my specimens.

772. *Hypacanthis spinoides.*—Common, up to 10,000 feet. It occurred in very large flocks at Ramni, in June (9,000 feet).

776. *Passer domesticus.*—Common about the hill stations and lower valleys. Yoshimath (6,000 feet) was the highest place in Garhwal where it was common.

780. *Passer cinnamomeus.*—This is the common Jungle-Sparrow in Kumaon: it is also found round the higher villages. Oates gives its range up to 7,000 feet, but I obtained specimens at Dungari, Garhwal, at well over 10,000 feet.

790. *Emberiza fucata.*—The Grey-headed Bunting is common throughout Garhwal.

794. *Emberiza stracheyi.*—Common up to 11,000 feet. It has a very soft, sibilant call-note; besides this, there is a long, disconnected sort of song.

803. *Melophus melanicterus.*—Common.

818. *Hirundo smithii.*—The Wire-tailed Swallow is generally distributed throughout the lower valleys.

822. *Hirundo nepalensis.*—Common.

826. *Motacilla alba.*—I shot a male, in full summer plumage, at Trijugi Narayan, Garhwal, on May 2nd. The testes were very small.

830. *Motacilla hodgsoni.*—Common towards the end of May, near Niti. It was very tame, and seems less restless than most Wagtails. I think that this species must breed in Garhwal, as the reproductive organs of those I shot were all fully developed. I was unable, however, to find any nests.

831. *Motacilla maderaspatensis.*—The only place at which I saw and procured the Large Pied-Wagtail was at Bageswar, to the east of Almora. It was very common about the river there.

832. *Motacilla melanope.*—Common and probably breeding in the higher ranges. Reproductive organs very large in May and June.

840. *Anthus trivialis.*—I procured several specimens in April, but did not see the bird later in the summer.

841. *Anthus maculatus.*—I shot some of this species at 10,000 feet at the end of May. The sexual organs were still very small.

844. *Anthus similis.*—Not very common, and only seen at low elevations.

847. *Anthus rufulus.*—Baijnath, in the Almora district, is the only place where I obtained this bird.

850. *Anthus vosaceus.*—Common on bare ground, at elevations of 10,000 feet and over. The day that I visited Kedarnath (May 4th) Hodgson's Pipit was in great force on the plain below the temple. It was a bitterly cold day and snowing hard, but the birds seemed quite cheerful. Its habits seem to be very similar to those of *A. pratensis*: it sings both on the wing and also when perched on some low bush or stone.

853. *Oreocorys sylvanus.*—The Upland Pipit is common on bare hill sides at moderate elevations. It has a very shrill call of two notes, frequently repeated, and soars, like a lark, to a height of twenty or thirty feet.

888. *Æthopyga gouldiæ.*—This Sun-bird was locally common, above 7,000 feet. At certain places, on a fine sunny day, one would notice numbers flitting about. On dull, overcast days one scarcely ever sees them.

890. *Æthopyga saturata.*—Rather less common than the preceding species.

946. *Gecinus squamatus.*—Very common in all well-wooded parts. I shot one specimen at above 11,000 feet. Like all the genus, it feeds a great deal on the ground. Small black ants are a very favourite food; the Woodpecker stands by the side of the ants' run, and picks them off as they come along. The bird also diligently hunts the rhododendrons.

950. *Gecinus occipitalis.*—Met with at moderate elevations, but it is, I think, nowhere very common.

951. *Gecinus chlorolophus.*—I only obtained one specimen in Garhwal of this species.

960. *Hypopicus hyperythrus.*—Very common indeed throughout the entire district, wherever there are trees. It has a very loud, harsh note, and taps the trees rapidly, making a loud rattle. The bill is pale yellow beneath.

961. *Dendrocopus himalayensis.*—The Western Himalayan Pied Woodpecker is also very common. It has a loud "clucking" note, and seems partial to very rotten trees: it feeds a great deal on the ground.

969. *Dendrocopus auriceps.*—Fairly common at moderate elevations.

974. *Iyngipicus pygmæus.*—I only procured this bird once, near Almora.

992. *Chrysocolaptes gutticristatus.*—Tickell's Golden-backed Woodpecker is not uncommon in the lower valleys of the eastern part of the Almora Division : I did not see it in Garhwal.

1006. *Megalæma marshallorum.*—Very common all over Kumaon, up to about 8,000 feet. 1 found this Barbet very wary indeed, and it was by no means easy to procure specimens. To start with, it is rather difficult to locate the particular tree from which the noisy chorus is proceeding. The Barbets keep a sharp look-out, and most of my attempts at stalking them ended in failure. However, the flocks seem to be very regular in their movements, frequenting a given tree at almost exactly the same time every day, as long as the fruit on it lasts. I obtained several specimens by taking up a position under a tree a little before the time that I had seen the birds the day before. They almost always kept the "appointment," and I got an easy shot.

1012. *Cyanops asiatica.*—Common in the low-lying valleys of the eastern part of Kumaon. I did not see or hear the bird in Garhwal.

1025. *Eurystomus orientalis.*—I got one specimen in Kumaon at an elevation of 4,000 feet. This was the only occasion on which I saw this bird.

1034. *Ceryle lugubris.*—The Himalayan Pied Kingfisher is common at moderate altitudes. It wanders about a good deal forsaking a stream as soon as the water gets at all thick. It is usually found in pairs and is rather wary. It perches indifferently on stones or branches.

1035. *Alcedo ispida.*—Not at all uncommon on the lower streams, both in Garhwal and in the Almora district. I only shot one specimen, at Bageswar. This, a male, has a shorter bill than my specimens from other parts of India, the distance from the gape being only 1·7″.

1067. *Upupa indica.*— Common up to about 6000 feet.

1068. *Cypselus melba.*—I saw two breeding places of the Alpine Swift in Garhwal; both on the Alaknanda river. They were quite inaccessible, being situated in high perpendicular cliffs at a considerable elevation above the river.

1072. *Cypselus leuconyx.*—At moderate elevations. This bird does not fly very fast, for a Swift. Two of my specimens, both hens, measure 6·8″; the wing of one being 6·5″, and of the other 6·55″, thus approaching the dimensions of *C. pacificus.* The feet of both were pale coloured and the claws almost black.

1073. *Cypselus affinis.*—Not seen above about 6000 feet.

1095. *Caprimulgus indicus.*—This was the only Nightjar I obtained in Garhwal : it is common.

1104. *Cuculus canorus.*—Very common.

1105. *Cuculus saturatus.*—Also common. I often heard the call of what I believed to be *C. poliocephalus*, but I obtained no specimens.

1107. *Cuculus micropterus.*—Very common. Its call notes are often heard on moonlight nights. It has a harsh alarm note, like "jŭg-jŭg-jŭg," frequently repeated.

1118. *Coccystes jacobinus.*—I shot one specimen, in the Almora district, at about 6000 feet, at the end of June.

1141. *Palæornis schisticeps.*—Very common indeed. I shot one bird in Garhwal, at 11,000 feet, in June.

1160. *Syrnium indrani.*

1184. *Glaucidium radiatum.*—These were the only two Owls that I procured, though I saw at least two other species that I was unable to identify.

1193. *Gyps himalayensis.*—I saw this fine Vulture on several occasions. There were five or six sitting on the rocks above Bampa, near the Niti Pass. They were not at all shy and allowed me to approach within thirty yards of them.

1199. *Gypaëtus barbatus.*—The Lämmergeyer is common, but the percentage of birds in fully adult plumage is very small.

1217. *Spilornis cheela.*—The Crested Serpent-Eagle is common, near running water, up to 6000 feet. It is very tame and has a loud cry.

1229. *Milvus govinda.*—The Common Pariah Kite is met with up to considerable elevations, wherever the country is open. I did not see *M. melanotis*, although constantly on the look out for it.

1248. *Accipiter virgatus.*—Fairly common.

1260. *Falco subbuteo.* —I saw either this species or *F. severus* many times, but could not procure a specimen.

1265. *Tinnunculus alaudarius.*—Very common.

1283. *Sphenocercus sphenurus.*—The Kokla Green Pigeon is very common. It has prolonged rather mournful notes, and a very swift and powerful flight. I saw one bird hanging almost head downwards from a slender branch, in a most un-pigeon-like attitude, as it attempted to reach some food. As I was within four yards of the bird, and watched it for some minutes, I am quite sure that I identified it correctly.

1292. *Columba intermedia.*—Common in cultivated districts, at moderate elevations.

1294. *Columba rupestris.*—I only saw the Blue Hill Pigeon at Badrinath, Garhwal. A few were feeding in company with *C. leuconota.* A hen bird, that I shot, had the irides orange, with a narrow inner circle of yellow. On May 20th, the ova were of the size of peas.

1296. *Columba leuconota.*—Very common at and above the snow-level, and not at all wild. They frequent the hill paths and the banks of streams, during the middle of the day, in pairs, or parties of three or four individuals. In the morning and evening they assemble in flocks of twenty to sixty birds and feed in the fields. I saw one of their breeding places at Mana, at the foot of the pass, on May 20th. It was in a cliff about eighty feet high, overhanging a cascade. The natives told me that the Snow-Pigeon always builds rather low down, near the water, to avoid the Choughs, which mostly frequent the higher parts of the cliff. I do not think that there were any eggs laid on the date of my visit, but I saw one bird carrying building materials to a ledge in the rock, and many courting, the males behaving just like domestic pigeons. I also saw one pair *in coitu.*

1297. *Dendrotreron hodgsoni.*—I saw this very handsome Wood Pigeon on several occasions, during the month of May, at about 7000 feet. Blanford says about this bird :—" A shy bird, usually seen in small flocks amongst the pine forests." I shot a pair at Gorikund, near Kedarnath, on May 2nd. They were feeding on low bushes, about thirty yards from my tent, and about fifty yards from the village, which was crowded with noisy pilgrims. They did not seem to be at all shy. I shot the hen first; the cock flew away for a short distance, and returned almost at once to the spot from which he had been disturbed. The reproductive organs were well developed, and the birds must soon have bred. I could get no information from the natives about their nidification. The claws are very bright yellow—not " pale yellow " as stated by Jerdon.

1305. *Turtur ferrago.*—The Indian Turtle Dove is common, and met with up to 10,000 feet.

1307. *Turtur suratensis.*—Very common, up to about 8000 feet.

1310. *Turtur risorius.*—Common at low elevations.

1333. *Catreus wallichi.*—I did not see the Cheer Pheasant myself, but I bought a skin said to have been obtained near Ranikhet.

1334. *Pucrasia macrolopha.*—Met with singly or in pairs in many parts of the district. At the time of my visit to Kumaon, all the pheasants were breeding, and I disturbed them as little as possible.

1336. *Gennæus albicristatus*—Very common. I saw a cock clapping his wings and making a great demonstration in front of an apparently indifferent hen.

1342. *Lophophorus refulgens*—Tolerably plentiful, and very wary.

1344. *Tragopan satyra.*—I did not come across this pheasant, but I bought a skin from a native. The bird was said to have been shot near the Pindari Glacier.

1372.   *Francolinus vulgaris.*—Very common up to 7000. I heard a cock calling at Ramni, Garhwal (about 9000 feet), just above the limit of cultivation.

· 1462.   *Totanus ochropus.*—I shot a Green Sandpiper in summer plumage, and saw a few others, at Adabadri, Garhwal, on April 20th. A few days later they disappeared.

⌇⌇⌇⌇⌇⌇⌇⌇

XII.—Noviciæ Indicæ XVII.   *Some new plants from Eastern India.—*
By D. PRAIN.

[Received 11th June ; Read 4th July, 1900.]

In this paper are contained descriptions of twelve previously undescribed species of plants from the north-eastern frontiers of India. A considerable number of these have been examined and compared at the Kew Herbarium by Sir George King, who has kindly undertaken, for some of them, the responsibility of joint authorship. The descriptions are, as usual, drawn up in such a way as to conform to the descriptions given in Sir J. D. Hooker's *Flora of British India.*

TILIACEÆ.

1.   GREWIA (Eugrewia) NAGENSIUM *Prain;* shrubby, *leaves* scabrous, ovate-lanceolate, acuminate, finely subequally serrate; *cymes* axillary, peduncled ; buds obovate, striate; *drupe* 2- or 1-lobed, subtesselately rugose with lenticular swellings, each crowned by a stellate hair.

ASSAM ; Eastern Naga Hills at Narazu, *J. W. Masters* 1263! Teock Ghat near Tingali Bam, *Prain's Collector* 128! 262! Margarita, *Prain's Collector !*

Young shoots scabrous with stellate hairs ; branches terete, sparsely stellately hairy. *Leaves* rather thick, 4·6 in. long, 2 5 in. wide, base rounded, 3-nerved, central nerve with 3-4 pairs of slightly arching nerves, sparsely stellately hairy above, rather densely stellately hairy, especially in the nerves, beneath ; stipules subulate as long as the petioles. *Cymes* axillary, umbellate, few-flowered, peduncles ·3–·5 in. long, pedicels as long, in fruit elongate and reaching ·6 in., bracts triangular-lanceolate, ·2 in. long, stellate-hairy outside, striate within. *Buds* ·25 in. long, ·2 in. wide. *Sepals* ·4 in. long, lanceolate. *Petals* linear, ·3 in. long. *Torus* densely addpressed-rusty-tomentose, ·15 in. long, cylindric. *Drupe* with 1 or 2 orbicular lobes, ·3 in. long and broad and ·25 in. thick.

The Calcutta Native Collector describes the flowers as yellow. The leaves most nearly resemble those of the Burmese species *G. microstemma ;* the margins are, however, more finely toothed. The flowers are quite unlike those of *G. microstemma* and most closely resemble those of *G. oppositifolia,* but the torus is very considerably longer than in that species. The fruits afford the most distinctive character; they approach most nearly to, though they are still widely different from, those of *G. umbellifera* Bedd. (*G. capitata* Dalz.), next to which species the systematic position of the present one is.

OLACINEÆ.

2. GOMPHANDRA SERRATA *King & Prain* ; *leaves* serrate, ovate or ovate-lanceolate, *cymes* axillary or also from leaf-scars lower down, rather longer than the petioles.

KACHIN HILLS ; Myitkyina, *Prain's Collector !*

A small tree, everywhere glabrous. *Leaves* chartaceous, glandular-serrate except at the cuneate base, apex acute, nerves about 9 pairs, prominent beneath; length usually about 6 in., breadth 2-3 in.; petiole ·5 in. *Cymes* finely puberulous. *Calyx* minute, 4-5-toothed. *Corolla* and stamens not seen. *Fruit* ·5 in. long, ovate, narrowed gradually to an acute tip, dark-brown, smooth, crowned by the remains of the stigma, pericarp thin, firmly coriaceous, the inner layer almost woody. *Seed* large, testa pale, thin.

Very different from any of the other species of *Gomphandra* either in Herb. Kew or Herb. Calcutta.

COMBRETACEÆ.

3. COMBRETUM KACHINENSE *King & Prain; leaves* ovate-lanceolate, caudate, glabrous above, finely rusty-pubescent beneath, only subopposite ; *flowers* in axillary, simple or sparingly branched, almost spicate lax racemes ; *calyx* not constricted above the ovary, densely rusty outside, glabrous within.

KACHIN HILLS ; near Sima, *Prain's Collector !*

A large scandent shrub, without thorns; branches densely rusty-pubescent. *Leaves* subopposite, 8-9 in. long, 2·25-3 in. wide, chartaceous, caudate apex ·5-·75 in. long, margin entire, base abruptly cuneate or almost rounded ; petiole densely rusty, ·35 in. long, stipules subulate ·25 in. long, subpersistent. Racemes 3-4 in. long, branches if present few, 1-1·5 in. long, rachis densely rusty ; bracteoles minute, rusty. *Flowers* subsessile. *Calyx* campanulate, 5-lobed, densely rusty externally throughout, ·2 in. long, lobes ovate-acute, glabrous within. *Petals* narrow obovate, exceeding the calyx, ·15 in. long, glabrous. *Stamens* 10, those of the antipetalous series with short but distinct filaments and lanceolate anthers, the others with short broad oblong anthers subsessile. *Ovary* sessile, quite glabrous, gradually narrowed into the glabrous simple style.

A very distinct species unlike any of the 5-merous Indian species. Among Indian *Combreta* it most resembles *C. dasystachyum* Kurz, which has however 4-merous flowers. In foliage it very closely resembles *Combretum ferrugineum* Schimper, from Abyssinia.

OLEACEÆ.

4. JASMINUM EXCELLENS *King & Prain;* glabrous ; *leaves* opposite pinnately 7-foliolate, less often 5-foliolate or occasionally close to the

inflorescence 3-foliolate, leaflets cordate- or ovate-lanceolate, glabrous, sub-3-nerved ; *cymes* axillary lax ; *calyx-teeth* short; *corolla* white, tube ·8 in. long.

KACHIN HILLS ; Shan Busti, Sadon, 5000 ft., *Prain's Collector !*

An extensive climber. *Terminal* leaflet 2-2·5 in. long, ·5-·75 in wide, lateral nearly as wide, usually 3, less often 2 pairs, only 1·25-1·5 in. long. *Cymes* 10-14-flowered, bracts subulate ·2-·25 in. long, pedicels ·5-1·25 in., slender. *Calyx* glabrous, teeth subulate ·1 in. long, as long as the tube. *Corolla* white, lobes elliptic, subacute, ·5 in. long, ·25-·3 in. wide. *Fruit* not seen.

Neither in the Kew Herbarium nor in that of Calcutta can we find anything like this very handsome Jasmine. The inflorescence reminds one most of that of *J. dispermum,* when its cymes are axillary, but the long slender pedicels, the calyx, and the leaves are very different. The species comes nearest *J. officinale* Linn., but again the calyx-teeth are different, being in our plant much shorter, while the cymes are axillary not terminal, and the pedicels are longer than in *J. officinale.*

ASCLEPIADACEÆ.

5. MARSDENIA (Eumarsdenia) LEIOCARPA *King & Prain;* quite glabrous except the finely puberulous rachis and pedicels ; *leaves* ovate-lanceolate or ovate caudate-acuminate, base rounded or cuneate ; *flowers* in rather lax axillary racemes ; *follicles* quite glabrous.

KACHIN HILLS ;· between Myitkynia and Sadon, *Prain's Collector !*

A large climber; *stem* stout, smooth as are the branches, petioles and leaves on both surfaces from a very early stage ; the youngest leaves very finely deciduously laxly puberulous. *Leaves* 3-6 in. by 1-3 in., nerves spreading, rather prominent beneath, 5-8 pairs, glandular above at the petiole ; petiole ·75-1·25 in. long. *Racemes* 4-6 in. long, lower portion of axillary rachis 1-2 in. long, uniform and glabrous (peduncle-like), the rest puberulous, slightly zig-zag, with small tumid minutely scaly and pubescent nodes in the retiring angles, about ·3 in. apart below, approximate above. *Flowers* at the nodes usually solitary, sometimes germinate, rarely more, on puberulous slender pedicels ·25 in. long. *Sepals* suborbicular, nearly glabrous, with faintly hyaline margins. *Corolla* not seen. *Follicles,* only one of a pair developed, when quite ripe 1·6 in. long, ·15 in. in diam., lanceolate, rather abruptly narrowed at the base, coriaceous, quite glabrous in the very youngest stage. *Seeds* narrowly ovoid, ·25 in. long.

Very nearly related to *M. tinctoria* R. Br., and *M. eriocarpa* Hook. f., but extremely distinct on account of its much smaller, quite glabrous follicles, and smaller seeds.

6. CEROPEGIA KACHINENSIS *Prain ; leaves* slightly puberulous above and on margins, glabrous glaucescent beneath, long petioled, ovate acuminate base rounded ; *corolla*-lobes subspathulate half as long as the slightly curved tube, their apices ciliate and connate so as to form a conical crown over the not greatly dilated throat; *coronal* lobes 10, triangular, ciliate, less than half as long as the linear, slightly clavate,

straight processes. C. pubescens *Prain in Rec. Bot. Surv. Ind.* i, 252 not of *Wall.*

KACHIN HILLS; Myitkyina, *Prain's Collector!*

A slender climber with glabrous stems, branches, peduncles and pedicels. *Leaves* 3–3·5 in. by 1·5–1·75 in.; petioles ·6–·75 in. *Peduncles* rather slender, as long as petioles, 8–12-fld., pedicels ·5–·6 in. slender. *Calyx*-segments lanceolate their tips pinkish, glabrous, ·2 in. long. *Corolla* slightly curved, 1·25 in. long, base slightly inflated, upper third of tube somewhat abruptly funnel-shaped; tube pale green with small purple spots in upper third and with pinkish lines below; lobes green in lower, yellow in upper half and the margins there purple-ciliate. *Follicles* slender, spreading horizontally, each 4 in. long, ·25 in. in diam., greenish with irregular red streaks. *Seeds* about 20 in each follicle, narrow ovate, compressed, ·4 in. long, ·12 in. wide, coma white, nearly twice as long as seeds.

Erroneously distributed in 1898 under the name *C. pubescens*, this is to be found in various collections. The species is perhaps most nearly related to *C. Thwaitesii* Hook., from Ceylon, but is abundantly distinct. The present description is drawn up from living specimens which have flowered in the Royal Botanic Garden, Calcutta.

ACANTHACEÆ.

7. GYMNOSTACHYUM (Cryptophragmium) LISTERI *Prain;* minutely puberulous; *leaves* large, short-petioled, oblanceolate; *panicles* mostly lateral, many-flowered; *corolla* ·75 in. long.

CHITTAGONG; Demagiri, in rocky places, *Lister* 162!

A small undershrub, under a foot high. *Leaves* attaining 8 in. by 3 in., widest at the junction of the anterior and middle third, acute or acuminate, tapering gradually in the basal half to a petiole ·5–·75 in. long, margins entire, veins slightly arched forward, 12–14 pairs, minutely puberulous on both surfaces. *Panicles* chiefly from the axils of the lower leaves and the leaf-scars below these, 1·5–2·5 in. long, branches subspicate; flowers solitary or clustered; bracts small, linear. *Sepals* ·2 in., linear. *Corolla* puberulous. *Anthers* linear-oblong. *Capsule* ·5 in., very narrow, glabrate. *Seeds* ovoid, compressed.

Very closely related to *Gymnostachyum latifolium* T. And. (*Cryptophragmium latifolium* Dalz.), and with that species standing, as regards habit, rather distinctly apart from the other Indian *Gymnostachya*. It is, however, very distinct from *G. latifolium* on account of its differently shaped leaves, its shorter flowers with different anthers, and its much smaller capsules. We are indebted to Mr. C. B. Clarke, for having very kindly compared our specimens with the material preserved in the Herbarium at Kew.

8. PERISTROPHE LONGIFOLIA *King & Prain;* *leaves* distinctly petioled, lanceolate, glabrous except for some adpressed hairs on midrib above and below; *bracts* lanceolate, faintly puberulous; *corolla* pink, 1·25 in. long.

KACHIN HILLS; Sadon, *Prain's Collector!* EASTERN NAGA HILLS; near Balijan, *Prain's Collector!*

Leaves 4–6 in. long, ·75–1 in. wide, margin undulate, gradually tapering from junction of middle and lower third to apex and to a slender petiole ·75 in. long; raphides very slender plentiful on both surfaces. *Bracts* ·75 in. long, ·25 in. wide, acute. *Filaments* sparsely pubescent; anther-cells linear, one superposed for half its length.

This is very like a plant at Kew from Ichang, (*Henry* n. 4153) as regards foliage, but the Ichang plant has broader bracts than the Kachin one though it has a similar condensed inflorescence. *Hypoestes salicifolia* O. Kuntze, represented at Kew by a flowerless scrap, named by its author, resembles our plant in leaves, inflorescence and bracts without however quite agreeing with it absolutely in these characters. Our plant has however the two-celled anthers that distinguish *Peristrophe* from *Hypoestes.* Henry's n. 4153 from Ichang, at Calcutta, differs considerably from our plant.

LABIATÆ.

9· GOMPHOSTEMMA (Pogosiphon) INOPINATUM *Prain;* ascending, leafy stems and flowering scapes distinct; *leaves* distinctly petioled; *spikes* erect, not interrupted even at the base.

KACHIN HILLS; Langkon, 3000 ft. elev., *Prain's Collector !*

Stems ascending, rooting below, 4-grooved, and with rounded angles, several from a woody rootstock with numerous tufted, woody, branching, slender roots; about a foot high; densely clothed with a close, ash-grey, stellate tomentum, intermixed with a copious pubescence of laxly spreading, long, white hairs. *Leaves* 4–6 pairs, the lowest small, the pairs 3-4 in. apart; petioles 1–1·5 in. (occasionally about 2 in.) long, pubescent like the stems but with fewer long, lax, white hairs in proportion to the stellate pubescence; lamina broadly ovate-acute, 3–5 in. long by 1·75–3 in. wide, the base of lower leaves slightly cuneate, of the upper rounded; margin finely crenate except at the basal fifth; nerves about 6 pairs, ascending; upper surface finely velvety with a soft, ash-grey, stellate pubescence interspersed with longer, simple, subadpressed tomentum, under surface softly velvety with a felted, whitish-grey, stellate pubescence. *Flowers* densely whorled, in radical spikes 2 in. long, 1·25 in. wide, on erect peduncles 3–6 in. long, with sometimes a pair of small foliaceous opposite bracts about ·25 below the spike; peduncles with pubescence exactly as on the stem, but themselves terete and more slender; floral bracts obovate, dentate, sparsely stellate-pubescent tinged with pink, the lowest ·5 in. long, ·25 in. wide. *Calyx* wide-campanulate, glabrous within, tube rather closely stellate externally, limb with 5 equal, wide-triangular claret or purple lobes, sparsely stellate on the strongish central and weaker marginal nerves; ·5 in. long, ·3 in. wide, the lobes ·2 in. long; bracteoles obovate-lanceolate ·2–·25 in. long, reddish. *Corolla* ·75 in. long, upper lip subentire, lower 3-lobed with slightly emarginate mid-lobe and inflated throat, apparently annulate within. *Stamens* exserted, filaments hirsute at their insertion. *Ovary* and style glabrous. *Nutlets* usually 4, sometimes 2–3, reddish, quite glabrous, wall very thickly coriaceous when dry.

Only one of our specimens had a few rather shrivelled corollas, from two of these, unsoaked as carefully as possible, the above description is given. Their colour is not particularly noted by the native collector, who simply remarks " flowers red," with reference doubtless to the purple or claret-coloured calyx. Further examination of less advanced specimens will be required in order to confirm the existence of a distinct annulus. Its other characters however amply justify its title to specific rank. It is not very like any of the hitherto described Indian *Gomphostemmata.* The fact that the flowers occur on independent leafless stems or scapes recalls the habit of *G. chinense* Oliv. and the fact that the calyx and less markedly the bracteoles are purple-coloured, recalls also *G. Curtisii,* and *G. pedunculatum* which are the other members of the group *Pedunculata* to which *G. chinense* belongs. The general facies of the species nevertheless rather recalls the *Strobilina* group of the § *Pogosiphon* to which, from the presence of hairs within the corolla tube, it must necessarily be referred. If, however, we are right in supposing that these hairs form a distinct annulus, instead of being scattered as in the other *Strobilina,* it must be considered in this respect as linking that group with the hitherto somewhat isolated *G. Hemsleyanum.*

This is the second new species recorded since the publication by the writer in 1891 of *An account of the Genus Gomphostemma* (Ann. Roy. Bot. Gard., Calcutta, iii. 227, *et seq.*). The other species, *Gomphostemma furfuracea* Hallier fil., has been very fully and accurately described and figured by its author, after comparison with the material in the Calcutta Herbarium, in *Bull. de l'herbier Boissier* vi. 351, 622 t. 9, f. 1 a–c (1898). It is a species of § *Eugomphostemma,* group *Melissifolia,* and as its bracts are not longer than the calyx it comes nearest to *G. velutinum* and *G. Mastersii.* The outer bracts are however in shape like those of *G. ovatum* and *G. melissifolium,* so that it stands, as its author has already indicated, intermediate between *G. ovatum* and *G. Mastersii.* It is a native of Eastern Sumatra.

Another point with reference to this genus may be noted in passing. In the account of the genus referred to above, the position of *Gomphostemma flavescens* Miq. was left doubtful. In the following year the writer was able to say that, judging from specimens of the plant (*Anthocoma flavescens* Zoll.) on which Miquel's species is based, kindly lent by Dr. Treub from the Buitenzorg Herbarium, this species was in reality *Cymaria acuminata* Dcne. In 1895 the writer was afforded, though the kindness of M. Drake del Castillo, an opportunity of examining the actual type specimen of *Anthocoma flavescens* and of thus confirming the accuracy of the identification published in *Annals of Botany* vi. 214 (1892).

CHLORANTHACEÆ.

10. CHLORANTHUS KACHINENSIS *King & Prain; leaves* subsessile, ovate, caudate-acuminate, margin finely gland-serrate except at the cuneate base; *anthers 3,* connate by their connective; *spikes* in terminal clusters.

KACHIN HILLS; Shan Busti, Sadon, near water, *Prain's Collector!*

An evergreen erect undershrub; *leaves* glabrous, shining above, dull and finely puberulous on the nerves beneath, nerves about 10 pairs doubly inarched within the margin, length 6–8 in., breadth 3–3·5 in., caudate apex ·75–1 in. long; petiole 1 in. long or 0. *Spikes* 3·5 in. long, 4–6 together, fascicled at the apex of the branches among linear bracts, surrounded by 2 closely approximated, distichous pairs of leaves.

The leaves most resemble those of *C. officinalis* Bl., but the fascicled instead of panicled spikes at once distinguish it. The inflorescence is like that of *C. nervosus* Coll. & Hemsl., from the Shan Plateau, which is however at once distinguished by its coarsely serrate, distinctly petioled and smaller leaves which are not caudate-acuminate at the apex.

LILIACEÆ.

11. Smilax (Eusmilax) Pottingeri *Prain; branches* terete, smooth or with few minute black verrucæ; *leaves* 6–8 in. by 4·5–5·5 in., ovate abruptly acuminate, the narrow tip ·6 in. long, thinly subcoriaceous or chartaceous, very dark green above, glaucescent beneath, 5-costate from the slightly cordate base, petiole 1·5 in, the basal portion ·6 in. long, narrowly sheathing; *peduncles* solitary, axillary, 3·5–4 in. long, very slender, rigidly wiry, terete, smooth, 25–30-flowered; *pedicels* rigid, slender, grooved, ·5–·6 in. long. S. macrophylla *Prain in Rec. Bot. Surv. Ind.* i. 275 (*not of Roxb.*).

Kachin Hills; Myaungjong, *Pottinger*; near Sadon, *Prain's Collector!*

An extensive climber; *leaves* somewhat shining above, with strong secondary reticulations; between the main-nerves on both surfaces; cirrhri slender, wiry, 4–5 in. long, springing from apex of sheathing part of petiole. *Peduncle* springing from a swelling ·2 in. above the petiole, bracts 0, bracteoles at base of pedicels shortly oblong obtuse, pale-brown, palea-like, persistent, making a small globose head ·25 in. across. ♂ flowers not seen. ♀ *Perianth* segments ovate obtuse under ·2 in. long; staminodes 3, style short with 3 stout recurved stigmas. *Fruit* small ·25 in. diam.

A very distinct species coming nearest S. *ferox* and its allies, but not very closely related to any hitherto described Indian species.

AROIDEÆ.

12. Cryptocoryne Cruddasiana *Prain*; *leaves* linear-lanceolate; tube of *spathe* narrow, longer than the limb, limb of spathe lanceolate acute not twisted, distinctly rather distantly transversely plicate within.

Kachin Hills; Keju river, near Sima, *Prain's Collector!*

Tuberous, stoloniferous. *Leaves* 5–8 in. long, ·25–·3 in. wide, rather abruptly acute, lower fourth to third sheathing; midrib distinct. *Scape* very short. *Tube* of spathe 3 in. long, limb 1·25 in. long, lanceolate acuminate, purple within, with transverse folds, ·1 in. apart, crossing its whole inner surface.

A very distinct species, in habit much resembling a small form of C. *ciliata* Fisch., and in this respect unlike any other Indian species of the genus. Its spathe has, however, a limb that is rather longer and much narrower in proportion to the tube than that of C. *ciliata*, while there are no fimbriæ but, instead, there are numerous transverse rugæ as in C. *spiralis* Fisch., which has however different leaves, a twisted limb to the spathe, with a tube much shorter than the limb. This species has been very kindly compared with the material preserved in Herb. Kew by Sir George King and Mr. N. E. Brown.

XIII.—*A list of the Asiatic species of* ORMOSIA.—*By* D. PRAIN.

[Received 21st June ; Read 4th July, 1900.]

On two previous occasions the writer has dealt with the genus *Ormosia* (N.O. *Leguminosæ*) in the Society's *Journal*.[1] The communication of three new forms from the Kachin Hills representing at least two new species, and the presence of two apparently undescribed forms among Chinese collections, renders it advisable to provide the requisite specific descriptions. It may therefore be as well to give at the same time a key to all the known Asiatic species as a preliminary to an exhaustive monograph of the genus.

According to Hooker and Jackson[2] the earliest name for the genus is TOULICHIBA[3] Adans. ; as however, the name ORMOSIA[4] Jacks. is in familiar use it is convenient to retain it. Other generic names have been from time to time applied to one or more species of *Ormosia*. These are LAYIA Hook. and Arn.,[5] CHÆNOLOBIUM Miq.,[6] and, in the writer's opinion, ARILLARIA Kurz.[7] To these Bentham has tentatively added MACROTROPIS DC.,[8] a genus founded by DeCandolle to include two plants from S. China and Cochin China that form the genus *Anagyris* Lour.[9] as opposed to the true *Anagyris* of Linnæus. This tentative reduction has been formally accepted in the *Index Kewensis* but it is not acceptable to the writer because the keel, Loureiro tells us, is in his two species longer than the standard ; this is not the case in any known *Ormosia* and as neither of Loureiro's plants are known to modern students it is better to keep *Macrotropis* separate. These objections, however, do not apply to the reduction of MACROTROPIS Miq. (not of DC.).[10] In the first place Miquel in dealing with the two species which he referred to *Macrotropis* found it necessary to provide for their reception a new section *Amacrotropis*, characterised by having the standard as long as the other petals, that is to say Miquel had to abandon the character that is most distinctive of the true *Macrotropis* before he could accommodate his two species in the genus. In the second place there are authentic examples of both Miquel's plants in Herb. Calcutta and both are true *Ormosias*.

1 J.A.S.B. lxvi. 2, 146 and 467 (1897).
2 Index Kewensis ii. 367.
3 Adans. Fam. ii. 326 (1763).
4 Jackson in Trans. Linn. Soc. x. 360 (1811).
5 Hooker and Arnott, Bot. Beech. Voy. 183, t. 38 (1833).
6 Miquel, Flor. Ind. Bat. Suppl. 302 (1860).
7 Kurz, J.A.S.B. xlii. 2. 70 (1873).
8 DeCandolle, Prodr. ii. 98 (1825).
9 Loureiro, Flor. Cochin-Chin. 260 (1790).
10 Miquel, Flor. Ind. Bat. Suppl. 294 (1860).

ARILLARIA Kurz, has not been accepted as a valid genus by Baker[1] or by the editors of the *Index Kewensis*. The species on which it is founded was treated by Roxburgh,[2] who has left a coloured drawing of the plant in Herb. Calcutta, as a *Sophora*. Wight has reproduced this figure[3] and in discussing it has suggested that the plant is nearer to *Ormosia* than to *Sophora* but that, owing to its having a fleshy pod, it is perhaps a distinct genus. This genus he refrained from founding because the account given by Roxburgh of the arillus was not clear to him. Kurz has confirmed and amplified Roxburgh's account of the arillus and has therefore provided the generic description that Wight did not venture to give. Taubert has adopted Kurz's genus, though his attitude may require to be discounted to some extent, for he at the same time retains among the *Ormosias* the species on which *Arillaria* is based.[4] In spite of the views expressed by Wight, Kurz and Taubert the writer agrees with Baker and Baillon[5] in thinking that the species may quite well be accomodated in *Ormosia*, though he nevertheless thinks the characters of the species (*Ormosia robusta*) are such as to entitle it to the rank of a subgenus.

Bentham has, for convenience, divided the Brazilian species of the genus into two groups,[6] *Concolores* or species with the leaflets glabrous to the naked eye on both sides except perhaps, the midrib, and with the leaves not much paler beneath than above, and *Discolores* with the leaves paler beneath and there manifestly puberulous silky or tomentose. Baker has also, in essence, adopted this method of subdividing the genus and Taubert has even formally adopted Bentham's groups as sections and applied them to the whole genus. This subdivision, however, does not always permit species that are naturally closely related to remain together and it is not improbable that a classification which depends more on the characters derived from fruit and seed and less on characters obtained from the shade of green and the degree of tomentum of the leaves will in future be found more satisfactory.

Below a purely tentative scheme of classification is briefly sketched :—

Pod with woody valves; seeds scarlet with or without a black spot near the hilum not enveloped in an aril; Sub-gen. TOULICHIBA.

Leaf-rachis bearing at its tip the distal pair of leaflets as well as the terminal leaflet; Sect. CHÆNOLOBIUM.

1 Hooker, Flor. Brit. Ind. ii. 252 (1878).
2 Roxburgh, Hortus Bengalensis 31 (1814).
3 Wight, Icones t. 245 (1840).
4 Engler Naturlich. Pflanzenfam. iii. 3. 194 (1894).
5 Baillon, Hist. des Plantes ii. 362 (1869).
6 Martius, Flora Brasil. xv. 1. 315 (1862).

Leaf-rachis prolonged beyond the distal pair of leaflets to support the terminal one ; Sect. ORMOSIA proper.

Pods with thickly woody valves not septate between the large seeds which are usually solitary ; Sub-sect. *Macrodisca.*

Pods with thickly woody valves septate between the small seeds which are usually several ; Sub-sect. *Layia.*

Pods with thinly woody valves and usually solitary always small seeds ; Sub-sect. *Amacrotropis.*

Pod with fleshy valves ; seeds black, enveloped in a fleshy arillus ; Sub-gen. ARILLARIA.

The Asiatic species of which sufficiently. complete material has been reported should be distributed as follows among these groups :—

I. TOULICHIBA.

1. CHÆNOLOBIUM. *O. pachycarpa, O. venosa, O. decemjuga, O. septemjuga, O. polita.*

2. ORMOSIA proper.

   (*a*) Macrodisca. *O. macrodisca, O. gracilis, O. travancorica.*

   (*b*) Layia. *O. emarginata, O. Henryi, O. inopinata, O. laxa, O. glauca, O. Balansae.*

   (*c*) Amacrotropis. *O. microsperma, O. parvifolia, O. sumatrana, O. yunnanensis.*

II. ARILLARIA. *O. robusta.*

The other species given in the subjoined key, which is more or less artificial, at all events in detail, are species of which the fruit is not yet known. In the account of *Ormosia* given in the Society's *Journal,* 1897, a previously undescribed species was there named *O. nitida.* There is however, a prior *O. nitida* Vogel,[1] which stands good ; it has therefore been necessary to rename the Malayan species.

Key to the Asiatic species of Ormosia.

Erect trees :—

Pod with fleshy valves ; seeds with complete arillus ; leaflets glabrous beneath ... ... 1. *robusta.*

Pod with woody valves :—

Seeds with a black adnate basal arillus, leaves minutely sparsely pubescent underneath :—

Panicles fastigiate, flowers white ; pod 8 cm. wide ; seed 2·5 cm. long ... ... 2. *macrodisca.*

Panicles lax, flowers yellow ; pod 3 cm. wide ; seed 2 cm. long ... ... 3. *gracilis.*

[1] Linnæa, xi. 405 (1837).

Seeds with a uniform pink testa and no arillus :—
Leaflets beneath glabrous or only downy along the
 midrib :—
Leaflets 3–5 :—
 Calyx glabrous ; leaflets obovate-oblong obtuse
  or emarginate, base cuneate ; stamens 9 ... **5.** *emarginata.*
 Calyx pubescent ; leaflets elliptic-oblong obtusely
  acuminate, base rounded ; stamens 5 ... **4.** *semicastrata.*
 Leaflets 7–9 ; base rounded ; calyx pubescent :—
 Leaflets narrowly oblong ; rachis prolonged
  beyond distal pair of leaflets :—
 Leaflets dark green gradually narrowed to an
  acute point ... ... **6.** *calavensis.*
 Leaflets pale grey-green, caudate-acuminate :—
 Pod broadly oblong, 5–6 cm. long, 3·5 cm. wide,
  seeds large 2·5 cm. long ... ... **7.** *travancorica.*
 Pod narrowly oblong, 5–7·5 cm. long, 3 cm.
  wide ; seeds small 1 cm. long ... **8.** *glauca.*
 Leaflets broadly oblong, apex rounded or shortly
  abruptly cuspidate ; rachis bearing distal
  pair of leaflets, as well as the terminal leaflet,
  at its tip ... ... **18.** *polita.*
Leaflets beneath more or less persistently hirsute or
 velvety :—
Leaflets with distinct petiolules and the leaf-rachis
 prolonged beyond the distal pair of leaflets :—
Pod large with thickly woody flattened valves ;
 pedicels long, ⅔rd to quite as long as calyx :—
Pods narrowly oblong, 6–7 cm. long, 2·25 cm.
 wide, seeds 1·25 cm. long or less :—
 Leaflets 7–9, thickly coriaceous, glabrous
  above, densely velvety beneath ... **9.** *Henryi.*
 Leaflets 15–17, chartaceous, deciduously
  puberulous above, softly pubescent beneath **12.** *laxa.*
Pods broadly oblong, 5–6 cm. long, 3·5 cm.
 wide ; seeds 1·5 cm. long or longer ; leaflets
 7–11, rarely 5 :—
 Pod glabrous ; racemes even in fruit much
  shorter than the leaves :—
 Corolla pink, leaflets persistently pubescent
  beneath ... ... **10.** *inopinata.*
 Corolla yellow, leaflets glabrescent with age **10b.** *inopinata*
           VAR. *dubia.*
 Pod pubescent ; racemes in fruit as long as
  the leaves ... ... **11.** *Balansae.*
Pod small with thinly woody convex valves, 1·5
 cm. wide ; pedicels less than half as long as
 calyx :—
 Leaflets small, 6 cm. long or shorter, 9–13,
  shortly acuminate ... ... **17.** *parvifolia.*

Leaflets large, 10 cm. long or longer :—

Leaflets thinly pubescent beneath ; panicles lax, bracts small :—

Leaflets 7–9, rarely 5, ovate, obovate or
elliptic, pale-green   ..   ...   14.   *sumatrana.*
Leaflets 13, narrow oblong dark-green   ...   15.   *yunnanensis.*
Leaflets densely pubescent beneath, dark-green 11–13; panicles fastigiate, bracts conspicuous :—

Pod glabrous   ...   .   16.   *microsperma.*
Pod hirsute   ...   .   16b. *microsperma*
VAR. *Ridleyi.*

Leaflets with short petiolules or subsessile, leaf-rachis bearing at its apex the distal pair of leaflets as well as the terminal leaflet : –

Pod with thinly woody valves, 2–2·5 cm. wide :—
Leaflets 13–15, ovate-acute   ...   ...   19.   *septemjuga.*
Leaflets 19–21, lanceolate.acuminate   ...   20.   *decemjuga.*
Pod with thickly woody valves, 3·5 cm. wide; leaflets ovate oblong :—

Leaflets abruptly shortly cuspidate; pod persistently woolly   ...   .   21.   *pachycarpa.*
Leaflets obtuse or subobtuse ; pod glabrous   ...   22.   *venosa.*
Climber ; leaves glabrous beneath, dark green   ...   13.   *scandens.*

ORMOSIA Jacks.

Subgenus 1.   ARILLARIA *Kurz* (pro genere) *Journ. As. Soc. Beng.* xlii. 2. 71.

1.   ORMOSIA ROBUSTA *Baker* in *Hook. fil. Flor. Brit. Ind.* ii. 252 (1878) ; *Taub.* in *Engl. Naturl. Pflanzenfam* iii. 3, 194 (1894).   O. floribunda *Wall. Cat.* 5337 (1832).   Sophora robusta *Roxb. Hort. Beng.* 31 (1814) ; *Wight Icones* t. 245 (1840). Arillaria robusta *Kurz Journ. As. Soc. Beng.* xlii. 2. 71 (1873) and xlv. 2, 224 (1876) and *For. Flor. Brit. Burma* i. 334 (1877) ; *Taub.* in *Engl. Naturl. Pflanzenfam.* iii. 3. 196 (1894).

ASSAM ; Brahmaputra Valley, near foot of Akha Hills, *King's Collector !* Silhet, *Roxburgh* (*Ic.* in Herb., *Calcutta) !* DeSilva (*Wall Cat.* 5337) ! CHITTAGONG ; Kodala Hill, *King's Collector !* BURMA ; Amherst, *Falconer !* Rangoon, *Kurz !* Pegu Yomah, *Kurz !*

Subgenus 2.   TOULICHIBA *Adans.* (pro genere) *Fam.* ii. 326 (1763).

§ EUORMOSIA.   ORMOSIA *Jacks.* (genus) *Trans. Linn. Soc.* x. 360 (1811).

¶ MACRODISCA.

2.   ORMOSIA MACRODISCA *Baker* in *Hook. fil. Flor. Brit. Ind.* ii. 253 (1878) ; *Prain, Journ. As. Soc. Beng.* lxvi. 2, 148 and 467 (1897).

MALAYAN PENINSULA ; Malacca, *Maingay !* Singapore, *Ridley !*

3. ORMOSIA GRACILIS *Prain, Journ. As. Soc. Beng.* lxvi. 2. 148 and 468 (1897).

MALAYAN PENINSULA ; Perak, *Scortechini ! Kunstler ! Wray !*

4. ORMOSIA SEMICASTRATA *Hance, Journ. Bot.* xx. 78 (1882) ; *Forbes & Hemsl.* in *Journ. Linn. Soc.* xxiii. 204 (1887).

CHINA ; Hongkong, *Ford,* fide Hance.

This species is not yet represented in Herb. Calcutta.

5. ORMOSIA EMARGINATA *Benth.* in *Hook. Kew. Journ.* iv. 77 (1852), and *Flor. Hong-Kong.* 96 (1861) ; *Forbes & Hemsl.* in *Journ. Linn. Soc.* xxiii. 204 (1887).

CHINA ; Hongkong, *Ford !*

6. ORMOSIA CALAVENSIS *Azaola* in *Blanco Flor. Filip.* ed. 2, 230 (1845) ; *Vid. Sinops.* t. 41, f. H(1883) and *Rev. Pl. Vasc. Filip.* 113 (1886).

PHILIPPINES ; Luzon, *Cuming* 1219 ! Alabat, *Vidal* 2617 !

Vidal y Soler suggests that this is the same as *Ormosia* (Arillaria) *robusta,* but the suggestion can only be explained on the assumption that Sen. Vidal had no good specimens of *O. robusta* before him. There are no fruits of this species in Herb. Calcutta ; if their structure is like that of *O. robusta* this species must be transferred to the subgenus *Arillaria.*

7. ORMOSIA TRAVANCORICA *Bedd. Flor. Sylvat.* i. t. 45 (1869) ; *Baker* in *Hook. fil. Flor. Brit. Ind.* ii. 253 (1878).

S. INDIA ; S. Canara, Tinivelly, Travancore, *Beddome* (Ic.)

This species is only represented at Calcutta by Beddome's figure.

¶¶ LAYIA *Hook. & Arn.* (pro genere) *Bot. Beech. Voy.* 183 t. 38 (1833).

8. ORMOSIA GLAUCA *Wall. Plant. As. Rar.* ii. 23. t, 125 (1831) and *Cat.* 5338 (1832) ; *Baker* in *Hook. fil. Flor. Brit. Ind.* ii. 253 (1878) ; *Gamble, Man. Ind. Timb.* xvii. (1881) and *Darjeel. List,* Ed. 2. 30 (1896) ; *Prain, Journ. As. Soc. Beng.* lxvi. 2. 467 (1897).

NEPAL ; Sonku, *Wallich !* SIKKIM ; Sivoke, 2500 ft., *Gamble !*

9. ORMOSIA HENRYI *Prain ;* leaflets 7–9, oblong, pale green, shortly stalked, thickly coriaceous, glabrous above, velvety beneath, pedicels as long as the calyx, pod narrow oblong, valves thick woody.

CHINA ; Hupeh, *Henry* 7577 !

A tree, with tawny-velvety branches. *Leaflets* usually 7, oblong lanceolate, very firmly coriaceous, 8–10 cm. long, 3–4·5 cm. wide, quite glabrous above, densely pale-buff velvety beneath, apex acute, base rounded, veins 8–9 pairs slender, somewhat prominent beneath ; petiolules 5 mm. and main rachis 8–9 cm., closely shortly tawny pubescent. *Flowers* in axillary racemes 8–9 cm. long, tawny pubescent as are the pedicels 1·25 cm. long, bracts and bracteoles deciduous. *Calyx* campanulate 6 mm. long, silky. *Corolla* and *Stamens* not seen. *Pod* hard thick, 5–7 cm. long, 2·5 cm.

wide, the valves black, smooth externally, slightly swollen opposite the ripe seeds, very faintly ribbed alongside the upper suture, seeds 2–5, bright scarlet, small, 1 cm. long, ·75 cm. wide, ·5 cm. thick, separated by partitions of the tawny suberous endocarp in which they are embedded, with no trace of arillus.

Nearest *O. glauca* Wall. but differing greatly in the velvety under-surface of the leaves.

10.   ORMOSIA INOPINATA *Prain;* leaflets 9, less often 11 or 7, rarely 5, ovate acuminate, beneath softly closely tawny pubescent on the midrib and veins, elsewhere sparsely pubescent, leaf-rachis and branchlets velvety, veins beneath prominent finely reticulate, large, distinctly stalked ; pedicels long ; pod compressed with thick woody valves.

VAR. TYPICA ; corolla reddish, leaflets persistently pubescent.

KACHIN HILLS ; Bansparao, near Sadon, *Prain's Collector !*

A large tree, with closely tawny-velvety sulcate branches. *Leaflets* rigidly subcoriaceous 15–16 cm. long, 5–6·5 cm. (the terminal leaflet sometimes 8 cm.) wide, above with midrib at first pubescent at length quite glabrous, rather pale-green shining, beneath persistently tomentose but the tomentum sparser with age, veins 7–9 pairs prominent beneath with a fine secondary reticulation visible also above especially on younger leaves, apex abruptly acuminate, base cuneate or rounded ; petiolules 8 mm and leaf-rachis 22 cm. long, closely velvety. *Flowers* in axillary racemes or few-branched panicles 20 cm. long closely velvety as are the pedicels 6 mm. (in fruit over 1 cm.) long, bracts and bracteoles minute deciduous, velvety. *Calyx* campanulate 9 mm. long, closely velvety both externally and within, teeth wide-triangular rather longer than the tube. *Corolla* reddish, twice as long as calyx. *Stamens* usually 9, all fertile, anthers oblong versatile. *Ovary* shortly stipitate, glabrous except for a few hairs on the dorsal and rather more on the ventral suture ; style glabrous, filiform, tip circinate ; stigma oblique ; ovules 4 or 3. *Pod* hard flattened, with woody valves, 6 cm. long, 3 cm. wide, 1·25 cm. thick, with faint depressions between the 3 or 4 seeds, obliquely ovate-oblong, with a distinct stipe 6 mm. long and minute tip at apex of diagonal axis remote from stipe, ventral suture with prominent parallel ridges 6 mm. apart projecting beyond level of line of dehiscence ; seeds cinnabar-red with a small white hilum and no arillus, ovate, 1·5 cm. long, 1 cm. across : sometimes slightly compressed and only 7 mm. thick.

10*b*.   VAR. DUBIA ; flowers yellow, leaves glabrescent with age on the under surface.

KACHIN HILLS ; Bomkatom, between Lashio and Sadon, *Prain's Collector !*

A large tree, branchlets faintly sulcate. *Leaflets* rigidly subcoriaceous 6–10 cm. long, 3–4·5 cm. wide, the terminal leaflet almost 5 cm. wide, light-green, glabrous shining above, pale beneath very sparsely persistently pubescent, veins 7–9 pairs prominent beneath as is the fine secondary venation which is hardly visible above, apex acuminate, base cuneate or rounded ; petiolules 6 mm. and leaf rachis 15 cm. long at first pubescent at length glabrous. *Flowers* in axillary racemes about 8 cm. long, rachis finely velvety as are the pedicels 6 mm. long not elongated in fruit. *Calyx* campanulate closely velvety both externally and within ; teeth wide-triangular, rather longer than the tube. *Corolla* yellowish white, twice as long as calyx.

Stamens and *ovary* as in *O. inopinata.* *Pod* hard flattened, with woody valves, 5 cm. long, 3 cm. wide, 1·25 cm. thick, somewhat swollen opposite the 1–2 seeds, ovate-acute with distinct stipe 6 mm. long and a prominent tip at apex of vertical axis remote from stipe, ventral suture with blunt parallel ridges 6 mm. apart not projecting beyond level of line of dchiscence; seeds cinnabar-red with a small white hilum and uo arillus, 1·25 cm. long, 1 cm. across, 8 mm. thick.

The foliage of the two trees here treated as varieties of one species is hardly distinguishable and the structure of their flowers is identical. The Native Collector who has communicated the specimens of both states, however, that besides the differences in colour of petals and in shape of pods and seeds, the two trees as they grow look very different. If this should turn out to be the case it may be necessary to treat the variety here described as a distinct species, to be known as *Ormosiu dubia.*

11. ORMOSIA BALANSAE *Drake del Castillo, Journ. de Botan.* v. 215 (1891).

TONKIN; near Ta-phap, in forests, *Balansa* 2178.

This species is not yet represented in Herb. Calcutta.

12. ORMOSIA LAXA *Prain;* leaflets 15, less often 17, lanceolate-acuminate, beneath and leaf-rachis and branchlets velvety, veins beneath inconspicuous, medium, distinctly stalked; pedicels long.

KACHIN HILLS; Shan Busti near Sadon, 5000 ft., *Prain's Collector !*

A tree, with tawny-velvety branches. *Leaflets* lanceolate, chartaceous, 6 cm. long, 2·5 cm. wide, at first finely deciduously puberulous above, densely softly tawny-velvety beneath, veins 5-6 pairs slender not prominent beneath, apex acuminate, tapering from the middle, base cuneate in the lower fourth; petiolules 5 mm. and main-rachis 20–25 cm. long, densely tawny-velvety. *Flowers* in axillary racemes or few-branched panicles 8–12 cm. long, densely tawny-velvety as are the pedicels 1·25–1 65 cm. long, bracts and bracteoles minute, deciduous, velvety. *Calyx* campanulate densely tawny-velvety outside, finely pubescent within, 1 cm. long, teeth wide-triangular almost as long as tube. *Corolla* twice as long as calyx. *Stamens* usually 5 fertile exserted in the open flower, sometimes 6 or 7, rarely 8 fertile—if 5, 6 or 7 fertile then with 3, 2, or 1 staminodes, always 2 stamens quite obsolete; anthers oblong, versatile. *Ovary* stipitate, silky with long tawny hairs especially on the sutures; style glabrous filiform, tip circinnate, stigma oblique; ovules 7. *Pod* 6–7 cm. long, 2·5 cm. wide, the valves black, smooth externally, slightly swollen opposite the ripe seeds, very faintly ribbed alongside both sutures; *seeds* 2-4, bright scarlet, 1 cm. long, 8 mm. wide, 6 mm. thick, separated by partitions of the pale woody endocarp in which they are embedded, with no arillus.

This very distinct species cannot be confounded with any of the hitherto described *Ormosias.*

13. ORMOSIA SCANDENS *Prain, Journ. As. Soc. Beng.* lxvi. 2, 147 and 467 (1897).

MALAYAN PENINSULA; Perak, *Kunstler !*

This species is distinguished from all the others by its climbing habit; as its fruit is not yet known its precise systematic position cannot be positively stated. It seems, however, as if it might prove to be a species of § *Layia.* It may be ultimately found advisible to subdivide § *Layia* into two groups; those with thick-walled large pods going into one and those with thin-walled short pods being placed in the other.

¶¶¶ AMACROTROPIS *Miq.* (pro sectione) *Flor. Ind. Bat. Suppl.* 294 (1860).

14. ORMOSIA SUMATRANA *Prain, Journ. As. Soc. Beng.* lxvi. 2, 150 and 469 (1897). Macrotropis sumatrana *Miq. Flor. Ind. Bat. Suppl.* 294 (1860).

MALAY ARCHIPELAGO ; Sumatra, *Teysmann* 3618 ! *Forbes* 2592 ! 2648 ! MALAYAN PENINSULA ; Malacca, *Holmberg !*

15. ORMOSIA YUNNANENSIS *Prain ;* leaflets 13, short-stalked, veins beneath distinctly raised, pedicels shorter than the calyx, pod sub-compressed with thin valves, seed ovate; racemes in rather close panicles.

CHINA ; Yunnan, mountains in western Szemao, 5,000 ft. elev., *Henry* 11,967 !

A small tree 20 ft. high, with rusty-pubescent branches. *Leaflets* oblong-lanceolate, firmly coriaceous, 10 cm. long, 3 cm. wide, glabrous above softly sparsely pubescent with longish adpressed ash-grey hairs beneath, veins about 10 pairs slender but prominent beneath, depressed above, secondary venation indistinct beneath not visible above, apex acute with a short finely acuminate sub-mucronulate tip, base shortly cuneate, petiolules 3·5 mm. and main-rachis 16 cm. long, rusty-pubescent. *Flowers* in axillary branched panicles 14 cm. long, rusty-pubescent as are the pedicels 2 mm. long, bracts and bracteoles ovate, 2·5 mm. long, 1·5 mm. wide, acute, deciduous, rusty pubescent. *Calyx* campanulate, rusty-pubescent both outside and inside, 8 mm. long, teeth triangular hardly as long as the tube. *Corolla* and *Stamens* not seen. *Pod* subsessile, irregularly orbicular if 1-seeded, oblong if 2-seeded, with a broadly triangular unilateral tip, 1·3 cm. wide, 2–3 25 cm. long, lineate between the seeds; valves thin, woody, rigid, black, glabrous, swollen opposite the seeds; *seeds* 1 or 2, bright scarlet, 8 mm. long, 6 mm. wide, 5 mm. thick, with white hilum and no arillus.

This species is most nearly related to *O. sumatrana ;* its chief interest lies in its being the most northerly representative of the ¶ *Amacrotropis*, all the other known members of which are Malayan.

16. ORMOSIA MICROSPERMA *Baker* in *Hook. fil. Flor. Brit. Ind.* ii. 253 (1878) ; *Prain, Journ. As. Soc. Beng.* lxvi. 2, 151 and 468 (1897). O. coarctata [*Benth. Mss.*] ; *Kurz, Journ. As. Soc. Beng.* xlii. 2, 71 (1871) *in part, not of Jacks.*

MALAYAN PENINSULA ; Malacca, *Griffith ! Maingay ! Derry !* Perak, *Kunstler !*

16b. VAR. RIDLEYI *Prain, Journ. As. Soç. Beng.* 2. 150 and 469 (1897).

MALAYAN PENINSULA ; Singapore, *Ridley !*

This " variety " is probably entitled to specific rank.

17. ORMOSIA PARVIFOLIA *Baker* in *Hook. fil. Flor. Brit. Ind.* ii. 253 (1878); *Prain, Journ As. Soc. Beng.* lxvi. 2. 149 and 469 (1897). Macrotropis? bancana *Miq. Flor. Ind. Bat. Suppl.* 295 (1860).

MALAY ARCHIPELAGO; Borneo, *Haviland* 57! Bangka, *Teysmann* 3405! MALAYAN PENINSULA; Singapore, *Ridley* 5929! 8096! Pahang, *Ridley* 1267! 5013! Malacca, *Griffith! Maingay! Goodenough!*

Besides being a very well characterised species this is much more widely distributed than most of the *Ormosias.* An authentic specimen of Miquel's *Macrotropis? bancana* in Herb. Calcutta shows that it is the same thing as Baker's *Ormosia parvifolia.* If the rule that the oldest specific epithet must under all circumstances be conserved is to be rigidly applied, then Mr. Baker's name must be abandoned in favour of the name *Ormosia bancana.*

§§ CHÆNOLOBIUM *Miq.* (pro genere) *Flor. Ind. Bat. Suppl.* 302 (1860).

18. ORMOSIA POLITA *Prain.* O. nitida *Prain, Journ. As. Soc. Beng.* lxvi. 2, 149 and 488 (1897) *not of Vogel.*

MALAYAN PENINSULA; Perak, *Kunstler!*

When a description was given of this very distinct species the fact was overlooked that there is already an *Ormosia nitida* Vogel, from Brazil; the name must therefore be replaced by another. This particular species is unlike the other *Chænolobia* is having perfectly glabrous dark-green shining leaves, and it moreover resembles the Malayan, as opposed to the Indo-Chinese *Layiæ* in having small pods. It also differs from the other *Chænolobia* in having well developed petiolules. Its agreement with *Chænolobium* lies in the fact that the leaf rachis is not prolonged beyond the last pair of leaflets which are attached along with the terminal leaflet.

19. ORMOSIA SÉPTEMJUGA *Prain, Journ. As. Soc. Beng.* lxvi. 2. 468 (1897). O. coarctata *Kurz, Journ. As. Soc. Beng.* xlii. 2. 71 (1872) *in part, not of Jacks.* Chaenolobium septemjugum *Miq. Flor. Ind. Bat. Suppl.* 302 (1860).

MALAYAN ARCHIPELAGO; Sumatra, *Diepenhorst* 2547!

An authentic specimen of Miquel's plant is preserved in the Calcutta Herbarium.

20. ORMOSIA DECEMJUGA *Prain, Journ. As. Soc. Beng.* lxvi. 2. 468 (1897). Chaenolobium decemjugum *Miq. Flor. Ind. Bat. Suppl.* 302 (1860). O. coarctata *Kurz, Journ. As. Soc. Beng.* xlii. 2, 71 (1872) *in part, not of Jacks.*

MALAYAN ARCHIPELAGO; Sumatra, *Teysmann* 3715!

An authentic specimen of Miquel's plant is in Herb. Calcutta. This specimen shows, in my opinion, that Kurz was not justified in supposing that this is the same as *Chenolobium septemjugum* and that further he was not justified in believing that either this or O. *septemjugum* is the same as *Ormosia microsperma* which he supposed to be the same thing as O. *coarctata* Jacks., a Guiana species.

21. ORMOSIA PACHYCARPA *Champ.* ex *Benth.* in *Hook. Kew. Journ.* iv. 76 (1852); *Benth., Flor. Hong Kong.* 96 (1861); *Forbes & Hemsl.* in *Journ. Linn. Soc.* xxiii. 204 (1887).

CHINA; Canton, *Reeves*, Hong-Kong, *Lamont, Champion, Ford!*

Bentham states that this species was found by Reeves at Canton as well as by various collectors in Hong-Kong. Hemsley, however, says that Reeves' specimens are without locality. This species is very closely related to the next, though the two are nevertheless specifically quite distinct.

22. ORMOSIA VENOSA *Baker* in *Hook. fil. Flor. Brit. Ind.* ii. 254 (1878) ; *Prain, Journ. As. Soc. Beng.* lxvi. 2. 152 (1897).

MALAY PENINSULA ; Malacca, *Maingay !*

A very distinct species, the one to which it is most closely related being the preceding, which comes from a very remote locality.

As regards distribution the most striking features connected with *Ormosia* are (1) the wide-spread occurrence of this genus throughout South Eastern Asia, from Hupeh in China to Bangka in the Malayan Archipelago, and from Travancore and Nepal to the Philippines : (2) the remarkably limited range of individual species with the exception of *O. parvifolia* (*O. bancana*) which extends from the Malay Peninsula to Bangka and Borneo, and to a less extent of *O. sumatrana* which occurs on both sides of the Straits of Malacca. *O. robusta* also has a wider range than most of the species for it extends from the valley of Assam through Silhet and Chittagong to Pegu and Tenasserim. It is interesting, however, to note that very closely related species such as *O. pachycarpa* and *O. venosä,* and again *O. yünnänensis* and *O. sumatrana,* may occur in widely separated localities. In the first instance one of the closely allied species is a native of Hong-Kong, the other is a native of Malacca; the specific names of the other pair indicate their respective habitats.

From the subjoined tabular statement it will be seen that of the 22 species enumerated one is S. Indian; one Himalayan ; two, but one of these with two quite distinct varieties, occur in the Kachin Hills ; five occur in China ; one in Tonkin ; one in the Philippines ; one in Borneo, though this species also occurs in Bangka and throughout the Malay Peninsula ; three in Sumatra, though one of these also occurs in Malacca ; and eight in the Malay Peninsula, though one of these extends to Bangka and Borneo and another extends to Sumatra. The remaining species, which forms a very distinct subgenus, is widely spread from Assam to Tenasserim in a region where no other species occurs—a region moreover which separates the two chief centres of the genus in South Eastern Asia, *viz.* :—the Kachin-S. China area, and the Malay Peninsula. It is further worthy of remark that, so far, no species has been recorded either from Java or from Ceylon.

Table of distribution of the South-Eastern-Asiatic species of Ormosia.

Species.	S. India.	C. and E. Himalaya.	Kachin Hills.	S. China.	Tonkin.	Assam and Chittagong.	Pegu and Tenasserim.	Sumatra.	Malay Peninsula.	Bangka.	Borneo.	Philippines.
O. travancorica	1	...	...	...	...	...	...	...	...	...	...	...
O. glauca	...	1	...	...	...	...	...	...	...	...	...	...
O. laxa	...	...	1	...	...	...	...	...	...	...	...	...
O. inopinata	...	...	1	...	...	...	...	...	...	...	...	...
O. Henryi	...	...	...	1	...	...	...	...	...	...	...	...
O. emarginata	...	...	...	1	...	...	...	...	...	...	...	...
O. semicastra	...	...	...	1	...	...	...	...	...	...	...	...
O. pachycarpa	...	...	...	1	...	...	...	...	...	...	...	...
O. yunnanensis	...	...	...	1	...	...	...	...	...	...	...	...
O. Balansae	...	...	...	...	1	...	...	...	...	...	...	...
O. robusta	...	...	...	...	...	1	1	...	...	...	...	...
O. septemjuga	...	...	...	...	...	...	...	1	...	...	...	...
O. decemjuga	...	...	...	...	...	...	...	1	...	...	...	...
O. sumatrana	...	...	...	...	...	...	...	1	1	...	...	...
O. microsperma	...	...	...	...	...	...	...	...	1	...	...	...
O. scandens	...	...	...	...	...	...	...	...	1	...	...	...
O. polita	...	...	...	...	...	...	...	...	1	...	...	...
O. gracilis	...	...	...	...	...	...	...	...	1	...	...	...
O. macrodisca	...	...	...	...	...	...	...	...	1	...	...	...
O. venosa	...	...	...	...	...	...	...	...	1	...	...	...
O. parvifolia	...	...	...	...	...	...	...	...	1	1	1	...
O. calavensis	...	...	...	...	...	...	...	...	...	...	...	1
	1	1	2	5	1	1	1	3	8	1	1	1

XIV.—*The Food-plants of the Butterflies of the Kanara District of the Bombay Presidency, with a Revision of the Species of Butterflies there occurring.*—By LIONEL DE NICÉVILLE, F.E.S., C.M.Z.S., &c.

[Received 23rd June; Read 4th July, 1900.]

In the Journal of the Bombay Natural History Society, vol. v, pp. 260–278, 349–375, plates A, B, C, D, E and F (1890), will be found a paper entitled "Notes on the Larvæ and Pupæ of some of the Butterflies of the Bombay Presidency," by J. Davidson, Bo. C.S., and E. H. Aitken. In the same Journal, vol. x, pp. 237–259, 372–393, 568–584, vol. xi, pp. 22–63, plates I, II, III, IV, V, VI, VII, and VIII (1896–98), appears a paper, which is practically a continuation of the same subject, under the title "The Butterflies of the North Canara District of the Bombay Presidency," by J. Davidson, T. R. Bell, and E. H. Aitken. The present paper is the third contribution to the subject, and is almost entirely based on the observations of Mr. T. R. D. Bell which have been placed at the disposal of the writer; but all the information contained in the above-cited papers regarding food-plants has been herein incorporated as well. The object of the paper is to give all the known food-plants of the butterflies bred by the three writers above mentioned in a compact form, and at the same time to give a revised list of the butterflies of the District of Kanara. In the first list the food-plants are arranged in botanical order, the order adopted being that of "The Flora of British India," in seven volumes, by Sir J. D. Hooker (1872–1897). In the second list the butterflies found in the Kanara district are arranged in order, with the food-plants of the larvæ given where known below each. Very large additions have been made to the known food-plants, nearly all of which have been discovered by Mr. Bell. It is probable that no single person has ever bred such a variety of species of butterflies in one tropical locality as Mr. Bell has done. The omissions are very few, and these Mr. Bell is trying to supply. In the list of the butterflies it will be noticed the *Aphnæus concanus*, Moore, which is a dry-season form of *A. lohita*, Horsfield, and *Baoris philotas*, de Nicéville, which Mr. Bell considers to be a small variety of *Baoris guttatus*, Bremer and Grey, caused by the larva having been starved, have been omitted. Their names appeared on pp. 386, n. 119, and 47, n. 207 of the second paper cited above; while *Nacaduba plumbeomicans*, Wood-Mason and de Nicéville, and *N. atrata*, Horsfield; *Halpe moorei*, Watson, and *H. ceylonica*, Moore; *Notocrypta feisthamelii,* Boisduval, and *N. restricta,* Moore, have been united, as these pairs of names are considered by the writer to represent but a single species in each instance. A considerable number of species of butterflies new to the lists have been added, bringing up the list from 233 (which includes the five species mentioned above as now omitted) to 245 species occurring in the Kanara district. The nomenclature has been brought up to date, the reasons for changing the published names being given in all cases. The importance of the Lists of Food-plants here presented need hardly be pointed out. No species of butter-fly can be said to be known otherwise than superficially by a study of its perfect or imago form alone; much more than this is wanted; its egg, larva in all stages, and pupa, should be studied, described, and if possible compared with the transformations of the allied species at all stages. In India, and probably in all tropical countries, seasonal dimorphism occurs to a very great and often unexpected extent, and this phenomenon can only be worked out fully by extensive breeding ex-periments at all seasons of the year, but more especially at the changes of the seasons, from dry to wet and from wet to dry. The lists here given will be of use not only to the student and collector of butterflies in the south-western littoral of India, but to students and collectors of butter-flies in most parts of India and to a less extent elsewhere, many of the species enumerated being very widely spread. It will even be of value as regards other species in the same genus or allied genera, as these allied species and genera will frequently be found to feed either on the same or on allied plants elsewhere. For breeding purposes a knowledge of the food-plant of the larva is half the battle, given a knowledge of that fact, success in breeding becomes almost a certainty. It is in many cases, probably in most, only necessary to shut up the female of any given species of butterfly in as natural conditions as possible with its food-plant for it to lay eggs; the rest requires only some care and attention. The writers of the series of papers on Kanarese butterflies have mainly relied on actually finding the larvæ of the different species to breed them successfully, while the present writer has been successful chiefly in breeding from eggs laid by captured butterflies. To find the food-plant of any particular butterfly often entails much patient watching of the females when ovipositing. Mr. Bell has discovered the transformations of several species not hitherto described by himself, Messrs. Davidson and Aitken. These new descriptions will be found when dealing with the butterflies of the Kanara district in the second list.

   The larvæ of all the butterflies enumerated feed on vegetable food, except that of *Spalgis epius,* Westwood, which feeds on *Coccidæ;* these are white, fluffy, onisciform insects commonly found on the young parts of many kinds of plants. They are often called "mealy-bugs" and "plant-lice," though the latter name is more properly applied to the *Aphidæ.* The larvæ of many kinds of butterflies will, when they cannot get vegetable food, eat each other or soft newly-formed pupæ. Mr. Bell has found that the greatest cannibals in this respect are the larvæ of certain *Lycænidæ,* and the worst amongst these again are the larvæ of *Zesius chrysomallus,* Hübner, for these will at times, even when plentifully supplied with their proper vegetable food, eat any larvæ which may be in a fit state to be eaten, *i.e.,* which are either on the point of casting their skins, have just cast them, or are just going to pupate. The lycænid larvæ which are most addicted, after that of *Z. chrysomallus,* to cannibalism, are those of the *Amblypodia* and *Tajuria* groups, those of *Arrhopala* and *Rapala* being nearly as bad. He has known one larva of *Tajuria cippus,* Fabricius, to eat up over a dozen young ones of its own species. In Kashmir Mr. Bell bred a single imago of *Hysudra selira,* Moore, from a larva which had been reared on the dead leaves and flowers of its food-plant, *Indigofera atropurpurea,* Hamilt. (Natural Order *Leguminosæ*), together with several newly-formed pupæ of its own species. The imago was a very fine, large specimen, so that the insect diet evidently agreed with the larva. Mr. Bell particularly noted this fact, as in all his previous experience he had been led to the conclusion that a cannibal diet was bad for the stomachs of the larvæ practising the habit of eating up its fellows, as they, as a general rule, have not been healthy, and have died before pupating. The tendency to cannibalism is not confined to the *Lycænidæ,* but exists also amongst the *Pierinæ*; the larvæ of *Appias* will eat each other and any other species of larvæ feeding on the same food-plant as themselves if forced to it by hunger. He has seen the larvæ of *Appias libythea,* Fabricius, and *Appias taprobana,* Moore, eat freshly-formed pupæ of their own species, as well as larvæ changing their skins, and also the larvæ and pupæ of *Leptosia xiphia,* Fabricius. Some of the caterpillars of the *Danainæ* will, when food is not to be had, eat individuals of their own species. Mr. Bell has never known a larva to eat another larva feeding on a food-plant of a species different from its own, so it is probable that all larvæ taste strongly of the plant they feed on, and it is also probable that cannibal larvæ are hardly conscious that they are eating up each other, being only guided to their proper food by the sense of taste, or possibly to a less extent by the sense of smell. None of the larvæ of the *Satyrinæ, Elymniinæ, Amathusiinæ, Acræinæ, Nymphalinæ, Libythæinæ, Nemeobiinæ, Papilioninæ,* or *Hesperiidæ,* have been found by Mr. Bell to eat anything but vegetable food. All rhopalocerous larvæ, however, with but very few exceptions, eat their own cast-off skins while these are still soft and moist; and the young larvæ on emerging from the egg will almost invariably under normal conditions make their first meal off the empty egg-shell. Mr. Bell notes that all the butterfly larvæ he has bred change their skins five times from the time they leave the egg to the time they turn to pupæ.

As regards the larvæ of the *Lycænidæ*, whether they are attended by ants or not, in may be noted that those which live in harmony with ants, and are probably largely dependant on their well-being on ants, the presence of the particular species of ant that lives with any particular species of butterfly larva often fixes the choice of the butterfly laying her eggs on a particular plant or not. If the right plant has no ants, or the ants on that plant are not the right species, the butterfly will lay no eggs on that plant. Some larvæ will certainly not live without the ants, and many larvæ are extremely uncomfortable when brought up away from their hosts or masters. In many cases it is just as important for breeding purposes to know the right species of ants as to know the right food-plant. In Kanara this is particularly noticeable in the cases of *Castalius ananda*, de Nicéville, *Zesius chrysomallus*, Hübner, *Aphnæus lohita*, Horsfield, and *Catapœcilma elegans*, Druce. *C. ananda* is "protected" by ants of the genus *Crēmastogaster*. On one occasion Mr. Bell was collecting larvæ at Katgal, and the ants were principally on *Zizyphus rugosa*, Lamk. (Natural Order *Rhamneæ*) but were also swarming all over six or seven species of different trees all round, and on all of these trees there were larvæ of *C. ananda* covered with ants and eating the leaves of the trees in all cases. Since then Mr. Bell has noticed the larva of *C. ananda* eating the leaves of many different plants and always in company with the same° species of ants. With regard to the *Zesius, Aphnæus* and *Catapœcilma* mentioned above, the female butterflies first look for the right species of ants, and the species of food-plant seems to be quite a secondary consideration, at any rate to a considerable extent. The larvæ of *Zesius* may be found on very nearly any plant that harbours the large red ant, *Œcophylla smaragdina*, Fabricius; so much so that Mr. Bell has often had a strong suspicion that the butterfly larvæ will occasionally eat the ant larvæ, although he has not actually seen them do so. The larva of this butterfly feeds on many species of plants not recorded in the lists, as Mr. Bell made no particular note of them, all these plants being affected by the large red ants. The larvæ of *Aphnæus* and *Catapœcilma* are only found on plants affected by ants of the genus *Cremastogaster*. As regards the four species of butterflies named above, the larvæ are often found in the ants' nests, and their pupæ also, but not invariably.

Mr. Bell has furnished me with the following detailed information on the subject of lycænid butterflies and ants :—

1. *Neopithecops zalmora*, Butler.   The larva but not the pupa is sometimes attended by ants, generally by a species of *Pheidole*.

2. *Cyaniris puspa*, Horsfield.   The larva is always attended by small ants of the genus *Cremastogaster*, but will live comfortably without them ; the pupa is not generally attended.

3. *Lycænesthes emolus*, Godart.   The larvæ, which are gregarious, are always attended by the common and large fierce red ant, *Œcophylla smaragdina*, Fabricius.

4. *Jamides bochus*, Cramer.   The larvæ are sometimes attended by ants of the genus *Cremastogaster*.

5. *Lampides celeno*, Cramer.   Sometimes attended by ants.

6. *Euchrysops pandava*, Horsfield.   Larva attended by ants generally.

7. *Castalius ananda*, de Nicéville.   Larva and pupa always strongly attended by ants of the genus *Cremastogaster*, and will not live well without them.

8. *Polyommatus bœticus*, Linnæus.   Larva sometimes attended by ants of the genus *Cremastogaster*.

9. *Surendra quercetorum*, Moore.   The imago may often be seen settled on branches of trees and bushes swarming with ants of the genus *Cremastogaster*, and being caressed by them ; the larva is attended by the same ants.

10. *Thaduka multicaudata*, Moore.   The larvæ and pupæ are gregarious, and are sometimes attended by ants of the genus *Cremasto-gaster*, and by *Œcophylla smaragdina*, Fabricius.

11. *Arrhopala centaurus*, Fabricius.   Both larva and pupa are always attended by *Œcophylla smaragdina*, Fabricius.

12. *Arrhopala amantes*, Hewitson.   Both larva and pupa are always attended by *Œcophylla smaragdina*, Fabricius.

13. *Arrhopala canaraica*, Moore.   Larva and pupa always attended by *Œcophylla smaragdina*, Fabricius, or by a small blackish ant.

14. *Arrhopala bazalus*, Hewitson.   Larva and pupa always attended by ants of the genus *Cremastogaster*.

15. *Curetis thetis*, Drury.   Mr. Bell has found hundreds of the larvæ of this species, but not one was attended by ants.   No doubt the long " bottle-brush " extrusible processes with which the larva is furnished are used to drive away ants as well as ichneumons.

16. *Zesius chrysomallus*, Hübner.   The larva is always attended by *Œcophylla smaragdina*, Fabricius, and will not live well without them.

17. *Aphnæus vulcanus*, Fabricius. Larva and pupa always attended by ants of the genus *Cremastogaster*.

18. *Aphnæus lohita*, Horsfield. Larva and pupa sometimes attended by *Œcophylla smaragdina*, Fabricius, and always by ants of the genus *Cremastogaster*.

19. *Tajuria indra*, Moore. Larva always attended by ants, but not to a very great extent.

20. *Tajuria cippus*, Fabricius. The larva is rarely attended by ants, although the female butterfly always lays her eggs on plants which are frequented by ants of the genus *Cremastogaster*.

21. *Catapœcilma elegans*, Druce. The larva and pupa are always attended by swarms of ants of the genus *Cremastogaster*, in fact they are found both in the permanent nests of those ants, and in small temporary nests formed by them on the branches, which latter are generally made to shelter scale insects.

22. *Deudorix epijarbas*, Moore. The larva is sometimes attended by ants of the genus *Cremastogaster*.

23. *Zinaspa todara*, Moore. Larva always attended by *Œcophylla smaragdina*, Fabricius, or ants of the genus *Cremastogaster*, and lives badly without them.

24. *Rapala schistacea*, Moore. The larva is desultorily attended by ants of the genus *Cremastogaster*.

25. *Rapala lankana*, Moore. Larva always attended by *Œcophylla smaragdina*, Fabricius.

26. *Virachola isocrates*, Fabricius. Ants are sometimes found in the fruits that contain larvæ of this butterfly, but do not seem to actually attend them.

27. *Virachola perse*, Hewitson. The same remarks apply to this species as to the last.

In many instances it will be found that the names of the plants given in the two lists below differ in spelling from that given in the two previously published papers on Kanarese butterflies. In the present paper all the names have been carefully revised, and the spelling herein given should be followed. It was not thought necessary also to draw particular attention to these variants, or to those cases in which the names of the butterflies were also incorrectly spelt.

At the end of Parts I and II will be found lists of the food-plants, and the butterflies whose larvæ feed on them, of a few butterflies discovered by Mr. Bell in the Western Himalayas and Kashmir.

PART I.

A List of the Food-plants arranged in the order of "The Flora of British India" on which the larvæ of the Butterflies of the Kanara District feed.

Order IV. ANONACEÆ.

1. UNONA DISCOLOR, Vahl.
Papilio agamemnon, Linnæus.

2. UNONA LAWII, Hook. f. and T.
Papilio eurypylus jason, Esper.
Papilio antiphates alcibiades, Fabricius.

3. POLYALTHIA LONGIFOLIA, Benth. and H. f.* The 'debdar,' or 'asoka.'
Papilio agamemnon, Linnæus.

4. ANONA SQUAMOSA, Linn. The custard-apple.
Virachola perse, Hewitson.
Papilio agamemnon, Linnæus.

5. ANONA RETICULATA, Linn.
Papilio agamemnon, Linnæus.

6. SACCOPETALUM TOMENTOSUM, Hook. f. and T.
Charaxes imna, Butler.
Papilio agamemnon, Linnæus.
Papilio eurypylus jason, Esper.
Papilio nomius, Esper.

Order XI. CAPPARIDEÆ.

7. CRATÆVA RELIGIOSA, Forst.
Leptosia xiphia, Fabricius.
Hebomoia australis, Butler.
Appias libythea, Fabricius.
Appias taprobana, Moore.

8. CADABA INDICA, Lamk.
Teracolus etrida, Boisduval.

---

* In the first Kanara paper (p. 263, n. 67) this plant is given under its synonymic name *Gualteria* [recte *Guatteria*] *longifolia*, Wall.

9. CAPPARIS HEYNEANA, Wall.   The plants of this genus are gene-
rally known as 'capers.'
Leptosia xiphia, Fabricius.
Hebomoia australis, Butler.
Nepheronia pingasa, Moore.
Nepheronia hippia, Fabricius.
Appias wardii, Moore.
Huphina remba, Moore.

10.   CAPPARIS DIVARICATA, Lamk.
Belenois mesentina, Cramer.

11.   CAPPARIS MOONII, Wight.
Hebomoia australis, Butler.

12.   CAPPARIS SEPIARIA, Linn.
Leptosia xiphia, Fabricius.
Teracolus eucharis, Fabricius.
Ixias pyrene, Linnæus.
Ixias marianne, Cramer.
Hebomoia australis, Butler.
Huphina nerissa, Fabricius.

13.   CAPPARIS HORRIDA, Linn. f.
Leptosia xiphia, Fabricius.
Nepheronia hippia, Fabricius.
Appias libythea, Fabricius.
Huphina nerissa, Fabricius.

14.   ? CAPPARIS TENERA, Dalz.
Prioneris sita, Felder.

Order XIII.   VIOLACEÆ.

15.   ALSODEIA ZEYLANICA, Thwaites.
Atella alcippe, Cramer.

Order XIV.   BIXINEÆ.

16.   FLACOURTIA MONTANA, Grah.
Cupha placida, Moore.
Atella phalantha, Drury.

17.   HYDNOCARPUS WIGHTIANA, Blume.
Cirrhochroa thais, Fabricius.

Order XIX. PORTULACEÆ.

18. PORTULACA OLERACEA, Linn.
Hypolimnas bolina, Linnæus.
Hypolimnas misippus, Linnæus.

Order XXV. DIPTEROCARPEÆ.

19. HOPEA WIGHTIANA, Wall.
Arrhopala centaurus, Fabricius.
Arrhopala amantes, Hewitson.
Arrhopala canaraica, Moore.
Arrhopala bazalus, Hewitson.
Rathinda amor, Fabricius.

Order XXVI. MALVACEÆ.

20. THESPESIA LAMPAS, Dalz. and Gibs.
Neptis kallaura, Moore.
Neptis jumbah, Moore.

21. KYDIA CALYCINA, Roxb.
Neptis jumbah, Moore.

22. BOMBAX MALABARICUM, DC.
Neptis jumbah, Moore.

Order XXVII. STERCULIACEÆ

23. HELICTERES ISORA, Linn.
Neptis jumbah, Moore.
Caprona ransonnetii, Felder.

24. WALTHERIA INDICA, Linn.
Hesperia galba, Fabricius.

Order XXVIII. TILIACEÆ.

25. GREWIA MICROCOS, Linn.
Neptis jumbah, Moore.
Coladenia indrani, Moore.

26. GREWIA sp.
Charaxes athamas, Drury.

Order XXXIII. RUTACEÆ.

27. RUTA GRAVEOLENS, Linn., var. ANGUSTIFOLIA, Pers.   The rue.
Papilio demoleus, Linnæus.

28. EVODIA ROXBURGHIANA, Benth.
Papilio demolion liomedon, Moore.
Papilio paris tamilana, Moore.

29. ZANTHOXYLUM RHETSA, DC.
Papilio polytes, Linnæus.
Papilio helenus daksha, Hampson.
Papilio buddha, Westwood.

30. ACRONYCHIA LAURIFOLIA, Blume.
Papilio demolion liomedon, Moore.

31. GLYCOSMIS PENTAPHYLLA, Correa.
Neopithecops zalmora, Butler.
Papilio demoleus, Linnæus.
Papilio polytes, Linnæus.
Papilio helenus daksha, Hampson.
Papilio abrisa, Kirby.

32. MURRAYA KŒNIGII, Spreng.
Papilio demoleus, Linnæus.

33. PARAMIGNYA MONOPHYLLA, Wight.
Papilio polymnestor, Cramer.

34. ATALANTIA MONOPHYLLA, Correa.   The wild lime.
Papilio polymnestor, Cramer.

35. CITRUS MEDICA, Linn.   The sour lime or citron or lemon or sweet lime.
Chilades laius, Cramer.
Papilio polytes, Linnæus.
Papilio helenus daksha, Hampson.

36. CITRUS DECUMANA, Linn.   The pomelo or shaddock.
Papilio demoleus, Linnæus.
Papilio polytes, Linnæus.
Papilio helenus daksha, Hampson.

37. ÆGLE MARMELOS, Correa. The bael tree.
Papilio demoleus, Linnæus.

Order XXXVII. MELIACEÆ.

38. AGLAIA ROXBURGHIANA, Miq.
Charaxes imna, Butler.

Order XXXIX. OLACINEÆ.

39. OLAX SCANDENS, Roxb.
Amblypodia anita, Hewitson.

Order XLI. CELASTRINEÆ.

40. SALACIA OBLONGA, Wall.
Bindahara sugriva, Horsfield.

Order XLII. RHAMNEÆ.

41. ZIZYPHUS JUJUBA, Lamk. The ' baer ' tree or ' jujube ' tree.
Neptis jumbah, Moore.
Tarucus theophrastus, Fabricius.
Castalius rosimon, Fabricius.
Castalius ananda, de Nicéville.
Castalius ethion, Doubleday and Hewitson.
Aphnæus vulcanus, Fabricius.

42. ZIZYPHUS XYLOPYRUS, Willd.
Neptis jumbah, Moore.
Castalius ananda, de Nicéville.
Castalius ethion, Doubleday and Hewitson.
Rapala varuna, Horsfield.

43. ZIZYPHUS RUGOSA, Lamk.
Neptis jumbah, Moore.
Castalius ananda, de Nicéville.
Castalius decidia, Hewitson.
Aphnæus vulcanus, Fabricius.
Aphnæus lohita, Horsfield.
Rapala varuna, Horsfield.
Rapala melampus, Cramer.

Order XLIV. SAPINDACEÆ.

44. ALLOPHYLUS COBBE, Blume.
Odontoptilum angulata, Felder.

45. SCHLEICHERA TRIJUGA, Willd.
Catochrysops strabo, Fabricius.
Rathinda amor, Fabricius.

Order XLVI. ANACARDIACEÆ.

46. MANGIFERA INDICA, Linn. The mangoe.
Euthalia garuda, Moore.

47. ANACARDIUM OCCIDENTALE, Linn. The cashewnut.
Euthalia garuda, Moore.

48. BUCHANANIA LATIFOLIA, Roxb.
Lycænesthes lycænina, Felder.

Order XLIX. CONNARACEÆ.

49. ROUREA SANTALOIDES, W. and A.
Charaxes wardii, Moore.

50. CONNARUS RITCHIEI, Hook. f.
Deudorix epijarbas, Moore.

Order L. LEGUMINOSÆ.

51. MILLETTIA RACEMOSA, Benth.
Neptis jumbah, Moore.
Hasora chabrona, Plötz.

52. SESBANIA ACULEATA, Pers.
Tarucus telicanus, Lang, in Bombay.
Terias hecabe, Linnæus.

53. ZORNIA DIPHYLLA, Pers. A vetch-like plant.
Vanessa cardui, Linnæus.
Plebeius trochilus, Freyer.
Zizera lysimon, Hübner.
Zizera otis, Fabricius.

54. OUGEINIA DALBERGIOIDES, Benth.
Catochrysops strabo, Fabricius.
Euchrysops cnejus, Fabricius.
Tarucus telicanus, Lang.
Curetis bulis, Doubleday and Hewitson.
Rapala melampus, Cramer.

55. ABRUS PRECATORIUS, Linn.
Lampides celeno, Cramer.
Curetis thetis, Drury.

56. BUTEA FRONDOSA, Roxb.   Bastard teak.
Jamides bochus, Cramer.
Polyommatus bœticus, Linnæus.

57. CAJANUS INDICUS, Spreng.
Polyommatus bœticus, Linnæus.

58. CYLISTA SCARIOSA, Ait.
Cyaniris puspa, Horsfield.
Everes argiades, Pallas.
Catochrysops strabo, Fabricius.
Euchrysops cnejus, Fabricius.

59. DALBERGIA LATIFOLIA, Roxb.   Blackwood or rosewood tree.
Neptis viraja, Moore.
Neptis jumbah, Moore.
Tapena thwaitesi, Moore.

60. DALBERGIA RUBIGINOSA, Roxb.
Tapena thwaitesi, Moore.

61. DALBERGIA CONFERTIFLORA, Benth.
Neptis kallaura, Moore.

62. DALBERGIA TAMARINDIFOLIA, Roxb.
Tapena thwaitesi, Moore.

63. DALBERGIA VOLUBILIS, Roxb.
Neptis viraja, Moore.*
Tapena thwaitesi, Moore.

64. PONGAMIA GLABRA, Vent.
Jamides bochus, Cramer.
Lampides celeno, Cramer.
Curetis thetis, Drury.
Hasora (Parata) alexis, Fabricius.

* In the second Kanara paper (p. 251, n. 41) *Dalbergia racemosa* appears as a food plant of *N. viraja*, Moore, but this is a wrong identification for *Dalbergia volubilis*.

65. DERRIS SCANDENS, Benth.
Hasora (Parata) butleri, Aurivillius.
Tapena thwaitesi, Moore.

66. DERRIS ULIGINOSA, Benth.
Hasora badra, Moore.

67. DERRIS HEYNEANA, Benth.
Curetis thetis, Drury.
Hasora (Parata) butleri, Aurivillius.

68. CÆSALPINIA MIMOSOIDES, Lam.
Neptis (Rahinda) hordonia, Stoll.
Charaxes athamas, Drury.

69. POINCIANA REGIA, Bojer.   The gold-mohar tree.
Charaxes athamas, Drury.
Terias silhetana, Wallace.

70. WAGATEA SPICATA, Dalz.
Neptis kallaura, Moore.
Neptis jumbah, Moore.
Charaxes wardii, Moore.
Charaxes fabius, Fabricius.
Lycænesthes emolus, Godart.
Lycænesthes lycænina, Felder.
Nacaduba atrata, Horsfield.
Euchrysops pandava, Horsfield.
Curetis thetis, Drury.
Aphnæus lohita, Horsfield.
Rapala lankana, Moore.
Terias silhetana, Wallace.

71. CASSIA FISTULA, Linn.   Indian Laburnum.
Catopsilia crocale, Cramer.

72. CASSIA OCCIDENTALIS, Linn.
Catopsilia pyranthe, Linnæus.
Terias hecabe, Linnæus.

73. CASSIA TORA, Linn.
Terias hecabe, Linnæus.

74. Cassia siamea, Lam.
Catopsilia crocale, Cramer.

75. Cassia glauca, Lam.
Terias hecabe, Linnæus.

76. Cassia pumila, Lam.
Terias libythea, Fabricius.

77. Saraca indica, Linn.
Lycænesthes emolus, Godart.
Lampides celeno, Cramer.
Cheritra jaffra, Butler.

78. Tamarindus indica, Linn.   The tamarind.
Charaxes fabius, Fabricius.
Virachola isocrates, Fabricius.

79. Xylia dolabriformis, Benth.   The Pegu iron-wood.
Neptis jumbah, Moore.
Cyaniris puspa, Horsfield.
Jamides bochus, Cramer.
Euchrysops pandava, Horsfield.
Arrhopala centaurus, Fabricius.
Arrhopala amantes, Hewitson.
Curetis thetis, Drury.
Zesius chrysomallus, Hübner.
Aphnæus lohita, Horsfield.
Cheritra jaffra, Butler.
Rapala varuna, Horsfield.
Coladenia indrani, Moore.

80. Acacia Intsia, Willd.
Neptis (Rahinda) hordonia, Stoll.   Form of larva with short processes.
Nacaduba noreia, Felder.
Surendra quercetorum; Moore.
Zinaspa todara, Moore.

81. Acacia Intsia, var. cæsia, W. and A.
Nacaduba noreia, Felder.
Zinaspa todara, Moore.
Rapala schistacea, Moore.
Rapala lankana, Moore.

J. ii. 26

82. Acacia pennata, Willd.
Neptis (Rahinda) hordonia, Stoll.   Form of larva with-long processes.
Charaxes athamas, Drury.
Nacaduba noreia, Felder.
Surendra quercetorum, Moore.
Zinaspa todara, Moore.
Rapala schistacea, Moore.
Rapala lankana, Moore.

83. Albizzia Lebbek, Benth.
Neptis viraja, Moore.
Charaxes athamas, Drury.

84.   Genus and Species unknown.
Neptis varmona, Moore.

Order LIII.   CRASSULACEÆ.

85. Bryophyllum calycinum, Salisb.
Talicada nyseus, Guérin.

Order LVIII.   COMBRETACEÆ.

86. Terminalia Bellerica, Roxb.
Badamia exclamationis, Fabricius.
Cupitha purreea, Moore.

87. Terminalia tomentosa, Bedd.
Arrhopala centaurus, Fabricius.
Zesius chrysomallus, Hübner.
Catapœcilma elegans, Druce.

88. Terminalia paniculata, Roth.
Lycænesthes emolus, Godart.
Castalius ananda, de Nicéville.
Arrhopala centaurus, Fabricius.
Arrhopala amantes, Hewitson.
Arrhopala bazalus, Hewitson.
Zesius chrysomallus, Hübner.
Aphnæus lohita, Horsfield.
Catapœcilma elegans, Druce.
Cupitha purreea, Moore.

89. Combretum ovalifolium, Roxb.
Cupitha purreea, Moore.

90. Combretum extensum, Roxb.
Lycænesthes emolus, Godart.
Ismene fergusonii, de Nicéville.
Bibasis sena, Moore.
Badamia exclamationis, Fabricius.

91. Quisqualis indica, Linn.
Rapala schistacea, Moore.
Rapala varuna, Horsfield.

Order LIX. MYRTACEÆ.

92. Psidium Guyava, Linn. The guava.
Zesius chrysomallus, Hübner.
Aphnæus lohita, Horsfield.

93. Eugenia zeylanica, Wight.
Rathinda amor, Fabricius.

94. Careya arborea, Roxb.
Euthalia (Cynitia) lepidea, Butler.
Rathinda amor, Fabricius.

Order LX. MELASTOMACEÆ.

95. Melastoma malabathricum, Linn.
Euthalia (Cynitia) lepidea, Butler.

Order LXI. LYTHRACEÆ.

96. Lagerstrœmia lanceolata, Wall.
Arrhopala centaurus, Fabricius.
Arrhopala amantes, Hewitson.
Aphnæus lohita, Horsfield.
Catapœcilma elegans, Druce.

97. Punica Granatum, Linn. The pomegranate.
Virachola isocrates, Fabricius.

Order LXIV. PASSIFLOREÆ.

98. MODECCA PALMATA, Lam. The wild passion-flower.
Telchinia violæ, Fabricius.
Cethosia mahratta, Moore.
Cynthia saloma, de Nicéville.

Order LXV. CUCURBITACEÆ.

99. ZEHNERIA UMBELLATA, Thwaites.
Parthenos virens, Moore.

Order LXXI. ARALIACEÆ.

100. HEPTAPLEURUM VENULOSUM, Seem. An ivy-like creeper.
Ismene gomata, Moore.

Order LXXV. RUBIACEÆ.

101. ADINA CORDIFOLIA, Hook. f.
Athyma selenophora, Kollar.

102. STEPHEGYNE PARVIFOLIA, Korth.
Limenitis (Moduza) procris, Cramer.

103. WENDLANDIA EXSERTA, DC.
Limenitis (Moduza) procris, Cramer.
Athyma inara, Doubleday and Hewitson.

104. MUSSÆNDA FRONDOSA, Linn.
Limenitis (Moduza) procris, Cramer.
Athyma inara, Doubleday and Hewitson.

105. RANDIA ULIGINOSA, DC.
Virachola isocrates, Fabricius.
Virachola perse, Hewitson.

106. RANDIA DUMETORUM, Lamk.
Virachola perse, Hewitson.

107. IXORA COCCINEA, Linn.
Rathinda amor, Fabricius.

Order LXXVIII.   COMPOSITÆ.

108.   BLUMEA sp.   A kind of thistle.
Vanessa cardui, Linnæus.

Order LXXXVIII.   MYRSINEÆ.

109.   EMBELIA ROBUSTA, Roxb.
Abisara fraterna, Moore.
Nacaduba atrata, Horsfield.

110.   ARDISIA HUMILIS, Vahl.
Abisara fraterna, Moore.
Nacaduba atrata, Horsfield.

Order XC.   EBENACEÆ.

111.   DIOSPYROS MELANOXYLON, Roxb.   Ebony.
Symphædra nais, Forster.
Euthalia (Dophla) laudabilis, Swinhoe.

112.   DIOSPYROS CANDOLLEANA, Wight.   Ebony.
Euthalia (Dophla) laudabilis, Swinhoe.

Order XCII.   OLEACEÆ.

113.   LINOCIERA MALABARICA, Wall.
Athyma ranga, Moore.

114.   OLEA DIOICA, Roxb.
Athyma ranga, Moore.

Order XCIII.   SALVADORACEÆ.

115.   SALVADORA PERSICA, Linn.
Teracolus amata, Fabricius.

Order XCIV.   APOCYNACEÆ.

116.   HOLARRHENA ANTIDYSENTERICA, Wall.
Euplœa (Crastia) core, Cramer.

117.   NERIUM ODORUM, Soland.   The oleander.
Euplœa (Crastia) core, Cramer.

118. Aganosma cymosa, G. Don.
Hestia malabarica, Moore.

119. Ichnocarpus frutescens, Br.
Euplœa (Crastia) core, Cramer.
Euplœa (Narmada) coreta, Godart.

Order XCV. ASCLEPIADEÆ.

120. Cryptolepis Buchanani, Roem. and Sch.
Danais (Parantica) aglea, Cramer.

121. Calotropis gigantea, Br. The 'Madar.'
Danais (Limnas) chrysippus, Linnæus.

122. Asclepias Curassavica, Linn.
Danais (Limnas) chrysippus, Linnæus.

123. Tylophora tenuis, Blume.
Danais (Parantica) aglea, Cramer.

124. Dregea volubilis, Benth. Wax plant.
Danais (Tirumala) limniace, Cramer.
Danais (Tirumala) septentrionis, Butler.

125. Genus and Species unknown.
Danais (Salatura) plexippus, Linnæus.

Order C. BORAGINEÆ.

126. Heliotropium strigosum, Willd.
Plebeius trochilus, Freyer.

Order CI. CONVOLVULACEÆ.

127. Argyreia speciosa, Sweet.
Zesius chrysomallus, Hübner.

128. Argyreia sericea, Dalz. and Gibs.
Aphnæus lohita, Horsfield.

Order CIX. ACANTHACEÆ.

129. Nelsonia campestris, Br.
Precis lemonias, Linnæus.
Zizera gaika, Trimen.

130. HYGROPHILA SPINOSA, T. Anders.
Precis almana, Linnæus.
Precis lemonias, Linnæus.
Precis hierta, Fabricius.
Precis orithyia, Linnæus.
Precis atlites, Linnæus.

131. DÆDALACANTHUS ROSEUS, T. Anders.
Celænorrhinus leucocera, Kollar.

132. STROBILANTHES CALLOSUS, Nees.
Precis iphita, Cramer.
Precis lemonias, Linnæus.
Kallima horsfieldii, Kollar.
Celænorrhinus ambareesa, Moore.
Celænorrhinus fusca, Hampson.

133. BLEPHARIS ASPERRIMA, Nees.
Sarangesa purendra, Moore.

134. BARLERIA PRIONITIS, Linn.
Precis lemonias, Linnæus.

135. BARLERIA sp.
Precis hierta, Fabricius.
Precis atlites, Linnæus.

136. ERANTHEMUM MALABARICUM, Clarke.
Kallima horsfieldii, Kollar.
Doleschallia polibete, Cramer.

137. ERANTHEMUM sp.
Celænorrhinus leucocera, Kollar.

138. Genus and Species unknown.
Sarangesa dasahara, Moore.

Order CXI. VERBENACEÆ.

139. ? LIPPIA NODIFLORA, Rich.
Precis almana, Linnæus. Mr. Bell thinks this food-plant is very doubtful for *P. almana.*

140. CLERODENDRON INFORTÚNATUM, Gaertn.
Zesius chrysomallus, Hübner.

Order CXVI. AMARANTACEÆ.

141. CYATHULA PROSTRATA, Blume.
Coladenia dan, Fabricius.

142. ACHYRANTHES ASPERA, Linn.
Coladenia dan, Fabricius.
Sarangesa dasahara, Moore.

143. ACHYRANTHES BIDENTATA, Blume.
Coladenia dan, Fabricius.

Order CXXIII. ARISTOLOCHIACEÆ.

144. ARISTOLOCHIA BRACTEATA, Retz.
Papilio aristolochiæ, Fabricius.

145. ARISTOLOCHIA INDICA, Linn.
Troides minos, Cramer.
Papilio hector, Linnæus.
Papilio aristolochiæ, Fabricius.

Order CXXVIII. LAURINEÆ.

146. CINNAMOMUM ZEYLANICUM, Breyn.   The wild cinnamon.
Cheritra jaffra, Butler.
Papilio sarpedon teredon, Felder.
Papilio clytia, Linnæus.

147. ALSEODAPHNE SEMECARPIFOLIA, Nees.
Papilio sarpedon teredon, Felder.
Papilio abrisa, Kirby.
Papilio clytia, Linnæus.

148. LITSÆA TOMENTOSA, Herb.
Papilio clytia, Linnæus.

149. LITSÆA SEBIFERA, Pers.
Papilio sarpedon teredon, Felder.
Papilio clytia, Linnæus.

Order CXXXII. LORANTHACEÆ.

150. LORANTHUS WALLICHIANUS, Schultz. All the plants of this Natural Order are often called 'Mistletoe' from the resemblance they bear to the European plant of that name.
Tajuria cippus, Fabricius.

151. LORANTHUS SCURRULA, Linn.
Euthalia garuda, Moore.
Euthalia lubentina, Cramer.
Camena deva, Moore.
Creon cleobis, Godart.
Tajuria cippus, Fabricius.
Ops melastigma, de Nicéville.

152. LORANTHUS TOMENTOSUS, Heyne.
Camena deva, Moore.
Ops melastigma, de Nicéville.

153. LORANTHUS LONGIFLORUS, Desrouss.
Castalius ananda, de Nicéville.
Zesius chrysomallus, Hübner.
Tajuria cippus, Fabricius.
Rathinda amor, Fabricius.
Delias eucharis, Drury.

154. LORANTHUS ELASTICUS, Desrouss.
Creon cleobis, Godart.
Tajuria indra, Moore.
Tajuria cippus, Fabricius.

Order CXXXV. EUPHORBIACEÆ.

155. GLOCHIDION LANCEOLARIUM, Dalz.
Athyma perius, Linnæus.

156. GLOCHIDION ZEYLANICUM, A. Juss.
Athyma inara, Doubleday and Hewitson.

157. GLOCHIDION VELUTINUM, Wight.
Athyma perius, Linnæus.
Athyma inara, Doubleday and Hewitson.

158. HEMICYCLIA VENUSTA, Thwaites.
Appias albina, Boisduval.

159. CROTON sp.
Rathinda amor, Fabricius.

160. TREWIA NUDIFLORA, Linn.
Thaduka multicaudata, Moore.

161. MALLOTUS PHILIPPINENSIS, Muell.
Coladenia indrani, Moore.

162. TRAGIA INVOLUCRATA, Linn.
Ergolis taprobana, Westwood.
Ergolis ariadne,'Linnæus.

163. TRAGIA INVOLUCRATA, var. CANNABINA, Linn.
Ergolis taprobana, Westwood.
Ergolis ariadne, Linnæus.
Byblia ilithyia, Drury, in Khandeish and the Deccan in the Bombay Presidency.

Order CXXXVI. URTICACEÆ.

164. CELTIS TETRANDRA, Roxb.
Apatura (Rohana) camiba, Moore.
Libythea rama, Moore.

165. TREMA ORIENTALIS, Blume.
Euripus consimilis, Westwood.
Neptis kallaura, Moore.
Neptis jumbah, Moore.

166. STREBLUS ASPER, Lour.
Euplœa (Crastia) core, Cramer.
Euplœa (Pademma) kollari, Felder.

167. FICUS BENGALENSIS, Linn. The banyan or 'bhor' tree.
Euplœa (Crastia) core, Cramer.
Cyrestis thyodamas, Boisduval.
Iraota timoleon, Stoll.

168. FICUS INDICA, Linn.
Cyrestis thyodamas, Boisduval.
Iraota timoleon, Stoll.

169. FICUS RELIGIOSA, Linn.   The pipal.
Euplœa (Crastia) core, Cramer.

170. FICUS GLOMERATA, Roxb.
Euplœa (Crastia) core, Cramer.
Cyrestis thyodamas, Boisduval.
Iraota timoleon, Stoll.

171. FLEURYA INTERRUPTA, Gaud.   A kind of nettle.
Hypolimnas bolina, Linnæus.

172. ELATOSTEMA CUNEATUM, Wight.
Hypolimnas bolina, Linnæus.

Order CXLVIII.   ORCHIDEÆ.

173. RHYNCHOSTYLIS RETUSA, Blume.
Chliaria othona, Hewitson.

174. SACCOLABIUM PAPILLOSUM, Lindl.
Chliaria othona, Hewitson.

Order CXLIX.   SCITAMINEÆ.

175. CURCUMA AMADA, Roxb.   Wild turmeric.
Udaspes folus, Cramer.
Notocrypta feisthamelii, Boisduval.

176. KÆMPFERIA PANDURATA, Roxb.
Lampides elpis, Godart.
Udaspes folus, Cramer.

177. HEDYCHIUM CORONARIUM, Kœnig.
Lampides elpis, Godart.
Udaspes folus, Cramer.
Notocrypta feisthamelii, Boisduval.

178. AMOMUM MICROSTEPHANUM, Baker.
Lampides elpis, Godart.
Udaspes folus, Cramer.
Notocrypta feithamelii, Boisduval.
Sancus pulligo, Mabille.

179. ELETTARIA CARDAMOMUM, Maton.   The cardamon.
Lampides elpis, Godart.

Order CLIV. DIOSCOREACEÆ.

180. DIOSCOREA PENTAPHYLLA, Linn.
Aphnæus lohita, Horsfield.
Loxura atymnus, Cramer.
Tagiades atticus, Fabricius.
Tagiades obscurus, Mabille.

Order CLVI. LILIACEÆ.

181. SMILAX MACROPHYLLA, Roxb.
Loxura atymnus, Cramer.
Tagiades atticus, Fabricius.

Order CLXIII. PALMEÆ.

182. ARECA CATECHU, Linn.   The Supari or Betel-nut.
Elymnias caudata, Butler.
Gangara thyrsis, Fabricius.
Suastus gremius, Fabricius.

183. CARYOTA URENS, Linn.   The Palmyra or fan palm.
Elymnias caudata, Butler.
Gangara thyrsis, Fabricius.
Suastus gremius, Fabricius.

184. PHŒNIX SYLVESTRIS, Roxb.   Wild date palm.
Elymnias caudata, Butler.
Gangara thyrsis, Fabricius.
Suastus gremius, Fabricius.
Hyarotis adrastus, Cramer.

185. CALAMUS PSEUDO-TENUIS, Becc. and Hook. f.   Rattan or cane.
Elymnias caudata, Butler.
Gangara thyrsis, Fabricius.
Suastus gremius, Fabricius.
Pedestes submaculata, Staudinger.
Hyarotis adrastus, Cramer.

186. COCOS NUCIFERA, Linn.   Coconut palm.
Elymnias caudata, Butler.

Gangara thyrsis, Fabricius.
Suastus gremius, Fabricius.

Order CLXXIII.   GRAMINEÆ.
187.   ORYZA SATIVA, Linn.   Rice.
Mycalesis (Orsotriæna) mandata, Moore.
Mycalesis (Calysisme) visala, Moore.
Mycalesis (Nissanga) junonia, Butler.
Ypthima baldus, Fabricius.
Melanitis ismene, Cramer.
Baoris (Parnara) guttatus, Bremer and Grey.
Ampittia dioscorides, Fabricius.
Baoris (Parnara) bevani, Moore.
Baoris (Chapra) subochracea, Moore
Baoris (Chapra) mathias, Fabricius.

188.   ZEA MAYS, Linn.   Maize or Indian corn.
Baoris (Parnara) conjuncta, Herrich-Schäffer.

189.   Grasses of different kinds.
Mycalesis (Orsotriæna) mandata, Moore.
Mycalesis (Calysisme) perseus, Fabricius.
Mycalesis (Calysisme) visala, Moore.
Mycalesis (Nissanga) junonia, Butler.
Ypthima baldus, Fabricius.
Ypthima huebneri, Kirby.
Melanitis ismene, Cramer.
Melanitis varaha, Moore.
Iambrix salsala, Moore.
Baoris (Parnara) guttatus, Bremer and Grey.
Baracus hampsoni, Elwes and Edwards.
Taractrocera ceramas, Hewitson.
Aëromachus indistinctus, Moore.
Padraona gola, Moore.
Baoris (Parnara) conjuncta, Herrich-Schäffer.
Baoris (Parnara) colaca, Moore.
Baoris (Chapra) subochracea, Moore.
Baoris (Chapra) mathias, Fabricius.

190.   BAMBUSA ARUNDINACEA, Willd.   Bamboo.
Lethe europa, Fabricius.

Lethe drypetis, Hewitson.
Melanitis gokala, Moore.
Discophora lepida, Moore.
Matapa aria, Moore
Padraona dara, Kollar.
Iambrix salsala, Moore.
Halpe ceylonica, Moore.
Telicota bambusæ, Moore.
Baoris oceia, Hewitson.
Baoris (Parnara) kumara, Moore

191. Bambusa sp.
Discophora indica, Staudinger.
Baoris (Parnara) philippina, Herrich-Schäffer.

192. Oxytenanthera monostigma, Beddome.
Discophora lepida, Moore.
Padraona dara, Kollar.
Halpe ceylonica, Moore.
Halpe hyrtacus, de Nicéville.
Telicota bambusæ, Moore.

193. Dendrocalamus strictus, Nees.
Discophora lepida, Moore.
Matapa aria, Moore.
Halpe honorei, de Nicéville.
Baoris oceia, Hewitson.

194. Ochlandra stridula, Thwaites.
Zipœtes saitis, Hewitson.
Discophora lepida, Moore.
Matapa aria, Moore.
Padraona dara, Kollar.
Halpe hyrtacus, de Nicéville.
Baoris oceia, Hewitson.
Baoris (Parnara) kumara, Moore.

Plants from the Western Himalayas and Kashmir, with the species of butterflies whose larvæ feed on them.

Order L. LEGUMINOSÆ.

1. Indigofera atropurpurea, Hamilt.
Thecla sassanides, Kollar.
Hysudra selira, Moore.

Order LXXVIII.   COMPOSITÆ.

2. CARDUUS sp.
Vanessa cardui, Linnæus.

Order CXXXVI.   URTICACEÆ.

3. URTICA PARVIFLORA, Roxb.
Vanessa caschmirensis, Kollar.

Order CXLI.   SALICINEÆ.

4. SALIX TETRASPERMA, Roxb.
Vanessa xanthomelas, Wiener Verzeichniss.

PART II.

A Revised List of the Butterflies of the Kanara District arranged in the order of Messrs. Davidson, Bell and Aitken's previous papers, with the names of the Food-plants on which the larvæ feed.

Family NYMPHALIDÆ.

Subfamily DANAINÆ.

1. HESTIA MALABARICA, Moore.
Aganosma cymosa, G. Don (*Apocynaceæ*).

H. lynceus, Drury, of the previous list, p. 239, is a species restricted by Dr. F. Moore in Lep. Ind., vol. i, p. 26 (1890), to Borneo.   It has the wings much longer and narrower and the coloration much darker than our South Indian species.   Dr. Moore (l.c., p. 18) does not record *H. mala-barica* from Kanara, but it undoubtedly occurs there.   But he describes (l.c., p. 21, pl. ii, figs. 2, *male*; 2a, *female*; I, *larva*; 1a, *pupa*) *H. kanarensis* as a new species from North Kanara and the South Konkan.   It is a smaller species than *H. malabarica,* 4·25 to 4·75 as against 5 0 to 5·5 inches in expanse.   " The markings similarly disposed and shaped but smaller, being about half the size of those in *H. malabarica.*"   This latter remark is an obvious exaggeration speaking of the markings as a whole, as can be verified by comparing Dr. Moore's figures of the two species in Lep. Ind.   I do not think that *H. kanarensis* can be retained as a species distinct from *H. malabarica.*   Its smaller size is its most distinctive character.

2. DANAIS (*Parantica*) AGLEA, Cramer.
Cryptolepis Buchanani, Roem. and Sch. (*Asclepiadeæ*).
Tylophora tenuis, Blume* (*Asclepiadeæ*).

3. DANAIS (*Tirumala*) LIMNIACE, Cramer.
Dregea volubilis, Benth.† (*Asclepiadeæ*).

4. DANAIS (*Tirumala*) SEPTENTRIONIS, Butler.
Dregea volubilis, Benth. (*Asclepiadeæ*).

Not recorded from Kanara in the lists, but undoubtedly occurs in
the district according to Mr. Bell, who says he has specimens of it
which agree with the typical form both in coloration and markings.
Specimens intermediate between *D. limniace*, Cramer, and *D. septen-*
trionis also occur, and for these Mr. Aitken in the first paper (p. 266)
suggested in joke the name *D. limnitrionis*. Mr. Fruhstorfer has
probably described this form as *Tirumala melissa dravidarum* in Berl.
Ent. Zeitsch., vol. xliv, p. 113 (1899).

5. DANAIS (*Limnas*) CHRYSIPPUS, Linnæus.
Calotropis gigantea, Br. (*Asclepiadeæ*).
Asclepius Curassavica, Linn. (*Asclepiadeæ*).

6. DANAIS (*Salatura*) PLEXIPPUS, Linnæus.
The *D. genutia*, Cramer, of the previous list, p. 240, but Linnæus'
name undoubtedly applies to this species. It has often been bred in
Kanara ; the larva feeds on an asclepiad, which has not been identified.

7. EUPLŒA (*Crastia*) CORE, Cramer.
Holarrhena antidysenterica, Wall. (*Apocynaceæ*).
Nerium odorum, Soland. (*Apocynaceæ*).
Ichnocarpus frutescens, Br. (*Apocynaceæ*).
Streblus asper, Lour. (*Urticaceæ*).
Ficus bengalensis, Linn. (*Urticaceæ*).
Ficus religiosa, Linn. (*Urticaceæ*).
Ficus glomerata, Roxb. (*Urticaceæ*).

8. EUPLŒA (*Pademma*) KOLLARI, Felder.
Streblis asper, Lour. (*Urticaceæ*).
The larva is said to also feed " On several species of *Ficus*."

* The *Tylophora carnosa*, Wall., of the first paper, p. 266, is a synonym of
T. *tenuis*, Blume.

† The *Hoya viridiflora*, Br., of the first paper, p. 266, is a synonym of *Dregea*
volubilis, Benth.

9. EUPLŒA (*Narmada*) CORETA, Godart.

Ichnocarpus frutescens, Br. (*Apocynaceæ*).

LARVA. The larva is of the type of that of *Euplœa* (*Crastia*) *core*, Cramer, except that the subdorsal tentacles are wanting on segment 6, and the two pairs on segments 3 and 4 are considerably longer. The head is round, smooth, shiny and black, with a narrow white band round the margin over the vertex, and a similar band down each side of the clypeus meeting at the apex of the clypeus; the labrum is white. The surface of the body is smooth and rather greasy looking. The spiracles are oval, black and shiny. The colour of the body varies somewhat, or rather the shade varies; segments 3 and 4 are dorsally always slightly yellowish. There is always a dark bluish, dorsal line or band. The general colour of the body is a light violet-green on the somites of the dorsal half, and a chocolate-green on the somites of the ventral half; there is a subspiracular yellow line dividing the two colours; a slight yellow shade around the spiracles, and the extreme base of all three pairs of tentacles is yellow; segment 2 is light yellow with a black subdorsal shiny spot, and the anal flap is yellow, with a very large shiny black mark, covering nearly three-fourths of the dorsum; tentacles dull indigo in colour; all the legs shiny black. Length 39 mm., breadth 6 mm., length of tentacles of segment 3 is 14 mm., of segment 4 is 9 mm., of segment 12 is 6 mm. All the tentacles are nearly straight.

PUPA. The pupa has nothing in any way to distinguish it in shape from that of *E. core*. The spiracles are light brown, oval, of the usual size. The colour of the pupa is silver, with one very broad subdorsal, and one very broad, spiracular, lightish brown band on the abdomen meeting on segment 6, which is entirely brown, as are also segments 4 and 5 except dorsally; the shoulders and the inner margin of the wing are broadly brown, as is also the vertex of the head; there is a narrow brown band along the outer margin of the wings; a broad, dorsal, thoracic, brown band forking from the apex to the hinder margin; a large, oblong, brown mark on the wing beyond the discoidal cell; the cremaster is strong, triangular, flattened dorsally and ventrally, with a rugose knob at the end, and a spherical tubercle at the base laterally : all shiny black, as are also the anal clasper scars; the costal margins of the wings and the haustellum are also brownish; the whole surface is shiny and quite smooth. Length 18·5 mm. including the cremaster, breadth at segment 7 is 8·5 mm., which is the broadest part.

HABITS. The habits are the same as those of *E. core* in every particular. The larva eats the young leaves, living on the underside of the leaves. It pupates under a leaf, hanging very freely. The larva is badly persecuted by ichneumons.

Subfamily SATYRINÆ.

10. MYCALESIS (*Orsotriæna*) MANDATA, Moore.
Oryza sativa, Linn. (*Gramineæ*).
Grasses (*Gramineæ*).

11. MYCALESIS (*Calysisme*) PERSEUS, Fabricius.
Grasses (*Gramineæ*).
Recorded without a number in the second Kanara paper, p. 242.

12. MYCALESIS (*Calysisme*) VISALA, Moore.
Oryza sativa, Linn. (*Gramineæ*).
Grasses (*Gramineæ*).
This appears in the Kanara papers as *M. minens.* Linnæus, a species
now restricted by Dr. F. Moore to N. and E. India, Burma, Siam, and
S.-E. China.

13. MYCALESIS (*Calysisme*) SUBDITA, Moore.
Not recorded in the Kanara papers, but Mr. Bell caught it in
one place, Tarimalapur, up the valley of the Kalinaddi river. Dr.
Moore restricts it to South India and Ceylon. It has not been bred in
the Kanara district.

14. MYCALESIS (*Nissanga*) JUNONIA, Butler.
Oryza sativa, Linn. (*Gramincæ*).
Grasses (*Gramineæ*).

15. LETHE EUROPA, Fabricius.
Bambusa arundinacea, Willd. (*Gramineæ*).

16. LETHE DRYPETIS, Hewitson.
Bambusa arundinacea, Willd. (*Gramineæ*).
The *L. drypetes* [sic] of Moore's " Lep. Ind.," and *L. todara,* Moore,
of the second Kanara paper, p. 243, that species being a synonym of
L. drypetis.

17. LETHE NEELGHERRIENSIS, Guérin.
The *L. neelgheriensis* [sic] of Moore's " Lep. Ind." It has not been
bred in Kanara, but the imago is found above the ghâts in Haliyal and
on the Dharwar frontier. In Ceylon the larva is said to feed on grasses.

18. YPTHIMA BALDUS, Fabricius.
Oryza sativa, Linn. (*Gramineæ*).

Grasses (*Gramineæ*).

The *Y. philomela*, Johanssen, of the second Kanara paper. It is quite possible that the true *Y. philomela*, Johanssen (I do not know what the *Y. philomela* of Linnæus is) may also occur in Kanara, as I have specimens from the Nilgiri Hills, the Wynaad (type of *Y. tabella*, Marshall), as well as from Henzada and Manlin in Burma. Dr. F. Moore in "Lep. Ind." restricts *Y. philomela*, Johansson [sic], to Java and Sumatra, and gives *Y. tabella* full specific rank, recording it from South India and Burma. In Lep. Ind., vol. ii, p. 58, he defines *Thymipa*, the genus in which he places *Ypthima philomela*, as having a prominent androconial patch in the male, while in describing *Thymipa tabella*, p. 73, he says it has "No glandular patch nor androconia." As a matter of fact the patch is sometimes absent, sometimes faintly present, and sometimes prominent, especially so in specimens from Java and Bali. In Sumatran specimens it is faint, in Burmese and South Indian examples apparently entirely absent.

19. YPTHIMA HUEBNERI, Kirby.

Grasses (*Gramineæ*).

Some Kanarese examples of this species have pure white cilia to the hindwing, while others have the outer half of the disc of the hindwing on the upperside suffused with white, thereby approximating in coloration to *Y. ceylonica*, Hewitson.

20. ZIPŒTES SAITIS, Hewitson.

Ochlandra stridula, Thwaites (*Gramineæ*).

21. MELANITIS ISMENE, Cramer.

Oryza sativa, Linn. (*Gramineæ*).

Grasses (*Gramineæ*).

The *M. leda*, Linnæus, of the first Kanara paper, p. 267.

22. MELANITIS VARAHA, Moore.

Grasses (*Gramineæ*).

Colonel C. Swinhoe's *Melanitis ampa*, described in Ann. and Mag. of Nat. Hist., sixth series, vol. v, p. 353, n. 1 (1890) from North Kanara, July, is a wet-season form of *M. varaha*.

23. MELANITIS GOKALA, Moore.

Bambusa arundinacea, Willd. (*Gramineæ*).

Subfamily ELYMNIINÆ.

24. ELYMNIAS CAUDATA, Butler.
Areca Catechu, Linn. (*Palmeæ*).
Caryota urens, Linn. (*Palmeæ*).
Phœnix sylvestris, Roxb. (*Palmeæ*).
Calamus pseudo-tenuis, Becc. and Hook. f. (*Palmeæ*).
Cocos nucifera, Linn. (*Palmeæ*).

Subfamily AMATHUSIINÆ.

25. DISCOPHORA LEPIDA, Moore.
Bambusa arundinacea, Willd. (*Gramineæ*).
Oxytenanthera monostigma, Beddome (*Gramineæ*).
Dendrocalamus strictus, Nees (*Gramineæ*).
Ochlandra stridula, Thwaites (*Gramineæ*).

26. DISCOPHORA INDICA, Staudinger.
Bamboos (*Gramineæ*).
This is a new record from Kanara.

LARVA. The larva is similar in shape and appearance to that of *D. lepida,* Moore. Head semi-elliptic, as seen from the front; clypeus black with a central white line, and bordered all round with yellowish; a narrow yellow line from over the vertex of the head down the middle of the face to the apex of the clypeus; eyes black; a bunch of long porrect hairs on the top of each lobe of the head; porrect hairs disposed moderately densely also down both cheeks; colour of the head red-brown, appearing lighter on the vertex of each lobe because of the light-coloured bunches of hair; surface dull. Spiracles oval, black, with white centres. Surface of the body dull, covered all over moderately densely with longish erect hairs of which. some few laterally are longer than the rest; there is a subdorsal bunch of spine-like hairs on segments 3 to 6, making these segments look brown over the dorsum; the other hairs are light mouse-coloured. Anal flap broadly rounded at the extremity, the two anal processes beneath it are short, conical, light yellow and hairy. Colour of the body light brownish-greyish, with a thin subdorsal white line (not a broad white dorsal band as in *D. lepida*); each segment with a latero-dorsal black mark near the front margin, which mark is very distinct on segments 2 to 7, and hardly perceptible on the rest; behind the black marks on segments 2 to 13 is a large light chocolate-brown mark or patch; legs rose-coloured. Size somewhat smaller than *D. lepida.*

PUPA. The pupa is in every way the same as that of *D. lepida,* and only differs from it in being smaller than the majority of pupæ of that species. In shade it is either bone-coloured or green according to whether it has been formed among dead leaves, &c., or green leaves; when green the edges of the pupa along the wings are yellow as in *D. lepida*, but when bone-coloured the black spots sprinkled over the surface in *D. lepida* are not present here ; the surface of the two contiguous head-points is also shiny, while in *D. lepida* it is dullish.

HABITS. In habits the larvæ and pupæ differ in no way from those of *D. lepida*. The larva, as is also the case in *D. lepida*, will suspend itself by the tail against a perpendicular surface with its ventral surface towards the perpendicular surface, and, when pupating, will turn itself round, so that, when the pupa is formed, the dorsal surface of the pupa will rest against the perpendicular surface.

Subfamily ACRÆINÆ.

27. TELCHINIA VIOLÆ, Fabricius.
Modecca palmata, Lam. (*Passifloreæ*).

Subfamily NYMPHALINÆ.

28. ERGOLIS TAPROBANA, Westwood.
Tragia involucrata, Linn. (*Euphorbiaceæ*).
Tragia involucrata, var. *cannabina*, Linn. (*Euphorbiaceæ*).

29. ERGOLIS ARIADNE, Linnæus.
Tragia involucrata, Linn. (*Euphorbiaceæ*).
Tragia involucrata, var. *cannabina*, Linn. (*Euphorbiaceæ*).
In Guzerat in the Bombay Presidency it has been bred on *Tragia cannabina*, Linn., which is given by Hooker as a var. of *T. involucrata.*

30. BYBLIA ILITHYIA, Drury.
This species has been bred in Khandeish and the Deccan, both in the Bombay Presidency, on *Tragia involucrata*, Linn., var. *cannabina*, Linn. (*Euphorbiaceæ*). It has not been bred in the Kanara District itself. There is a very interesting paper in Ann. and Mag. of Nat. Hist., sixth series, vol. xviii, p. 333 (1896) by Mr. Guy A. K. Marshall, entitled "Notes on the Genus *Byblia* (=*Hypanis*)," in which he notes that he is convinced "That all the Asiatic and continental African forms of *Byblia* are referable to a single species." Dr. Chr. Aurivillius in Kongl. Svens. Vet.-Akad. Handl., vol. xxxi, n. 5, p. 158 (1898), records *B. ilithyia*, Drury, from Africa, the Cape Verde Islands, Arabia,

and India, and *B. goetzius*, Herbst, from Africa only, giving a woodcut of
the upperside of the forewing of each to allow of their easy identification.

31. EURIPUS CONSIMILIS, Westwood.
Trema orientalis, Blume (*Urticaceæ*).

EGG. The egg is green, shiny, and spherical in shape, though
slightly higher than broad, and has twenty-two prominent ridges from
top to bottom, all parallel to each other in the manner of meridional
lines: these ridges merge into the level of the surface of the egg
towards its top. The breadth is 1·3 mm.

LARVA. · The larva is of the type of *Ergolis*, but has no spines on the
body. Body cylindrical, thickest in the middle, and gradually decreasing
in width to the very narrow hinder end, which terminates in two nearly
parallel, conical processes about 2 mm. in length bearing short hairs; the
body also decreases from the middle of segment 2 (where it is
about the same breadth as at segment 11/12). The head is nearly
square, slightly higher than broad however; the face almost flat,
and each lobe is surmounted by a stout cylindrical process or horn,
which horn is slightly longer than the head is high; it has three
or four short, yellow spines before the middle, and is shortly bifurcated at the top; the front face of the horn is nearly in the same
plane as the face; these two horns diverge at an angle of about 35°,
and are rather widely and squarely separated at the base; on the hind
vertex of the head between the horns are two small conical red-brown
spines; along the side margin of the head in continuation with the
outside edge of the horn are three sharp spines; the surface of the
head is otherwise smooth and shiny; the colour of the head is dark
green, with a long narrow white clypeus, a white band from the base of
each horn running down each side of the central line of the face and
along the sides of the clypeus to the jaws, as also a white band on each
cheek separating the face from the cheek; the horn-spines and the two
vertex spines are tipped with black: the head is higher than segment 2
but about the same breadth. The surface of the body is dull and rough,
each segment being set with seven transverse rows of minute, conical,
yellow tubercles; the colour of the body is dark green with a red spot
in the spiracular region of segments 3, 7, 10, and 12/13, those on
segments 7 and 10 being larger than the other two; the yellow tubercles are each surmounted by a short hair; there is a small brown spot
or two next each spiracle. Spiracles flush, oval, rather large, light
green, with a very narrow, shiny black border, and a central thin white
slit. Length 42 mm. altogether, breadth 7 mm., length of horn 5 mm.,
breadth of head 4 mm. The breadth across the base of the anal

processes or points (that is the breadth of the larva at that base) is about 1·5 mm.

Pupa. The pupa is similar to that of *Apatura* (*Rohana*) *camiba*, Moore, in general facies and mode of suspension. Looked at sideways the pupa is a section of a circle : about a quarter moon-crescent, the ventral line being nearly straight, and the dorsal line highly and evenly curved, the abdominal segments being much compressed laterally and highly carinated in the dorsal line, the edge of the carination being sharp, *i.e.*, thin. The pupa is the same breadth from the shoulders to segment 8, and is in the middle twice as high as broad ; segments 4, 5 and 6 are separated slightly in the dorsal carinated edge, and segment 6 is the highest part of the carination ; the transverse dorsal section in the middle is pear-shaped, the same section across the ventrum from spiracle to spiracle being a compressed semicircle ; the head has two strong, slightly diverging, conical processes, narrowly separated at the base and about 1·5 mm. in length : these head-points are about half as far apart at the tips as the pupa is broad in the middle : the pupa increases evenly in width to the shoulders, which have each two small smooth tubercles ; the thorax is convex transversely ; the cremaster is stout, triangular, flattened above and below. The surface of the pupa is dull, and is transversely wrinkled all over as seen under a lens ; there is a low indistinct ridge from each head-point running back on to the thorax, and the wing-edge is a ridge from the shoulder to segment 4. The spiracles are depressed, oval, and the colour of the pupa. The pupa is green, densely streaked with white on the thorax and head, more obscurely on the rest of the surface ; the dorsal ridge and wing-edge ridge is yellowish ; and there is a rather prominent brown-yellow rugosity on the spiracular line at the hinder margin of segment 7. Length 29 mm. over all, height at middle 12 mm., breadth at middle over 8 mm.

Habits. The egg is laid on the upperside of the leaf, or near the edge on the underside. The larva, on emerging, eats the egg-shell partially, and then makes a bed of silk anywhere on the upperside of the leaf. Having grown somewhat it betakes itself to the middle of the leaf and lies along the midrib near the point, covering the surface of the leaf with a thick carpet of silk ; over this carpet it weaves a network of silk which is free of the surface of the leaf, and on the top of this network the larva rests with its face in the same plane as the ventral surface, so that the horns are resting on the web. When about to pupate the larva wanders, and finally finishes up on the under surface of some leaf, where it undergoes its transformation. The pupa is stoutly attached by the tail only, so that the ventral surface is parallel to the under surface of the leaf.

32. CUPHA PLACIDA, Moore.
Flacourtia montana, Grah. (*Bixineæ*).

Herr H. Fruhstorfer in Berl. Ent. Zeitsch., vol. xliii, p. 198 (1898), has named this species *C. erymanthis maja* from Karwar in North Kanara.

33. ATELLA PHALANTHA, Drury.
Flacourtia montana, Grah. (*Bixineæ*).

34. ATELLA ALCIPPE, Cramer.
Alsodeia zeylanica, Thwaites (*Violaceæ*).

In the second paper, p. 248, it is noted that the larva was found " On a tree, which we believe to be a very local species of *Hydnocarpus* [*Bixineæ*, the Natural Order which next follows the *Violaceæ*], but this required verification." This identification is incorrect, the food-plant being *Alsodeia zeylanica*.

35. CETHOSIA MAHRATTA, Moore.
Modecca palmata, Lam. (*Passifloreæ*).

36. CYNTHIA SALOMA, de Nicéville.
Modecca palmata, Lam. (*Passifloreæ*).

37. APATURA (*Rohana*) CAMIBA, Moore.
Celtis tetrandra, Roxb. (*Urticaceæ*).

Dr. F. Moore in Lep. Ind., vol. iii, p. 3 (1896), gives the genus *Apatura* (part), Fabricius, as a synonym of *Potamis*, Hübner. He uses *Apatura* for the species placed in this paper under *Hypolimnas*, Hübner.

38. PRECIS IPHITA, Cramer.
Strobilanthes callosus, Nees (*Acanthaceæ*).

Dr. F. Moore in Lep. Ind., vol. iv, p. 62 (1899) notes that this species is not a true *Precis*, the type of that genus being the *Papilio octavia* of Cramer, an African species, which has the discoidal cell of the forewing closed, while *Junonia* (in which *iphita* is best placed) has it open. Dr. Chr. Aurivillius in Kongl. Svens. Vet.-Akad. Hand., vol. xxxi, n. 5, p. 131 (1898), gives *Junonia* as a synonym of *Precis*, the latter being the older name. Dr. A. G. Butler in Ann. and Mag. of Nat. Hist, seventh series, vol. iv, p. 373 (1899), notes that " Prof. Aurivillius shows that *Precis* has priority over *Junonia* ; therefore, although the latter is a far more satisfactory name for the genus (became more descriptive), I suppose it will have to go."

39. PRECIS ALMANA, Linnæus.
Hygrophila spinosa, T. Anders. *(Acanthaceæ).*

Mr. Bell notes that the larva feeds on some other unidentified acanthads. In the first paper, p. 272, *Asteracantha longifolia,* Nees[*] *(Acanthaceæ),* and *Lippia nodiflora,* Rich. *(Verbenaceæ)* are recorded : the latter food-plant Mr. Bell considers to be doubtful.

40. PRECIS LEMONIAS, Linnæus.
Nelsonia campestris, Br. *(Acanthaceæ).*
Hygrophila spinosa, T. Anders. *(Acanthaceæ).*
Strobilanthes callosus, Nees *(Acanthaceæ).*
Barleria Prionitis, Linn. *(Acanthaceæ).*

41. PRECIS HIERTA, Fabricius.
Hygrophila spinosa, T. Anders. *(Acanthaceæ).*
? *Barleria* sp. *(Acanthaceæ).*

42. PRECIS ORITHYIA, Linnæus.
Hygrophila spinosa, T. Anders. *(Acanthaceæ).*

43. PRECIS ATLITES, Linnæus.
Hygrophila spinosa, T. Anders. *(Acanthaceæ).*
Barleria sp. *(Acanthaceæ).*

EGG. The egg is barrel-shaped, there being thirteen longitudinal ridges from top to bottom, parallel to each other, and not continued on to the flat top ; these ridges under the lens are finely beaded and are thin, being one-fifth as broad as the interspace at the middle of the egg ; the flattish top of the egg has a small white ring in the centre— the micropyle ; the surface of the egg is shiny and smooth ; the colour is dark green with all the ridges white.

LARVA. The larva resembles in shape those of the larvæ of all the species of *Precis* found in the Kanara district, and the disposition of the spinous processes is the same. There are two such processes, one above the other, on segment 2 below the spiracle on the base of the leg ; two in a horizontal line on the base of the legs of segments 3 and 4 ; a triangle of three on segments 5 and 6 where the leg would be were there one ; two in a horizontal line above and one below, and two in a line on the base of the proleg of segments 7 to 10 ; two, one below the other, on segment 11 ; one subspiracular on segments 3 to 12 ; all except the subspiracular ones dirty watery-white in colour, and set with

---

[*] *Asteracantha* is given by Hooker as a subgenus of *Hygrophila,* and *longifolia* Nees, as a synonym of *Hygrophila spinosa,* T. Anders.

fine white hairs as long as the processes in the subspiracular line, these subspiracular spines being one-fourth as long again as the ones beneath that line ; besides these processes there are also the following :—a dorsal, dorso-lateral, and supra-spiracular process on segments 5 to 11 ; on segments 3 and 4 a dorso-lateral and supra-spiracular process ; on segments 13 and 14 a dorso-lateral process ; on segment 12 two dorsal processes, one in front of the other, as well as a dorso-lateral and supra-spiracular process ; all these processes from the subspiracular ones upwards are of the same length, shiny blackish in colour, set with two whorls of dark yellow-brown spinelets nearly as long as the processes ; the processes are just under 2 mm. in length ; segment 2 has ten simple, slightly-curved, spinous hairs along the front margin, black and rather long ; the surface of the body is covered besides, (and herein lies its difference from other larvæ of *Precis*), with 1 mm. long fine pure white hairs, each hair springing from a minute circular pure white tubercle ; a narrow dorsal line and the whole of the dorsum of segment 3 have no white hairs.　　The anal flap is nearly semicircular in outline and somewhat thickened at the extremity, where it is yellow-ochreous in colour.　　The spiracles are oval, black, with shiny black borders.　　Head rather small, squarish, with the vertex indented, giving an appearance of being bilobed ; the vertex of each lobe bears a conical, shining, ochreous tubercle surrounded by three or four small ones each bearing a hair, the hair of the large tubercle always long and white ; another small tubercle in the middle of each lobe ; some stiff black hairs on the upper part of the face, some soft white hairs about the base ; colour of the head dark bronzy-blackish-brown, with a rather large triangular black clypeus ; colourless labrum ; ochreous basal antennal joint, and blackish second joint.　　Neck dull greenish-black.　　Colour of body velvety black, looking, under the lens, slightly shiny greenish-black ; the abdomen lighter blackish ; a subspiracular band, sending a short shoot up and forwards before each spiracle, legs and prolegs, all brownish-ochreous. The whole larva appears frosted with white on account of the presence of the small white hairs.　　Length 40 mm., breadth 5 mm.

Pupa.　The pupa is almost exactly the same shape as that of *Precis almana*, Linnæus, in every way ; it differs mainly in the colour, which is a dull light brown throughout, with the front faces of the tubercles slightly darker, and the hinder faces somewhat lighter, than the colour of the body ; the head-points, the apex of the thorax, and the sides of the cremaster, dark brown ; the hinder half of segment 8 lighter than the body ; the apex of the thorax is more pointed than that of *P. almana*, the apex being a conical point ; the head-points are much more pronounced, being conical ; the front slope from the apex of the

thorax to the front of the pupa is straight, instead of convex as in
P. almana; the cremaster is smoothly-triangular, and has no tubercles;
the whole pupa is slighter. Spiracles of segment 2 indicated by a small
semicircle of a light red-brown colour on the surface of segment 3; the
other spiracles are narrowish, black, somewhat raised ovals. Length
17·5 mm., breadth 6·25 mm. at the shoulders and at segment 8; between
these points the pupa is somewhat constricted.

HABITS. The egg is laid on a stalk of grass, on the dead stem of
any plant, in fact anywhere; the larva on emergence easily finds its
food-plant, which generally grows in great abundance all around; it lies
on the underside of the leaves and low down on the plant, drops to the
ground curled up when touched, and remains a long time thus. The
food-plant grows chiefly in damp places and always in great quantities.
The pupa as a rule is affixed to a stem or leaf in some thick place, and,
like the larva, is not easy to find. The butterfly is hardly ever found in
jungle, but is very plentiful along the coast in open cultivation,
especially about rice fields and on the banks of tanks or ponds.

44. NEPTIS (*Rahinda*) HORDONIA, Stoll.
Cæsalpinia mimosoides, Lam. (*Leguminosæ*).
Acacia Intsia, Willd. (*Leguminosæ*). Form of larva with short
processes.
Acacia pennata, Willd. (*Leguminosæ*). Form of larva with long
processes.
The form of *N. hordonia* with short processes in the larva is said
to feed also on "Several species of *Albizzia*" (*Leguminosæ*).

45. NEPTIS VIRAJA, Moore.
Dalbergia latifolia, Roxb. (*Leguminosæ*).
Dalbergia volubilis, Roxb.* (*Leguminosæ*).
Albizzia Lebbek, Benth. (*Leguminosæ*).

46. NEPTIS VARMONA, Moore.
Peas of several kinds (*Leguminosæ*).
The *N. leucothoë*, Cramer, of the second Kanara paper, p. 251, that
species = *N. matuta*, Hübner, according to Dr. Moore, who restricts it to
Java and Borneo.

47. NEPTIS COLUMELLA, Cramer.
The *N. ophiana*, Moore, of the second Kanara paper, which is a
synonym of *N. columella*, Cramer. Though so widely-spread, this species

---

* This food-plant has been wrongly identified—there is no such species—as
Dalbergia racemosa, in the second Kanara paper, p. 251.

has never been bred, though Messrs. Davidson, Bell and Aitken once obtained eggs and young larva, but did not record the food-plant.

48. NEPTIS KALLAURA, Moore.

Thespesia Lampas, Dalz. and Gibs. (*Malvaceæ*).
Dalbergia confertiflora, Benth. (*Leguminosæ*).
Wagatea spicata, Dalz. (*Leguminosæ*).
Trema orientalis, Blume (*Urticaceæ*).

LARVA. The larva is very similar to that of *N. jumbah,* Moore, except that it is furred-looking (minutely spined) all over. The processes are exactly the same as in *N. jumbah,* except that the subdorsal ones of segment 6 are shortly conical, and segments 7 and 8 have a small subdorsal tubercle, lacking in *N. jumbah*; all the processes are covered with cylindrical, rather long tubercles each bearing a rather long bristle in continuation; the head has the points on the vertex of each lobe rather more accentuated than in *N. jumbah,* and the surface is entirely covered with the same cylindrical hair-bearing tubercles as the rest of the body. The spiracles are roundly oval, broadly black-bordered, with light brown centres reaching from top to bottom, that is, the black border is thin at the top and the bottom. The arrangement of the coloration is exactly as in *N. jumbah*: it is very light brown-pink all over, including the head, except laterally on segments 9 to 14 where it is deep olive-moss-green strongly suffused with rusty orange, except along the borders, and a short dark moss-green diagonal line in the centre of each segment 10 to 12; there is the merest tinge of the same olive-moss-green dorsally on segments 5 to 9, with a dorsal thin lighter line; the head has a thin dark border to the clypeus, and a slightly dark surface towards the vertex, which dark colour, however, is only visible between the light tubercles. Length 19 mm., breadth at middle 5·5 mm., the middle being the broadest part.

PUPA. The pupa is very like that of *N. jumbah.* The two conical head-points are wide apart, the points at the base of the antennæ (one at the base of each) are half the distance apart and about the same size, the point on the shoulder is small and sharp, the subdorsal conical tubercle on segment 6 is rather large; and there is a low, not accentuated, ridge from spiracle to spiracle on segment 7, which is slightly curved convexly towards segment 8; the thorax has a rather highly peaked, laterally flattened apex, and this apex is rounded in outline seen from the side, the hinder slope being at an angle of about 75° to the longitudinal axis of the pupa; the wing outline, that is the expansion outline (the inner margin of the forewing), is very highly curved: nearly a semicircle, and its apex rises to nearly the same height as the apex of, the thorax; the dorsal line of the abdomen is slightly toothed at the hinder margin of segments 8 to 11; the cremaster is triangular, with strong prominent waved sustensor ridges; the spiracles are rather small, oval, shiny and black.   The colour of the pupa is always the same: a light fresh bone-colour with a brown-pink suffusion on the dorsal region of segments 10 to 12, a brown-gold streak along the ridge of segment 7, a brown-gold semicircle in front of the tubercles of segment 6 on the dorsum, and a similar streak along the hinder margin of the thorax; a large subdorsal mother-of-pearl gold patch on segment 4, a smaller one on segment 5, some gold at the base of the tubercle on segment 6, a silver-golden glean on the sides of the thorax and on the whole wing-surface in certain lights, the whole of segment 2 silvery, a thin dorsal yellow line on the abdominal segments.   Length 16 mm, breadth at segment 6 from apex to apex of wing expansion 8 mm., distance between head-points nearly 3·5 mm.

HABITS.   The habits of the larva are the same as those of *N. jumbah* in every way.   The pupa is of course suspended by the tail and hangs down perpendicularly, and is always attached to the underside of a leaf.

49.   NEPTIS JUMBAH, Moore.
Thespesia Lampas, Dalz. and Gibs. (*Malvaceæ*).
Kydia calycina, Roxb. (*Malvaceæ*).
Bombax malabaricum, DC. (*Malvaceæ*).
Helicteres Isora, Linn. (*Sterculiaceæ*).
Grewia Microcos, Linn. (*Tiliaceæ*).
Zizyphus Jujuba, Lamk. (*Rhamneæ*).
Zizyphus Xylopyrus, Willd. (*Rhamneæ*).
Zizyphus rugosa, Lamk. (*Rhamneæ*).
Millettia racemosa, Benth. (*Leguminosæ*).
Dalbergia latifolia, Roxb. (*Leguminosæ*).
Wagatea spicata, Dalz. (*Leguminosæ*).
Xylia dolabriformis, Benth. (*Leguminosæ*).
Trema orientalis, Blume (*Urticaceæ*).

50.   CIRRHOCHROA THAIS, Fabricius.
Hydnocarpus Wightiana, Blume (*Bixineæ*).

51.   HYPOLIMNAS BOLINA, Linnæus.
Portulaca oleracea, Linn. (*Portulaceæ*).
Fleurya interrupta, Gaud. (*Urticaceæ*).
Elatostema cuneatum, Wight (*Urticaceæ*).

Dr. F. Moore in Lep. Ind., vol. iv, p. 135   (1899), gives the genus *Hypolimnas*, Hübner, as a synonym of *Apatura*, Fabricius.

52. HYPOLIMNAS MISIPPUS, Linnæus.
Portulaca oleracea, Linn. (*Portulaceæ*).

53. PARTHENOS VIRENS, Moore.
Zehneria umbellata, Thwaites (*Cucurbitaceæ*).

54. LIMENITIS (*Moduza*) PROCRIS, Cramer.
Stephegyne parvifolia, Korth. (*Rubiaceæ*).
Wendlandia exserta, DC. (*Rubiaceæ*).
Mussænda frondosa, Linn. (*Rubiaceæ*).

55. ATHYMA PERIUS, Linnæus.
Glochidion lanceolarium, Dalz. (*Euphorbiaceæ*).
Glochidion velutinum, Wight (*Euphorbiaceæ*).

56. ATHYMA RANGA, Moore.
Linociera malabarica, Wall. (*Oleaceæ*).
Olea dioica, Roxb. (*Oleaceæ*).
Athyma ranga and *A. mahesa*, both of Moore, being synonymous, the older name has to be used for this species. In the second Kanara paper, p. 254, it appears under the latter name.

57. ATHYMA INARA, Doubleday and Hewitson.
Wendlandia exserta, DC. (*Rubiaceæ*).
Mussænda frondosa, Linn. (*Rubiaceæ*).
Glochidion zeylanicum, A. Juss. (*Euphorbiaceæ*).
Glochidion velutinum, Wight (*Euphorbiaceæ*).

58. ATHYMA SELENOPHORA, Kollar.
Adina cordifolia, Hook. f. (*Rubiaceæ*).

59. SYMPHÆDRA NAIS, Forster.
Diospyros melanoxylon, Roxb. (*Ebenaceæ*).

60. EUTHALIA (*Dophla*) LAUDABILIS, Swinhoe.
Diospyros melanoxylon, Roxb. (*Ebenaceæ*).
Diospyros Candolleana, Wight (*Ebenaceæ*).

61. EUTHALIA (*Cynitia*) LEPIDEA, Butler.
Careya arborea, Roxb. (*Myrtaceæ*).
Melastoma malabathricum, Linn. (*Melastomaceæ*).

62. EUTHALIA GARUDA, Moore.
Mangifera indica, Linn. (*Anacardiaceæ*).

Anacardium occidentale, Linn. (*Anacardiaceæ*).
Loranthus scurrula, Linn. (*Loranthaceæ*).

" Larva commonly feeds on the mango and cashewnut tree, also on the mulberry and the rose, and on *Loranthus*" (First Kanara paper, p. 276).

63. EUTHALIA LUBENTINA, Cramer.
Loranthus scurrula, Linn. (*Loranthaceæ*).

64. VANESSA * CARDUI, Linnæus.
Zornia diphylla, Pers. (*Leguminosæ*).
Blumea sp. (*Compositæ*).

65. CYRESTIS THYODAMAS, Boisduval.
Ficus bengalensis, Linn. (*Urticaceæ*).
Ficus indica, Linn. (*Urticaceæ*).
Ficus glomerata, Roxb. (*Urticaceæ*).

66. KALLIMA HORSFIELDII, Kollar.
Strobilanthes callosus, Nees (*Acanthaceæ*).
Eranthemum malabaricum, Clarke (*Acanthaceæ*).

67. DOLESCHALLIA POLIBETE, Cramer.†
Eranthemum malabaricum, Clarke (*Acanthaceæ*).

68. CHARAXES WARDII, Moore.
Rourea santaloides, W. and A. (*Connaraceæ*).
Wagatea spicata, Dalz. (*Leguminosæ*).

Dr. F. Moore in Lep. Ind., vol. ii, p. 262, pl. clxxxviii, figs. 2, 2*a*, male; 2*b*, larva and pupa (1896), has named the South Indian form of *Charaxes schreiber*, Godart—*Eulepis wardii.* The Hon. Walter Rothschild in Nov. Zool., vol. vi, p. 222 (1899), refers to it as local race *a*, *Eulepis schreiber wardi.* The butterfly appears as *C. schreiberi*, Godart, in the second Kanara paper, p. 257.

69. CHARAXES ATHAMAS, Drury.
Grewia sp. (*Tiliaceæ*).
Cæsalpinia mimosoides, Lam. (*Leguminosæ*).

* *Pyrameis cardui* of previous papers. The types of both *Vanessa* and *Pyrameis* is the *Papilio atalanta* of Linnæus. *Vanessa* is the older name of the two. *Pyrameis* therefore becomes a synonym of *Vanessa.*

† In Lep. Ind., vol. iv, p. 155 (1900), Dr. F. Moore has named the *Doleschallia* from N.-E. and S. India, Ceylon and Burma—*D. indica.* He restricts *D. polibete* to the Malayan Islands, Amboina, Waigiou and Batchian.

Poinciana regia, Bojer (*Leguminosæ*).
Acacia pennata, Willd. (*Leguminosæ*).
Albizzia Lebbek, Benth. (*Leguminosæ*).

Dr. F. Moore in Lep. Ind., vol. ii, p. 252 (1896) records *Eulepis athamas* from South India, while the Hon. W. Rothschild records in Nov. Zool., vol. vi, p. 249 (1899) " a¹. *Eulepis athamas agrarius* f. (temp. ?) *madeus,* Rothschild" from "Karwar, September and October; N. Canara, September."

70. CHARAXES FABIUS, Fabricius.
Wagatea spicata, Dalz. (*Leguminosæ*).
Tamarindus indica, Linn. (*Leguminosæ*).

71. CHARAXES IMNA, Butler.
Saccopetalum tomentosum, Hook. f. and T. (*Anonaceæ*).
Aglaia Roxburghiana, Miq. (*Meliaceæ*).

Family LEMONIIDÆ.
Subfamily LIBYTHÆINÆ.

72. LIBYTHEA RAMA, Moore.
Celtis tetrandra, Roxb. (*Urticaceæ*).

LARVA. At first sight the larva reminds one rather of the larva of a species of *Catopsilia* (*Pierinæ*). It is the same thickness from segment 4 to segment 10, narrowing to the head and to segment 14; segment 2 is about the same breadth as the head at the front margin, but is wider behind, and has the front margin very slightly produced in the dorsal line; the anal flap is rounded behind, and its dorsal slope is nearly a quarter of a circle, the extremity nearly touching the resting surface; it has a depressed dorsal oval mark two-thirds the width of the segment reaching from its hinder extremity towards the front margin, which surface or mark is covered with brown streaks and has no hairs on it, as has the rest of the segment; the prolegs are rather long. The head is small, being only about half as broad as the larva is at the centre; it has a dull smooth surface set with minute rather sparse dark bristles; a rather large clypeus; it is round in shape, with a shallow broad curved depression on the vertex; is green in colour with brown markings as seen under a lens, antennæ reddish, labrum green, eyes black, and some light hairs about the jaws. The surface of the body is dull, each segment has four broad ridges, that is, each segment is divided into four by thin depressed transverse lines, and on these ridges are rows of minute bristle-like black hairs all over, some even on the ventrum. The spiracles are light yellow, black-rimmed, oval, and flush to the surface, of ordinary size. Colour dark green, sometimes with a brownish tinge, with a thin, dorsal, light yellow line from segments 4 to 12, and a narrow, yellow, supra-spiracular band from the head to the anal end. Length 26 mm., breadth at middle 4·5 mm., breadth at head 2·25 mm.

PUPA. Unlike that of any butterfly from the Kanara district, though it somewhat resembles that of *Ergolis*, but is fixed with its longitudinal axis parallel to the surface to which it is attached like that of *Elymnias caudata*, Butler; the ventral line is therefore straight from the head to segment 10, and thence the rest of the pupa is at right-angles to its longitudinal axis; the front of the pupa seen from above is absolutely square, the head ending in a broad straight edge; the head and segment 2 form a trapezoidal piece, which is broadest transversely to the pupa-length; the sides or lateral outline of this trapeze being absolutely straight lines; the dorsal line of the pupa in segments 1 and 2 is slightly convex; the thorax forms at the shoulders the broadest part of the pupa, sloping out suddenly laterally at an angle of 135° with the lateral line of the head-piece; the thorax is somewhat convex and highly carinated along the dorsal line, this carination starting from the front margin in—seen laterally—an absolute straight line to just before the hinder margin, where it ends abruptly in a somewhat rounded peak; the dorsal outline falling thus abruptly from the peak to the hinder margin of the thorax; the dorsal outline of the abdomen starts from segment 3/4 and ascends to a small sharp peak at the margin of segment 5/6, whence it descends gradually in a very slight carination to segment 8/9, and then in a curve of a quarter circle to the cremaster; a thin linear low carination connects the point of the shoulder with the abdominal peak, and the wings are slightly thickened at and behind the shoulders; the transverse section of the abdomen after the peak is nearly circular; the abdominal peak is somewhat higher than the apex of the thoracic carination, and the straight top of the thoracic carination is at an angle of 45° to the longitudinal axis of the pupa. The spiracles of segment 2 are depressed narrow slits, the other spiracles are light, nearly white, ovals facing somewhat forwards. The cremaster is dorsally triangular, and embraces somewhat the last segment, its attachment-surface being considerably longer in the sense of the length of the pupa than its breadth. Colour of pupa light green, with the tops of all the carinations yellow, with a black speck on the abdominal peak; the surface of the pupa is smooth and somewhat shiny. Length 12·5 mm., breadth at the shoulders 5 mm., breadth at the front of the head just over 3 mm., height at the abdominal peak 6 mm., height at the apex of the thorax 5 mm.

HABITS. The eggs are laid on the young shoots and leaves, generally

J. II. 30

on a low shrub near a nulla with water in it and open to the sunlight. The larvæ live generally on the underside of the leaves, eating all but the ribs or veins to which one finds them hanging. They emit much web and fall by a silk thread when disturbed, but only when touched or otherwise violently molested ; the larva rests with its true legs off the surface, and its head curved down and often turned to one side. The pupa is formed always on the underside of a leaf, and rests quite parallel to the surface of the leaf. The larva reminds one forcibly of a pierine larvæ of the *Oatopsilia, Ixias* or *Teracolus* type. The butterfly, which is rare in the Kanara district, appears to be found only in the neighbourhood of its food-plant.

Subfamily NEMEOBIINÆ.

73.   ABISARA FRATERNA, Moore.
Embelia robusta, Roxb. (*Myrsineæ*).
Ardisia humilis, Vahl (*Myrsineæ*).

Family LYCÆNIDÆ.

74.   NEOPITHECOPS ZALMORA, Butler.
Glycosmis pentaphylla, Correa (*Rutaceæ*).

75.   SPALGIS EPIUS, Westwood.
The larva is wholly carnivorous, feeding on *Coccidæ,* and not touching vegetable food at all.

76.   MEGISBA MALAYA, Horsfield.
This species has never been bred.

77.   PLEBEIUS* TROCHILUS, Freyer.
Zornia diphylla, Pers. (*Leguminosæ*).
Heliotropium strigonum, Willd. (*Boragineæ*).
LARVA. Of the ordinary lycænid shape, but rather narrowed at both ends, broadest a little before the middle, somewhat narrow. Head small, hidden under segment 2 in repose, having a rather long neck ; nearly black with a whitish labrum ; shiny, glabrous, and round. Segment 2 swollen on the front margin, somewhat depressed in the

---

* Dr. A. G. Butler in Ann. and Mag. of Nat. Hist., seventh series, vol. v, p. 61, n. 23 (1900), records *Plebeius trochilus,* Freyer, from Nyasaland, noting that " According to de Nicéville this is a *Chilades.*" In this I followed Dr. F. Moore, who placed the *Lycæna putli* of Kollar, which is a synonym of *Lycæna trochilus,* Freyer, in the genus *Chilades.* I do not know what the type of *Plebeius* is.

centre of the dorsum, in shape a perfect semicircle. Anal end rounded, the posterior segments sloping and flattened dorsally ; the usual gland, and protrusible organs on segment 12, present. Spiracles very minute, round, edged with dark colour, above the marginal red band. The surface of the body covered moderately densely with small, cylindrical, white tubercles, from the top of each of which springs a rather short, brownish hair. The colour of the body is green, with a dorsal red band edged with white, and a marginal (subspiracular) similar band edged broadly below with white; two parallel diagonal white lines on each segment, laterally, between the dorsal and marginal red bands ; on the dorsal red band near the front margin of each segment 3 to 10 is a small, green spot, which, on segments 3 to 5 at least, is a distinct depression, circular in shape. Length 7 mm., breadth nearly 2·5 mm.

PUPA. The pupa is distinctly lycænid in shape, but is rather long and narrow, is slightly constricted dorsally behind the thorax, though not laterally ; the thorax is somewhat humped ; the front of the pupa is blunt, squarish ; the hinder end narrow and rounded ; thickest about segment 8 and highest about segment 7. The spiracles of segment 2 are indicated by an oval, white mark. The other spiracles are small, nearly round, white. The surface of the body is rather shiny, covered with small, cylindrical, white tubercles which are each surmounted by a rather long, somewhat curved, hair, these hairs are longest on the front portion of the pupa, and are nearly colourless. The colour of the pupa is green, very watery-coloured on the wings and anterior parts. Length 6 mm., breadth 2·75 mm.

HABITS. The egg is laid in the axil of a leaf, on the flower, or on a pod, and the larva, on emerging, gets inside the flower or pod as the case may be and feeds therein, changing the flowers, pods, &c., as necessary. It changes to a pupa on a leaf or in some such place, and the pupa has a lax tail-attachment and a body-band.

78. CHILADES LAIUS, Cramer.
Citrus medica, Linn. (*Rutaceæ*).
Now recorded for the first time from Anshi and Gairsoppa in the Kanara District.

79. CYANIRIS PUSPA, Horsfield.
Cylista scariosa, Ait. (*Leguminosæ*).
Xylia dolabriformis, Benth. (*Leguminosæ*).

80. CYANIRIS LIMBATUS, Moore.
Has not been bred.

81. ZIZERA LYSIMON, Hübner.
Zornia diphylla, Pers. (*Leguminosæ*)

82. ZIZERA GAIKA, Trimen.
Nelsonia campestris, Br. (*Acanthaceæ*).

83. ZIZERA OTIS, Fabricius.
Zornia diphylla, Pers. (*Leguminosæ*).

LARVA. The larva is of the ordinary limaciform shape like those of the genera *Cyaniris*, *Jamides* and *Polyommatus*. Segment 2 is semicircular with a dorsal triangular depression as usual; the anal end is dorsally flattish, with the dorsal line inclined to the axis of the length of the larva at an angle of about 30°, and semicircularly rounded in outline, with the two cylindrical, protrusible organs and a slit-shaped gland; the body is broadest from segment 5 to segment 7, and highest at segment 5; segment 3 overhangs segment 2 as usual. The head is round, shiny, smooth, yellow, the jaws dark brown, the labrum white. Before the last moult the head is entirely dark shiny red-brown coloured and hidden under segment 2. The surface of the body is dull, covered with minute white tubercles each surmounted by a minute, sharp, white spine or hair; there is a subdorsal row of three or four curved, shiny, dark brown, rather large (for the size of the larva) hairs on segments 3 to 10, only two hairs on each side on segments 7 to 10; the margin of the body is rather densely hairy, especially on segments 2 and 14. Spiracles shiny, minute, round (or rather hemispherical), yellow. Colour grass green, with a subspiracular yellow band from segments 5 to 12; an indistinct dark dorsal line. Length 9 mm., breadth 3 mm.

PUPA. The pupa is of the ordinary shape of those of the genera *Cyaniris*, *Jamides* and *Polyommatus*, broadest at the middle, highest at the thorax, constricted in the dorsal line behind the thorax, rounded narrowly behind, narrowly square in front, the vertex of the head is in a plane perpendicular to the longitudinal axis of the pupa and rather large; segment 2 is straight on the front margin, with a slight triangular shallow and wide indentation in the dorsal line, and a hind margin curved concavely towards the front of the pupa: its dorsal line is in a plane at an angle of nearly 45° to the length of the axis, and the segment is very slightly convex transversely to the length of the pupa; the thorax is humped, and the line joining the front and hind margins is in a plane at an angle of about 30° to the longitudinal axis; the dorsal line of segments 4 to 14 is convex; the ventral line of the pupa is straight; the wings are slightly expanded laterally in parallel lines. The surface of the body is smooth, slightly shiny, and covered with erect, stiff, minute, light hairs, especially along the front margin of segment 2 and about the anal end; these hairs are simple and pointed. The spiracles of segment 2 are indicated by smooth, oval, yellow surfaces, the other spiracles are minute, shiny, convex, white surfaces. The colour of the pupa is light green with a black, dorsal stripe to segment 2; a dark, dorsal, thoracic line, a black smudge along the border of the wing at segments 4 and 5, two supra-spiracular spots. on segments 7 to 12, and a dorsal, dark green line on the abdominal segments; the wings and shoulders are slightly blotched with brownish. Length 7 mm., breadth 3 mm.

HABITS. The eggs are laid anywhere on the plant, on the leaves, stalks or flowers; the larva eats the flowers, pods, &c., and is difficult to find. The pupa, attached by the tail and by a body-band, is formed anywhere convenient, on the upperside or underside of any leaf, either dead or alive.

84. LYCÆNESTHES EMOLUS, Godart.
Wagatea spicata, Dalz. (*Leguminosæ*).
Saraca indica, Linn. (*Leguminosæ*).
Terminalia paniculata, Roth (*Combretaceæ*).
Combretum extensum, Roxb. (*Combretaceæ*).

85. LYCÆNESTHES LYCÆNINA, Felder.
Buchanania latifolia, Roxb. (*Anacardiaceæ*).
Wagatea spicata, Dalz. (*Leguminosæ*).

LARVA. The larva is of the usual onisciform shape, is rather broad and stout; the segments, being somewhat swollen dorsally, are very distinctly marked, so that the lateral view of the dorsal line shows considerable constrictions between the segments; the larva is broadest and highest at segment 7, segment 2 has the usual dorsal large depression and is semicircular, hiding the head as usual; the anal segment is thickened round the margin, and slopes dorsally at about an angle of 45° to the longitudinal axis, and is semicircular in the hinder outline; the gland, which is transverse and mouth-shaped, and the usual circular-mouthed, white, protrusible organs on segment 12, are present. Head small, shiny, smooth, yellowish; the eyes black. Spiracles of the ordinary size, round, white. Surface of the body dull, covered with small, simple, reddish hairs, which are not very densely disposed, and are for the most part nearly adpressed to the surface. The colour is dark green, with a deep rose-coloured, rather fine, but very distinct, dorsal line, and a large, triangular, greenish-yellow, subdorsal patch touching the dorsal line in one basal angle (the base of the triangle‐being the hinder margin of the segment) on each segment 3 to 10, each triangle being bounded narrowly with deep rose colour exteriorly (towards the spiracles); there is a marginal (underneath the spiracles on the margin of the larva) yellow band, interrupted at the margins of the segments by a deep rose-coloured mark, along segments 4 to 10. The dorsal line extends from segment 3 to the anal end. There is a small round depression on the dorsum of segments 3, 4 and 5, also one on the front margin of segment 2, and a lateral longitudinal depression parallel to the margins of the segments on segments 3 to 6. There is a fringe of porrect hairs round the margin of segment 2. Length 11 mm., breadth 5 mm.

PUPA. The pupa is nearly exactly like that of *Lycænesthes emolus,* Godart, except that it is slightly more robust, *i.e.,* stouter and more compact. The diamond-shaped dorsal mark at the hinder margin of the thorax is also present, it is this marking indeed that makes the pupa so similar to that of *L. emolus.* The pupa is blunt in front, the vertex of the head is flat, and in a plane at right-angles to the longitudinal axis of the pupa; the head is not visible from above; segment 2 is more or less semicircular in outline as to the front margin, though somewhat squarish to fit the flat head-surface; no dorsal constrictions, except that segment 4 is slightly lower than the apex of the thorax; no lateral constriction; the thoracic dorsal slope is gradual and in the same line with that of segment 2, which is at an angle of less than 45° with the longitudinal axis of the pupa; the apex of the thorax, which is just near its hinder margin, is the highest point of the pupa, and the broadest part is at segment 7, though the breadth varies little from the shoulders to that segment; the anal end of the pupa is somewhat broadly rounded. The spiracles of segment 2 are indicated by narrow white slits; the other spiracles are raised, oval and white. The surface of the body is covered with very minute, white tubercles which are not very densely disposed. The colour is green with a dorsal thoracic yellow line, the thoracic dorsal diamond being yellow margined with brown; there is a lateral interrupted yellowish line; the ventrum and underside are whitish; the margins of segments 1/2 and 2/3 and the wings show whitish-yellow; and the whole pupa is more or less spotted-looking. Length 10 mm., breadth at centre 4·5 mm., height at thoracic apex 4 mm.

HABITS. The egg is laid on a flower or in an axil of a flower stalk, and the caterpillar at first bores into a flower bud, but afterwards lives outside curled round the flowers generally on which it feeds, being very difficult to see owing to its patchy coloration. It pupates amongst the flowers, or on a flower stem, or on a leaf, &c., fixing itself by a body-band and by the tail.

86. TALICADA NYSEUS, Guérin.
Bryophyllum calycinum, Salisb. (*Crassulaceæ*).

87. EVERES ARGIADES, Pallas.
Cylista scariosa, Ait. (*Leguminosæ*).

88. NACADUBA MACROPHTHALMA, Felder.
Has never been bred.

89. NACADUBA HERMUS, Felder.
Has never been bred.

90. NACADUBA NOREIA, Felder.
Acacia Intsia, Willd. (*Leguminosæ*).
Acacia Intsia, var. *cæsia*, W. and A. (*Leguminosæ*).
Acacia pennata, Willd. (*Leguminosæ*).

91. NACADUBA ATRATA, Horsfield.
Wagatea spicata, Dalz. (*Leguminosæ*), *N. plumbeomicans.*
Embelia robusta, Roxb. (*Myrsineæ*), *N. atrata.*
Ardisia humilis, Vahl (*Myrsineæ*), *N. atrata.*
Under this name I have included *N. plumbeomicans*, Wood-Mason and de Nicéville, a species originally described from the Andamans, as it is doubtfully distinct from *N. atrata*, Horsfield, although in the second Kanara paper separate descriptions are given of the larva and pupa of both.   Mr. Bell thinks that the two butterflies may be distinct.

92. NACADUBA DANA, de Nicéville.
This species has never been bred.

93. JAMIDES BOCHUS, Cramer.
Butea frondosa, Roxb. (*Leguminosæ*).
Pongamia glabra, Vent. (*Leguminosæ*).
Xylia dolabriformis, Benth. (*Leguminosæ*).

94. LAMPIDES ELPIS, Godart.
Kæmpferia pandurata, Roxb. (*Scitamineæ*).
Hedychium coronarium, Kœnig (*Scitamineæ*).
Amomum microstephanum, Baker (*Scitamineæ*).
Elettaria Cardamomum, Maton (*Scitamineæ*), the cultivated cardamom.

95. LAMPIDES CELENO, Cramer.
Abrus precatorius, Linn. (*Leguminosæ*).

Pongamia glabra, Vent. (*Leguminosæ*).
Saraca indica, Linn. (*Leguminosæ*).

96. CATOCHRYSOPS STRABO, Fabricius.
Schleichera trijuga, Willd. (*Sapindaceæ*).
Ougeinia dalbergioides, Benth. (*Leguminosæ*).
Cylista scariosa, Ait. (*Leguminosæ*).

97. EUCHRYSOPS CNEJUS, Fabricius.
Ougeinia dalbergioides, Benth. (*Leguminosæ*).
Cylista scariosa, Ait. (*Leguminosæ*).
In "The Entomologist," vol. xxxiii, p. 1 (1900), Dr. A. G. Butler
describes the genus *Euchrysops,* which differs from the genus *Catochrysops,* Boisduval (which has hairy eyes), by having the eyes of the
imago " Quite smooth instead of hairy."

98. EUCHRYSOPS PANDAVA, Horsfield.
Wagatea spicata, Dalz. (*Leguminosæ*).
Xylia dolabriformis, Benth. (*Leguminosæ*).

99. TARUCUS THEOPHRASTUS, Fabricius.
Zizyphus Jujuba, Lamk. (*Rhamneæ*).

100. TARUCUS TELICANUS, Lang.
Ougeinia dalbergioides, Benth. (*Leguminosæ*).
This is an older name for *Tarucus plinius,* Fabricius, of the Kanara
lists. In the first Kanara list, p. 353, it is noted that the larva in Bombay feeds on *Sesbania aculeata,* Pers. (*Leguminosæ*).

101. CASTALIUS ROSIMON, Fabricius.
Zizyphus Jujuba, Lamk. (*Rhamneæ*).

102. CASTALIUS ANANDA, de Nicéville.
Zizyphus Jujuba, Lamk. (*Rhamneæ*).
Zizyphus Xylopyrus, Willd. (*Rhamneæ*).
Zizyphus rugosa, Lamk. (*Rhamneæ*).
Terminalia paniculata, Roth (*Combretaceæ*).
Loranthus longiflorus, Desrouss. (*Loranthaceæ*).

103. CASTALIUS ETHION, Doubleday and Hewitson.
Zizyphus Jujuba, Lamk. (*Rhamneæ*).
Zizyphus Xylopyrus, Willd. (*Rhamneæ*).

104. CASTALIUS DECIDIA, Hewitson.
Zizyphus rugosa, Lamk. (*Rhamneæ*).

105. POLYOMMATUS BŒTICUS, Linnæus.
Butea frondosa, Roxb. (*Leguminosæ*).
Cajanus indicus, Spreng. (*Leguminosæ*).

106. AMBLYPODIA ANITA, Hewitson.
Olax scandens, Roxb. (*Olacineæ*).

107. IRAOTA TIMOLEON, Stoll.
Ficus bengalensis, Linn. (*Urticaceæ*).
Ficus indica, Linn. (*Urticaceæ*).
Ficus glomerata, Roxb. (*Urticaceæ*).

108. SURENDRA QUERCETORUM, Moore.
Acacia Intsia, Willd. (*Leguminosæ*).
Acacia pennata, Willd. (*Leguminosæ*).

109. THADUKA MULTICAUDATA, Moore.
Trewia nudiflora, Linn. (*Euphorbiaceæ*).

EGG. The egg is similar in shape to that of the species of the genus *Arrhopala*, *i.e.*, it is dome-shaped, but is broadest above the base; it looks, however, to be turban-shaped, *i.e.*, flat on the top. The reason of this is that there are two rows of long delicate feathery-looking spikes, finely bifurcated at the top, placed at right-angles to the polar axis of the egg and slightly converging, one row to the other, at the points. There are two-and-a-half cells from the base to near the summit of the egg, and these spikes are situated where the walls of the middle row of cells intersect with the walls of the top perfect cells and the bottom demi-cells. The cells are large, nearly regularly quadrilateral, with fine rather high walls, and are flat-bottomed. On the summit the egg is punctuated, and has a rather large, central, circular depression (micropyle). The colour is finely granulated green, the walls of the cells and the spikes being white. The cells round the "equator" of the egg are ten in number. The breadth of the egg is 0·6 mm., and the height 0·4 mm.

LARVA. The larva in shape and habits agrees in all respects absolutely with that of the species of the genus *Arrhopala*. The head is hidden beneath the second segment, is shiny, rather large, and black. Segment 2 is semi-circular, very slightly indented in the middle of the front margin; the middle dorsal depression is semi-elliptical (round end anterior), velvety black, with a dorsal green line. Segment 3 is suddenly somewhat higher and broader than segment 2. Segments 4 to 11 nearly of the same breadth and length ; segments 12 to 14 decrease in breadth, the anal segment being rather flat, thickened round the outer margin, and broadly rounded at the end, with a dorsal square velvety black patch bisected by a dorsal green line. The whole larva is depressed, being of one height from segment 4 to segment 10, both inclusive. The spiracles are plainly visible, rather long ovals in shape, and yellow in colour. The surface of the body is covered with minute, short, star-like hairs, light-coloured and sparse ; on the black patches they are black and denser than anywhere else ; the surface is also laterally corrugated on each segment, with a few latero-ventral deep punctuations ; the whole margin of the larva bears long simple hairs placed somewhat far apart. The gland on segment 11 is large and conspicuous, surrounded by an oval, deep black patch, which patch has a thin green line all round just within its margin ; the longer axis of the oval is transverse to the body length. The organs on segment 12 are circular-mouthed, protruded as white cylinders. The colour of the larva is light green, with a dorsal dark green line flanked on either side by a white line ; a latero-dorsal white line ; a lateral white line : all six white lines commence on segment 3, and end just in front of the gland on segment 11. The segments are distinct. The space on the dorsum between the lateral white line and the white line flanking the dorsal green line is obscure rose-coloured. The larva changes to a brown-pink before turning into a pupa. The total length of the full-grown larva is 19 mm., the breadth is 7·5 mm.

PUPA. The pupa resembles that of the species of the genus *Arrhopala*. The head is hidden, bowed. Segment 2 is large, very convex, with a semi-circular front margin, ascending in the dorsal line at the same angle as the front of the thorax. Thorax very evenly curved to the apex, then evenly descending to segment 5 from the rounded apex. The pupa is slightly constricted dorsally behind the thorax, not at all laterally ; dorsal outline straight from segments 5 to 8, then descending gradually to the front margin of segment 10, after which the surface is perpendicular to the longitudinal axis of the pupa. In lateral outline the pupa increases from the head to the slightly angular shoulders, then still more, though slightly, to segments 7 and 8, after which it decreases gradually to the end ; the end is rounded, though not broadened out like a hoof, and is applied round the margin closely to the surface of suspension. The shoulder has a small tubercular swelling. The spiracular expansion of segment 2 is small, long, facing forwards. Spiracles with swollen lips, oval, conspicuous, light brown in colour. Gland scar and marks of organs of segment 12 conspicuous. Body surface very finely covered with minute tubercular granules which sometimes coalesce into lines. Colour very dark rosy-brown, lighter on the abdomen and dorsum; a dorsal light brown line on segments 2 to 4; a row of two or three light brown spots parallel to the margins of the segments on each side of the dorsal line on segments 6 to 10. Underparts of pupa light rosy-brownish-yellow. The pupa is fastened by the tail and a median band. Total length 14 mm., breadth at segment 7 is 6·5 mm., height at the apex of the thorax 5·3 mm., breath at the shoulders 5 mm.

HABITS. The eggs are laid singly or in tows and threes on leaves, leaf stalks, stems and twigs, even on the trunk of the tree, generally in cracks, crevices, or axils. One female lays many on the same tree. The butterfly is fond of the sun, and sits for long periods on one leaf basking with closed wings, sometimes on a twig, stem, or trunk of a tree; with care it can be caught in the fingers, but once on the wing its flight is extremely rapid though not sustained. The larva from the first moult makes a house or shelter for itself by turning over a bit of the edge of a leaf, fixing it and lining it with silk, and eating holes all round through both the layers of the leaf except on the outer side; it makes new nests as required, feeding always on the tender leaf on which is its house. It wanders off to some crevice in the bark, hole in the tree, or even down to the ground, to pupate, getting under a dead leaf, or clod of earth, or into a hole in the ground, in the latter case. A dozen pupæ are sometimes found together. The butterfly is difficult to kill by squeezing. Some of the larvæ are attended by ants of the genus *Cremastogaster*, some are not: at any rate the ants do not appear to care much for them, as they will leave them on the slightest alarm. The pupæ are sometimes attended by these same ants. The reason the butterfly is so rare is most probably because the tree on which the larva feeds is, as a general rule, about 150 feet in height, with a clear stem of some 60 feet, and the butterflies keep to the top. The reason of the success in obtaining so many larvæ and pupæ was that extensive cuttings of this tree had taken place, and there were large areas covered with young stool-shoots. Generally, at other times, and in other places the butterflies—or what was presumed to be this butterfly—have been noticed flying round and basking on the leaves of the tops of high trees of *Trewia nudiflora*, Linn. The known range of this butterfly is curious, as it is recorded only from Tenasserim in Burma and from the Nilgiri Hills and North Kanara in South-Western India. The female has hitherto only been recognised: Mr. Bell has sexed when freshly caught and newly emerged all the specimens (a large number) in his collection. The male can hardly be said to differ superficially from the female, it is usually somewhat smaller, and the bright smalt-blue of both wings on the upperside is slightly lighter and more silvery in shade.

110. ARRHOPALA CENTAURUS, Fabricius.
Hopea Wightiana, Wall. (*Dipterocarpeæ*).
Xylia dolabriformis, Benth. (*Leguminosæ*).
Terminalia tomentosa, Bedd. (*Combretaceæ*).
Terminalia paniculata, Roth (*Combretaceæ*).
Lagerstrœmia lanceolata, Wall.* (*Lythraceæ*).

111. ARRHOPALA AMANTES, Hewitson.
Hopea Wightiana, Wall. (*Dipterocarpeæ*).
Xylia dolabriformis, Benth. (*Leguminosæ*).
Terminalia paniculata, Roth (*Combretaceæ*).
Lagerstrœmia lanceolata, Wall. (*Lythraceæ*).

112. ARRHOPALA ABSEUS, Hewitson.
Has not been bred in Kanara.

113. ARRHOPALA CANARAICA, Moore.
Hopea Wightiana, Wall. (*Dipterocarpeæ*).

114. ARRHOPALA BAZALUS, Hewitson.
Hopea Wightiana, Wall. (*Dipterocarpeæ*).
Terminalia paniculata, Roth (*Combretaceæ*).

115. CURETIS THETIS, Drury.
Abrus precatorius, Linn. (*Leguminosæ*).
Pongamia glabra, Vent. (*Leguminosæ*).
Derris Heyneana, Benth. (*Leguminosæ*).
Wagatea spicata, Dalz. (*Leguminosæ*).
Xylia dolabriformis, Benth. (*Leguminosæ*).
Mr. Bell notes that the larva of this species feeds on other species of *Leguminosæ* than those named above.

116. CURETIS BULIS, Doubleday and Hewitson.
Ougeinia dalbergioides, Benth. (*Leguminosæ*).

117. ZESIUS CHRYSOMALLUS, Hübner.
Xylia dolabriformis, Benth. (*Leguminosæ*).

* This plant in the second Kanara paper, p. 382, appears under the name of *Lagerstrœmia microcarpa*, Wight, which is given by Sir J. D. Hooker as a synonym of *L. lanceolata.*

Terminalia tomentosa, Bedd. (*Combretaceæ*).
Terminalia paniculata, Roth (*Combretaceæ*).
Psidium Guyava, Linn. (*Myrtaceæ*).
Clerodendron infortunatum, Gaertn. (*Verbenaceæ*).
Argyreia speciosa, Sweet (*Convolvulaceæ*).
Loranthus longiflorus, Desrouss. (*Loranthaceæ*).

Mr. Bell notes that the larva of this butterfly is found on many other plants than those given above frequented by the ferocious red, or yellow, tree ant, *Œcophylla smaragdina*, Fabricius. The larvæ are so persistently carnivorous that each one has to be bred by itself.

118. CAMENA ARGENTEA, Aurivillius.

Camena (1865) is perhaps too near to *Camæna* (1850), the latter name having priority, in which case the former name must give way to *Pratapa*. This butterfly appears in the second Kanara paper, p. 384, as *Camena cippus*, Fabricius, but Dr. Chr. Aurivillius in Ent. Tids., vol. xviii, p. 146, n. 48 (1897), has shown that the true "*Hesperia*" *cippus* of Fabricius is the same as "*Hesperia*" (*Tajuria*) *longinus*, Fabricius, and that the latter name must sink as a synonym to the former, which is the older. Dr. Aurivillius has renamed the *Camena cippus* of authors, but not of Fabricius, *Pratapa argentea*. It has never been bred.

119. CAMENA DEVA, Moore.
Loranthus scurrula, Linn. (*Loranthaceæ*).
Loranthus tomentosus, Heyne (*Loranthaceæ*).

120. CREON CLEOBIS, Godart.
Loranthus scurrula, Linn. (*Loranthaceæ*).
Loranthus elasticus, Desrouss. (*Loranthaceæ*).

121. APHNÆUS VULCANUS, Fabricius.
Zizyphus Jujuba, Lamk. (*Rhamneæ*).
Zizyphus rugosa, Lamk. (*Rhamneæ*).

122. APHNÆUS LOHITA, Horsfield.
Zizyphus rugosa, Lamk. (*Rhamneæ*).
Wagatea spicata, Dalz. (*Leguminosæ*).
Xylia dolabriformis, Benth. (*Leguminosæ*).
Terminalia paniculata, Roth (*Combretaceæ*).
Psidium Guyava, Linn. (*Myrtaceæ*).
Lagerstrœmia lanceolata, Wall. (*Lythraceæ*).
Argyreia sericea, Dalz. and Gibs. (*Convolvulaceæ*).
Dioscorea pentaphylla, Linn. (*Dioscoreaceæ*).

Under *A. lohita* the *A. concanus* of Moore, which is recorded (p. 386) doubtfully as a species distinct from *A. lohita* in the second Kanara paper, is included.

123. APHNÆUS ABNORMIS, Moore.
A single specimen has been obtained in the Kanara district at Jaggalbett, above the ghâts. It has never been bred.

124. APHNÆUS ICTIS, Hewitson.
There is a single male of this species in Mr. Bell's collection which I have examined. It has never been bred. This species and the last are new to the Kanara list.

125. TAJURIA INDRA, Moore.
Loranthus elasticus, Desrouss. (*Loranthaceæ*).

126. TAJURIA CIPPUS, Fabricius.
Loranthus Wallichianus, Schultz. (*Loranthaceæ*).
Loranthus scurrula, Linn. (*Loranthaceæ*).
Loranthus longiflorus, Desrouss. (*Loranthaceæ*).
Loranthus elasticus, Desrouss. (*Loranthaceæ*).
This species appears in the second Kanara paper as *Tajuria longinus*, Fabricius. With regard to the change of name see No. 118 *ante*.

127. TAJURIA JEHANA, Moore.
Mr. Bell has a male of this species taken at Karwar in the North Kanara district. It is new to the list of Kanarese butterflies, and has never been bred.

128. OPS MELASTIGMA, de Nicéville.
Loranthus scurrula, Linn. (*Loranthaceæ*).
Loranthus tomentosus, Heyne (*Loranthaceæ*).

129. CHLIARIA OTHONA, Hewitson.
Rhynchostylis retusa, Blume (*Orchideæ*).
Saccolabium papillosum, Lindl. (*Orchideæ*).

130. ZELTUS ETOLUS, Fabricius.
This species has never been bred.

131. CHERITRA JAFFRA, Butler.
Saraca indica, Linn. (*Leguminosæ*).

Xylia dolabriformis, Benth. (*Leguminosæ*).
Cinnamomum zeylanicum, Breyn (*Laurineæ*).

132. RATHINDA AMOR, Fabricius.
Hopea Wightiana, Wall. (*Dipterocarpeæ*).
Schleichera trijuga, Willd. (*Sapindaceæ*).
Eugenia zeylanica, Wight (*Myrtaceæ*).
Careya arborea, Roxb. (*Myrtaceæ*).
Ixora coccinea, Linn (*Rubiaceæ*).
Loranthus longiflorus, Desrouss. (*Loranthaceæ*).
Croton sp. (*Euphorbiaceæ*).

133. HORAGA ONYX, Moore.
This species has never been bred.

134. CATAPŒCILMA ELEGANS, Druce.
Terminalia tomentosa, Bedd. (*Combretaceæ*).
Terminalia paniculata, Roth (*Combretaceæ*).
Lagerstrœmia lanceolata, Wall. (*Lythraceæ*).
Mr. H. H. Druce in Proc. Zool. Soc. Lond., 1895, p. 612, suggests
that if the Sumatran, Indian and Ceylonese species of *Catapœcilma* allied
to *C. elegans* should prove to be distinct from the typical Bornean form
it may be called *C. major*.

135. LOXURA ATYMNUS, Cramer.
Dioscorea pentaphylla, Linn. (*Dioscoreaceæ*).
Smilax macrophylla, Roxb. (*Liliaceæ*).
The pupa of this species is suspended by the tail only, with no
median band. In the second Kanara paper, p. 390, the latter is inad.
vertently given as being present.
As the eight genera, *Camena*, *Creon*, *Aphnæus*, *Tajuria*, *Ops*, *Cheritra*,
Rathinda and *Loxura* have the pupa suspended by the cremaster only
with no median girth, they would seem to form one very natural group,
this character being an extremely aberrant one in the *Lycænidæ*. The
larvæ and pupæ in these genera are also very similar. In the genus
Spalgis also the pupa is attached by the cremaster only with no median
band.

136. DEUDORIX EPIJARBAS, Moore.
Connarus Ritchiei, Hook. f. (*Connaraceæ*).

137. ZINAPSA TODARA, Moore.
Acacia Intsia, Willd. (*Leguminosæ*).

Acacia Intsia, var. *cæsia*, W. and A. (*Leguminosæ*).
Acacia pennata, Willd. (*Leguminosæ*).

138. RAPALA SCHISTACEA, Moore.

Acacia Intsia, var. *cæsia*, W. and A. (*Leguminosæ*).
Acacia pennata, Willd. (*Leguminosæ*):
Quisqualis indica, Linn. (*Combretaceæ*).

139. RAPALA LANKANA, Moore.

Wagatea spicata, Dalz. (*Leguminosæ*).
Acacia Intsia, var. *cæsia*, W. and A. (*Leguminosæ*).
Acacia pennata, Willd. (*Leguminosæ*).

LARVA. The larva is similar in shape to the larvæ of other species of *Rapala*. Head shiny, smooth, very light yellow, with the jaws and the basal joint of the antennæ white. The " teeth " or processes on the segments are round-topped, with a slight constriction before the end of each tooth, giving the teeth the appearance of being ball-topped; the anal teeth and subdorsal teeth of segment 2 are smaller than the rest, being little more than knobs. There is a subdorsal tooth to each segment except segment 13, a subspiracular tooth to each segment except segments 13 and 14; the subdorsal teeth of segments 11 and 12 are much further apart than on any other segment; the teeth of segment 2 are on the front margin, those of segment 14 on the hinder margin, all the rest in the middle of their respective segments. The surface of the body is smooth, oily-looking, each tooth having about eight golden-coloured hairs proceeding from the extremity, which are about the same length as the tooth itself. The spiracles are oval and black; the gland on segment 11, and the cylindrical protrusible organs of segment 12, are present. The colour of the larva is a light oily yellow-greenish, with a diagonal white band along the base of each subdorsal tooth, and below it, from the front margin near the subdorsal line down and back to the hinder margin to just behind the spiracle; the area on each segment in front of this line is suffused with brown; a light brown patch with a white cross on it dorsally on segment 2, and a small subdorsal green rising on the hinder margin of that segment; the dorsal parts of segments 11 and 12 are pink-greenish; there is a dorsal bluish line. The colour of the larva may be dark green with the diagonal lines yellow, and the areas in front of these lines deep rich brown-red when it is feeding on the red flowers of *Wagatia spicata*, Dalz. Length 19 mm. if moving, breadth 5 mm. omitting the teeth, 7 mm. if the teeth be included.

PUPA. The pupa agrees in shape with those of the other species of *Rapala* found in North Kanara, the length from segment 4/5 to the front being about equal to the length from the same point to the anal end; the breadth is greatest at segment 7, though but little more than at the shoulders; the height is the same at segment 7 and at just before the hinder margin of the thorax; the vertex of the head is inclined towards the ventral line of the pupa; the front margin of segment 2 slightly overhangs the head, and its dorsal line is inclined at an angle of about 45° to the length axis of the pupa; this segment is large, rather square in the front outline (which is the front of the pupa), and slightly pinched laterally behind the spiracles on the margin of segment 2/3; the thorax is very large, and increases in breadth from the front margin to the shoulders, is slightly pinched in front dorsally, and has the hinder margin coming to a point in the dorsal line; its dorsal line is evenly curved, is at an angle of 45° to the length-axis of the pupa in its anterior half, and curves through a plane parallel to that axis as far as the margin of segment 3/4; segment 4 has its dorsal line inclined from the front margin to the hinder margin towards the length-axis, so that the pupa is constricted somewhat dorsally, its dorsal line rising again from segment 4/5 to segment 7; the margin of segment 9/10 is raised, as is usual with the pupæ of *Rapala*, there being a thin interval between the margins; the hinder end of the pupa is rounded; the ventral line is straight. The surface of the pupa is finely reticulated, and has some minute erect hairs on segment 2, on the thorax, and on the dorsum of the other segments; these hairs are sparse. The spiracles of segment 2 are raised, oval, small, golden-yellow surfaces which face forwards, the other spiracles are of the same colour, small, oval and raised. The colour of the pupa is light rose-brown, segments 4 and 5 are generally dark, the margin of segment 9/10 is dark; there is a dorsal darkish line and a darkish lateral spot on each abdominal segment; the wings, segment 2, and the sides of the thorax, are darkish. Length 12·25 mm., breadth 6·25 mm. at segment 7, at the shoulders 5·75 mm.

HABITS. The larva lives on the flowers and young parts of the plant on which it feeds, and always on such plants as are frequented by the large red ants *Œcophylla smaragdina*, Fabricius, which tend the larvæ. Pupation takes place anywhere, amongst the flowers, in the crevices of the stems, &c., and the pupa is attached by the tail and by a body-band. The pupa when disturbed makes the usual creaking noise which is so common amongst lycænid pupæ.

140. RAPALA VARUNA, Horsfield.

Zizyphus Xylopyrus, Willd. (*Rhamneæ*).
Zizyphus rugosa, Lamk. (*Rhamneæ*).

J. II. 32

Xylia dolabriformis, Benth. (*Leguminosæ*).
Quisqualis indica, Linn. (*Combretaceæ*).

141. RAPALA MELAMPUS, Cramer.
Zizyphus rugosa, Lamk. (*Rhamneæ*).
Ougeinia dalbergioides, Benth. (*Leguminosæ*).

142. BINDAHARA SUGRIVA, Horsfield.
Salacia oblonga, Wall. (*Celastrineæ*).

143. VIRACHOLA ISOCRATES, Fabricius.
Tamarindus indica, Linn. (*Leguminosæ*).
Punica Granatum, Linn. (*Lythraceæ*).
Randia uliginosa, DC. (*Rubiaceæ*).

144. VIRACHOLA PERSE, Hewitson.
Anona squamosa, Linn. (*Anonaceæ*).
Randia uliginosa, DC. (*Rubiaceæ*).
Randia dumetorum, Lamk. (*Rubiaceæ*).

Family PAPILIONIDÆ.

Subfamily PIERINÆ.

145. LEPTOSIA XIPHIA, Fabricius.
Cratæva religiosa, Forst. (*Capparideæ*).
Capparis Heyneana, Wall. (*Capparideæ*).
Capparis sepiaria, Linn. (*Capparideæ*).
Capparis horrida, Linn. f. (*Capparideæ*).
This butterfly appears in the second Kanara list under the synonymic generic name *Nychitona*.

146. DELIAS EUCHARIS, Drury.
Loranthus longiflorus, Desrouss. (*Loranthaceæ*).

147. PRIONERIS SITA, Felder.
This rare butterfly has not been bred in Kanara, but Mr. Bell has seen a female laying eggs on a plant which is probably *Capparis tenera*, Dalz. (*Capparideæ*). The eggs hatched out, but the larva failed to reach maturity. In Ceylon Mr. E. E. Green says that the larva feeds on *Capparis*.

148. CATOPSILIA PYRANTHE, Linnæus.
Cassia occidentalis, Linn. (*Leguminosæ*).

149. Catopsilia crocale, Cramer.

Cassia Fistula, Linn. (*Leguminosæ*).

Cassia siamea, Lam.* (*Leguminosæ*).

Larva. The larva [of *C. catilla*, Cramer] is very similar in every way to that of *Catopsilia crocale*, Cramer [these two species are in de Nicéville's opinion one and the same species]. The head is round, green, the clypeus edged with brown, covered with small, shiny, black tubercles which are not very large and do not hide the colour of the head; the anal flap is rounded, but looks square at the extremity, and is covered with small tubercles, not black but green, each bearing a short hair; the body is covered with rows of small black tubercles as in *C. crocale*, of which only the row along the spiracular line is conspicuous. The spiracles are oval, shiny and white. The colour is green, with a spiracular white band touched with bright yellow on segments 2 to 5, and these segments, especially 3 and 4, are distinctly flanged on the spiracular line as in the larva of *Hebomoia australis*, Butler, though not to so great an extent. Length 51 mm., height 7 mm.

Pupa. The pupa is the same as that of *C. crocale* at first sight, but the dorsal line of the thorax is absolutely parallel to the longitudinal axis of the pupa for two-thirds of its length; consequently the hinder part just before the margin is perpendicular to this parallel part, i.e., is raised suddenly though very slightly above the front margin of segment 4, and the front end of this parallel dorsal line is at an angle, and a sharpish angle, with the front slope of the thorax; the shoulder too is distinctly angled, i.e., the point where the lateral line of the head and segment 2 meets that of the wings; the front margins of segments 9 and 10 in the dorsal line when looked at sideways show a minute peak overhanging the hinder margins of segments 8 and 9 respectively; the cremaster is distinctly bifid at the extremity, and has some shiny, very short, black suspensory hooks dorsally as well as at the extremity. There is a dorsal rugose black tip to the snout terminating the head, which snout is cylindrical in its apical half; there is no black line round the eye as in *C. crocale*, and there is a dark green-blue dorsal line, which is yellow on the thorax, as well as the supra-spiracular yellow line. Length 34 mm., length of snout 3 mm., breadth at segment 7 is 9 mm., height at apex of curve of wings (segment 6) 10 mm., height at the apex of the thorax 8 mm.

[—— Habits. The habits are the same as in *C. crocale* in every particular both as to the larva and the pupa. Mr. Bell notes that until the day (30th July, 1898), on which he wrote this description he always

* This plant appears in the first Kanara paper, p. 360, under its synonymic name *Cassia sumatrana*, Roxb.

considered *C. crocale* and *C. catilla* were one and the same butterfly, but it always struck him as somewhat anomalous that nearly all the butterflies caught below the hills on the North Kanara coast should be of the former form, while the great majority of those caught on the tops of the hills and in the heavy jungles should be of the latter form. There is but little doubt, he says, that the former is a more or less open-country butterfly, while the latter keeps nearly altogether to the jungles.

150. TERIAS HECABE, Linnæus.
Sesbania aculeata, Pers. (*Leguminosæ*).
Cassia occidentalis, Linn. (*Leguminosæ*).
Cassia Tora, Linn. (*Leguminosæ*).
Cassia glauca, Lam. (*Leguminosæ*).

151. TERIAS SILHETANA, Wallace.
Poinciana regia, Bojer (*Leguminosæ*).
Wagatea spicata, Dalz. (*Leguminosæ*).

152. TERIAS LIBYTHEA, Fabricius.
Cassia pumila, Lam. (*Leguminosæ*).

153. TERIAS LÆTA, Boisduval.
This butterfly has not been bred in Kanara.

154. TERIAS VENATA, Moore.
There are specimens of this species in Mr. Bell's and my own collection taken in Karwar in October. It is new to the Kanara list, and it has not been bred.

155. TERACOLUS AMATA, Fabricius.
Salvadora persica, Linn. (*Salvadoraceæ*).

156. TERACOLUS ETRIDA, Boisduval.
Cadaba indica, Lamk. (*Capparideæ*) in Bombay.
Mr. Bell has caught a single specimen only of this butterfly in Kanara, from whence Dr. Butler also records it.

157. TERACOLUS EUCHARIS, Fabricius.
Capparis sepiaria, Linn. (*Capparideæ*) at Bijapur.
This species is new to the Kanara list. Dr. F. Moore has recorded it from North Canara as *T. pallens*, Moore, which is a synonym

of *T. eucharis.* Mr. Bell has received it from Bijapur, which lies N.-E. of Kanara, but has never taken it in Kanara itself.

158. IXIAS PYRENE, Linnæus.

Capparis sepiaria, Linn. (*Capparideæ*) at Bijapur.

Dr. A. G. Butler in his latest revision* of the genus *Ixias* restricts *I. pyrene,* Linnæus, to China, and records *I. frequens,* Butler, from "India generally," *I. dharmsalæ,* Butler, from "India, from Darjiling to the Western Provinces and southwards to the Neilgherries," and *I. pirenassa,* Wallace, from "Western India southwards to Depalpur." It is unknown to the writer to which of these species, if any, Dr. Butler would assign the butterfly that is placed here under the parent form. *I. cingalensis,* Moore, is restricted by Dr. Butler to Ceylon. Mr. Bell doubts the occurrence of *I. pyrene* in Kanara, though it certainly occurs commonly elsewhere in the Bombay Presidency, and at Bijapur, N.-E. of Kanara, but it has been recorded from "A place half-way up the ghât on the road to the Gairsoppa Falls."

159. IXIAS MARIANNE, Cramer.

Capparis sepiaria, Linn. (*Capparideæ*).

New to the Kanara list. It is common on the Malemani or Gairsoppa Ghât in Kanara. It has not been bred in Kanara.

160. HEBOMOIA AUSTRALIS, Butler.

Cratæva religiosa, Forst. (*Capparideæ*).
Capparis Heyneana, Wall. (*Capparideæ*).
Capparis Moonii, Wight (*Capparideæ*).
Capparis sepiaria, Linn. (*Capparideæ*).

This species appears in the second Kanara list as *H. glaucippe,* Linnæus, but Dr. Butler has recently separated off the South Indian and Ceylonese form under the name *H. australis* from the North-East Indian, Burmese, Malayan Peninsula and Chinese *H. glaucippe.*†

161. NEPHERONIA PINGASA, Moore.

Capparis Heyneana, Wall. (*Capparideæ*).

162. NEPHERONIA HIPPIA, Fabricius.

Capparis Heyneana, Wall. (*Capparideæ*).
Capparis horrida, Linn. f. (*Capparideæ*).

---

* Ann. and Mag. of Nat. Hist., seventh series, vol. i, pp. 133-143 (1898).
† Ann. and Mag. of Nat. Hist., seventh series, vol. i, pp. 289-293 (1898). ...

163. APPIAS LIBYTHEA, Fabricius.
Cratæva religiosa, Forst. (*Capparideæ*).
Capparis horrida, Linn. f. (*Capparideæ*).

This species is entirely omitted from Dr. A. G. Butler's recent monograph of the genus *Catophaga* (Ann. and Mag. of Nat. Hist., seventh series, vol. ii, pp. 392–401, 458–467 (1898). Possibly with *A. zelmira*, Cramer, he considers it to be a true *Appias*, and generically distinct from the genus *Catophaga*, Hübner.

164. APPIAS TAPROBANA, Moore.
Cratæva religiosa, Forst. (*Capparideæ*).

This species is given in the second Kanara list as *A. hippoides*, Moore, that species being a synonym of *A. hippo*, Cramer, found in North India, Burma, the Malay Peninsula, Indo-China, China, and many islands of the Malay Archipelago. *A. hippo* is distinct from *A. taprobana*, the latter occurring in South India and Ceylon only.

165. APPIAS ALBINA, Boisduval.
Hemicyclia venusta, Thwaites (*Euphorbiaceæ*).

This species is given as *A. neombo*, Boisduval, in the second Kanara list. *A. neombo* is a species which cannot be satisfactorily identified, though I have some specimens of *Appias* from North Kanara that agree fairly well with the original description, but these in my opinion do not represent a distinct species, but are probably a dry-season form of *A. wardii*, Moore, or possibly of *A. albina*, Boisduval. They were all caught or bred in December. Dr. A. G. Butler in Ann. and Mag. of Nat. Hist., seventh series, vol. ii, p. 397, n. 11 (1898), places *Pieris neombo* as a synonym of *Catophaga albina*.

LARVA. The larva is very like that of *Appias taprobana*, Moore, in appearance, but is as a rule more thickly covered with black tubercles. The body is more or less cylindrical, but narrows somewhat at segments 2 and 3, and still less so at the anal end; the anal flap is thick, semicircularly rounded, and inclined at an angle of 45° to the length-axis of the larva, and has a small, very slightly developed, conical tubercle before the extremity on each side of the dorsal line; the front half of the flap is shiny and black, and has some conical tubercles of different sizes all over it, each surmounted by a single fine hair, there being one subdorsal tubercle larger than the rest; the posterior half is green and smooth except for the tubercle above the extremity just mentioned; the body is somewhat stouter in the middle; the head is broader than the body at segment 2. Head round, shiny, oily yellow all over, with a rather large and rather narrow triangular clypeus, the labrum and antennæ coloured like the head ; the surface covered with small, conical, setiferous, black tubercles, three on each side of the dorsal line on the vertex, two or the border of the clypeus at the apex on each side, and one above the apex on each side of the central line, about eight on each lobe besides, in addition to which there are several small cylindrical points, all, tubercles and points, with a surmounting fine hair. Surface of body rugose with six transverse rows, from above the spiracular region over the dorsum, of small, shiny, conical, setiferous, black tubercles to each segment ; segments 2, 12 and 13 have only a few transverse rows of such tubercles ; the front row of each segment is generally composed of larger tubercles than the others, and especially the subdorsal tubercle of that first row is generally large, the tubercles of segment 2 nearly render the whole segment black ; the surface is shiny as well as the tubercles, and has besides a few cylindrical, setiferous, black points. Spiracles of the ordinary size, flush, oval and white. The colour of the body is a rather light green, sometimes with a tinge of lilac, with a yellow-white, spiracular, narrow band from segment 2/3 to segment 12, where the band expands somewhat. The black tubercles may sometimes be very small, just black specks. The eyes are only four in number on each side, and are arranged in an arc above the base of the antennæ, they are shiny, of the same colour as the head, and are generally bordered with black. Length 30 mm., breadth 3·75 mm.

PUPA. The pupa is very like that of *A. taprobana* ; the head-process from between the eyes is long, flattened at the sides, slightly curved, pointed at the extremity and directed upwards and forwards, sometimes straight out in a line with the axis, sometimes inclined to it ; it is as long as segments 4 and 5 (in the dorsal line) together ; and the edges on the ventral surface are minutely serrated. The front margin of segment 2 is produced into a small subdorsal tooth, and the dorsal line is rather strongly carinated ; the thorax is rather highly carinated on the dorsal line, the lateral outline of this carina being a curve which is slightly broken at the apex, just before which the carina is double, and the edges somewhat minutely serrate ; the lateral teeth of segments 6, 7 and 8 are all the same size and pointed ; the dorsal line of the abdomen from segment 6/7 somewhat carinated, the carination splitting down the sides of the cremaster on segment 14 ; the cremaster is rather small, square as seen from above, slightly bifid, with the lateral carina, which is continued forwards on to segments 11, 12 and 13. The spiracles of segment 2 are thin, yellow lines, the other spiracles are oval, flush, and white. The surface of the pupa is shiny, smooth except for a superficial wrinkling, the carinæ, and the teeth; there is a round, blunt, low production of the shoulder. The colour of the pupa is dirty whitish with a pink shade on segments 4 to 14; the same but transparent. looking on the rest; the head-production, the points on segment 2, the teeth of segments 5 to 7 (sometimes), and the extremity of the cremaster are black—only the top and lower edge of the point of the head however; there is a black spot on the hinder edge of segment 2 dorsally, one just behind the shoulder, one lateral on segments 3 and 4 to 12, and one dorsal on the front margin of segments 9 and 10. There is always a semicircle of six darkish spots dorsally on segments 6 and 7. The colour of the pupa when formed under a leaf is probably green, with the markings as above. The underside or ventrum of the pupa is always whitish. Length 21·25 mm., of which the process on the head is 2·75 mm., breadth at shoulders 5 mm., breadth at segment 7 from tip to tip of teeth 7·5 mm.

HABITS. The habits of the larva are those of *A. taprobana*. The pupa is formed on the underside of a leaf, on the trunk of the tree, or on any flat surface, and is attached by the tail and by a body-band.

166. APPIAS WARDII, Moore.
Capparis Heyneana, Wall. (*Capparideæ*).

Recorded by Dr. Butler, l.c., p. 398, n. 12, from the Nilgiris, Mysore, and Rangoon. The latter locality is certainly erroneous. The species is confined to South India.

167. HUPHINA NERISSA, Fabricius.
Capparis sepiaria, Linn. (*Capparideæ*).
Capparis horrida, Linn. f. (*Caparideæ*).

H. nerissa is the parent form of this group of this genus occurring in India, but typically it is not found in South India, being represented there by *H. phryne*, Fabricius, under which name it appears in the second Kanara list, p. 574, n. 158.

168. HUPHINA REMBA, Moore.
Capparis Heyneana, Wall. (*Capparideæ*).

169. BELENOIS MESENTINA, Cramer.
Capparis divaricata, Lamk. (*Capparideæ*).

Subfamily PAPILIONINÆ.

170. TROIDES MINOS, Cramer.
Aristolochia indica, Linn. (*Aristolochiaceæ*).

The generic name Ornithoptera, under which this species is given in the Kanara papers, is a synonym of *Troides.*

171. PAPILIO HECTOR, Linnæus.
Aristolochia indica, Linn. (*Aristolochiaceæ*).

172. PAPILIO ARISTOLOCHIÆ, Fabricius.
Aristolochia bracteata, Retz. (*Aristolochiaceæ*).
Aristolochia indica, Linn. (*Aristolochiaceæ*).

The food-plant of this butterfly, *Aristolochia,* Linnæus, must apparently have been known to Fabricius in Europe in 1775, more than a century and a quarter ago, when he described the insect, and probably named it after the pabulum of the larva.

173. PAPILIO AGAMEMNON, Linnæus.
Unona discolor, Vahl (*Anonaceæ*).
Polyalthia longifolia, Benth. and H. f.* (*Anonaceæ*).
Anona squamosa, Linn. (*Anonaceæ*).
Anona reticulata, Linn. (*Anonaceæ*).
Saccopetalum tomentosum, Hook. f. and T. (*Anonaceæ*).

174. PAPILIO SARPEDON, TEREDON, Felder.
Cinnamomum zeylanicum, Breyn (*Laurineæ*).
Alseodaphne semecarpifolia, Nees (*Laurineæ*).
Litsæa sebifera, Pers. (*Laurineæ*).

An aberration of this species has been described by Colonel Swinhoe from Matheran in South India as a distinct species under the name of *Delchina* [sic] *thermodusa.* It has the anterior blue spot of the median band of the forewing absent. This aberration is found also in Ceylon.

175. PAPILIO EURYPYLUS JASON, Esper.
Unona Lawii, Hook. f. and T. (*Anonaceæ*).
Saccopetalum tomentosum, Hook. f. and T. (*Anonaceæ*).

In the first Kanara paper, p. 364, this species appears under the name *Papilio doson,* Felder, and in the second, p. 578, as *Papilio telephus,* Felder, these two species being given by the Hon. Walter Rothschild in Nov. Zool., vol. ii, p. 432 (1895) as synonyms of *Papilio eurypylus jason,* Esper.

176. PAPILIO NOMIUS, Esper.
Saccopetalum tomentosum, Hook. f. and T. (*Anonaceæ*).

---

* In the first Kanara paper, p. 363, this plant is given under its synonymic name *Gualteria* [recte *Guatteria*] *longifolia,* Wall.

177. PAPILIO ANTIPHATES ALCIBIADES, Fabricius.
Unona Lawii, Hook. f. and T. (*Anonaceæ*).

178. PAPILIO DEMOLEUS, Linnæus.
Ruta graveolens, Linn., var. *angustifolia,* Pers. (*Rutaceæ*).
Glycosmis pentaphylla, Correa (*Rutaceæ*).
Murraya Kœnigii, Spreng. (*Rutaceæ*).
Citrus decumana, Linn. (*Rutaceæ*).
Ægle Marmelos, Correa (*Rutaceæ*).
In the Kanara papers this species is given under its synonymic name *P. erithonius,* Cramer.

179. PAPILIO POLYTES, Linnæus.
Zanthoxylum Rhetsa, DC. (*Rutaceæ*).
Glycosmis pentaphylla, Correa (*Rutaceæ*).
Citrus medica, Linn. (*Rutaceæ*).
Citrus decumana, Linn. (*Rutaceæ*).

180. PAPILIO POLYMNESTOR, Cramer.*
Paramignya monophylla, Wight (*Rutaceæ*).
Atalantia monophylla, Correa (*Rutaceæ*).

181. PAPILIO HELENUS DAKSHA, Hampson.
Zanthoxylum Rhetsa, DC. (*Rutaceæ*).
Glycosmis pentaphylla, Correa (*Rutaceæ*).
Citrus medica, Linn. (*Rutaceæ*).
Citrus decumana, Linn. (*Rutaceæ*).

182. PAPILIO DEMOLION LIOMEDON, Moore.
Evodia Roxburghiana, Benth. (*Rutaceæ*).
Acronychia laurifolia, Blume (*Rutaceæ*).

183. PAPILIO PARIS TAMILANA, Moore.
Evodia Roxburghiana, Benth. (*Rutaceæ*).
Though Mr. Bell have given me the name of the food-plant of the larva of this splendid butterfly, the largest of its group, he has not furnished me with a description of its transformations. He has only seen the female laying eggs on the plant named.

* In the first Kanara paper, p. 367, one of the food-plants of this butterfly is given as *Garcinia Xanthochymus,* Hook. f. (*Guttiferæ*), but Mr. Bell considers this record to be incorrect.

184.   Papilio buddha, Westwood.
Zanthoxylum Rhetsa, DC. (*Rutaceæ*).

185.   Papilio abrisa, Kirby.
Glycosmis pentaphylla, Correa (*Rutaceæ*).
Alseodaphne semecarpifolia, Nees (*Laurineæ*).

186.   Papilio clytia, Linnæus.
Cinnamomum zeylanicum, Breyn (*Laurineæ*).
Alseodaphne semecarpifolia, Nees (*Laurineæ*).
Litsæa tomentosa, Herb.* (*Laurineæ*), in Bombay.
Litsæa sebifera, Pers. (*Laurineæ*).
In the two Kanara papers this butterfly appears as *Papilio dissimilis*, Linnæus, or *P. panope*, Linnæus, but the Hon. W. Rothschild has recently shewn that *P. clytia* is the oldest name for it.

187.   Papilio pandiyana, Moore.
This is the only *Papilio* in Kanara of which the food-plant has not been discovered.   The allied *P. jophon*, Gray, of Ceylon, has been bred, but its food-plant has not been recorded.   The larva shews that this butterfly comes into the first group of the genus, being very similar to that of *P. hector*, Linnæus, and *P. aristolochiæ*, Fabricius, and probably feeds on the same plants.

Family HESPERIIDÆ.

188.   Ismene gomata, Moore.
Heptapleurum venulosum, Seem. (*Araliaceæ*).

189.   Ismene fergusonii, de Nicéville.
Combretum extensum, Roxb. (*Combretaceæ*).

190.   Bibasis sena, Moore.
Combretum extensum, Roxb. (*Combretaceæ*).

191.   Hasora (*Parata*) alexis, Fabricius.
Pongamia glabra, Vent. (*Leguminosæ*).
Dr. Chr. Aurivillius in Ent. Tids., vol. xviii, p. 150, n. 68 (1897), has recently shewn that "*Hesperia*" *alexis*, Fabricius, is an older name

---

* In the first Kanara paper, p. 369, this plant is mentioned under *Tetranthera apetala*, Dalz. and Gibs., which is given by Sir J. Hooker as a synonym of *Litsæa tomentosa*, Herb.

for the "*Papilio*" *chromus* of Cramer, under which this species appears in the Kanara papers.

LARVA. Head squarely rounded as seen from the front, moderately thick through; covered with rather long, erect, light hairs; the colour is yellow or red-fuscous; when yellow a black spot (the eyes) at the base of each lobe just above the jaws; head slightly bilobed. Segment 2 is smaller than the head, and has a dorsal, broad, black collar; it is often greenish when it has two black, lateral spots. The shape is cylindrical, the section being circular, the anal end slightly sloping and finishing off round; the last segment has a shiny, dorsal, black shield at the end. The spiracles are rather long ovals, large and white. The body is sparsely covered with rather long, erect, white hair. The colour is a more or less dark mauve on the dorsal half-segments, suffused with whitey-yellow dorsally; there is a dorsal, pure mauve line, and a more or less indistinct subdorsal pure mauve line; as also a broad, latero-marginal band of yellowish-green bordered above and below by a white line. Ventrum greenish-yellow. The larva is oily looking. There may be a lateral, black spot on each or any of segments 5 to 9.

PUPA. Head high, somewhat bowed, with a conical boss on the vertex, pointing upwards and forwards, between the eyes; the eyes are very prominent. Segment 2 broad. Thorax stout, convex, humped in the usual way; shoulders somewhat narrower than the head. Section of body circular. The pupa decreases evenly in diameter from the shoulders to the end, with a slight dorsal constriction to the cremaster, which is small and nearly cubical. The last segment before the end is broad dorsally, but disappears laterally, and is raised on the front margin above the margin of the next segment, with a triangular indentation dorsally. Segment 14 shows as a semicircular dorsal shield-like piece, deeply indented on the dorsal line. Spiracles rather large, rather long ovals, light brown in colour, the spiracle of segment 2 linear. Surface shiny, widely and finely wrinkled, covered with more or less numerous hairs, erect on the anterior part of the body, adpressed on the posterior part; ventrally on abdomen the hairs are erect. Colour green, generally sprinkled with white powder, with a yellowish tinge on the abdomen; the depressions of segments 13 and 14 edged with shiny black. The pupa is attached by the tail and a body-band.

HABITS. The larva feeds on young leaves, and makes a loose cell by bringing the two edges of a leaflet together laxly. It pupates in such a cell. It is very moth-like in its habits: the larva runs out of its cell when disturbed. The pupa wriggles considerably when touched.

192. HASORA (*Parata*) BUTLERI, Aurivillius.
Derris scandens, Benth. (*Leguminosæ*).

Derris Heyneana, Benth. (*Leguminosæ*).

Dr. Chr. Aurivillius in the above-cited paper renames the *Hasora alexis* of Butler, but not of Fabricius—*Hasora butleri*. Messrs. Elwes and Edwards in Trans. Zool. Soc. Lond., vol. xiv, p. 301 (1897), cite Moore's figure of *Parata alexis* in Lep. Cey. as a synonym of *Hasora chromus*, Cramer, = *H. alexis*, Fabricius, but make no reference to Fabricius' original description or Butler's figure of *H. alexis*. The very broad, clearly defined, discal, white band of the hindwing on the underside will separate *H. butleri* from *H. alexis*.

LARVA. The exact type of *Hasora (Parata) alexis*, Fabricius. Head from in front nearly circular, slightly, though distinctly, indented on the vertex; under a lens the surface is irregularly rugose, covered sparsely, with the exception of the upper part of the face and the vertex, with long, fine, erect, white hairs; colour very dark brown, with reddish jaws; the head is shiny, small for the body as compared with other hesperids. Segment 2 is narrower than the head, and is shiny dark brown, with a double, dorsal, greenish line and some long hairs as on the head. Segment 3 as broad as the head from the front margin. The larva is fat and greasy-looking, thickest about segments 6 to 9, after which its diameter decreases very gradually to the broadly rounded, sloping, anal segment. The transverse section of the body is circular. The spiracles are very small, oval in shape, yellow. The surface of the body is sparsely · hairy, the hairs are fine, long, erect and white, most thickly disposed round the margin of the larva; otherwise the larva is quite smooth. Colour green, suffused dorsally as far as a lateral yellow line with rather dark violet; a dorsal, dark green line bordered by a yellow line on each side, and a subdorsal line of the same colour, so that there are four parallel dorsal lines altogether; a marginal yellow line; all these lines are not continued on to the yellowish-green anal segment; a black spot laterally just outside the subdorsal line on segments 6, 8, 10 and 12; this spot under a lens is velvety-looking. The ventrum is green. Length 32 mm. when the larva is walking, breadth 6·3 mm. at the broadest part, height 6·3 mm.

PUPA. The exact counterpart of that of *H. alexis* in shape and colour. A very slight boss between the eyes, a small oblong space just above the boss and between the bases of the antennæ, just touching these bases and not reaching the front margin of segment 2, dark brown-green in colour consequent on being free from the white powder which covers the whole pupa. Spiracles oval and black. The whole surface of the body is rather sparsely pitted and covered with rather long, fine, white hairs, which are semi-adpressed to the surface, these hairs spring from the pits, one from each, and are densest on the eyes. The cremaster is short, stout, oblong, slightly curved and black. The last segment before the cremaster is roundly indented in the middle of the front margin, and the indentation is lined with black. The colour through the white powder is green on the thorax and pink on the abdomen. The only black markings are the spiracles, the markings above mentioned on the last segments, and a dorsal black line from the front margin of the thorax a third of the whole length of the thorax towards the hinder margin. Length 23 mm., breadth at shoulders (which are slightly angulated and the broadest part of the pupa) 6·3 mm., the height of the thorax (which is the highest part of the pupa) 6·3 mm.

HABITS. The habits of the larva are exactly those of *H. alexis.* The cell is composed of a few tender, soft leaves joined together loosely by an irregular web, which web is also generally spun over the mouth of the cell or shelter; the pupa is formed in the cell. The larvæ are much eaten by spiders, and are greatly attacked by parasitic *Diptera* and *Hymenoptera*. The eggs are laid on the young white leaves (the leaves of the food-plant are sometimes rose or rose-brown in colour) in a shady place, very often high up amongst the foliage of the trees amongst which the creeper climbs, often to a great height. The larva has always been found at an elevation of 900 to 1000 feet above sea level.

193. HASORA CHABRONA, Plötz.
Millettia racemosa, Benth. (*Leguminosæ*).
EGG. The egg is very small for so large a hesperid, and is red or pink when first laid.
LARVA. Head rather square, broader than high, thick through, not very large for the body, rather flat on the face; the top quarter of the head red-brown in colour, the rest black, the jaws yellow and black. Segment 2 smaller than the head, shiny, smooth, swollen-looking, white in colour, with a dorsal and a lateral black band, and a marginal brown spot laterally. Body cylindrical, increasing rapidly in width from the collar to segment 4, then gradually to the middle which is the broadest part; the anal segment rather narrow, overhanging the legs, sloping, with the extremity rounded, and dorsally black and shiny. The surface of the body covered throughout with long, very fine, white hairs, each segment with four transverse fine yellowish lines. Colour of larva dirty bluish-green, with four subdorsal, broad, yellow lines, two on each side, besides a marginal and submarginal yellow line; the dirty ground-colour is spotted finely with yellow. Length 33 mm.
PUPA. Very like that of *Hasora* (*Parata*) *alexis*, Fabricius, but light pink on the abdomen, dirty green-white on the wings, thorax and head. The point on the head between the eyes is short and sharply

conical. Cremaster oblong, stout, thicker than· broad. The spiracle of segment 2 with a flush, rather· large, oval, black surface near it. Spiracles oval, large and black. Surface pitted all over, an erect ·short hair being placed in each pit, these hairs are not very short, and are longest on the head, eyes, and segment 2. Thorax stout, the constriction behind the thorax slight. A long, dorsal, black streak on the front slope of the thorax reaching its front margin, and a short, dorsal, black mark on the hinder margin ; a dorsal,·black mark on segment 13 ; cremaster black ; two black spots on the inner margin of each eye ; the point on the head black. The entire pupa is covered with a white powder. Length 25·3 mm., breadth at the thorax or at segment 7 is 7 mm., height at the apex of the thorax 7 mm.

HABITS. The habits are similar to those of *H. alexis* in. nearly all particulars.

194. HASORA BADRA, Moore.
Derris uliginosa, Benth. (*Leguminosæ*).

LARVA. The larva is most like that of *Hasora chabrona*, Plötz, but differs in markings and coloration. It is circular in transverse section, thickest in the middle and rather stout. The head seen from the front is nearly round and somewhat bilobed, there being a depressed line· over the vertex to the apex of the clypeus ; it is finely rugose, and has a covering of fine, rather long, erect, white hairs which are not very densely disposed ; the colour is dark rose-red with the clypeus black, and a black patch, varying a good deal in size, on the front face of each lobe at the base and reaching as far as the centre ; the·eyes are also black ; the jaws tipped with white ; the head is of the ordinary size for the larva of the genus *Hasora*. Segment 2 is shiny black. Segments 13 and 14 are dirty white, the anal segment slopes in the dorsal line at an angle of more than 45° with the longitudinal axis, and is broadly rounded at the extremity. The surface of the body is dull, covered all over, but not very densely, with fine, moderately long, erect, white hairs. Spiracles oval, of the ordinary size, and yellowish. The colour is light neutral tint, with a double subdorsal line on each side running from segment 3 to segment 12, both rather ill-defined as to outline and running together in places ; five or six thin lines on each segment running across transversely to the length of the body from just above the spiracle on each side ; all these lines yellow ; on segments 3, 4, 6, 10 and 12 is a large lateral deep purple spot or patch interrupting the anterior pair of the transverse yellow lines ; on segments 5, 7, 9 and 11 there is a deep purple line between the anterior pair of transverse yellow lines in the same position as the deep purple patch on segments 3, 4, 6, 10 and 12 ;

there is a subspiracular white line; the ventrum is white with a slight, indigo-blue wash in the colour. Length 38 mm. when at rest, breadth 6 mm.

PUPA. The pupa is nearly exactly the same as that of *Hasora* (*Parata*) *alexis*, Fabricius; it has, however, a longer snout. The eyes are prominent, and the front of the head (the front of the pupa) is square, except that the vertex is produced into a triangular piece as seen from above, really a short cone surmounted by a porrect, rather long, cylindrical, blunt-topped snout directed slightly upwards, which is as long as segment 2 is broad (breadth in the direction of the longitudinal axis of the pupa); segment 2 is broad and convex, with its dorsal line at an angle of 45° to the longitudinal axis; the thorax is humped, and the front slope is at a greater angle to the longitudinal axis than segment 2, the back slope being nearly parallel to that axis; the body is slightly constricted before the shoulders in lateral outline, and also at segment 4; the pupa is thickest in the middle, though only slightly more so than at the thorax, the shoulders are evenly rounded and not prominent; the pupa is circular in section decreasing in diameter from the middle to the anal end, where the stout parallelo-pipedal cremaster runs up on to the anal segment in a lateral pear-shaped piece; segment 12 has a dorsal convex shield, with a dorsal semicircular notch in it. The spiracles of segment 2 are indicated by a conical black tubercle on the front margin of the thorax; and the other spiracles are oval and black. The surface of the body is covered all over with a white powder, and is shiny beneath this powder where visible. The colour is a very pale green, with a black patch round each spiracle; there is an oval, blob-like, black patch at the front and hinder margins of the thorax in the dorsal line; a black spot above and below on the front of each eye; a black border to the lateral pieces of the cremaster and to the dorsal shield of segment 12; and the snout is rugose and black. Length 28 mm. over all, length of snout 2·25 mm., breadth at centre 7 mm., height at centre 7 mm., breadth of front of head 5 mm.

HABITS. The larva makes a cell, similar to that of all the other species of *Hasora*, of young leaves loosely bound together with silk web, and feeds, whilst young, on the young leaves, though, when full grown, it will feed on the fully matured leaves. The pupa is formed in the cell, which is thinly covered inside with silk. It is attached by the tail, and by a body-band.

195. BADAMIA EXCLAMATIONIS, Fabricius.
Terminalia Bellerica, Roxb. (*Combretaceæ*).
Combretum extensum, Roxb. (*Combretaceæ*).

196. Hesperia galba, Fabricius.
Waltheria indica, Linn. (*Sterculiaceæ*).

197. Caprona ransonnetii, Felder.
Helicteres Isora, Linn. (*Sterculiaceæ*).

198. Odontoptilum angulata, Felder.
Allophylus Cobbe, Blume (*Sapindaceæ*).

199. Coladenia indrani, Moore.
Grewia Microcos, Linn. (*Tiliaceæ*).
Xylia dolabriformis, Benth. (*Leguminosæ*).
Mallotus philippinensis, Muell. (*Euphorbiaceæ*).
This species is given under *Coladenia tissa,* Moore, in the second
Kanara paper. I have given my reasons in Journ. A.S.B., vol. lxviii,
pt. 2' p. 225, n. 190 (1899) for considering *C. indrani* and *C. tissa* to
represent a single species.

200. Coladenia dan, Fabricius.
Cyathula prostrata, Blume (*Amarantaceæ*).
Achyranthes aspera, Linn. (*Amarantaceæ*).
Achyranthes bidentata, Blume (*Amarantaceæ*).

201. Satarupa bhagava, Moore.
Mr. Bell obtained a single male at Anshi on 27th December, 1898.
It is new to the Kanara list, and has not been bred. Messrs. Elwes and
Edwards do not record it from South India at all, but it occurs in
Orissa and the Nilgiri Hills.

202. Sarangesa purendra, Moore.
Blepharis asperrima, Nees (*Acanthaceæ*).

203. Sarangesa dasahara, Moore.
Achyranthes aspera, Linn. (*Amarantaceæ*).
An unidentified Acanthad.
Mr. Bell notes that *Coladenia dan,* Fabricius, and *S. dasahara* are
so much alike in the larval and pupal states that they should certainly
not be generically separated. The larva of *Sarangesa purendra,* Moore,
only differs from them in colour, but feeds on a different food-plant to
the other species mentioned above.

204. Tapena thwaitesi, Moore.
Dalbergia latifolia, Roxb. (*Leguminosæ*).

J. ii. 34

Dalbergia rubiginosa, Roxb. (*Leguminosæ*).
Dalbergia tamarindifolia, Roxb. (*Leguminosæ*).
Dalbergia volubilis, Roxb. (*Leguminosæ*).
Derris scandens, Benth. (*Leguminosæ*).

Messrs. Elwes and Edwards (l.c., p. 147, pls. xviii, fig. 19, *male*; xxii, fig. 16, *inner face of left clasp of male*) describe *Tapena hampsoni* as a species distinct from *T. thwaitesi*, from the Nilgiris and N. Canara. As, however, they say they have never seen typical *T. thwaitesi* from Ceylon, it is probable that their *T. hampsoni* is a synonym of that species.

205. CELÆNORRHINUS LEUCOCERA, Kollar.
Dædalacanthus roseus, T. Anders. (*Acanthaceæ*).
Eranthemum sp. (*Acanthaceæ*).

206. CELÆNORRHINUS AMBAREESA, Moore.
Strobilanthes callosus, Nees (*Acanthaceæ*).

207. CELÆNORRHINUS FUSCA, Hampson.
Strobilanthes callosus, Nees (*Acanthaceæ*).
Messrs. Elwes and Edwards place "*Plesioneura*" *fusca* as a synonym of *Celænorrhinus spilothyrus*, Felder, with a query, and record it from North Canara.

208. TAGIADES ATTICUS, Fabricius.
Dioscorea pentaphylla, Linn. (*Dioscoreaceæ*).
Smilax macrophylla, Roxb. (*Liliaceæ*).

209. TAGIADES OBSCURUS, Mabille.
Dioscorea pentaphylla, Linn. (*Dioscoreaceæ*).
1 have no specimens of this species from Kanara.

210. TAGIADES ALICA, Moore.
Messrs. Elwes and Edwards (l.c., p. 140), record this species from N. Canara, this being the only locality in South India given by them for it ; elsewhere they record it from Burma, the Andaman Isles, the Malay Peninsula, and Pulo Laut near Borneo. In my collection there is a good series of both sexes of this species from Kanara. It is probable that the *T. obscurus*, Mabille, recorded from Kanara, is wrongly identified. It has not been bred, unless, as is probable, the transformations recorded for *T. obscurus* really apply to this species.

211. CUPITHA PURREEA, Moore.
Terminalia Bellerica, Roxb. (*Combretaceæ*).

Terminaliu paniculata, Roth. (*Combretaceæ*).
Combretum ovalifolium, Roxb. (*Combretaceæ*).

This curious little butterfly has a wide distribution, being found in South India (Kanara and the Nilgiri Hills), in Orissa, the Eastern Himalayas, Assam, Burma, the Andaman Isles, Sumatra, Nias, Java, Bali, Borneo, Celebes, and the Philippine Isles.

212. MATAPA ARIA, Moore.*
Bambusa arundinacea, Willd. (*Gramineæ*).
Dendrocalamus strictus, Nees (*Gramineæ*).
Ochlandra stridula, Thwaites (*Gramineæ*).

213. GANGARA THYRSIS, Fabricius.†
Areca Catechu, Linn. (*Palmeæ*).
Caryota urens, Linn. (*Palmeæ*).
Phœnix sylvestris, Roxb. (*Palmeæ*).
Calamus pseudo-tenuis, Becc. and Hook. f. (*Palmeæ*).
Cocos nucifera, Linn. (*Palmeæ*).

214. PADRAONA DARA, Kollar.*
Bambusa arundinacea, Willd. (*Gramineæ*).
Oxytenanthera monostigma, Beddome (*Gramineæ*).
Ochlandra stridula, Thwaites (*Gramineæ*).

215. IAMBRIX SALSALA, Moore.
Grasses (*Gramineæ*).
Bambusa arundinacea, Willd. (*Gramineæ*).

216. BAORIS (*Parnara*) GUTTATUS, Bremer and Grey.
Oryza sativa, Linn. (*Gramineæ*).
Grasses (*Gramineæ*).

This is the *Baoris bada*, Moore, of the second Kanara list, p. 45, n. 204, placed by Messrs. Elwes and Edwards (l.c., p. 281) as a synonym of *Parnara guttatus*. *Baoris* (*Parnara*) *philotas*, de Nicéville, of the second list, p. 47, n. 207, is believed to be a synonym of this species by Mr. Bell, who says that starved larvæ of *B. guttatus* produce *B. philotas*. See also *Baoris* (*Parnara*) *bevani*, Moore, No. 239, *infra*.

---

* In the second Kanara paper, pp. 42, 44, the food-plant of this species is given as *Teinostachyum*. It should be *Ochlandra*.

† In the second Kanara list, p. 43, *Calamus Rotang*, Linn. (*Palmeæ*), is given as the food-plant of this species, but Mr. Bell informs me that this cane is not found in the district, though Sir J. Hooker gives it from the Deccan Peninsula and Ceylon.

217. SUASTUS GREMIUS, Fabricius.
Areca Catechu, Linn. (*Palmeæ*).
Caryota urens, Linn. (*Palmeæ*).
Phœnix sylvestris, Roxb. (*Palmeæ*).
Calamus pseudo-tenuis, Becc. and Hook. f. (*Palmeæ*).
Cocos nucifera, Linn. (*Palmeæ*).

218. PEDESTES SUBMACULATA, Staudinger.
Calamus pseudo-tenuis, Becc. and Hook. f. (*Palmeæ*).

This butterfly appeals in the second Kanara list, p. 47, n. 206, as *Isma submaculata*, Staudinger. In Trans. Zool. Soc. Lond., vol. xiv, p. 230 (1897), Messrs. Elwes and Edwards place it in the genus *Plastingia*, in which it was originally described, but say that they have not seen a specimen of it. On page 193 they describe *Pedestes maculicornis* as a new species from Pulo Laut, near Borneo. In my opinion this species s a synonym of *P. submaculata*. In my collection there are specimens of it from Kanara, Cachar, the Daunat Range of Middle Tenasserim in Burma, Perak in the Malay Peninsula, and Pulo Laut. It was originally described from Palawan in the Philippines. In spite of the key to the species of the genus *Pedestes* given by Messrs. Elwes and Edwards on p. 193, I am unable to separate *P. fuscicornis*, described by those gentlemen also from Pulo Laut, from their *P. maculicornis.*

EGG. The egg is laid on the underside of the leaves. It is dome-shaped, standing on a narrow band. It has sixteen "meridians" which start from the top of the band and run towards the top of the dome; these meridians are thin and raised above the surface. The surface is very finely lined transversely to the meridians. The colour is greenish, with the meridians brown. As the egg found was empty, the larva having eaten its way out of the top and made a large hole, the fact as to whether the meridians meet at the top of the egg or not cannot be stated. Breadth 0·8 mm.

LARVA. The head is semi-elliptical in shape, slightly rough as to surface, somewhat shiny, very light yellow-brown in colour, with a dark brown band round the back (not visible from the front view) just reaching the jaws; a brown medial line splitting down the sides of the clypeus, and a medial brownish line, broadest in the middle of its length, on each lobe of the face, starting from the clypeus and diverging from the medial line of the face and nearly reaching the vertex of the lobe; the jaws and the lower parts of the clypeus dark brown; the head is much larger than the second segment. Body broadest at segment 5, sub-cylindrical in shape; the last segment, ending semicircularly, is sloping and rather large, slightly corrugated on the dorsum towards the posterior margin, and with a small, lateral, round, tubercular, light yellow-brown spot. All the segments are clearly distinguished. The spiracles are of the ordinary size, rather round (actually slightly oval), light yellowish-brown in colour. Each segment has a good many colourless glassy-looking spots towards its front margin; these spots are small but are clearly visible under a lens. The surface of the body is finely frosted and dull, destitute of hairs except round the margin of segment 14. General colour of the larva bluish-greeny-white, beneath yellowish-green. Total length 22 mm.

Pupa. Like that of *Suastus gremius*, Fabricius. Head as broad, if not broader, than at the shoulders, and is, together with the segment 2, very large for the pupa, and is slightly bowed. Thorax rather short, strong, convex, but only slightly humped. Body of the same breadth from the shoulders to segment 8 and then tapering to the end; constricted between segments 2 and 3; the body is circular in transverse section from segment 3 to segment 13. Spiracular expansion of segment 2 large, nearly flush with the thorax, semicircular in shape (the straight side facing forwards); dark red-brown in colour. Spiracles small, linear, dark red-brown. Surface of the body finely rugose. Eyes and cremaster covered with short, erect hairs; surface of pupa bearing short, erect hairs as seen under a lens, with slightly longer hairs on the posterior portion. Cremaster hexagonal, small, with next to no suspensory hairs, brown. The pupa has the head green with a shade of brown, the thorax is green, and the abdomen waxy yellow; the surface is covered with a white powder. The cell in which it is formed is tightly closed, and the pupa is attached very slightly by the tail only.

- Habits. The habits of the larva are similar to those of *S. gremius*, the cell being made tightly, clothed with silk inside, and the edges eaten in crenulations. The larva eats above the cell towards the point of the leaf leaving the midrib, and pupates in the cell. Great quantities of old cells are found, pointing to the fact that the larvæ are very liable to the attack of enemies. The food-plants were always found in dark shady evergreen jungle. The pupa-cell is cut free by the larva before pupation and falls to the ground, and no pupa is therefore ever found except among rubbish at the foot of the plant.

219. Halpe astigmata, Swinhoe.
This species has never been bred.

220· Halpe ceylonica, Moore.
Bambusa arundinacea, Willd. (*Gramineæ*).
Oxytenanthera monostigma, Beddome (*Gramineæ*).

This, butterfly, is given in the second Kanara list as *Halpe moorei*, Watson, as well as *H. ceylonica*, the former appears to me to be a, synonym of the latter. In describing *H. moorei*, Watson does not refer to *H. ceylonica*, except to give it in his list of the species of the genus as a distinct species.

221. HALPE HONOREI, de, Nicéville.
Dendrocalamus strictus, Nees (*Gramineæ*).

222. HALPE HYRTACUS, de Nicéville.
Oxytenanthera monostigma, Beddome. (*Gramineæ*).
Ochlandra stridula, Thwaites (*Gramineæ*).

LARVA. The head of the larva is the same shape as those of the genus *Baoris*; it is nearly round, slightly indented on the top, convex on the face, thick through, rugose as to the surface, and finely hairy all over, the hairs rather short; the clypeus and about the jaws, the whole margin of the head as well as all the hinder part, and a central broad band—all very dark brown; the rest of the head dirty yellow. Body of the usual shape of *Baoris*, the anal segment rounded at the extremity; that segment covered all over with star-shaped reddish-brown spots, from each of which springs a short seta. Spiracles small, oval, a little darker than the colour of the body. Surface of the body covered with short, erect, fine, colourless hairs, which are rather longer on the anal margin than elsewhere. The colour of the larva is a transparent greenish dirty yellow, with a brown tinge on the hinder segments; a dorsal dark green line. The body is finely folded at the margins of the segments. Length 28 to 38 mm.

PUPA. The pupa has the head bowed, square in front, perfectly parallel-sided, much broader transversely to the length of the pupa than in the direction of that length, nearly as broad as the thorax at the shoulders; there is a slight boss between the eyes, with erect, rather long, light hairs in front and around the eyes. Segment 2 narrow, parallel-sided. Thorax only slightly humped, its front slope in the same line of ascent as the head, and segment. 2, twice the diameter of segment 2 (in the sense of the pupal height) at the apex, evenly convex, rounded at the shoulders; the apex is the highest and the shoulders the broadest part of the body. The transverse section of the body is circular from the shoulders to the anal end, which narrows off into the rather short and triangular cremaster; the cremaster has a rounded extremity, is perfectly flat beneath, nearly perfectly wedge-shaped, with feebly developed dorsal sustensor ridges, and a tuft of suspensory hairs on the upperside of the rounded extremity; the cremaster is hollowed out at each side at its base as in the pupa of *Baracus hampsoni*, Elwes and Edwards. Spiracles narrow, oval, small, of a darker yellow colour than the body. The spiracular expansions of segment 2 large, kidney-shaped, with the round side backwards, the edge slightly raised above the body, strainer-shaped as to the hollow area, rugose on the surface, facing forwards and outwards, very conspicuous, dark brown. The surface of the body is covered thickly all over with semi-erect, short, light hairs; the surface is irregularly and finely rugose. The colour of the pupa is a brownish-red, light, dirty yellow, with a lateral dark smudge along each side of the thorax, as well as some dark spots; each abdominal segment from 6 to 12 having two transverse rows of small, dark spots. The pupa is stout, very similar in shape to that of *Telicota bambusæ*, Moore. The proboscis does not extend in the least beyond the wing-cases. It is suspended by the tail only. Length 22 mm., breadth 5·5 mm.

HABITS. The habits of the larva are those of *T. bambusæ* in as far as the cell-making is concerned. It is just as sluggish, or even more so, in its movements; and does not excrete any cereous matter in the cell prior to changing to a pupa.

223. TELICOTA BAMBUSÆ, Moore.
Bambusa arundinacea, Willd. (*Gramineæ*).
Oxytenanthera monostigma, Beddome (*Gramineæ*).

224. BARACUS HAMPSONI, Elwes and Edwards.
A very long-leafed soft grass (*Gramineæ*).
This butterfly is given in the second Kanara list as *B. septentrionum*, Wood-Mason and de Nicéville, a species restricted by Messrs. Elwes and Edwards to Sikkim and the Shan hills of Upper Burma, but it is found also in Cachar, and in Middle Tenasserim of Lower Burma. *B. hampsoni* was described from N. Canara, but it occurs also on the Nilgiri Hills.

225. TARACTROCERA CERAMAS, Hewitson.
Grasses (*Gramineæ*).

226. TARACTROCERA MÆVIUS, Fabricius.
This butterfly has never been bred.

227. ZOGRAPHETUS OGYGIA, Hewitson.
This species also has not been bred.

228. Hyarotis adrastus, Cramer.

Phœnix sylvestris, Roxb. (*Palmeæ*).

Calamus pseudo-tenuis, Becc. and Hook. f. (*Palmeæ*).

Egg. The egg is dome-shaped, widest at the bottom, the base being narrowly flanged. There are about thirty-two very fine, low meridian-like ridges not extending on to the top of the egg, reaching only to two-thirds of the height of the egg from the base; the top surface is finely frosted, all the surface is semi-satiny and frosted-looking. Colour dirty very light brown. The egg is always laid on the top of a leaf near the edge, and sometimes there are five or six eggs placed in a row.

Larva. The larva is of the type of *Baoris oceia*, Hewitson, with which it agrees in every way. The body is long and parallel-sided, flat beneath, convex over the dorsum, narrowing somewhat towards the head in segments 2 and 3; segment 12 somewhat dilated behind the spiracles; the anal segments sloping to a thin edge, which lies flat on the resting surface, and a broadly rounded extremity. Head large, broadly oval from the front view, face slightly convex, somewhat bilobed, with a narrow clypeus, the surface pitted, dull, light brown-yellow in colour, with a dark brown medial band over the vertex to the apex of the clypeus; clypeus red-brown, with a dark medial line bordered with whitish; a few short hairs about the mouth. Spiracles oval, light yellow, with a fine brown border. Surface of the body dull, the segments well marked though not constricted in any way, and with some transverse lines; a fringe of straight, fine, colourless hairs round the margin of the anal segment, but none elsewhere. Colour blue-green-white, with a dorsal, dark indigo line from segment 3 to segment 11; segment 2 yellowish; an indistinct lateral white line. Length 33 mm., breadth 4 mm.

Pupa. The pupa is also of the type of *B. oceia*. The thorax is slightly humped; the abdomen is depressed, being broader than high; the process or snout between the eyes is long and sharp, turned up very slightly at the extremity; the proboscis is free from the ends of the wings and reaches to the end of the cremaster; the cremaster is broad, thin, spatulate, and rather long, with the edges dorsally swollen. The spiracles of segment 2 are indicated by a kidney-shaped swelling with a frosted-looking surface. The other spiracles are oval, flush, and nearly white. The surface of the body is very finely corrugated, except the frontal process which is rather coarsely ringed. The thorax has a dorsal double series of small tooth-like tubercles. The pupa is bone-coloured, with a pinkish dorsal suffusion from segments 4 to 11, a subdorsal blackish band, with a spot touching it on the outside in the middle of each segment 3 to 11 ; a dorsal pink line, with a black dot on the margin of each segment on this dorsal line ; a dorsal, thin, black line on the cremaster, and a dorsal darkish line on the anterior process. The pupa is about the same breadth from the shoulders to segment 9, and is highest at the thorax.   Length 31 mm. over all, of which the frontal process is 4 mm., and the cremaster is 3 mm., breadth 5 mm., height at segment 9 is 4 mm., height at thorax 4·75 mm.

HABITS.   The young larva on emerging from the egg makes a small semi-tube for itself on the underside of a leaf, covering the hollow with silk and stretching bands of silk across at intervals.   Having grown somewhat the larva fastens one leaf on top of another and lives between them, eating the leaf lying under the top one.   When about to pupate the larva wanders, finally pupating on the underside of a leaf with or without a covering.   The pupa is fixed to the leaf by the tail and by a body-band.   All the sheaths, cells, houses, nests or shelters of the larva, by whatever name we choose to call them, are thickly and evenly lined with silk.   The pupæ are very often found on the dry dead leaves round the base of the palms (*Phœnix sylvestris*).

229.   AMPITTIA DIOSCORIDES, Fabricius.
Oryza sativa, Linn. (*Gramineæ*).
Grasses (*Gramineæ*).

This is the *Ampittia maro*, Fabricius, of the second Kanara list, p. 54, but Dr. Chr. Aurivillius has recently shewn (Ent. Tids., vol. xviii, p. 150, n. 65 (1897) that *A. dioscorides* is an older name for it.

230.   AËROMACHUS INDISTINCTUS, Moore.
Grasses (*Gramineæ*).

231.   PADRAONA GOLA, Moore.
Soft grasses (*Gramineæ*).

232.   BAORIS (*Parnara*) CONJUNCTA, Herrich-Schäffer.
Zea Mays, Linn.* (*Gramineæ*).
Coarse broad-leaved grasses (*Gramineæ*).

Mr. Bell says that this species is the *Chapra promnens* [recte *prominens*, Moore = *Baoris* (*Chapra*) *sinensis*, Mabille], of the first paper, p. 371, n. 82.

233.   BAORIS OCEIA, Hewitson.
Bambusa arundinacea, Willd. (*Gramineæ*).

* Given as *Zea mais* in the second Kanara paper, p. 57'

Dendrocalamus strictus, Nees (*Gramineæ*).
Ochlandra stridula, Thwaites (*Gramineæ*).

234. BAORIS (*Parnara*) KUMARA, Moore.
Bambusa arundinacea, Willd. (*Gramineæ*).
Ochlandra stridula, Thwaites (*Gramineæ*).

In the first Kanara paper (p. 370, n. 80) it is noted that this species feeds on rice, *Oryza sativa,* Linn., Natural Order *Gramineæ,* and that the figure of the larva (pl. F, fig. 4) is represented on a bamboo leaf by mistake. Mr. Bell has only bred the larva on bamboos, and doubts that it feeds on rice or on any true grasses. Messrs. Elwes and Edwards (l.c., p. 276) record *P. kumara* in South India only from the Nilgiris, in North India only from Sikkim, and from Java and Borneo. In spite of the elaborate keys given by these gentlemen to distinguish the various species of the genus, I find it extremely difficult to differ- entiate many of the species given as distinct from coloration and mark- ings only. I have not studied the prehensores, which appear to be the only safe test by which they can be satisfactorily distinguished, and that test will apply to the males only. I possess no specimens of *B. kumara* from Kanara.

235. BAORIS (*Parnara*) PHILIPPINA, Herrich-Schäffer.
Bamboo (*Gramineæ*).

Recorded by Messrs. Elwes and Edwards (l.c., p. 276) from N. Canara (E. H. Aitken), but omitted from their two Kanara lists by Messrs. Davidson, Bell and Aitken. They have probably failed to identify it, placing it under *B. kumara,* Moore, to which species it is so closely allied that in spite of Messrs. Elwes and Edwards key to separate them, I am often in doubt as to which of the two species I should apportion certain specimens. *B. philippina* is said to have in the forewing a white spot in the submedian interspace touching the submedian nervure just beyond its middle, which *B. kumara* lacks. All my Kanarese specimens possess this spot. Many specimens have this spot very faint indeed, obsolescent in fact, so that one is in doubt as to which species these specimens belong.

LARVA. The larva has the head heart-shaped, but it is rather narrow at the top and slightly indented, the vertices of the lobes being rounded; the face is shiny and pitted all over, without hairs except about the mouth where there are a few; the colour is white, with a broad, black band round the head, ending at the eyes, a black line down the centre of the face, splitting into two just before the apex of the clypeus, the two parts running parallel to the sides of the clypeus for a very short distance, when they stop, and are met by a brown line parallel to the central one commencing one-third the length of the head from the vertex, and ending where it meets the furcation of the central line; the sides of the clypeus are lined with black, and there is a black line down the centre of the clypeus; the head is not large for the body. The colour of the head varies from that described above to pure white without any markings. The body is more or less cylindrical, lying with the ventrum flat on the leaf; the neck is considerably narrower than the head; the anal segment is flattened, roughened on top by pittings, it lies flat on the surface of the leaf, and is rounded at the end; segment 12 is slightly broader than segment 11, being swollen at the sides; the larva is very little broader and higher at segment 5 than anywhere else. Spiracles small, round, yellow. The surface of the larva is smooth, not shiny; the colour is a pure, opaque, bluish-white all over, with a slight yellowish tinge on the margins of the front segments, which are finely folded. Length 42 mm.

PUPA. The pupa is very similar in shape to that of *Baoris conjuncta*, Herrich-Schäffer, also in mode of suspension and in colour. The "beak" (long head-process) is slightly curved downwards, with a blunt tip, and a small piece stuck on to the top at the point with a slight point directed backwards. The proboscis is free, reaching to the hinder margin of segment 9. The cremaster is long, thin, triangular, scolloped out on the top, transparent. The surface of the pupa is shiny, with minute, erect, thick hairs covering it sparsely. The spiracular marks of segment 2 are narrowly oval in shallow depressions. Spiracles linear, white; the marks on segment 2 are the same colour as the body. Colour of body very transparent darkish green, with a rather broad, double, dorsal, white line on the abdomen. Length 32 mm., breadth 6 mm., length of snout or beak alone 3 mm. The pupa is fastened by the tail and by a band round the body.

HABITS. The larva lives in a more or less laxly-made cylindrical cell. It pupates free on the underside of a leaf, drawing the sides together slightly at either end of the leaf.

236. BAORIS (*Parnara*) PLEBEIA, de Nicéville.

This is the only species in the genus which has in the male a tuft of long hairs towards the base of the forewing on the underside attached to the inner margin of the wing and turned under and forwards. By this character alone it is quite easy to distinguish the male. Messrs. Elwes and Edwards (l.c., p. 274) record it from Sikkim, from Java on M. Paul Mabille's authority, from Kina Balu mountain in Borneo, and from Pulo Laut island near Borneo. In my collection are specimens from Sikkim, and from Taungoo in Upper Tenasserim of Burma. Its occurrence in Kanara is I think doubtful, and Mr. Bell tells me that he has no specimens of it in his collection. It has never been bred, though Mr. Bell says that the larva feeds on bamboo in the second paper. The insect bred was not *B plebeia.*

237. Baoris (*Parnara*) canaraica, Moore.
This species has never been bred.

238. Baoris (*Parnara*) colaca, Moore.
Soft, small grasses (*Gramineæ*).

239. Baoris (*Parnara*) bevani, Moore.
Oryza sativa, Linn. (*Gramineæ*).
Recorded in the first Kanara paper, p. 370, n. 79, but omitted by Mr. Bell from the second paper. It is doubtless a wrong identification, the specimens referred to being *B. guttatus,* Bremer and Grey, n. 216 *ante,* though it may occur in Kanara, as Messrs. Elwes and Edwards (l.c., p. 283) record it from Bombay. I have no specimens of *B. bevani* from any part of South India. In Bombay Mr. Aitken says he has bred it on grass. It is doubtful if he knew the species when he wrote.

240. Baoris (*Chapra*) sinensis, Mabille.
Recorded as *Chapra promnens* [recte *prominens*], Moore, in the first Kanara paper, (p. 371, n. 82), which is synonym of *B. sinensis,* the larva feeding on "Some species of Arum." The species is omitted altogether from the second paper. Mr. Bell says it is a wrong identification, and that the specimens so recorded were *Baoris* (*Parnara*) *conjuncta,* Herrich-Schäffer.

241. Baoris (*Chapra*) subochracea, Moore.
Oryza sativa, Linn. (*Gramineæ*).
Grasses (*Gramineæ*).
Mr. Bell thinks that this species and the following are one and the same, in which I am inclined to follow him, as I have never been able to separate them satisfactorily. The form with the underside grey, typical *B. subochracea,* is never found in Kanara in the rains, and is probably a dry-season form of *B. mathias.* Messrs. Elwes and Edwards (l.c., p. 275) keep them distinct. They say that the form of the male genitalia is different in the two species,

242. BAORIS (*Chapra*) MATHIAS, Fabricius.
Oryza sativa, Linn. (*Gramineæ*).
Grasses (*Gramineæ*).

243. UDASPES FOLUS, Cramer.
Curcuma Amada, Roxb.* (*Scitamineæ*).
Kæmpferia pandurata, Roxb. (*Scitamineæ*).
Hedychium coronarium, Kœnig (*Scitamineæ*).
Amomum microstephanum, Baker (*Scitamineæ*).

244. NOTOCRYPTA FEISTHAMELII, Boisduval.
Curcuma Amada, Roxb. (*Scitamineæ*).
Hedychium coronarium, Kœnig (*Scitamineæ*).
Amomum microstephanum, Baker (*Scitamineæ*).

In the second Kanara paper (p. 62) the larva of this species is said to feed on "*Maranta.*" That genus of plants is not kept distinct by Sir J. D. Hooker, but appears in the synonymy of several genera of the *Scitamineæ* in conjunction with various species of plants Messrs. Davidson, Bell and Aitken give *N. restricta*, Moore, as a species distinct from *N. feisthamelii*, and describe the transformations of both. From these descriptions the larvæ would appear to differ considerably in the two species, and feed on different plants, *N. feisthamalii* on *Amomum*, and *N. restricta* on *Curcuma*. Further observations on the subject are very desirable, as Messrs. Elwes and Edwards (l.c., p. 239) unite the two. Typically the imagines are quite distinct, but intermediate examples are apparently common. Mr. Bell maintains stoutly that the two species are distinct.

245. SANCUS PULLIGO, Mabille
Amomum microstephanum, Baker (*Scitamineæ*).
Mr. Bell says that the larvæ of *Sancus pulligo* and *Notocrypta feisthamelii*, Boisduval, can hardly be distinguished, but that the larvæ of *N. feisthamelii* and *N. restricta*, Moore, cannot be, with care, mistaken the one for the other.

Butterflies from the Western Himalayas and Kashmir, with the food-plants of their larvæ.

1. VANESSA CARDUI, Linnæus.
Carduus sp. (*Compositæ*).

---

* In the second Kanara paper, p. 62, the food-plant of this species is said to be *Curcuma aromatica*, Salisb., which is an incorrect identification.

2. VANESSA CASCHMIRENSIS, Kollar.
Urtica parviflora, Roxb. (*Urticaceæ*).

3. VANESSA XANTHOMELAS, Wiener Verzeichniss.
Salix tetrasperma, Roxb. (*Salicineæ*).

4. THECLA SASSANIDES, Kollar.
Indigofera atropurpurea, Hamilt. (*Leguminosæ*).

5. HYSUDRA SELIRA, Moore.
Indigofera atropurpurea, Hamilt. (*Leguminosæ*).

I may note that in the descriptions of larvæ above fourteen segments are always reckoned; all fourteen are very obvious in some larvæ,
the head being the first, the anal flap the fourteenth; all fourteen
somites are very distinct also in some pupæ, hardly distinguishable as
regards 13 and 14 in others.   Segment 2/3, &c., is the margin common
to segment 2 and segment 3.

XV.—*Note on the avian genus* Harpactes, *Swainson.*—*By* LIONEL DE
NICÉVILLE, *Natural History Secretary.*

In a footnote on p. 130 *ante* Colonel C. T. Bingham points out that
the genus *Harpactes* of Swainson cannot be used for a genus of birds,
being preoccupied.   This was first pointed out by Cabanis and Heine in
Mus. Hein., n. iv, pt. 1, p. 154 (1863), and referred to by Mr. Charles
W. Richmond in Proc. United States National Museum, vol. xvii, p. 602,
footnote (1894) ; by Mr. Harry C. Oberholser in Proc. Acad. Nat.
Sciences Philadelphia, 1899, p. 206 ; and in the Ibis, seventh series,
vol. vi, p. 555 (1900).   Messrs. Cabanis and Heine, Richmond, and
Oberholser specify *Pyrotrogon*, Bonaparte, as the name by which these
birds should be generically known, the typical species being *Trogon
ardens*, Temminck.

JOURNAL

OF THE

ASIATIC SOCIETY OF BENGAL.

◦◦●◦●◦◦

Vol. LXIX. Part II.—NATURAL SCIENCE.

No. III.—1900.

XVI.—*Materials for a Carcinological Fauna of India. No. 6. The Brachyura Catometopa, or Grapsoidea.*—By ·A. ALCOCK, M.B., C.M.Z.S., *Superintendent of the Indian Museum.*

[Received 25th June; Read 4th July.]

In treating the Catometopes I have in the main followed the scheme of Milne Edwards (*Annales des Sciences Naturelles* for 1852 and 1853) as modified by Dana, and I may introduce this paper with a statement of the points at which it deviates from the former of those classical works.

In the first place, following Dana and most subsequent authors, I have evicted the *Telphusidæ.* With them must also go *Gecarcinucus,* which is an undoubted Telphusoid, although it is persistently ranked with the *Geocarcinidæ.*

Again I have followed the lead of Dana in his treatment of the *Gonoplacæa* of Milne Edwards, the genera of which are distributed among the *Ocypodidæ* and the Sesarmine *Grapsidæ,* while *Gonoplax* itself is relegated to the *Carcinoplacidæ.*

This step necessitates a considerable enlargement of Milne Edwards' group of *Carcinoplacinæ,* and a reconstruction of his *Ocypodinæ,* and in carrying this out I have in the main followed Dana's admirable system.

The isolation of *Myctiris* as an independent family, which was first suggested by Dana, is here accented, but at the same time I fully agree with Milne Edwards estimate of this singular form as a " satellite " of the Ocypodoids.

J. II. 37

In grouping the genera of the *Grapsidæ* I have departed very little from the arrangement of Milne Edwards, who recognized—though his successors have ignored it—the independence of the *Varuna* group.

I have adopted Dana's family of *Geocarcinidæ*, but with some hesitation, for Milne Edwards' estimation of the group as a subfamily of *Grapsidæ* has much to recommend it.

I gladly follow Milne Edwards in recognizing the *Hymenosoma* group as a *tribu principale* not distantly related to the Ocypodes and quite distinct from and independent of the *Pinnoteres* group.

As regards additions to the *Catometopa* as known to and recognized by Milne Edwards, I may mention the *Rhizopinæ* (Stimpson, Miers), the *Hexapodinæ*, the *Palicidæ* (which include *Cymopolia* formerly classed with the Dorippidæ), and the new family *Ptenoplacidæ*.

From the system of Dana I would dissent only in separating the *Hymenosoma*-group from the *Pinnoteridæ*; in enlarging the *Scopimerinæ* (=*Dotinæ*) at the expense of the *Ocypodidæ*; in splitting the *Grapsinæ* into two equal groups,—one round *Grapsus*, the other round *Varuna*; and in removing *Gecarcinucus* from the *Geocarcinidæ*.

The scheme of classification proposed by Miers seems to me to, too often, disregard natural relations without facilitating the recognition of species by way of compensation.

The most conspicuous instance is the family *Pinnoteridæ*, in which we find *Pinnoteres* and its kindred included with such undoubted Ocypodoids as *Dotilla* and *Scopimera*, with *Mictyris*, with *Hymenosoma* and its allies, and finally with *Hexapus* whose affinities are quite clearly with the *Rhizopinæ*.

Again by the exclusion of *Scopimera* and *Dotilla* and by the inclusion of the *Gonoplacidæ*, Miers family of *Ocypodidæ* becomes unnatural and incomplete.

I follow Miers in treating the *Rhizopinæ* as a subfamily of *Carcino-placidæ*.

Ortmann obviates some difficulties by separating *Gonoplax* and *Ommatocarcinus* from the *Carcinoplacidæ* as a distinct family, and by altogether removing the *Hymenosomidæ* from the Catometopes. By the latter step his *Pinnoteridæ* gain in natural value, as they further do by the restoration of *Scopimera* and *Dotilla* to their place among the

Ocypodoids ; so that both his *Pinnoteridæ* and *Ocypodidæ* are far more natural families than those of Miers. I am doubtful, however, whether Ortmann has assigned its full rank to *Mictyris*, or their proper place to the *Hexapodinæ*.

The Catometope crabs of the Indian fauna number about 140, of which 136 are noticed in the present paper. Of these, 31 are new to science, and include 2 species of *Libystes*, 1 of *Psopheticus*, 2 of *Litochira*, 1 of *Notonyx*, 1 of *Ceratoplax*, 1 of *Typhlocarcinus*, 2 of *Pinnoteres*, 3 of *Dotilla*, 2 of *Scopimera*, 1 of *Clistostoma*, 1 of *Tylodiplax*, 1 of *Elamena*, 2 of *Hymenicus*, 2 of *Ptychognathus*, 1 of *Pyxidognathus*, 3 of *Sesarma*, 2 of *Palicus* (*Cymopolia*), and 1 of each of the following new genera, *Typhlocarcinodes* (Rhizopinæ), *Lambdophallus* (Hexapodinæ), and *Chasmocarcinops* (Asthenognathinæ).

The new species are, for the most part, either little crabs that are liable to be overlooked, or inhabitants of depths which, though moderate, are inaccessible to ordinary collectors.

As heretofore, most of the new species come from the copious collections of the " Investigator " and will be duly figured in the *Illustrations of the Zoology of the R.I.M.S. Investigator.*

Tribe CATOMETOPA.

Quadrilatera, Latreille (pt.), Fam. Nat. du Règne Anim. p. 269.

Catomètopes, Milne Edwards (pt.), Hist. Nat. Crust. II. p. 1.

Cancri (pt.), *Ocypodes*, *Grapsi*, *Pinnotheridea*, De Haan, Faun. Japon. Crust.

Ocypodidæ, Milne Edwards (pt.), Ann. Sci. Nat., Zool., (3) XVIII. 1852, pp. 128, 140.

Grapsoidea, Dana, U. S. Expl. Exp. Crust. pt. I. pp. 67, 306.

Catometopa, Miers, Challenger Brachyura, p. 216.

Catometopa, Ortmann, Zool. Jahrb., Syst. VII. 1893–94, pp. 411, 683, plus *Majoidea Hymenosomidæ*, p. 31 : and in Bronn's Thier Reich, V. ii. *Arthropoda*, pp. 1165, 1168, 1175.

The carapace is variable, but commonly and typically it is transverse, more or less quadrate, with large branchial and small and indistinct hepatic regions and a broad front. The front also is variable in form, but typically it is much deflexed.

The orbits, typically, occupy the whole or the greater part of the anterior border of the carapace on either side of the front. The typical fold of the antennules is transverse ; but it may be oblique, or nearly vertical, and in a few cases there are no distinct fossæ at all into which these appendages can fold.

The epistome, typically, is extremely short, but occasionally it is of considerable length. The buccal orifice is typically, but by no means always, square cut.

The palp of the external maxillipeds usually articulates either at the summit, or at or near the external angle, of the merus; but often, as in almost the whole family *Gonoplacidæ*, it articulates distinctly at the antero-internal angle.

The genital ducts of the male usually perforate the sternum opposite the last pair of legs: if, as happens in the family *Gonoplacidæ,* they perforate the bases of the last pair of legs, *they pass forwards to their destination in a groove in the sternum.*

The abdomen of the male is very often narrow at its base and so does not cover all the space between the last pair of legs.

The branchiæ are often fewer than 9—from 8 to 6—on either side: their efferent channels open on either side of the palate.

The *Catometopa* may be divided into 9 families. One of these, the *Gonoplacidæ*, so closely approaches the Cyclometope family *Xanthidæ* that such Xanthoid forms as *Geryon* and *Camptoplax* have by some authors been included in it, while, on the other hand, some of its constituent genera, such as *Gonoplax* and *Carcinoplax*, have been ranged among the Cyclometopes.

Three other families, namely, the *Grapsidæ*, the *Geocarcinidæ*, and the *Ocypodidæ*, include the typical Catometopes, upon which our general conception of the group is founded.

The remaining five families are more or less aberrant, they are the *Pinnoteridæ*, the *Mictyridæ*, the *Hymenosomidæ*, the *Palicidæ*, and the *Ptenoplacidæ*.

Of these aberrant families, the *Pinnoteridæ* are probably most nearly related to the *Gonoplacidæ*, the *Mictyridæ* to the *Ocypodidæ*, and the *Palicidæ* to the *Grapsidæ*.

The true position of the *Hymenosomidæ* appears to me to be still doubtful. Many authors place them near the *Pinnoteridæ* and *Mictyridæ*, and I think that their most natural place is alongside the *Mictyridæ*. Ortmann alone boldly removes them from the Catometope grade altogether and unites them with the Oxyrhynchs, which I think is a decided mistake.

There remains the family *Ptenoplacidæ,* which includes the single species *Ptenoplax notopus*. This, though it has a superficial resemblance to *Macrophthalmus*, is remote from that genus in many important characters, and, though it has no look of *Hexapus*, yet shows an attraction to *Hexapus* and *Lambdophallus* that can hardly be accidental.

The 9 families may be characterized as follows, their compass in relation to the schemes of other authors will be noted in the sequel:—

Family GONOPLACIDÆ.   Marine Catometopes closely resembling Cyclometopes.   The palp of the external maxillipeds articulates at or near the antero-internal angle of the merus, never at the antero-external angle or at the middle of the anterior border : the exognath of the external maxillipeds is of normal size and is not concealed.   The interantennular septum is a thin plate.   The division of the orbit into two fossæ is not accented.

Family GRAPSIDÆ.   Littoral (rock-haunting), or pelagic (drift-weed and timber-haunting), or estuarine and paludine, or fluviatile, or rarely terrene Catometopes.   The palp of the external maxillipeds articulates either at the antero-external angle, or at the summit, or at the middle of the anterior border of the merus : the exognath is either abnormally slender or abnormally broad.   The interantennular septum is very broad. The division of the orbit into two fossæ is accented.   [Front of great breadth : carapace usually quadrilateral, with the lateral borders either straight or very slightly arched, and the orbits at or very near the antero-lateral angles : the buccal cavern is square and there is generally a gap, which is often large and rhomboidal, between the external maxillipeds].   Male openings sternal.

Family GEOCARCINIDÆ.   Terrene Catometopes (Land-crabs).   The palp of the external maxillipeds articulates either at the antero-external angle or at the middle of the anterior border of the merus (but is sometimes, though never in any Indian species, completely hidden behind the merus) : the exognath is slender and inconspicuous (some-times more or less concealed) and sometimes carries no flagellum. The interantennular septum is very broad and the antennular fossæ are narrow.   The front is of moderate breadth and always strongly deflexed : the carapace is more or less transversely oval, *the antero-lateral borders being strongly arched and the fronto-orbital border being very much less than the greatest breadth of the carapace.*   In all the Indian forms there is a wide rhomboidal gap between the external maxillipeds.   Male openings sternal.

Family. OCYPODIDÆ.   Amphibious littoral and estuarine crabs, burrowing, and commonly gregarious.   The palp of the external maxillipeds is coarse, and articulates at or near the antero-external angle of the merus : the exognath is generally slender and often more or less concealed.   The interantennular septum is generally broad, but in one subfamily (*Macrophthalminæ*) is a thin plate. The front is usually of no great breadth, and is often a narrow lobe more or less deflexed. The orbits occupy the whole anterior border of the carapace outside the front, and their outer wall (between the far ends of the upper and lower borders) is often defective. The buccal cavern is usually large and a little narrower in front than behind, the external maxillipeds are foliaceous and usually completely close it, but if they do not they never leave between them a wide rhomboidal space exposing the mandibles. The abdomen of the male is narrow. Male openings sternal.

Family PINNOTERIDÆ. Small crabs, usually living as commensals in the mantle-cavity of Bivalve Mollusks or Ascidians, in the cloaca of Holothurians, in worm-tubes, or in coral-stocks, and hence often exhibiting degeneration of some of the organs of special sense. The external maxillipeds vary : the merus, though often very large, is never quadrilateral, and never carries the palp distinctly at the antero-internal angle : the ischium is often small, and is sometimes absent or indistinguishably fused with the merus, in which case the merus lies with its long axis directed obliquely or almost transversely inwards : the exognath is small and more or less concealed. The interantennular septum, when distinguishable, is a thin plate. [The front is narrow, the eyes and orbits very small, the corneæ sometimes obsolescent : the antennules and antennæ are usually very small and cramped. The buccal cavern is short and of great breadth, being commonly semi-circular in outline. The male abdomen is very narrow]. Male openings sternal.

Family MICTYRIDÆ. Amphibious Catometopes resembling the *Ocypodidæ* in habits. The buccal cavern is of enormous size and is completely closed by the enormous foliaceous convex external maxillipeds, whose coarse palp articulates with the antero-external angle of the merus, and whose short slender exognath is entirely concealed and carries no flagellum. The interantennular septum is narrow. The orbits are represented by a small post-ocular spine, the eyes being quite unconcealed. [Carapace elongate-globose : front a narrow declivous lobe : the rudimentary antennular flagella fold nearly vertically, and are a good deal concealed by the front : the abdomen of the male resembles that of the female and covers the greater part of the sternum. No membranous spaces (*tympana*) on the meropodites of the legs or on the sternum]. Male openings sternal.

Family HYMENOSOMIDÆ. Small marine and estuarine Catometopes having a curious superficial resemblance to some of the Oxyrhynch crabs of the Inachine subfamily, a resemblance heightened by the fact that the epistome is sometimes nearly as long as broad. The palp of the external maxillipeds articulates near the antero-external angle of the merus, but as the antero-internal angle of the merus is sometimes truncated the true relations of the palp are often not quite clear : the exognath is slender and partly or entirely concealed. There are no orbits and the eyes are exposed and little retractile. [Carapace thin, flat, triangular or subcircular, not very well calcified, usually produced to form a horizontal rostrum. Antennular fossæ shallow and ill defined. Antennal peduncle slender. Buccal cavern square, the ischium of the external maxillipeds well developed]. Male openings sternal.

Family PALICIDÆ. Small Catometopes having a sort of *Dorippe* appearance. The Indian members of the family are found among coral- and shell-shingle, at a moderate depth, and have a kind of protective resemblance to an eroded flake of coral rock. The external maxillipeds close the buccal cavern ventrally but not anteriorly : their merus is a very small joint articulating with the retreating antero-external angle of the ischium, and carrying the palp at the middle of the oblique-lying anterior (or inner) border, their exognath is not concealed and is rather broad. The interantennular septum is a thin plate. The orbit has 2 or 3 deep gaps in the upper border. Front of moderate breadth, little or not at all deflexed : antennal flagella of good length : epistome absent : abdomen of male narrow. Compared with the other 3 pairs, *the 4th (last) pair of legs, which are dorsally situated, are rudimentary* in all the Indian species. Male openings sternal : *female openings placed far forward on the sternal segment corresponding with the first pair of ambulatory legs* (2nd peræopods).

Family PTENOPLACIDÆ. Represented by an aberrant Catometope found only in Indian Seas at a depth of 100 to 250 fathoms. The external maxillipeds are slender and subpediform, not nearly covering the buccal cavity : their palp articulates with the summit of the slender merus : their exognath is of normal size and form, and is not concealed. The interantennular septum is a thin rudimentary plate. The orbits are very incomplete below. The front is a narrow, little deflexed lobe. No distinct antennular fossæ. Antennal flagella of good length. No epistome. Abdomen of male narrow. Compared with the other 3 pairs, *the last (fourth) pair of legs are rudimentary, being also placed close together dorsally : the last segment of the sternum is also rudimentary.*

The male openings are in the bases of the last pair of legs but the ducts run forward in a sternal groove.

Most of these families can be further split into subfamilies, as is shown in the following scheme :—

Family GONOPLACIDÆ, Dana.

Gonoplaciens, Milne Edwards (pt.), Hist. Nat. Crust. II. 56.

Gonoplacés Cancéroides plus *Carcinoplacinæ*, Milne Edwards, Ann. Sci. Nat. Zool. (3) XVIII. 1852, pp. 162, 164.

Gonoplacidæ, Dana, U. S. Expl. Exp. Crust. pt. I. pp. 308, 310.

Carcinoplacinæ plus *Gonoplacinæ* plus *Hexapodinæ*, Miers, Challenger Brachyura, pp. 222, 237, 275.

Carcinoplacini, Ortmann, Zool. Jahrb., Syst., VII. 1893-94, p. 683.

Carcinoplacidæ plus *Gonoplacidæ* plus *Hexapodinæ*, Ortmann in Bronn's Thier Reich, tom. cit. pp. 1175, 1176, 1177.

This family may be divided into the 5 following subfamilies :—

Subfamily I. PSEUDORHOMBILINÆ (*Carcinoplacinæ* Miers, *Carcinoplacidæ* Ortmann). Carapace Xanthoid, the regions seldom well defined : front usually of good breadth and square cut, often little deflexed : eyes and orbits of normal size and form, the eyes well pigmented and the eyestalks normally movable except in certain deep-sea genera : the antennules fold transversely : antennal flagella of fair length. Epistome well defined : buccal cavern square-cut and usually completely closed by the external maxillipeds, which have a subquadrate merus. The base of the male abdomen covers the whole space between the last pair of legs. Male openings not sternal.

Subfamily II. GONOPLACINÆ (*Gonoplacinæ* Miers, *Gonoplacidæ* Ortmann). The anterior border of the subquadrate carapace is entirely occupied by the square-cut front and orbits, the front being either narrow or of fair breadth, and the orbits being long narrow trenches for the elongate eyestalks. In other respects similar to the *Pseudorhombilinæ*.

Subfamily III. PRIONOPLACINÆ (not represented in India). Differs from *Pseudorhombilinæ* only in the form of the male abdomen, which is not broad enough at base to cover all the space between the last pair of legs.

Subfamily IV. RHIZOPINE (*Rhizopinæ* Miers, Ortmann). With the exception of one species (*Notonyx nitidus*) the eyestalks are fixed, and very often the "cornea" is minute or obsolete : the lower border of the orbit has a tendency to run downwards towards the epistome. The carapace usually has its antero-lateral corners cut away and rounded off : the front may be square-cut and broad, but is more often narrow and more or less distinctly bilobed and deflexed. The antennules may be of fair size and transversely folded, but more often, owing to the narrowness of the front, they are cramped, and fold obliquely : sometimes they cannot be folded in their fossæ at all. Antennal flagella usually short. The epistome may either be well defined and prominent, or ill defined and sunken. The buccal cavern may be squarish, but it often is decreased in breadth anteriorly : the external maxillipeds have a square merus and may completely close the buccal cavern, or there may be a gap between them. The male abdomen does not nearly cover the space between the last pair of ambulatory legs. Male openings sternal.

Subfamily V. HEXAPODINÆ (*Pinnoteridæ-Hexapodinæ* Miers, Ortmann). *Only three pairs of legs besides the chelipeds, the last segment of the sternum also aborted.* Carapace much broader than long with the antero-lateral corners cut away and rounded off. Front narrow : eyes, orbits and antennæ small : the antennules fold transversely. Epistome well defined : buccal cavern with the sides a little anteriorly-convergent, or not, nearly closed by the external maxillipeds, whose merus is either quadrate or has the antero-external angle rounded off. The male abdomen does not nearly fill the space between the last pair of ambulatory legs. Male openings sternal.

Family PINNOTERIDÆ, Edw.

Pinnotheridæ, De Haan (part), Faun. Japon., Crust., pp. 5,'34.
Pinnothériens, Milne Edwards (part), Hist. Nat. Crust. II. 28.
Pinnotherinæ, Milne Edwards, Ann. Sci. Nat. Zool. (3) XVIII. 1852, p. 138, and XX. 1853, p. 216 : Dana, U. S. Expl. Exp., Crust. pt. I. pp. 378, 379 : Miers, Challenger Brachyura, p. 274 : Ortmann, Zool. Jahrb., Syst., VII. 1893-94, p. 691 ; and in Bronn's Thier Reich, *tom. cit.* p. 1177.

I propose, with some diffidence, as I have not examined enough of the forms included, to divide this family into 4 subfamilies :—

Subfamily I. PINNOTERINÆ. Ischium of the external maxillipeds either rudimentary, or indistinguishably fused with the merus to form a single piece which is usually oblique, sometimes transverse. Usually the carapace is not transverse and the palp of the external maxillipeds not so large as the merus-ischium.

J. II. 38

Subfamily II. PINNOTHERELINÆ. Ischium of the external maxillipeds distinct and independent, but smaller than the merus, the latter joint little oblique. Usually the carapace is broadly transverse, and often the palp of the external maxillipeds is the largest part of these appendages.

Subfamily III. XENOPHTHALMINÆ. Ischium of the external maxillipeds distinct, as large as or larger than the merus, the latter joint little oblique, the palp of ordinary size. *The orbits are narrow chinks situated dorsally with their long axis at right angles to the anterior border of the carapace.*

Subfamily IV. ASTHENOGNATHINÆ (*Asthenognathidæ* Stimpson). External maxillipeds weak and slender, not nearly meeting across the buccal cavern, the ischium distinct and larger than the merus, the palp of ordinary size. Eyes in the normal position.

Family GRAPSIDÆ, Dana.

Grapsoidiens, Milne Edwards, Hist. Nat. Crust. II. 68.

Grapsinæ, Milne Edwards, Ann. Sci. Nat., Zool., (3) XVIII. p. 136 and XX. p. 163.

Grapsidæ, Dana, U. S. Expl. Exp., Crust. p. 329 : Miers, Challenger Brachyura, p. 252 : Ortmann, Zool. Jahrb., Syst., VII. 1893-94, p. 699, and in Bronn's Thier Reich, *tom. cit.* p. 1177.

This family can be divided into four well characterized subfamilies as follows :—

Subfamily I. GRAPSINÆ (*Grapsacea*, Edw., *Grapsinæ* in part, Dana, Kingsley, Miers, Ortmann). Front strongly deflexed : the lower border of the orbit runs downwards towards the buccal cavern : antennal flagellum very short : the external maxillipeds leave a wide rhomboidal gap between them, they are not traversed by any oblique hairy crest, their palp articulates at the antero-external angle of the merus, and their exognath is very slender and is exposed throughout. The male abdomen fills all the space between the last pair of ambulatory legs.

Subfamily II. VARUNINÆ (*Varunacea* and *Cyclograpsacea* part, Milne Edwards; *Grapsinæ* in part, Dana, Kingsley, Miers, Ortmann). Front moderately or little deflexed, sometimes sublaminar : the suborbital crest, which supplements the defective lower border of the orbit, is rather distant from the orbit and usually runs nearly in a line with the anterior border of the epistome : antennal flagellum usually of good length : the external maxillipeds do not often gape widely, though usually there is something of a gap, they are not traversed by any oblique hairy crest, their palp articulates with the middle of the anterior border of the merus, and their exognath is generally broad and is exposed throughout. The male abdomen, though not narrow, rarely covers all the space between the last pair of ambulatory legs.

Subfamily III. SESARMINÆ. (*Sesarmacea* and *Cyclograpsacea* part, Milne Edwards; *Sesarminæ*, Dana, Kingsley, Miers, Ortmann). Front strongly deflexed : the lower border of the orbit commonly runs downwards towards the angle of the buccal cavern : the external maxillipeds leave a` wide rhomboidal gap between them, *an oblique hairy crest traverses them from a point near the antero-external angle of the ischium to a point near the antero-internal angle of the merus,* their palp articulates either at the summit or near the antero-external angle of the merus, and their exognath is slender and either partly or almost entirely concealed. The male abdomen either fills or does not quite fill all the space between the last pair of ambulatory legs. Antennal flagella variable.

Subfamily IV. PLAGUSIINÆ. (*Plagusiacea*, Milne Edwards; *Plagusiinæ*, Dana, Kingsley, Miers, Ortmann). The front is cut into lobes or teeth by the antennular fossæ, which are visible in a dorsal view as deep clefts : the lower border of the orbit curves down into line with the prominent anterior border of the buccal cavern : the external maxillipeds do not completely close the buccal cavern but they do not leave a wide rhomboidal gap, they are not traversed by any oblique hairy crest, their palp articulates near the antero-external angle of the merus, and their slender exposed exognath has no flagellum. The antennal flagella are short. The male abdomen fills all the space between the last pair of legs.

Family GEOCARCINIDÆ, Dana.

Gécarciniens, Milne Edwards, Hist. Nat. Crust. II. 16.
Gecarcinacea, Milne Edwards (pt.), Ann. Sci. Nat., Zool., (3) XX. 1853, p. 200.
Gecarcinidæ, Dana (pt.), U. S. Expl. Exp. Crust. pt. I. p. 374.
Geocarcinidæ, Miers, Challenger Brachyura, p. 216.
Gecarcinidæ, Ortmann (pt.), Zool. Jahrb. Syst. VII. 1893.94, pp. 699, 732, and in Bronn's Thier Reich, *tom. cit.* p. 1178.

I think it inadvisable to subdivide this small group, which Milne Edwards, with more justice, regarded as itself only a subfamily of the *Grapsidæ*.

Gecarcinucus is a Telphusoid and should not be referred here. *Epigrapsus* and *Grapsodes*, if they are distinct from one another, belong here rather than to the *Grapsidæ*.

Family PALICIDÆ (*vel* CYMOPOLIDÆ).

This little and aberrant family is probably best treated as an appendage to the *Grapsidæ*.

Family OCYPODIDÆ, Ortmann (pt.).

Ocypodiens, Milne Edwards, Hist. Nat. Crust. II. p. 39.
Ocypodinæ, Milne Edwards, Ann. Sci. Nat. Zool. (3) XVIII. 1852, p. 140, plus
Gonoplacés Vigils (pt.), p. 155.
Macrophthalmidæ, Dana, U. S. Expl. Exp. Crust. pt. I. pp. 308, 312.
Ocypodinæ, Miers (pt.), Challenger Brachyura, p. 236, and *Myctirinæ* (pt.), p. 275.
Ocypodidæ, Ortmann (pt.), Zool. Jahrb., Syst., VII. 1893-94, pp. 700, 741; and
in Bronn's Thier Reich, *tom. cit.*, p. 1179.

In the treatment of this family nothing can be added to the scheme of Dana, where they are divided into 3 sub-families as follows :—

Subfamily I. OCYPODINÆ (*Ocypodiacés Ordinaires* Edw., *Ocypodinæ* Dana (pt.), Miers (pt.), Ortmann). Carapace deep, subquadrilateral, the regions seldom well defined : front narrow deflexed, commonly a mere lobe between the long eyestalks : antennular flagellum small, folding obliquely or almost vertically, the interantennular septum broad : the external maxillipeds completely close the buccal cavern, their exognath is inconspicuous but is not, or not entirely, concealed, and may either have, or be destitute of, a flagellum : chelipeds remarkably unequal either in both sexes or in the male only. *There is an orifice or recess, the edge of which is thickly fringed with hair, between the bases of the 2nd and 3rd pairs of true legs.*

Subfamily II. SCOPIMERINÆ (*Ocypodiacés Globulaires* Edw., *Dotinæ* Dana, *Myctirinæ* (pt.) Miers, Ortmann). Carapace very deep, cuboidal or globose : front narrow deflexed, commonly a mere lobe : antennular flagellum rudimentary, folding nearly vertically and hidden beneath the front, interantennular septum broad : buccal cavity large, sometimes enormous, completely closed by the external maxillipeds which are commonly very prominent and have small linear concealed exognaths with or without a flagellum : chelipeds equal or subequal in both sexes. *Orbits shallow. Curious membranous spaces known as "tympana" exist on the meropodites of the legs and often of the chelipeds also ; and sometimes on some of the segments of the sternum.* No hairy recesses between the bases of the 2nd and 3rd pairs of true legs.

Subfamily III. MACROPHTHALMINÆ. (*Gonoplacés Vigils* pt. Edw., *Macrophthalminæ* Dana, Miers, Ortmann). Carapace usually quadrilateral, broader than long (sometimes more than twice as broad as long),

flattish and not very deep, the regions usually well defined : front
variable, but never very broad : antennules with a well developed flagellum that folds transversely, interantennular septum very narrow :
eyestalks usually elongate: the external maxillipeds do not always meet
across the buccal cavern, though the gap between them is never very
wide, their exognath is not, or not entirely, concealed and has a flagellum : chelipeds usually subequal.   No special recess between the bases
of any of the legs.

Family MICTYRIDÆ, Dana.

Pinnothériens, Milne Edwards (pt.), Hist. Nat. Crust. II. 39.
Myctiroidea, Milne Edwards, Ann. Sci. Nat., Zool., (3) XVIII. 1852, p. 154.
Mictyridæ, Dana, U. S. Expl. Exp., Crust. pt. I, pp. 309, 389.
Pinnotheridæ-Myctirinæ, Miers (pt.), Challenger Brachyura, p. 275 ; Ortmann
(pt.) in Bronn's Thier Reich, *tom. cit.,* p. 1179.
Ocypodidæ-Myctirinæ, Ortmann (pt.), Zool. Jahrb., Syst. VII. 1893-94, pp. 742,
747.

There can be little question that Milne Edwards was right in
reckoning *Mictyris* as a " satellite " of the *Ocypodidæ,* or that Dana's
plan of separating them as a distinct family is fully justified.   The
affinities which several authors find between *Mictyris* and the *Pinnoteridæ*
are by no means easy to recognize.

Family HYMENOSOMIDÆ, Ortmann.

Pinnothériens, Milne Edwards (pt.), Hist. Nat. Crust. II. 39.
Hymenosominæ, Milne Edwards, Ann. Sci. Nat., Zool. (3) XX. 1853, p. 221.
Pinnotheridæ-Hymenicinæ, Dana, U. S. Expl. Exp., Crust. pt. I. pp. 379, 384.
Pinnotheridæ-Hymenosominæ, Miers, Challenger Brachyura, p. 275.
Majoidea-Hymenosomidæ, Ortmann, in Bronn's Thier Reich, *tom. cit.,* p. 1168.

Three types seem to be distinguishable in this family : in one (*e.g.*
Hymenosoma) there is no epistome and the external maxillipeds almost
encroach on the bases of the antennules, which appendages are not
concealed by the front; in the second (*e.g. Halicarcinus*) there is an
epistome of considerable length, but the antennules are still unconcealed by the front; in the third (*e.g. Hymenicus*) there is a long
epistome and the antennules are quite concealed by the front.

Family PTENOPLACIDÆ.

This family has no very close connexions with any of the others
although it is an undoubted Catometope.

The following is a list of all the Catometope genera known to me arranged according to the foregoing scheme. As in previous papers, the genera known to me by autopsy are marked with an asterisk, and all the Indian genera are printed in roman type.

Family GONOPLACIDÆ, Dana.

Subfamily I. PSEUDORHOMBILINÆ, nov.

? Brachygrapsus, J. S. Kingsley, Proc. Ac. Nat. Sci. Philad. 1880, p. 203.

Bathyplax, A. Milne Edwards, Bull. Mus. Comp. Zool., VIII, 1880-81, p. 16 : Miers, Challenger Brachyura, p. 230.

? Camptandrium, Stimpson, Proc. Ac. Nat. Sci. Philad. 1858, p. 106.

* Carcinoplax (=Curtonotus).

* Catoptrus (= Goniocaphyra).

? Cryptocœloma, Miers, Zool. H. M. S. Alert, p. 227.

* Eucrate.

Freyvillea, A. Milne Edwards, Bull. Mus. Comp. Zool. VIII, 1880-81, p. 15.

Heteroplax, Stimpson, Proc. Ac. Nat. Sci. Philad. 1858, p. 94.

* Libystes.

* Litochira.

* Pilumnoplax.

* ? Platypilumnus.

* Pseudorhombila.

* Psopheticus.

Subfamily II. PRIONOPLACINÆ, nov.

Eucratoplax, A. Milne Edwards, Bull. Mus. Comp. Zool. VIII, 1880-81, p. 17.

Eucratopsis, Smith, Amer. Journ. Sci. XLVIII, 1869, p. 391, and Trans. Connect. Acad. II. 1871-73, p. 35.

Euryplax, Stimpson, Ann. Lyc. Nat. Hist., New York, VII, 1862, p. 60.

Glyptoplax, Smith, Trans. Connect. Acad. II, 1871-73, p. 164.

Oediplax, Mary J. Rathbun, P. U. S. Nat. Mus. XVI, 1893, p. 241.

Panoplax, Stimpson, Bull. Mus. Comp. Zool. II, 1870-71, p. 151.

Prionoplax, Milne Edwards, Ann. Sci. Nat., Zool., (3) XVIII, 1852, p. 163.

Speocarcinus, Stimpson, Ann. Lyc. Nat. Hist., York, VII, 1862, p. 58.

Subfamily III. GONOPLACINÆ, Miers.

* Gonoplax, Leach, Trans. Linn. Soc. XI, 1815, p. 323: Miers, Challenger Brachyura, p. 245.

Ommatocarcinus, White, in Voy. H. M. S. Rattlesnake, II, p. 393; Miers, Challenger Brachyura, p. 246.

Subfamily IV. RHIZOPINÆ, Stimpson, Miers.

* Camatopsis.

* Ceratoplax.

? *Chasmocarcinus,* Mary J. Rathbun, Bull. Nat. Hist. Iowa, 1898, p. 284.

* Hephthopelta.

* Notonyx.

Rhizopa, Stimpson, Proc. Ac. Nat. Sci. Philad. 1858, p. 95.

* Scolopidia (=Hypophthalmus).

* Typhlocarcinus.

* Typhlocarcinodes.

* Xenophthalmodes.

Subfamily V. HEXAPODINÆ, Miers.

Amorphopus, Bell, Journ. Linn. Soc., Zool., III, 1859, p. 27.

* *Hexapus,* De Haan, Faun. Japon. Crust., p. 35.

* Lambdophallus.

Thaumastoplax, Miers, Ann. Mag. Nat. Hist. (5) VIII, 1881, p. 261.

Family PINNOTERIDÆ, Edw.

[? Subfamily I. PINNOTERINÆ, *nov.*

Cryptophrys, Mary J. Rathbun, P. U. S. Nat. Mus. XVI, 1893, p. 250.

Dissodactylus, S. I. Smith, Trans. Connect. Acad. II, 1871–73, p. 172.

Durckheimia, de Man, Zool. Jahrb., Syst., 1889, p. 442.

Fabia, Dana, Proc. Ac. Nat. Sci. Philad. 1851, p. 253, and U. S. Expl. Exp., Crust. pt. I. p. 382.

? *Holothuriophilus,* Nauck, Zeits. Wiss. Zool. XXXIV, 1880, pp. 24, 66.

Ostracoteres, Milne Edwards, Ann. Sci. Nat., Zool., (3) XX, 1853, p. 219.

? *Parapinnixa,* Holmes, Proc. Calif. Acad. IV, 1893-94, pp. 565, 587.

Pinnaxodes, Heller, Novara Crust., p. 67.

* Pinnoteres.

? *Scleroplax*, Mary J. Rathbun, P. U. S. Nat. Mus. XVI, 1893, p. 250.

* Xanthasia.

? Subfamily II. PINNOTHERELINÆ, *nov.*

? *Malacosoma*, de Man, Notes Leyden Mus. I, 1879, p. 67.

Opisthopus, Mary J. Rathbun, P. U. S. Nat. Mus. XVI, 1893, p. 251.

Pinnixa, White, Ann. Mag. Nat. Hist. XVIII. 1846, p. 177 (= *Tubicola*, Lockington, Proc. Calif. Acad. VII. 1876, p. 55).

Pinnotherelia, Milne Edwards and Lucas, in Voy. Amér. Mérid., Crust. p. 24 (1843).

Pseudopinnixa, Ortmann (*nec* Holmes), Zool. Jahrb., Syst. VII, 1894, p. 694.

* Tetrias.

? *Tritodynamia*, Ortmann, Zool. Jahrb., Syst. VII, 1194, p. 692.

? Subfamily III. XENOPHTHALMINÆ, nov.

* Xenophthalmus.

? Subfamily IV. ASTHENOGNATHINÆ, Stimpson.

Asthenognathus, Stimpson, Proc. Acad. Nat. Sci. Philad. 1858, p. 107.

* Chasmocarcinops.

Family OCYPODIDÆ, Ortmann, emend.

Subfamily I. OCYPODINÆ, Dana.

Acanthoplax, Milne Edwards, Ann. Sci. Nat. Zool. (3) XVIII, 1852, p. 151.

* Gelasimus.

*Heloecius, Dana, Amer. Journ. Sci., (2) XII, 1851, p. 286, and U. S. Expl. Exp., Crust. pt. I. p. 319.

* Ocypoda.

Subfamily II. MACROPHTHALMINÆ, Dana.

* Clistostoma.

Chænostoma, Stimpson, Proc. Ac. Nat. Sci. Philad. 1858, p. 97.

Euplax, Milne Edwards, Ann. Sci. Nat., Zool., (3) XVIII, 1852, p. 160 ; Miers, Challenger Brachyura, p. 251.

Hemiplax, Heller, Novara Crust. p. 40: Miers, Challenger Brachyura, p. 250.

*Macrophthalmus.

Paraclistostoma, de Man, Zool. Jahrb., Syst., VIII, 1895, p. 580.

*Tylodiplax.

Subfamily III. Scopimerinæ.

*Dotilla (= *Doto*, De Haan).

Ilyoplax, Stimpson, Proc. Ac. Nat. Sci. Philad. 1858, p. 98.

*Scopimera.

*Tympanomerus (= *Dioxippe*, de Man).

Family MICTYRIDÆ, Dana.

*Mictyris.

Family HYMENOSOMIDÆ, Ortmann.

*Elamene.

?? *Elamenopsis*, A. Milne Edwards, Nouv. Archiv. du Mus. IX, 1873, p. 324.

Halicarcinus, White, Ann. Mag. Nat. Hist. XVIII, 1846, p. 178: Miers, Challenger Brachyura, p. 280 (= *Liriopea*, Gay, Hist. Fis. Chile, pt. III. Zool. p. 158.

*Hymenicus.

Hymenosoma, Leach, Milne Edwards, Hist. Nat. Crust. II, 35 : Miers, Challenger Brachyura, p. 279.

Rhynchoplax, Stimpson, Proc. Ac. Nat. Sci. Philad. 1858, p. 109.

*Trigonoplax.

Family GRAPSIDÆ, Dana.

Subfamily I. Grapsinæ, Dana (pt.).

*??? Epigrapsus.

*Geograpsus.

Goniopsis, De Haan, Faun. Japon. Crust., p. 33 (pt.) : Miers, Challenger Brachyura, p. 266.

*Grapsus.

Leptograpsus, Milne Edwards, Ann. Sci. Nat., Zool., (3) XX, 1853, p. 171 : Miers, Challenger Brachyura, p. 257 (sub-genus of *Grapsus*).

*Metopograpsus.

Orthograpsus, Kingsley, Proc. Ac. Nat. Sci. Philad. 1880, p. 194 (sub-genus of *Grapsus*).

*Pachygrapsus.

Perigrapsus, Heller, Verh. zool.-bot. Ges. Wien, XII, 1862, p. 522, and Novara Crust. p. 48.

Subfamily II. VARUNINÆ, nov.

? *Acmæopleura,* Stimpson, Proc. Ac. Nat. Sci. Philad., 1858, p. 105.

Brachynotus, De Haan, Faun. Japon., Crust., p. 34, 1835: Miers, Challenger Brachyura, p. 264 (= *Heterograpsus,* Lucas, Expl. Sci. Algerie, Anim. Artic. I, p. 18, 1849 : = *Hemigrapsus,* Dana, Amer. Journ. Sci. (2) XII, 1851, p. 288, and U. S. Expl. Exp., Crust., pt. I. p. 348).

Cyrtograpsus, Dana, Amer. Journ. Sci. (2) XII, 1851, p. 288, and U. S. Expl. Exp., Crust., pt. I, p. 351.

Eriochir, De Haan, Faun. Japon. Crust. p. 32.

Euchirograpsus, A. Milne Edwards, Bull. Mus. Comp. Zool., VIII, 1880-81, p. 18 : and Milne Edwards and Bouvier " Hirondelle " (Monaco) Crust., Brachyures et Anomures, p. 46.

Glyptograpsus, S. I. Smith, Trans. Connect. Acad. II, 1871–73, p. 153.

Planes, Leach, Malac. Pod. Brit., Expl. of pl. xxvii, figs. 1–3, 1815 (= *Nautilograpsus,* Milne Edwards, Hist. Nat. Crust. II, 89, 1837.)

Platychirograpsus, de Man, Zool. Anz. 1896, p. 292, and Mitteil. Nat. Mus. Hamburg, XIII, 1896, p. 95.

Platygrapsus, Stimpson, Proc. Ac. Nat. Sci. Philad. 1858, p. 104 : Miers, Challenger Brachyura, p. 263 (= *Platynotus,* De Haan, Faun. Japon., Crust., p. 34).

Pseudograpsus, Milne Edwards, Hist. Nat. Crust. II, 81 : Miers, Challenger Brachyura, p. 261 (= *Pachystomum,* Nauck, Zeits. Wiss. Zool. XXXIV, 1880, p. 67).

*Ptychognathus (= *Gnathograpsus,* A. M. Edw. = *Cœlochirus,* Nauck).

*Pyxidognathus.

Utica, White, P. Z. S. 1847, p. 85, and Ann. Mag. Nat. Hist., XX, 1847, p. 206.

*Varuna (= *Trichopus,* De Haan).

Subfamily III. SESARMINÆ, Dana.

Aratus, Milne Edwards, Ann. Sci. Nat., Zool., (3) XX, 1853, p. 187.

Chasmagnathus, De Haan, Faun. Japon., Crust., p. 27 (= *Paragrapsus,* Milne Edwards, Ann Sci. Nat., Zool., (3) XX, 1853, p. 195).

*Clistocœloma.

Cyclograpsus, Milne Edwards, Hist. Nat. Crust., II, 77, 1837 (= *Gnathochasmus*, MacLeay, in Smith's Ill. Ann. S. Afr. p. 65, 1838).

Helice, De Haan, Faun. Japon., Crust, p. 28 : Miers, Challenger Brachyura, p. 268.

Metaplax (= *Rhaconotus*, Gerst.).

Metasesarma.

Metopaulias, Mary J. Rathbun, P. U. S. Nat. Mus. XIX, 1897, p. 144.

Sarmatium (= *Metagrapsus*, Edw.).

Sesarma (= Holometopus, Edw.).

Subfamily IV. PLAGUSIINÆ, Dana.

Liolophus (= *Acanthopus*, De Haan).

Plagusia.

Family GEOCARCINIDÆ, Dana.

Cardiosoma (= *Discoplax*, A. M. Edw.).

Epigrapsus.

Gecarcinus, Leach, Trans. Linn. Soc. XI, 1815, p. 322 : Miers, Challenger Brachyura, p. 217.

Pelocarcinus (= Gecarcoidea, Edw., = Hylæocarcinus, W.-M., = Limnocarcinus, de Man).

Uca, Latr., Encycl. Méthod. X, p. 685 : Milne Edwards, Hist. Nat. Crust. II, 21.

Family PALICIDÆ, Rathbun (name only).

Palicus (= *Cymopolia*).

Crossotonotus, A. Milne Edwards, Nouv. Archiv. du Mus. IX, 1873, p. 282, and Journ. Mus. Godeffroy, I, 1873, p. 258.

Family PTENOPLACIDÆ, Alcock.

Ptenoplax.

Family I. GONOPLACIDÆ, Dana.

Subfamily i. PSEUDORHOMBILINÆ, Alcock.

Key to the Indian Genera.

I. Front with the edge cut straight and square, never curved, often prominent :—

    1. The fronto-orbital border, though extensive, is much less than the greatest breadth of the carapace, so that the antero-lateral borders of the carapace have

a distinctly Cancroid arch : the carapace is usually
much broader than long :—

    i.   Dactyli of last pair of legs styliform .... PSEUDORHOMBILA.
    ii.   Dactyli of last pair of legs compressed and
       ciliated :—
         a.   Antero-external angle of merus of
            external maxillipeds not particularly
            produced :—
            α.   Carapace transversely quadri-
               lateral, its antero-lateral borders
               with few teeth .... CARCINOPLAX.
            β.   Carapace transversely elliptical,
               its antero-lateral borders with 5
               or 6 teeth .... CATOPTRUS.
         b.   Antero-external angle of merus of exter-
            nal maxillipeds strongly produced out-
            wards : last pair of legs sometimes
            paddle-like.................................. LIBYSTES.
    2.   The fronto-orbital border is not so very much less
       than the greatest breadth of the carapace in extent,
       so that the antero-lateral borders of the carapace
       are either slightly arched or nearly straight : the
       carapace is broader than long but is not conspicu-
       ously transverse :—
       i.   The antennal flagellum stands loosely in
         orbital hiatus :—
         a.   Carapace deepish, rather markedly trans-
            verse : the meri of the legs with a
            spine or spines on the anterior border ... PSOPHETICUS.
         b.   Carapace shallow, depressed, and flat,
            little broader than long :—
            a.   Legs spiny .... PLATYPILUMNUS.
            β.   Legs unarmed......................... PILUMNOPLAX.
       ii.   A process of the basal antenna-joint com-
         pletely fills up and closes the orbital hiatus,
         entirely excluding the antennal flagellum ...... EUCRATE.
II.   Front with the edge slightly but distinctly curved, never
    cut straight and square ; carapace and appendages in all
    the Indian species tomentose and hairy......................... LITOCHIRA.

EUCRATE, De Haan.

Eucrate, De Haan, Faun. Japon. Crust. p. 36: de Man, Journ. Linn. Soc., Zool.,
1887–88, p. 88 : Ortmann, Zool. Jahrb., Syst. VII. 1893-94, p. 685.
    Heteroplax, Stimpson, Proc. Ac. Nat. Sci. Philad. 1858 (1859) p. 94.

Carapace deepish, subquadrilateral, a little broader than long,
smooth and with little or no distinction of regions, convex fore and aft,
very slightly so from side to side.

The extent of the fronto-orbital border is not much less than the greatest breadth of the carapace, the antero-lateral borders therefore, which are toothed, are short and but slightly arched. Front square-cut and straight, well delimited from the well-defined supra-orbital angles, usually notched or grooved in the middle line, about a third the breadth of the carapace.

Upper border of orbit with two distinct sutures. The orbital hiatus is compactly filled and closed by a process of the basal antenna-jóint, so that the antennal flagellum, which is of good length, lies entirely outside the hiatus. The antennules fold transversely.

Buccal cavern square, completely closed by the external maxillipeds, the flagellum of which articulates with the inner angle of the merus. Efferent branchial channels of palate well defined.

Chelipeds subequal, much more massive and shorter, or not much longer, than the legs.

Legs slender, unarmed ; the propodite and dactylus of the last pair are compressed and are usually, but not always, somewhat broadened.

In both sexes all seven abdominal segments are distinct, and in the male the third segment covers the whole width of the sternum between the bases of the last pair of legs.

Distribution : Indo-Pacific (Indian, Australian and Japanese).

Following de Man and Ortmann, I restrict the genus *Eucrate* to those species in which the orbital hiatus is completely stopped-up by a process of the basal antenna-joint.

Key to the Indian species of the genus Eucrate.

I. Antero-lateral borders of the carapace cut into four teeth (including the outer orbital angle) all of which are distinct: dactylus of last pair of legs distinctly palmulate : front grooved or notched in the middle line :—

    1. Carapace nearly smooth........................... *E. crenata.*

    2. Carapace with some short transverse ridges in its antero-lateral part...................... *E. crenata* var. *affinis.*

II. Antero-lateral borders cut into four teeth (including the orbital angle) of which the 2nd and 4th are hardly distinguishable : front with the median notch almost obsolete : dactylus of last pair of legs palmulate .... *E. crenata* var. *dentata.*

III. Antero-lateral borders cut into three teeth (including the orbital angle) : dactylus of last pair of legs almost styliform .... *E. sexdentata.*

1. *Eucrate' crenata*, De Haan.

Cancer (Eucrate) crenatus, De Haan, Faun. Japon. Crust. p. 51, pl. xv. fig. 1.

Eucrate crenata, Ortmann, Zool. Jahrb , Syst., VII. 1893-94, p. 688.

? *Pilumnoplax sulcatifrons*, Stimpson, Proc. Ac. Nat. Sci. Philad. 1858 (1859), p. 93 : Tozzetti, 'Magenta' Crust. p. 102, pl. vii. fig. 2.

Carapace smooth, its length about five-sixths of its breadth. Front not quite a third the breadth of the carapace, notched and groved in the middle line. Major diameter of orbit about half the width of the front.

Antero-lateral borders of carapace cut into 4 bluntish teeth, the middle two of which are the largest : a short ridge runs on to the dorsum of the carapace from the last tooth.

Chelipeds less than twice the length of the carapace, not much longer than the legs, especially in the female : one or two teeth at the far end of the upper border of the arm, and one at the inner angle of the wrist : hand rather short and squat, the fingers, which are stout, are a little longer than the palm : there is a characteristic patch of fur at the far end of the upper surface of the wrist.

Legs smooth, the last 3 joints more or less ciliated : in the 4th (last) pair the propodite and dactylus are broader and more compressed than in the other legs.

In the Indian Museum are 3 specimens from the Andamans and 1 from Madras (besides 3 from Hongkong).

The carapace of the largest specimen is 10 millim. long and 12 millim. broad.

2. *Eucrate crenata* var. *affinis*, Haswell.

Eucrate affinis, Haswell, P. L. S., N. S. Wales, VI. 1881-82, p. 547 and Cat. Austral. Crust. p. 86 : de Man, Journ. Linn. Soc., Zool. XXII. 1887-88, p. 89, pl. v. fig. 5.

? *Pseudorhombila sulcatifrons*, var. *australiensis*, Miers, Zool. H. M. S. Alert, p. 242, pl. xxiv. fig. c.

Differs from typical *E. crenata*, specimens of the same sex and of approximately the same size compared, only in the following characters :—

(1) the carapace is more sculptured, for besides the short transverse ridge on the dorsum of the carapace that runs from the last tooth of either antero-lateral border, there are similar ridges running (*a*) from the 2nd tooth of either antero-lateral border, parallel with the orbit, and (*b*) parallel with the front, near the anterior limit of the gastric region ; there is also a beaded ridge running parallel with either postero-lateral border :

(2) the patch of fur on the wrist may be smaller :

A single specimen from Mergui (Anderson collection) has the carapace 12 millim. long and 15 millim. broad.

In a large series of specimens these distinctions would probably fail.

3.　*Eucrate crenata* var. *dentata.*

? *Heteroplax dentatus,* Stimpson, Proc. Ac. Nat. Sci. Philad. 1858, (1859), p. 94 : A. O. Walker, Journ. Linn. Soc. Zool. XX. 1886-1890, p. 110.

Differs from the typical *E. crenata,* only in the following particulars :—

(1) the front is entire, the median notch being inconspicuous or absent :

(2) the outer orbital angle and the third tooth of the antero-lateral border are large and acute, while the 2nd and 4th teeth are quite inconspicuous.

In the Indian Museum are two small specimens, one from Palk Strait (the other from Hongkong).

4.　*Eucrate sexdentata,* Haswell.

Eucrate sexdentata, Haswell, P. L. S., N. S. Wales, VI. 1881-82, p. 548, and Cat. Austral. Crust. p. 86.

? *Pseudorhombila vestita* var. *sexdentata,* Miers, Zool. ' Alert,' p. 240, pl. xxiv. fig. B, and Challenger Brachyura, p. 229.

Differs from *E. crenata* in the following particulars :—

(1) the only ridges on the carapace are two exceeding faint ones running parallel with the postero-lateral borders :

(2) the antero-lateral borders are cut into 3 teeth, of which the last is spine-like :

(3) the median emargination of the front is much less distinct :

(4) the chelipeds are about $1\frac{3}{4}$ times the length of the carapace and are decidedly shorter than the legs : there is only one distinct tooth near the far end of the upper border of the arm : the tooth at the inner angle of the wrist is very large and acute :

(5) the propodite and dactylus of the last pair of legs are not broader than those of the other legs.

In the Indian Museum is a single male from the Gulf of Martaban, 20 fms. The carapace is 11·5 millim. long and 13·5 millim. broad.

CARCINOPLAX, Edw.

Carcinoplax, Milne Edwards, Hist. Nat. Crust. II. 60, and Ann. Sci. Nat., Zool., (3) XVIII. 1852, p. 164 : Ortmann, Zool. Jahrb , Syst., &c., VII. 1893-94, p. 685.

Curtonotus, De Haan, Faun. Japon. Crust., p. 20 (*nom. preocc.*).

The chief differences between this genus and *Eucrate* are that (1) the carapace is very much broader, and its antero-lateral borders are much more arched, the fronto-orbital border being relatively much less extensive; (2) the supra-orbital angles are almost merged in the front, and the median notch of the front is almost obsolete; and (3) the orbital hiatus is not stopped up by any process of the basal antenna-joint.

Carapace deepish, subquadrilateral, usually much broader than long, smooth and with little or no distinction of regions, convex fore and aft, very slightly so from side to side.

The extent of the fronto-orbital border is much less than two-thirds the greatest breadth of the carapace, and the antero-lateral borders, which are toothed, are well arched. Front square-cut and straight, faintly notched or longitudinally grooved in the middle line, not very distinctly demarcated from the supra-orbital angles, from a third to a fourth, or less, the width of the carapace.

The upper border of the orbit is sinuous and may, or may not, be marked 'by a single faint suture line. The basal antenna joint is short and the antennal flagellum stands loosely in the open orbital hiatus. The antennules fold transversely.

Buccal cavern, palate, and external maxillipeds as in *Eucrate*.

Chelipeds subequal, much more massive and sometimes, in the adult, much longer than the legs.

Legs slender, unarmed; in the last pair the propodite and dactylus are compressed and decidedly broadened for swimming.

In both sexes all seven abdominal segments are distinct, and in the male the third segment covers the whole width of the sternum between the bases of the last pair of legs.

Distribution : Indo-Pacific (Indian, Japanese, Californian).

I exclude from the genus *Carcinoplax* those species, e.g., *setosa* and *integra*, which have the edge of the front turned down and arched: these it seems to me are better associated with *Litochira*.

Key to the Indian species of the genus Carcinoplax.

I. The long diameter of the orbit is nearly three-fourths the width of the inter-orbital space : a spine or tooth at the outer angle of the wrist. Chelipeds in the adult male very much longer than the legs ... ... *C. longimanus.*

II. The long diameter of the orbit is about half the width of the inter-orbital space: no spine or tooth at the outer angle of the wrist. Chelipeds rather shorter than the legs ... ... *C. longipes.*

5. *Carcinoplax longimanus*, De Haan.

Cancer (Curtonotus) longimanus, De Haan, Faun. Japon. Crust. p. 50, pl. vi. fig. 1.
Carcinoplax longimanus, Milne Edwards, Ann. Sci. Nat. Zool. (3) XVIII. 1852,
p. 164: Ortmann, Zool. Jahrb., Syst., VII. 1893-94, p. 688.

Carapace, length a little more than two-thirds its breadth, its surface (like that of the chelipeds) finely frosted : in the young the hepatic
are obscurely delimited from the branchial and gastric regions and are
very slightly tumescent.

Front proper about two-ninths the greatest breadth of the carapace,
very faintly notched in the middle line, its free edge longitudinally
grooved.

Orbits shallow, their major diameter more than two-thirds the
width of the front: borders of orbit finely beaded, the upper border
sinuous but entire.

Antero-lateral borders of carapace not much more than half the
length of the postero-lateral, well arched, armed with 3 teeth or tubercles
(including the outer orbital angle) which become much worn away
in adults.

Chelipeds subequal, massive, varying in length with increase in
age—from 2 or 2½ times the length of the carapace in females and
young males to 4 times and more the length of the carapace in old
males, the palm being the principal joint in which the lengthening
takes place. There is a spine or tooth in the distal half of the upper
surface of the arm, and one at either angle (inner and outer) of the
wrist: a blunt crest, ending in a blunt tooth, traverses the inner surface
of the palm.

The legs are long : the 3rd pair, which are slightly the longest, are
a little more than twice the length of the carapace. The last two
joints—as also the anterior border of the carpus—of all the legs are
plumose.

In the Indian Museum are 2 specimens from the Gulf of Martaban
and the Andaman Sea 53 and 60 fathoms, (besides a large male from
Japan).

In spirit the colour is a light reddish ochre, the fingers uncoloured.

6. *Carcinoplax longipes* (Wood-Mason).

Nectopanope longipes, Wood-Mason, Ann. Mag. Nat. Hist., March, 1891, p. 262 :
Alcock and Anderson, Ill. Zool. Investigator, Crust. pl. xiv. fig. 7.
Carcinoplax longipes, Alcock, Investigator Deep-Sea Brachyura, p. 71.

Carapace, length more than three-quarters its breadth, the regions
barely indicated.

Front proper about a third the greatest breadth of the carapace, remarkably prominent, as faintly as possible notched in the middle line.

Orbits shallow, their upper border sinuous but entire, their major diameter about half the width of the front. Eyes small.

Antero-lateral borders of carapace not two-thirds the length of the postero-lateral, moderately arched, armed with two pro-curved spine-like teeth, and with a small blunt denticle just behind the ill-defined orbital angle.

Chelipeds twice the length of the carapace; the arm has a denticle beyond the middle of the upper border, and there is a strong spine—with sometimes a secondary spinule at its base—at the inner angle only of the wrist.

The legs are long and have the dactylus well plumed and the 2 preceding joints more scantily hairy: the third pair, which are slightly the longest, are nearly $2\frac{1}{2}$ times the length of the carapace: though the terminal joints of the fourth (last) pair are compressed they are not so subfoliaceous as those of *C. longimanus.*

In the Indian Museum are 20 specimens from the Andamans 220 to 290 fathoms and off Travancore, 430 fathoms.

In the largest specimen the carapace is 14 millim. long and 17 millim. broad.

In spirit the colour is white with a faint pink tinge, the fingers blackish-brown.

7. PSEUDORHOMBILA, Edw.

Pseudorhombila, Milne Edwards, Hist. Nat. Crust. II. 59, and Ann. Sci. Nat., Zool., (3) XVIII. 1852, p. 164.

The only particulars in which *Pseudorhombila* differs from *Carcinoplax* are that the regions of the carapace are better defined, that the square-cut front is more distinctly bilobed, that the supra-orbital border has two distinct sutures, and that the dactyli of the last pair of legs are styliform.

The only specimen in the Indian Museum that is perhaps referable to this genus is too small and too much damaged for description: it is from the Andamans.

LIBYSTES, A. M. Edw.

Libystes, A. Milne Edwards, Ann. Soc. Entom. France, (4) VII. 1867, p. 285, and Nouv. Archiv. du Mus. IV. 1868, p. 84.

This genus unites *Carcinoplax* with *Catoptrus.* It chiefly differs from *Carcinoplax* in having (1) a much shorter and broader carapace, (2) a much shorter and broader buccal cavern, with external maxillipeds that have the antero-external angle of the merus remarkably produced outwards, and (3) the 3rd to 5th abdominal terga of the male fused together. From *Catoptrus* it chiefly differs (1) in having the carapace more subquadrilateral than elliptical, and (2) in the curious *Amphitrite*-like form of the external maxillipeds.

. Carapace deepish, subquadrilateral or subelliptical, vastly broader than long, with little or no distinction of regions, convex fore and aft, slightly so from side to side.

The extent of the fronto-orbital border is vastly less than the greatest breadth of the carapace, so that the antero-lateral borders, which may be toothed or entire, have a Cancroid-like curve. Front square-cut and quite straight, not well separated from the supra-orbital angles, slightly notched in the middle line, a third or less the greatest breadth of the carapace.

Orbits shallow, their upper border entire The basal antenna-joint is short, and the antennal flagellum stands loosely in the orbital hiatus. The antennules fold transversely.

Buccal cavern square-cut, much broader than long; the efferent branchial canals of the palate very well defined. The merus of the external maxillipeds is short and broad and has the external angle much produced, as in many species of *Neptunus*.

Chelipeds subequal, much more massive and longer than the legs; the hands however, which are somewhat tumid, are unequal in the adult.

Legs slender, unarmed: in the Indian species the last pair are almost as paddle-like as those of the typical swimming-crabs of the Portunid family.

In the male the abdomen covers the whole width of the sternum between the last pair of legs, and the 3rd–5th abdominal terga are fused together.

The sternal canals of the male are more perfect than in any other Gonaplacoid known to me.

Key to the Indian species of Libystes.

I.   Antero-lateral borders of the carapace serrated almost
exactly like those of *Catoptrus nitidus*   ...   .   *L. Edwardsi.*

II. Antero-lateral borders of the carapace entire ...   ...   *L. Alphonsi.*

Distribution : Indo-Pacific (Madagascar to Sandwich Is.).

8. *Libystes Edwardsi*, n. sp.

Carapace, length about four-sevenths of the breadth, finely pitted under lens, somewhat granular near the antero-lateral borders : an angular eminence near either posterior angle and a slight concavity of the postero-lateral part of the lateral epibranchial regions give the carapace a somewhat quadrilateral cast.

Front a good deal less than a third the breadth of the carapace, perfectly straight, faintly notched in the middle line. Eyes small.

Antero-lateral borders of the carapace with 5 or 6 granular denticles followed by a sharp procurved spine.

The chelipeds have the hands unequal in the adult. They are more than three times the length of the carapace and are smooth and unarmed. The fingers are slender and hooked at tip, especially in the smaller hand : they are a good deal longer than the palm in the smaller hand, and about as long as the palm in the larger hand. On the immobile finger of the smaller hand there are several irregular enlarged teeth. [In the young, as in *Catoptrus*, the hands are nearly equal, and the fingers of both hands are equally long and slender].

The legs are slender and the longest pair are not much more than twice the length of the carapace. The last 3 joints of the last pair form typical swimming paddles.

An apparently adult specimen from the Persian Gulf and 3 young from the Andamans are in the Indian Museum.

The carapace of the large specimen is 8 millim. long and 14 millim. broad.

9. *Libystes Alphonsi*, n. sp.

Differs from *L. Edwardsi* in the following particulars :—

(1) the carapace, though of the same proportions, is more quadrilateral and more convex fore and aft, and the eminences at the posterior angles are wanting :

(2) the antero-lateral borders of the carapace are smooth and entire :

(3) the front is more deflexed and more distinctly divided in the middle line :

(4) the chelipeds (in the young) are about $2\frac{1}{2}$ times the length of the carapace and are nearly equal and similar : the fingers are hardly as long as the palm :

(5) The last 3 joints of the last pair of legs are much broadened and compressed, but are not such unmistakeable paddles as those of *L. Edwardsi*.

In the Indian Museum is a single specimen from the Andamans : its carapace is 4 millim. long and 7 millim. broad.

This species differs but little, except in the sub-quadrilateral shape of the thorax, from the *Libystes nitidus* described and figured by M. A. Milne-Edwards.

CATOPTRUS, A. M. Edw.

Catoptrus, A. Milne Edwards, Ann. Sci. Nat. Zool. (5) XIII. 1870, p. 82 : Ortmann, Zool. Jahrb., Syst. VII. 1893-94, p. 685.

Goniocaphyra, de Man, Archiv fur Naturges. LIII. 1887, i. p. 339.

Carapace transversely elongate-elliptical, without distinction of regions, moderately convex in both directions.

The extent of the fronto-orbital border is vastly less than the greatest breadth of the carapace, the antero-lateral borders, which are serrated, are therefore well curved. Front straight, slightly notched in the middle line, not distinctly separated from the supra-orbital angles, less than a third the greatest breadth of the carapace.

Orbits shallow, their upper border entire. The antennal flagellum, which is of good length, stands in the orbital hiatus. The antennules fold transversely.

Buccal cavern, palate, and external maxillipeds as in *Eucrate*.

Chelipeds much as in *Libystes*. Legs as in *Libystes*, except that the last pair, though they have the dactylus compressed and ciliated, are never paddle-like.

Abdomen as in *Libystes*.

Distribution : Indo-Pacific (Mauritius to Samoa).

Catoptrus really differs from *Libystes* only in the form of the merus of the external maxillipeds and of the last pair of legs, which are not paddle-like as they are in one species of *Libystes*.

10. *Catoptrus nitidus*, A. M. Edw.

Catoptrus nitidus, A. Milne Edwards, Ann. Sci. Nat., Zool., (5) XIII. 1870, p. 82 : de Man, Notes Leyden Mus. XII. 1890, p. 67 : Ortmann, Zool. Jahrb., Syst., VII. 1893-94, p. 687.

Goniocaphyra truncatifrons, de Man, Archiv fur Nat. LIII. 1887, p. 339, pl. xiv. fig. 1, and Notes Leyden Mus. XII. 1890, p. 67.

Goniocaphyra sp., Zehntner, Rev. Suisse Zool. II. 1894, p. 163, pl. viii. fig. 12, 12a.

Carapace, length less than two-thirds its breadth, perfectly smooth and shining except for some fine granulation near the antero-lateral borders.

Front about a third the greatest breadth of the carapace, faintly notched and grooved in the middle line.

Antero-lateral borders cut into five teeth followed by a procurved spine.

Merus of external maxillipeds having the external angle very slightly produced.

Chelipeds unequal, much longer and more massive than the legs, the larger one about three times the length of the carapace : they are smooth and unarmed, except that the anterior border of the arm is finely serrulate and that one of the serrations at either the near or far end (rarely at both) is enlarged to form a spine. In the smaller cheliped the fingers are slender hooked and finely toothed, and are rather longer than the slightly swollen palm : in the larger cheliped they are stouter and more coarsely toothed and are shorter than the swollen palm.

Legs slender, the longest pair are hardly more than twice the length of the carapace; the dactylus of all, though compressed, is slender.

In the Indian Museum are 16 specimens from off Ceylon 34 fathoms (besides 3 from Mauritius and 2 from Samoa).

In the largest specimen (from Mauritius) the carapace is 9·5 millim. long and 14·5 millim. broad. The Indian specimens, though they include egg-laden females, are much smaller.

PSOPHETICUS, Wood-Mason.

Psopheticus, Wood-Mason, Admin. Rep. Marine Survey of India, 1890-91, p. 20 (name only) : Alcock, Investigator Deep-Sea Brachyura, p. 72.

Psopheticus in several respects connects *Carcinoplax* and *Pseudorhombila* with *Eucrate*, and hence serves to emphasize the opinion of Miers as to the closeness of the ties that connect the three latter genera.

As in *Pseudorhombila* and *Carcinoplax*, the carapace is much broader than long and the orbital hiatus is open. As in *Pseudorhombila*, the dactylus of the last pair of legs is styliform. As in *Eucrate*, the fronto-orbital border occupies almost all the breadth of the carapace.

Carapace deepish, quadrilateral or subquadrilateral, a good deal broader than long, with the regions hardly defined, moderately convex fore and aft, flat from side to side.

Fronto-orbital border little, if at all, less than the greatest breadth of the carapace, the antero-lateral borders of the carapace therefore— which are short—are either very slightly arched or are in the same straight line with the postero-lateral borders. Front square-cut, straight, prominent, entire, not well delimited from the supra-orbital angles, a third the breadth of the carapace, or a little less.

Upper border of orbit very sinuous and with a single faint short suture line. The antennal flagellum, which is of good length, stands loosely in the orbital hiatus. The antennules fold transversely.

Mouth and external maxillipeds as in *Eucrate.*

Chelipeds much stouter than the legs. The legs end in a slender styliform dactylus, and have one or many spines on the anterior border of the merus.

In both sexes the abdomen consists of seven separate segments, and in the male the third segment covers the whole width of the sternum between the last pair of legs.

Distribution : Andaman Sea.

Key to the (*Indian*) species of Psopheticus.

I. Carapace quite quadrilateral, the fronto-orbital border being equal to the greatest breadth of the carapace : meropodites of legs with numerous spines ... ... *P. stridulans.*

II. Carapace subquadrilateral, the fronto-orbital border being about three-fourths its greatest breadth : meropodites of legs with a single spine ... ... *P. insignis.*

11. *Psopheticus stridulans,* Wood-Mason.

Psopheticus stridulans, Wood-Mason, Illustrations of the Zoology of the Investigator, Crustacea, pl. v, fig. 1. (1892): Alcock, Ann. Mag. Nat. Hist., May 1894, p. 402 ; and Investigator Deep-Sea Brachyura, p. 73.

Carapace quite quadrilateral, three-fourths as long as broad, smooth and polished, crossed transversely in its posterior half by a broad groove which is continued obliquely across the pterygostomian regions to the angles of the mouth.

Owing to the large size of the eye and orbit, the extent of the fronto-orbital border is equal to the greatest breadth of the carapace.

A thin sharp prominent tooth at the outer orbital angle, and an obliquely-prominent spine at the junction of the antero-lateral and postero-lateral borders.

The subocular and subhepatic regions are inflated, and together form a granular eminence against which a strong spine on the upper border of the arm can be brought to play, producing a sound. Hence the names *Psopheticus* and *stridulans.*

The major diameter of the reniform eye is between a sixth and a seventh the breadth of the carapace ; though the orbit does not conceal the eye its edges are well and cleanly cut.

The chelipeds in the adult male are a little more, in the adult female a little less, than twice the length of the carapace, but are slightly shorter than the legs: they are smooth and polished, as also are the legs. The arm has a strong upstanding claw-like tooth near the middle of its upper border, one or two spinules near the far end of the outer border, and a spinule near the far end of the inner border : the wrist has both the inner and the outer angles spiniform.

The third pair of legs, which are slightly the longest of the four, are rather more than two-and-a-half times the length of the carapace. In all, the anterior edge of the meropodites is armed with spines and the same edge of the carpopodites with spinules—these being least numerous and least distinct in the case of the first pair.

Colours in glycerine : chelipeds and legs rather dusky red ; carapace dusky red behind the transverse groove—which forms a very sharply-defined red band—livid red, or almost violet, in front of it ; eyestalks almost purple, eyes purplish-black. Eggs in life magenta.

The carapace of the largest male is 15 millim. long and 20 millim. broad.

Only known, so far, from the Andaman Sea : 2 males and a female from 173 fms., 2 males and a female (Types of the species and genus) from 188–220 fms , 7 females (3 with eggs) from 185 fms., a male and 4 females from 370–419 fms.

12. *Psopheticus insignis,* n. sp.

Carapace subquadrilateral, the antero-lateral borders being slightly arched, about three-fourths as long as broad, smooth, crossed transversely by two very low and indistinct ridges—one (convex forwards) between the lateral epibranchial spines, the other at the level of the post-cardiac region. The extent of the fronto-orbital border is about three-fourths the greatest breadth of the carapace.

There is a bluntish tooth at the outer orbital angle, and an obliquely prominent spine at the junction of the antero-lateral and postero-lateral borders, the edge of the carapace between the two being granular.

Eye small, subglobular, its diameter being hardly a tenth the greatest breadth of the carapace.

Chelipeds more than $2\frac{1}{2}$ times as long as the carapace and decidedly longer than the legs : they are unarmed except for a small tooth or spinule at the outer angle of the wrist.

The meropodites of the legs have the anterior border sharply granular, and in the case of the last three pair of legs there is a spine near the far end of this border. The longest pair of legs are hardly $2\frac{1}{3}$ times as long as the carapace.

Two specimens, from the Gulf of Martaban, 60 and 67 fms.

The carapace of the largest is 13 millim. long and 19 millim. broad.

Colours in glycerine, reddish: in the middle of the carapace is a large deep-red shield with a milk-white edge and centre.

This species closely connects *Psopheticus* with *Carcinoplax*.

PILUMNOPLAX, Stimpson restr.

Pilumnoplax, Stimpson, Proc. Ac Nat. Sci. Philad. 1858 (1859) p. 93: Miers, Challenger Brachyura, p. 225: Alcock, Investigator Deep Sea Brachyura, p. 74.

Carapace depressed, flat, a little broader than long, the regions very faintly indicated. Fronto-orbital border two-thirds, or more, the greatest breadth of the carapace: the antero-lateral borders, which are toothed, are slightly arched or oblique. Front square-cut, straight, rather prominent, more or less confluent with the supra-orbital angles, often notched or grooved in the middle line.

Supra-orbital border often with two fissures. The antennal flagellum, which is of good length, stands in the orbital hiatus. The antennules fold transversely, or nearly so.

Mouth and mouth-parts as in *Eucrate.*

Chelipeds either subequal or unequal, much more massive than the legs. Legs slender, their dactyli compressed.

The abdomen in both sexes is seven-jointed: in the male the 3rd segment covers the whole width of the sternum between the last pair of legs.

Distribution: Tropical and S. Atlantic (deep sea), Arabian Sea (deep), Japan, Fiji.

The species of *Pilumnoplax* are characterized by the flat, depressed carapace, which is also comparatively narrow and, owing to the prominence of the perfectly straight front, is subhexagonal in shape.

13.   *Pilumnoplax americana*, Rathbun.

Pilumnoplax americanus, Mary J. Rathbun, Bull. Lab. Nat. Hist. Iowa, 1898, p. 283, pl. vii figs. 1, 2.

Pilumnoplax Sinclairi, Alcock, Investigator Deep Sea Brachyura, p. 74, pl. iii. fig. 1.

Carapace subquadrilateral, much depressed, a little more than three-quarters as broad as long, very finely frosted, perfectly bare, the regions fairly indicated.

Front horizontal, slightly prominent, square cut, grooved but not distinctly notched in the middle, more than a third the greatest breadth

J. II. 41

of the carapace; its free edge is turned vertically, downwards and rather deeply grooved from side to side.

The antero-lateral borders are not much more than half the length of the postero-lateral: they are thin and sharp, and are cut into three teeth, of which the first is broad and bicuspid and the other two are acute. On the postero-lateral borders, just behind the junction with the antero-lateral, is a denticle.

The eyes are small but well-formed, and are freely movable. The orbits conceal the retracted eyes to dorsal view : their upper margin is fissured near the middle, and the lower margin is slightly excavated just below the outer angle : the inner angle of the lower margin is not prominent, though dentiform.

The chelipeds in both sexes are very unequal, the larger one being not quite twice as long as the carapace; their surface, under the lens, is finely frosted : the inner angle of the wrist is strongly pronounced and is capped by a pair of acute teeth.

Legs moderately stout, unarmed, smooth, almost hairless : the third pair, which are somewhat the longest, are about two-and-a-half-times the length of the carapace. The dactyli are compressed-styliform.

Colours in spirit french-grey, fingers much darker grey.

A single female specimen, from off the Travancore coast 430 fms., has the carapace 13 millim. long and 16 millim. broad.

This species is closely related to *Pilumnoplax heterochir* (Studer) Miers, but is distinguished from it by the entire and more prominent front, by the absence of transverse markings on the carapace, by the longer legs, and by the smoothness of the chelipeds and legs.

From *Pilumnoplax abyssicola* Miers, which it also closely resembles, it is distinguished by the smooth carapace (to the naked eye), by the turned-down milled edge of the front, by the spinule on the postero-lateral border, by the fissured upper-margin of the orbit, and by the double spine at the inner angle of the wrist.

Distribution : Off Atlantic coasts of North America (Florida and Georgia) 440 and 70 to about 200 fms. Off Travancore coast 430 fms.

A single specimen from the latter locality is in the Indian Museum collection.

[PLATYPILUMNUS, Wood-Mason.

Platypilumnus, Wood-Mason MS., Alcock, Ann. Mag. Nat. Hist., May, 1894,: p. 401: Journ. Asiatic Soc. Bengal, Vol. LXVII. pt. 2, 1898, p. 232: Investigator, Deep Sea Brachyura, p. 62.

This genus, like so many of the preceding, has strong affinities with the *Xanthidæ*: it may prove to belong to that family, where I have already, with reserve, placed it.

I may here, however, state that it closely resembles *Pilumnoplax*, having a flat, depressed, slightly transverse carapace. It differs from *Pilumnoplax* in the following particulars :—

(1) the front is more prominent, so that the carapace is more decidedly hexagonal :

(2) the fronto-orbital border is sharply serrated and the chelipeds and legs are profusely spiny :

(3) the external maxillipeds do not completely close the buccal cavern, but leave a wide gap between their anterior margin and the edge of the epistome :

(4) the dactyli of the legs are styliform.

Distribution : Andaman Sea.]

[*Platypilumnus gracilipes,* Wood-Mason.

Platypilumnus gracilipes, Wood-Mason MS., Alcock, Ann. Mag., Nat. Hist., May, 1894, p. 401 : Ill. Zool. Investigator, Crust., pl. xiv. fig. 6 : J.A.S.B. Vol. LXVII, pt. 2, 1898, p. 232 : Investigator Deep Sea Brachyura, p. 63.

A description of the female (which is the only sex known) has been already given in this *Journal* (*loc. cit.*)].·

Litochira, Kinahan.

Litochira, Kinahan, Journ. Roy. Soc. Dublin, I. 1858, p. 121 : Miers, Challenger Brachyura, p. 231.
? Brachygrapsus, Kingsley, Proc. Ac. Nat. Sci. Philad. 1880 (1881) p. 203.

Carapace and appendages in all the Indian species thickly tomentose and hairy.

Carapace deepish, either subquadrilateral and a good deal broader than long, or almost square, smooth, with little or no distinction of regions, flat, but declivous anteriorly.   Fronto-orbital border not much less than, if not equal to, the greatest breadth of the carapace : antero-lateral borders short and if arched at all, very slightly so, and usually, but not always, with 2 or 3 teeth or spines.

Front not well delimited from the supra-orbital angles, its free edge deflexed and somewhat arched, never square-cut and laminar; more or less distinctly bilobed.

Upper border of orbit entire.   The antennal flagellum, which is of good length, stands in the orbital hiatus.   The antennules fold transversely, or nearly so.

Mouth and external maxillipeds as in *Eucrate,* &c.

Chelipeds subequal, more massive and usually shorter than the legs. The legs, including the dactyli, are compressed.

The abdomen of the male occupies the whole width of the sternum between the last pair of legs : in both sexes it consists of 7 segments.

I restrict the genus *Litochira* to those species which have the edge of the front turned down and distinctly arched as is shown in Kinahan's figure. These species fall into two groups, in one of which the carapace is a good deal broader than long, as in Kinahan's type, while in the other it is nearly square. Perhaps these two groups should be separa ted, though I do not recommend this course.

Distribution : S. Atlantic and Indo-Pacific (Cape to Australia).

Key to the Indian species of Litochira.

I. Length of carapace about two-thirds the greatest breadth of the carapace and equal to the extent of the fronto-orbital border ; the antero-lateral borders distinctly arched :—

    1. Antero-lateral borders of the carapace with three truncated teeth, exclusive of the orbital angle .. *L. angustifrons.*

    2. Antero-lateral borders with two distinct, though blunt, teeth .... *L. setosa.*

    3. Antero-lateral borders with hardly any trace of lobulation—almost entire .... *L. integra.*

II. Carapace more nearly square, the fronto-orbital border almost equal to its greatest breadth, so that the antero-lateral borders are almost in the same straight line with the postero-lateral borders or a very little curved :—

    1. Antero-lateral borders with two spines and one at the orbital angle : legs unarmed .... *L. Beaumontii.*

    2. Antero-lateral borders with two spines : no spine at the orbital angle : meropodites of the legs with some spines .... *L. quadrispinosa.*

14. *Litochira integra* (Miers).

Carcinoplax integra, Miers, Zool. H. M. S. Alert, p. 543, pl. xlviii. fig. C : de Man, Journ. Linn. Soc., Zool., XXII, 1887-88, p. 93.

Length of the carapace about two-thirds its breadth and equal to the extent of the fronto-orbital border.

Antero-lateral borders arched, without spines, though when completely denuded they are granular and show faint but quite distinguishable traces of division into two lobules besides the orbital angle.

Chelipeds less than twice the length of the carapace and shorter than the legs, unarmed except for an indistinct blunt tooth near the far end of the upper border of the arm : inner angle of wrist dentiform. Legs unarmed.

A single female from Mergui : its carapace is 6 millim. long and 9 millim. broad.

15.   *Litochira setosa* (A. M. Edw.).

Carcinoplax setosa, A. Milne Edwards, Nouv. Archiv. du Mus. IX. 1873, p. 267, pl. xii. fig. 2 : de Man, Archiv f. Naturges. LIII. 1887, i. p. 349, and Journ. Linn. Soc., Zool., XXII. 1887–88, p. 93.

The only essential difference between this species and the preceding is that the carapace here is a little more depressed and that the antero-lateral borders are cut into 2 blunt teeth besides the blunt orbital angle. The size is about the same.

In the Indian Museum are 16 specimens, from the Andamans and Mergui.

16.   *Litochira angustifrons,* n. sp.

Carapace, length a little more than two-thirds the breadth. Fronto-orbital border nearly five-ninths the breadth of the carapace in extent. Antero-lateral borders arched, cut into 4 teeth (including the outer orbital angle) the edges and dorsal surface of which are granular : the first 3 teeth are sharply truncated, the fourth is subacute.

Chelipeds, in the adult male, nearly twice the length of the carapace and hardly shorter than the legs ; in the female much less than twice the length of the carapace and markedly shorter than the legs. There is a lobule near the far end of the upper border of the arm, and the inner angle of the wrist is subacute.

Two specimens, from Bombay and Karachi. The carapace of the larger is 13 millim. long and 18 millim. broad.

This species appears to be closely related to *Pilumnoplax ciliatus* Stimpson.

17.   *Litochira Beaumontii,* n. sp.

Carapace, length more than two-thirds the greatest breadth, nearly square. The extent of the fronto-orbital border is hardly less than the breadth of the carapace. The antero-lateral borders are hardly arched and are armed with 3 sharp spinules—including one at the outer orbital angle.

The chelipeds are much shorter than the legs and, like them, are unarmed, except that the inner angle of the wrist is dentiform. The longest (penultimate) pair of legs are more than $2\frac{1}{2}$ times as long as the carapace.

In the Indian Museum are 4 specimens, from the Andamans and from off Ceylon 34 fms. The carapace of the type specimen is 5 millim. long and 7 millim. broad.

Colour in spirit, uniform yellow.

18. *Litochira quadrispinosa*, Zehntder.

Litochira quadrispinosa, Zehntner, Rev. Suisse de Zool. II. 1894, p. 171, pl' viii. figs. 11, 11b.

Differs from *L. Beaumontii* in the following particulars only :—

(1) the carapace is still more nearly square :

(2) there are 2 spines on the antero-lateral borders but none at the outer orbital angles :

(3) the inner border of the ischium and arm of the chelipeds is serrated, and the meropodites of the legs are armed with spines.

(4) the colouration is yellow, with a large purplish-brown horse-shoe behind the front, and with sinuous markings of the same colour on the lateral subfrontal and suborbital regions of the carapace : the greater part of the antennal flagella is of the same purplish-brown colour.

In the Indian Museum is a single specimen from the Andamans : the carapace is 4 millim. long and 5 millim. broad.

Subfamily ii. GONOPLACINÆ.

19. GONOPLAX, Leach.

Gonoplax, Leach, Trans. Linn. Soc. XI. 1815, pp. 309,-323, and Malac. Pod. Brit. : Desmarest, Consid. Gen. Crust. p. 124, and Dict. Sci. Nat. XXVIII. p. 243 : De Haan, Faun. Japon. Crust., p. 19 : Milne Edwards, Hist. Nat. Crust. II. 60, and Ann. Sci. Nat. Zool. (3) XVIII. 1852, p. 162 : Dana, U. S. Expl. Exp. Crust. pt. I. p. 310 : Bell, Brit. Stalk-eyed Crust. p. 129 : Heller, Crust. Sudl. Europ. p. 102 : Miers, Challenger Brachyura, p. 245.

Rhombilia, Lamarck (part), Hist. Nat. Anim. sans Vert. (2) V. p. 466 : Latreille, Encyc. Méthod. X. p. 292.

Carapace subquadrilateral, with the antero-lateral angles acute and the lateral borders posteriorly convergent, a good deal broader than long, moderately convex, the regions but faintly indicated.

The front and orbits occupy the whole anterior border of the carapace : the front is square cut, laminar, and obliquely deflexed, and takes up between a third and a fourth of the anterior border of the carapace, the rest being taken up by the trench-like orbits.

Eyestalks long and slender : the antennules fold quite transversely beneath the front : the antennæ have a short basal joint and a slender flagellum of good length, standing in the orbital hiatus.

The buccal cavern is square and is well separated from the prominent epistome : the efferent branchial channels are not well defined. The external maxillipeds completely close the buccal cavern : their merus is square and carries the flagellum at the antero-internal angle.

Chelipeds in both sexes much more massive, and in the male very much longer, than the legs, which are long and slender.

The abdomen in both sexes consists of 7 separate segments : in the male the 3rd segment nearly but not quite covers the sternum between the last pair of legs.

Distribution : North-Eastern Atlantic coasts, Mediterranean basin ; Persian Gulf ; East Indian Archipelago.

In the Indian Museum there is a young female, lately received by myself from the Persian Gulf, of a species of *Gonoplax.* Apart from the shortness of the chelipeds it differs from *G. angulata,* of which we have several good specimens from Europe, only in wanting the terminal spine to the upper border of the meropodites of the legs.

Subfamilies iii. & iv. Rhizopinæ & Hexapodinæ.
Key to the Indian Genera.

A. Four pairs of legs, besides the chelipeds (*Rhizopinæ*) : —

   I. The antennulary flagella can be completely retracted within the antennulary fossæ :—

      1. The epistome is of good length fore and aft, it is not in any way confused with the palate but is commonly prominent and almost vertical :—

         i. Eyes well formed, rarely deficient in pigment :—

            a. Eyes in all respects perfect : front straight, entire, from two-fifths to half the greatest breadth of the carapace : merus of the external maxillipeds nearly square...... Notonyx.

            b. Eyes either quite perfect or deficient in pigment : front slightly curved and notched in the middle, about a third the greatest breadth of the carapace : antero-external angle of the merus of the external maxillipeds much produced... Ceratoplax.

         ii. Eyes obsolete or nearly so :—

            a. Carapace much broader than long, the postero-lateral borders parallel ... . Typhlocarcinus.

            b. Carapace a little broader than long, the postero-lateral borders anteriorly-convergent .... Xenophthalmodes.

2. The epistome is short, sunken, and not boldly separated from the palate :—

    i. Eyes minute, orbits concealed beneath the anterior border of the carapace : merus of external maxillipeds with a sharp antero-external angle ... .   SCALOPIDIA.

    ii. Eyes obsolete or nearly so, orbits visible from above : antero-external angle of merus of external maxillipeds rounded off ....   TYPHLOCARCINODES.

II. The basal joint of the antennules completely fills its fossa, into which the flagellum cannot therefore be retracted :—

    1. Eyes small, but perfect : outer border of merus of external maxillipeds almost straight ....   HEPHTHOPELTA.

    2. Eyes reduced to a speck of pigment : outer border of merus of external maxillipeds with a strongly convex bulge outwards ....   CAMATOPSIS.

B. Only three pairs of legs besides the chelipeds, the last pair of other crabs not being represented even by a rudiment. The *vasa efferentia* of the male open on the 4th sternal segment (*Hexapodinæ*) ....   LAMBDOPHALLUS.

Subfamily iii. RHIZOPINÆ, Stimps.

NOTONYX, A. M. Edw.

Notonyx, A. Milne Edwards, Nouv. Archiv. du Mus. IX. 1873, p. 268 : Miers, Challenger Brachyura, p. 235.

Carapace deepish, subquadrilateral with the antero-lateral angles rounded off, broader than long, perfectly nude smooth and polished, without any indication of regions, convex fore and aft and anteriorly declivous.

Fronto-orbital border a good deal more than three-fourths the greatest breadth of the carapace : antero-lateral borders short, entire, curved. Front straight, sublaminar, from two-fifths to half the breadth of the carapace.

Eyes small but well developed, the eyestalks movable, obpiriform : orbits in the usual marginal position. The antennules fold transversely in well formed pits. Basal antenna-joint short ; the flagellum, which is of fair length, stands in the orbital hiatus.

Epistome well formed, nearly vertical : buccal cavern a little wider in front than behind. A slight hiatus between the external maxillipeds, the merus of which appendages is square and carries the flagellum at the antero-internal angle.

   Chelipeds subequal, or a little unequal, smooth and polished, much more massive and but little shorter than the legs: palm short and rather deep, with the lower border sharply carinate.

Legs smooth, unarmed, with a very few scattered lank hairs: dactyli styliform.

The abdomen in both sexes consists of 7 separate segments and does not nearly conceal the sternum between the last pair of legs.

Distribution: Indo-Pacific, from Fiji to the Persian Gulf.

Key to the Indian species of Notonyx.

I. Carapace, length about three-fourths the breadth : merus of external maxillipeds about as long as the ischium   ...   *N. nitidus.*

II. Carapace, length about five-sixths the breadth : merus of external maxillipeds much shorter than the ischium   ...   *N. vitreus.*

20.   *Notonyx nitidus*, A. M. Edw.

Notonyx nitidus, A. Milne Edwards, Nouv. Archiv. du Mus. IX, 1873, p. 269, pl. xii. fig. 3.: Miers, Challenger Brachyura, p. 236.

Carapace, length a little more than three-fourths the greatest breadth. Front between a third and two-fifths the breadth of the carapace. Orbits elongate. Merus of the external maxillipeds as long as the ischium.

A small denticle near the far end of the upper border of the arm : inner angle of wrist pronounced, but not acute.

Legs with some scattered hairs along the edges, the 3rd pair, which are slightly the longest, are about $2\frac{1}{2}$ times the length of the carapace and nearly half again as long as the chelipeds.

In the Indian Museum is a single specimen from the Persian Gulf : its carapace is 8·5 millim. long and 11 millim. broad.

21.   *Notonyx vitreus*, n. sp.

Carapace, length about five-sixths the greatest breadth, rather tumid. Front nearly half the breadth of the carapace. Merus of the external maxillipeds shorter than the ischium.

No denticle on the arm: inner angle of wrist blunt. Legs with hardly any hairs, otherwise resembling those of *N. nitidus.*

In the Indian Museum is a single specimen from the Andaman Sea, 53 fathoms : its carapace is 5 millim. long and 6 millim. broad.

CERATOPLAX, Stimpson.

Ceratoplax, Stimpson, Proc. Ac. Nat. Sci. Philad. 1858, p. 96 : Miers, Challenger Brachyura, p. 233.

Carapace deep, subquadrilateral with the antero-lateral angles rounded off, a good deal broader than long, the regions very indistinctly

J. II. 42

and incompletely indicated, strongly convex fore and aft and anteriorly declivous.

Fronto-orbital border about two-thirds the greatest breadth of the carapace: antero-lateral borders sharp, entire, curved: postero-lateral borders parallel.

Front about a third the greatest breadth of the carapace, its free edge slightly arched, notched in the middle line.

The orbits are in the usual position and the eyestalks are immovably fixed in them, but the eyes are fairly well formed, though they may be deficient in pigment. The antennules fold transversely in proper pits. The basal antenna-joint is short: the flagellum, which is of good length, stands in the orbital hiatus.

Epistome well formed and prominent: buccal cavern quadrilateral, slightly increasing in breadth from behind forwards, almost completely closed by the external maxillipeds, the merus of which has *the antero-external angle much produced* and carries the flagellum at the antero-internal angle.

Chelipeds subequal, more massive but decidedly shorter than the legs; the palm short, deep, and compressed.

Legs slender, unarmed, the 3rd pair the longest: dactyli styliform.

The abdomen in both sexes consists of 7 separate segments and does not nearly occupy the space between the last pair of legs.

Distribution: Indo-Pacific from the Bay of Bengal to Ecuador.

Key to the Indian species of Ceratoplax.

I. Surface of carapace nude, eyes well pigmented: outer surface of palm polished and nearly smooth ...   . *C. ciliata.*

II. Surface of carapace tomentose, eyes deficient in pigment: rows of vesiculous granules on the outer surface of the palm   ...   . *C. hispida.*

22.   *Ceratoplax ciliata,* Stimpson.

Ceratoplax ciliatus, Stimpson, Proc. Ac. Nat. Sci. Philad. 1858, p. 96: A. O. Walker, Journ. Linn. Soc., Zool., XX. 1890, p. 110.

Ceratoplax ciliata, Miers, Challenger Brachyura, p. 234, pl. xix. fig. 3: Cano, Boll. Soc. Nat. Napol. III. 1889, p. 229.

Carapace, chelipeds and legs rather scantily fringed with hairs, but with a nude surface.

Carapace, length a little more than three-fourths the greatest breadth, sparsely punctate, the regions not distinguishable. Front about a third the greatest breadth of the carapace, its free edge slightly arched and notched in the middle line. Eyes well pigmented. Chelipeds decidedly shorter than the legs: inner angle of wrist sharp, but not produced: outer surface of palm smooth and polished, except for a few depressed granules inferiorly. Third pair of legs not twice the length of the carapace.

In the Indian Museum is a single specimen from the Andaman Sea, 53 fms.

23. *Ceratoplax hispida*, n. sp.

Carapace, chelipeds and legs with a tomentose surface, and fringed with longer silky hairs.

Carapace, length a little less than three-fourths the breadth, when denuded its régions (and three gastric subregions) are just distinguishable, and its surface is pitted and its lateral margins granular. Front a little more than a third the greatest breadth of the carapace, its free edge decidedly arched and notched in the middle line. Eyes very deficient in pigment. Chelipeds (in the female—male unknown) much shorter than the legs: inner angle of wrist sharply dentiform; outer surface of palm with numerous rows of vesiculous granules. Third pair of legs two-and-a-half times the length of the carapace.

In the Indian Museum is a single specimen from Palk Straits: its carapace is 9 millim. long and 13 millim. broad.

TYPHLOCARCINUS, Stimpson.

Typhlocarcinus, Stimpson, Proc. Ac. Nat. Sci. Philad. 1858, p. 95.

Carapace as in *Ceratoplax.* Fronto-orbital border about half the greatest breadth of the carapace. Front less than a fourth the breadth of the carapace, more or less distinctly bilobed. Antero-lateral borders well curved, often emarginate in places: postero-lateral borders parallel.

Orbits in the usual position, completely filled by the immovable eye-stalks: eyes obsolete, or nearly so. The antennules fold nearly transversely, in proper pits. Basal antenna-joint short; the flagellum, which is short, stands in the orbital hiatus.

Epistome well formed and prominent: buccal cavern completely, or almost completely, closed by the external maxillipeds, the flagellum of which articulates with the antero-internal angle of the merus; the outer angle of the merus not produced.

Chelipeds subequal or unequal, much more massive than the legs from which they do not much differ in length: palm short deep and compressed, with sharp upper and lower borders.

Legs slender, unarmed, the 3rd pair slightly the longest: dactyli styliform.

The abdomen in both sexes consists of 7 separate segments and does not nearly occupy all the sternum between the last pair of legs.

Distribution: Indo-Pacific, from the Persian Gulf to Hongkong.

From *Rhizopa*, of which we possess specimens from Hongkong, this genus differs only in having the eyes obsolete and the external maxillipeds more closely opposed to each other. It may well be doubted whether these differences are of generic value.

Key to the Indian species of Typhlocarcinus.

I. Antero-lateral borders with 2 or 3 emarginations:—
    1. Buccal cavern decreasing in size from behind forwards: antero-external angle of merus of external maxillipeds obsolete and rounded off ...   . *T. nudus.*
    2. Buccal cavern quite square: antero-external angle of merus of external maxillipeds sharp ...   . *T. villosus.*
II. Antero-lateral borders of carapace entire: buccal cavern quite square   ...   . *T. rubidus.*

24. *Typhlocarcinus nudus*, Stimpson.

Typhlocarcinus nudus, Stimpson, Proc. Ac. Nat. Sci. Philad. 1858, p. 96.

Carapace much transverse, its length only about five-eighths its greatest breadth, its surface smooth and bare, the regions hardly distinguishable. The posterior part of the antero-lateral border has two or three obscure notches.

The front, which is about a fifth the greatest breadth of the carapace, is grooved in the middle line—almost bilobed. Orbits broadly oval, almost subcircular.

Buccal cavern considerably decreasing in breadth from behind forwards: merus of the external maxillipeds with the antero-external angle obsolete and rounded off; the exognath very narrow.

Chelipeds and legs smooth, with only a few scant hairs on the margin. Chelipeds, in the male about twice the length of the carapace, a little longer than any of the legs: inner angle of wrist sharp, but not produced: palms unequal, smooth and polished, the upper border smooth and crest-like, the lower border with a distinct moulding.

In the Indian Museum are 25 specimens, from Karachi and the Mekrán coast, Madras coast and Sandheads, and the Andamans.

In this species a tiny speck of pigment denotes an eye.

25. *Typhlocarcinus villosus*, Stimpson.

Typhlocarcinus villosus, Stimpson, Proc. Ac. Nat. Sci. Philad. 1858, p. 96 Miers, P. Z. S., 1879, pp. 20, 40: Walker, Journ. Linn. Soc. Zool. XX. 1890, p. 110, pl. ix. figs. 6–8: Ortmann, Zool. Jahrb. Syst. VII. 1893–94, p. 689.

Carapace and appendages everywhere covered with velvet. Carapace about three-fourths as long as broad, its greatest breadth across the middle: when denuded it is granular in places and the regions are hardly distinguishable. Three blunt granular teeth on the lateral borders, two of which are antero-lateral, the third being postero-lateral.

Front between a fourth and a fifth the breadth of the carapace, bilobed: orbits piriform.

Buccal cavern quite square: antero-external angle of merus of external maxillipeds well marked but not produced, the exognath normal.

Chelipeds about twice as long as the carapace, and nearly the same length as the 3rd (longest) pair of legs, their outer surface, especially that of the palm, is granular: inner angle of the wrist produced, dentiform. The legs are fringed with coarsish hairs.

In the Indian Museum, besides a specimen from Hongkong, are 6 from various parts of the coast of the Bay of Bengal.

The carapace of the best specimen is 6 millim. long and 8 millim. broad.

In this species also there is a tiny speck of pigment for an eye.

26. *Typhlocarcinus rubidus,* n. sp.

Carapace perfectly smooth and nude, except for a few hairs on the anterior and antero-lateral margins, its length a little over three-fifths its breadth, the regions hardly distinguishable, though the epibranchial regions have a decided dorsal bulge.

The antero-lateral borders, which, like the postero-lateral are blunt and granular, are quite entire.

Front about a fifth the breadth of the carapace, bilobed, the median groove very deep. Orbits piriform. Buccal cavern and external maxillipeds as in *T. villosus.*

Chelipeds and legs rather hairy, but there is always a large smooth bare space on the outer surface of the wrist and palm. Chelipeds about as long as the longest legs, less than twice the length of the carapace: inner angle of wrist produced, dentiform: below and above the bare patch on the wrist and hand the surface, when denuded, is granular.

The colour is a rich ruddy brown.

In the Indian Museum are 18 specimens from the Bay of Bengal, 20 to 65 fms.

The largest specimen has the carapace nearly 7 millim. long and 10 millim. broad, but there are egg-laden females smaller than this.

There is no pigment speck to represent an eye in this species.

XENOPHTHALMODES, Richters.

Xenophthalmodes, Richters, in Möbius Meeresf. Maurit. p. 155, 1880.

Carapace rudely semicircular in outline, the posterior border being the longest, and the postero-lateral borders being anteriorly-convergent to form a common curve with the well-arched anterior and antero-lateral borders: it is but little broader than long, is convex fore and aft and strongly declivous anteriorly, and shows the regions indistinctly and incompletely.

Fronto-orbital border less than half, front less than a fifth, the greatest breadth of the carapace, the front being prominent and bilobed.

Orbits in the usual position, completely filled by the immovable eye-stalks: eyes obsolete. The antennules are small, and fold obliquely rather than transversely in proper pits. Basal antenna-joint short: the flagellum, which also is short, stands in the orbital hiatus.

Epistome and mouth parts, as also the abdomen, as in *Typhlocarcinus*.

Chelipeds a little unequal, much more massive and rather longer than the legs, of which the 3rd pair is slightly the longest. Palm short deep and compressed, with sharp edges.

Legs slender, unarmed : dactyli styliform.

Distribution : Indian Ocean, from Mauritius and the Red Sea to the Andamans.

This genus differs from *Typhlocarcinus* in having the carapace more elongate and more semicircular in outline, the front more prominent and narrower, and the antennules more cramped in consequence.

27.   *Xenopthalmodes moebii,* Richters.

Xenophthalmodes moebii, Richters, in Möbius, Meeresf. Maurit. p. 155, pl. xvi. fig. 29 and pl. xvii. figs. 1-5 1880 : Miers, P. Z. S. 1884, pp. 10, 12 : de Man, Notes Leyden Mus. XII. 1890, p. 68, pl. iii. fig. 5.

The carapace has rather a lop-sided look and is practically smooth, except for two rather deep semilunar impressions that incompletely separate the gastro-cardiac from the epibranchial regions : its surface is bare, but its free edges, like the edges of the chelipeds and legs, are thickly fringed with longish silky hairs : its length is about five-sixths the greatest breadth, which is quite posterior. Front very decidedly bilobed. Orbits oval. Buccal cavern very slightly decreasing in breadth anteriorly : the merus of the external maxillipeds has the antero-external angle rounded off.

Chelipeds in the male a little longer than the legs, and with the hands decidedly unequal : the inner angle of the wrist is acuminate : the upper edge of the palm is sharp and crest-like, the lower edge has a low granular crest or moulding, the surface of the palm is smooth and polished. The larger cheliped, measured along its convexities, is about twice the length of the carapace.

In the Indian Museum are 13 specimens, from the Persian Gulf, Malabar coast, Coromandel coast, Gulf of Martaban, and the Andamans. The carapace of the largest specimen is 10 millim. long and 12 millim. broad.

In one very young specimen the eye is represented by a tiny speck of pigment, as shown in de Man's figure, but in large specimens there is no trace of this speck.

SCALOPIDIA, Stimpson.

Scalopidia, Stimpson, Proc. Ac. Nat. Sci. Philad. 1858, p. 95 : Miers, Challenger Brachyura p. 223.

Hypophthalmus, Richters, Abh. Senck. Nat. Ges. Frankfurt, XII. 1881, p. 429.

Carapace of but moderate depth, moderately convex fore and aft and but moderately declivous anteriorly : it is a good deal broader than long and inclines somewhat to a semicircular outline, the greatest breadth being quite posterior, the postero-lateral borders being anteriorly convergent, and the antero-lateral borders being nicely curved : the regions are distinctly mapped out by fine grooves.

Fronto-orbital border about two-fifths, front about a fourth the greatest breadth of the carapace : front rather obscurely bilobed, antero-lateral borders acute.

Eyes minute, eyestalks fixed in small orbits which lie entirely beneath the anterior border of the carapace. The antennules fold transversely in shallow and rather inadequate pits. Basal antenna-joint short; the flagellum, which is of moderate length, stands quite clear of the orbital hiatus.

Epistome sunken, not well demarcated from the edge of the buccal cavern : the latter is squarish and broader in front than behind. There is a considerable gap between the external maxillipeds, the merus of which is square and has a sharp antero-external angle and carries the flagellum at the antero-internal angle.

Chelipeds a little unequal, much shorter and not much more massive (except as regards the larger palm) than the third pair of legs : palm short and compressed, with sharp edges.

The legs have the merus broadened, especially in the case of the 2nd and 3rd pair : the 3rd pair is considerably the longest.

The abdomen consists of 7 separate segments, and does not nearly occupy all the sternum between the last pair of legs.

Distribution : Indo-Pacific, from Madagascar to China.

28. *Scalopidia spinosipes*, Stimpson.

Scalopidia spinosipes, Stimpson, Proc. Ac. Nat. Sci. Philad. 1858, p. 95 : J. R. Henderson, Trans. Linn. Soc., Zool., (2) V. 1893, p. 379.

Carapace and appendages downy. Carapace, length about two-thirds the greatest breadth, its surface closely punctate: all the regions are quite plainly defined by grooves, which also subdivide the gastric into three subregions, and the epibranchial into two—an anterior and a posterior; and the cardiac region has a distinct bulge. The sharp-cut antero-lateral borders are, like the anterior border, very finely serrated, and are marked off from the blunt postero-lateral borders by a minute spine.

The larger cheliped is barely half again as long as the carapace: both chelipeds have the lower edge of the arm finely serrated, have a spinule near the far end of the upper border of the arm and one at the outer angle of the wrist, and have the inner angle of the wrist strongly dentiform.

The legs have their edges, except in the case of the dactyli, closely and evenly spinulate, but there is a tendency for the spines to fail on the posterior edge of the carpus and propodite. The 3rd pair, which are considerably the longest, are much more than $2\frac{1}{2}$ times the length of the carapace. The legs increase remarkably in length from the 1st to the 3rd, and the 4th are about the same length as the first. The dactyli are sharp, strong, styliform and ciliated: those of the last pair are curved, those of the other pairs are straight.

Henderson records this species from the Gulf of Martaban: the only specimens in the Indian Museum are from Hongkong.

29. Typhlocarcinodes, n. gen.

Apparently one of the links between *Typhlocarcinus* and its allies on the one hand and *Scalopidia* on the other.

Carapace moderately deep, shaped much as in *Typhlocarcinus*, but slightly more elongate, the free edges hairy. Fronto-orbital border about three-fifths, front about a third, the greatest breadth of the carapace: front prominent, its free edge convex and entire.

Orbits in the normal position, narrow, button-hole shaped; eye-stalks tapering, immovable; eyes obsolete or nearly so. *Antennules cramped, folding very obliquely—nearly longitudinally—*in proper pits. Antennal peduncle small and cramped, the flagellum standing in the orbital hiatus.

Epistome sunken, linear : buccal cavern square, its anterior angles, like the antero-external angles of the merus of the external maxillipeds, rounded off: the external maxillipeds completely close the buccal cavern and have the flagellum articulated to the antero-internal angle of the merus.

The abdomen does not nearly occupy all the space between the last pair of legs.

The above diagnosis is framed on a broken specimen, without chelipeds or legs, in the Indian Museum. In the form of the front and shape of the carapace this specimen has a strong resemblance to the *Typhlocarcinus integrifrons* described and figured by Miers in Ann. Mag. Nat. Hist. (5) VIII. 1881, p. 260, ‘pl. xiv. fig. 1. Miers himself was doubtful about referring his species to *Typhlocarcinus.*

Our specimen is too much damaged to furnish a useful specific diagnosis.

HEPHTHOPELTA, Alcock.

Hephthopelta, Alcock, Investigator Deep Sea Brachyura, p. 76.

Carapace very deep, inflated, rudely semicircular, about as long as broad, convex fore and aft and vertically deflexed anteriorly, all its borders entire and all, except the posterior, tumid, the cardiac and branchial regions well delimited.

Front considerably less than a third the greatest breadth of the carapace, bilobed, vertically deflexed ; the whole extent of the fronto-orbital border is more than half the greatest breadth of the carapace.

Orbits small, shallow, excavated in the vertically-deflexed anterior border of the carapace, not concealing the eyes. Though the eyes are small and their stalks immovably fixed, they are *well formed, well defined and well pigmented.*

The antennulary fossæ are completely filled by the basal antennulary joint, to the exclusion of the flagella.

The basal antenna-joint is small, slender, and does not nearly reach the front ; the flagellum, which arises in the orbital hiatus, is hardly longer than the orbit.

The epistome is of considerable width fore and aft and, though sunken, is well defined from the palate. The buccal cavern is square, though very slightly narrower in front than behind : the excurrent branchial canals are well defined. The external maxillipeds, which completely cover the buccal cavern, have the merus shorter and slightly narrower than the ischium and somewhat oval in shape, and the palp jointed to the antero-internal angle of the merus and of good size.

The legs are all long and slender and end in a slender dactylus : the third pair are slightly the longest.

The chelipeds are lost in the single specimen obtained, which is a female.

30. *Hephthopelta lugubris,* Alcock.

Hephthopelta lugubris, Alcock, Investigator Deep Sea Brachyura, p. 77. pl. iv, fig. 2.

Carapace as long as broad, roughly semicircular or semiglobose, of thin texture, its surface very finely frosted and somewhat pubescent,

The fronto-orbital region is vertically deflexed and almost invisible in a dorsal view.

Epibranchial and cardiac regions tumid, circumscribed by deepish grooves.

Legs subcylindrical, with a finely frosted and pubescent surface: the third pair, which are slightly the longest, are about $2\frac{3}{4}$ times the length of the carapace: the posterior (lower) border of the merus of the first two pairs is spinulose.

Colours in spirit, light yellow, eyes black.

A single female, without chelipeds, from the Andaman Sea, 490 fms. The carapace is 8 millim. long, and the same in breadth.

CAMATOPSIS, Alcock.

Camatopsis, Alcock, Investigator Deep Sea Brachyura, p. 75.

Carapace deep, rudely sub-semicircular, hardly broader than long, strongly convex fore and aft and declivous anteriorly : its antero-lateral borders short sharp and entire, its postero-lateral borders long sharpish and slightly convergent anteriorly : its only markings are two longitudinal grooves hardly visible on the undenuded carapace, that mark off the epibranchial regions.

Front considerably less than a fourth the greatest breadth of the carapace, obscurely bilobed ; the whole fronto-orbital border is about half the greatest breadth of the carapace.

Orbits large, deep, and normally cut in the anterior border of the carapace : eyestalks large, tumid, conical, almost immovably fixed in the orbits : eyes reduced to a speck of pigment placed on the under surface of the tip of their stalks.

Antennulary fossæ small, and filled entirely by the basal antennulary joint, *to the complete exclusion of the large flagellum.*

The small basal antenna-joint is wedged in between and beneath the eyestalk and antennule, the second joint hardly reaches to the front, the flagellum is large and considerably longer than the orbit.

The epistome is of considerable width fore and aft, especially at its middle, and though sunken, is well separated from the palate. The buccal cavern is square, though rather broader in front than behind, and is almost entirely covered by the external maxillipeds. These have the *merus as long as, and markedly broader than the ischium, owing to the strongly convex bulge of the outer border of the merus :* the palp, which is of good size, is jointed to the antero-internal angle of the merus.

The chelipeds are moderately massive and in the male the hands are unequal. The arm is short and trigonal, the wrist rather long narrow and crooked.

Legs sufficiently long and stout, the penultimate pair being the longest; their dactyli are sharply trigonal and elegantly plumose : the last pair have the dactylus slightly curved and compressed.

.. The abdomen of the male, which is four-jointed, does not nearly fill the space between the last pair of legs.

Between the 4th and 5th segments of the sternum, in the male, is intercalated a long narrow plate that covers the external genital ducts.

31. *Camatopsis rubida,* Alcock and Anderson.

Camatopsis rubida, Alcock and Anderson, Ann. Mag. Nat. Hist. Jan. 1899, p. 13 :: Alcock, Investigator Deep Sea Brachyura, p. 76, pl. iv. fig. 3.

Carapace very finely granular when denuded of the short velvet that covers it and all parts of the body and appendages. The narrow front and the antero-lateral borders form a semicircular curve : the postero-lateral borders are anteriorly convergent, the greatest breadth of the carapace being between the bases of the penultimate pair of legs. The tumid anterior (true inner) borders of the eyestalks bulge beyond the orbital concavities of the anterior border of the carapace.

The efferent branchial canals cause an angular bulging or carina-tion of the pterygostomian regions.

The chelipeds are unequal in the male (female unknown), the longer one being about $1\frac{3}{4}$ times the length of the carapace. They are unarmed. In the larger hand the fingers meet only at tip and are finely toothed in the distal half only, being rather deeply notched in the basal half, while on the inner surface of the movable finger is a curious truncated spine. In the smaller hand the fingers meet through-out their extent and only the immovable finger is distinctly toothed, one or two of its teeth being enlarged.

The first and last pair of legs are about $1\frac{3}{8}$ times, the second and third pair are about twice, the length of the carapace. In the last pair of legs the terminal joints are more strongly ciliated, and the dactylus is slightly curved and compressed as for swimming.

Colours in spirit rich chocolate brown. Animal entirely covered with velvet.

Three males from the Andaman Sea, 194 fathoms. The carapace of the largest is 9 millim. long and 10 millim. broad.

Subfamily iv. HEXAPODINÆ, Miers.

LAMBDOPHALLUS, nov. gen.

Near *Hexapus,* De Haan, from which it chiefly differs in the form of the anterior pair of male sexual appendages, which are rigidly bent into the form of an **L,** the horizontal limb of which is lodged in a special trench in the first segment of the sternum.

Carapace much broader than long, broadest behind. Front narrow, nearly vertically deflexed. Orbits small, circular, widely communicating with the antennular fossæ. The antennules fold transversely. Antennæ small, standing in the orbital hiatus.

Epistome well-defined. Buccal cavern with the sides slightly convergent anteriorly. The external maxillipeds have coarse palps, which, when folded, fill the rather broad space that exists between the ischiopodites: the merus is subquadrilateral, with the antero-external angle rounded off, and the palp articulates with its antero-internal angle: the exognath is not concealed.

Chelipeds unequal in the male, shorter but more massive than the legs.

Only three pairs of legs, the fourth pair entirely absent.

Sternum extremely broad. Abdomen of the male very narrow. The efferent ducts of the male sex open on the 4th sternal segment inside the fossa into which the abdomen fits.

82.   *Lambdophallus sexpes,* n. sp.

Resembles *Hexapus sexpus,* De Haan, with a specimen of which I have compared it, but differs in numerous important characters.

Carapace subquadrilateral with the anterior angles broadly rounded off, much broader than long, convex fore and aft and anteriorly deflexed, nearly flat from side to side, the gastric and cardiac regions well defined, the surface uniformly finely granular under a lens.

Front nearly vertically deflexed, its edge square-cut but grooved or notched in the middle line, its breadth about a fifth the greatest breadth of the carapace.

Orbits freely communicating with the antennular fossæ: eyestalks immovable and very short, eyes small but well pigmented.

Antennules large, folding transversely ; the inter-antennular septum narrow if complete.

Epistome lozenge-shaped, well defined: the sides of the buccal cavern converge slightly from behind forwards : the ischiopodites of the external maxillipeds are rather narrow and leave between them a widish gap, which, however, is filled by the flagella.

There is a deep crescentic groove across the pterygostomian region, just in front of the bases of the chelipeds, and there are several close-set oblique scorings near the antero-lateral angles of the buccal cavern.

Chelipeds in the male unequal, more massive than the legs, the larger one not $1\frac{1}{2}$ times the length of the carapace : under the lens their outer surface is very finely and uniformly granular: the fingers are short, especially in the larger hand, and meet only at tip, and at the base of the dactylus of the larger hand is a molariform tooth.

Legs tomentose: only 3 pairs are present, the 4th pair not being represented even by a rudiment. The first pair, which are not much longer than the chelipeds, are the shortest and slenderest: the next two pairs, which are about equal in size, are not quite twice the length of the carapace.

Sternum very broad, finely and uniformly granular: in the male, in the first sternal segment, on either side of the last abdominal tergum, is a long narrow oblique trench, in which the ends of the modified abdominal appendages are lodged.

Male abdomen very narrow, not a fifth the breadth of the sternum at its base. The first tergum is short fore and aft, the second is linear and has a somewhat trilobed form, the 3rd 4th and 5th are fused to form a sort of hexagonal plate with the distal end narrowed, the 6th and 7th are separate.

The anterior of the two pairs of male abdominal appendages are most curiously modified: they are very long and stiff and are L shaped, and the proximal limb of the L lies beneath and parallel with the abdomen, while the distal limb of the L emerges at right angles to the abdominal tergum, and, instead of being free, lies in the special sternal canal before mentioned.

In the Indian Museum are 2 specimens, from the Bay of Bengal, 65 fathoms. The carapace is 4·5 millim. long and 7 millim. broad.

Family II. PINNOTERIDÆ, Edw.

Key to the Indian genera of Pinnoteridæ.

I. Carapace ill-calcified: the ischium of the external maxillipeds is indistinguishably fused with the much enlarged merus :—

    1. Edges of the carapace swollen and ill-defined: dactylus of the external maxillipeds small and often abnormally placed, but present .... PINNOTERES.

    2. The edge of the carapace, in all but its short fronto-orbital portion, forms a thin upturned crest: dactylus of the external maxillipeds wanting, or represented by a tiny pencil of hairs XANTHASIA.

II. Carapace well calcified: the ischium of the external maxillipeds is distinct and independent :—

    1. Ischium of the external maxillipeds much smaller than the merus: dactylus of the external maxillipeds very large, spathulate. Orbits and eyes normal, the orbits circular .... TETRIAS.

2. Ischium of the external maxillipeds as well deve.
loped as the merus, the dactylus not enlarged.
The orbits. are narrow slits situated dorsally
with their long axis almost at right angles with
the anterior border of the carapace, and the
eyes are minute or obsolescent .... XENOPHTHALMUS.

3. Ischium of the external maxillipeds very much
larger than the merus, the appendages as a
whole being slender and not nearly closing the
buccal cavern. The orbits are in the usual
marginal position.................... .... CHASMOCARCINOPS.

Subfamily XENOPHTHALMINÆ, nov.

XENOPHTHALMUS, White.

Xenophthalmus, White, Ann. Mag. Nat. Hist. XVIII. 1846, p. 177 : Milne
Edwards, Ann. Sci. Nat., Zool., (3) XX. 1853, p. 220: Burger, Zool. Jahrb., Syst.
VIII. 1894-95, p. 386.

Carapace broader than long and broadest behind, arched antero-
laterally, the regions faintly indicated. Front narrow, strongly deflexed.

*The orbits are small, oblique or nearly longitudinal, button-hole like
slits, placed dorsally almost at right angles to the frontal border*, and the
eyestalks are immovably embedded in them. The eyes are, at most,
minute specks of pigment. The antennules and antennæ are extremely
small, the antennules folding nearly vertically beneath the front.

Epistome not defined. Buccal cavern almost semicircular, completely closed by the external maxillipeds. The external maxillipeds
have the ischium and merus equally well developed (the ischium being
nearly square and the merus about a quadrant of a circle) and the
palp articulated at the antero-external angle of the merus. Exognath
small and concealed.

Chelipeds in the male " with the hands somewhat elongated and
thickened," in the female short and very slender.

Legs fairly stout, the third pair the longest.

The abdomen in both sexes consists of seven separate segments.

Key to the Indian species of Xenophthalmus.

I. The legs are ciliated and the third (longest) pair are not
twice the length of the carapace ... . *X. pinnoteroides.*

II. The legs are ciliated towards the tip only, and the third.
(longest) pair are more than twice the length of the
carapace .... *X. obscurus.*

33. *Xenophthalmus pinnoteroides*, White.

Xenophthalmus pinnotheroides, White, Ann. Mag. Nat. Hist. XVIII. 1846, p. 178,
pl. ii. fig. 2, and Samarang Crust. p. 68, pl. xii. fig. 3: Milne Edwards, Ann. Sci.

Nat., Zool., (3) XX. 1853, p. 221: Stimpson, *Proc. Ac. Nat. Sci.* Philad. 1858, p. 107: Sluiter, Tijds. Nederl. Ind. XL. 1881, p. 162: J. R. Henderson, Trans. Linn. Soc., Zool., (2) V. 1893, p. 394.

This species is included in the Indian Fauna on the authority of Professor J. R. Henderson. It seems to be characterized by having the ischium and merus of the external maxillipeds deeply grooved, longitudinally, near the outer margin; the legs stout and hairy, the third pair barely twice as long as the carapace; and the three terminal joints of the first pair of legs broadened so that their edges are almost carinate: the lateral borders of the carapace are granular or finely denticulate.

34. *Xenophthalmus obscurus*, Henderson.

Xenophthalmus obscurus, J. R. Henderson, Trans. Linn. Soc., Zool., (2) V. 1893, p. 394, pl. xxxvi. figs. 18, 19.

Carapace glabrous and shiny, but its surface is somewhat creased: the median regions are separated from the branchial regions by grooves or depressions, and each branchial region is traversed obliquely in its posterior part by a low ridge.

The rounded-off antero-lateral corners of the carapace are traversed by three low fine ridges, nearly parallel with one another: one of these defines the pterygostomian region, the next appears to be the true antero-lateral border, while the most dorsal one runs from the angle of the orbit to the junction of the antero-lateral and postero-lateral borders.

Front narrow, nearly vertically deflexed, longitudinally grooved in the middle line, its free edge square-cut but faintly sinuous. The eyes are just visible as minute linear specks, placed posteriorly.

No epistome. The ischium and merus of the external maxillipeds are not deeply grooved near the outer border.

Chelipeds in the female shorter and much slenderer than the first and last pair of legs.

The 3rd pair of legs are the longest, being about $2\frac{1}{4}$ times the length of the carapace: the second pair, though a little shorter than the 3rd, are equally stout. The first and last pairs are about equal to one another in size (in the female) being hardly longer than the carapace, and slenderer than the other legs. The terminal joints of all the legs are hairy: the posterior borders of the meropodites of the first three pairs are spiny, the anterior border being very finely serrulate.

In the Indian Museum are two females, one, with eggs, from off the Ganjam coast, 20 fathoms, the other from the Andamans. The carapace in the larger female is 6 millim. long and 8 millim. in greatest breath.

Subfamily ASTHENOGNATHINÆ, Stimps.

CHASMOCARCINOPS, n. gen.

Carapace deep, convex fore and aft and declivous anteriorly : its greatest breadth is quite posterior, so that the postero-lateral borders, which are blunt, are anteriorly-convergent, though slightly so : the antero-lateral borders are sharp and form an elegant curve with the anterior border : the regions are nearly as well defined as they are in *Scalopidia* : its length is hardly less than its breadth.

The fronto-orbital border is considerably more than a third, but the front (which is bilobed) is only about a sixth, the greatest breadth of the carapace.

· The orbits, which are in the usual marginal position, are small, and *the eyestalks,* which are immovable, *are shrunk within them :* the eyes are minute.

The antennulary flagella are large and cannot be retracted into the antennular pits, which are filled entirely by the basal joint.

The antennal flagella are long—considerably more than a third the length of the carapace—and stand in the orbital hiatus.

Epistome sunken and not altogether well demarcated from the palate. The buccal cavern has its antero-external angles rounded off, *and is not nearly closed by the external maxillipeds :* these have the merus *much shorter and narrower than the ischium,* oval and somewhat oblique, and the flagellum appears to articulate with the summit of the merus.

The chelipeds are about as long as the legs and are very unequal in the male.

The third pair of legs are slightly the longest. As in *Scolopidia* the dactylus of the last pair of legs is recurved.

The abdomen in both sexes is narrow, not nearly occupying all the space between the last pair of legs, and in the male consists of 5 pieces, the 3rd–5th segments being fused. In the male also, as in *Camatopsis,* there is, on either side, a narrow plate intercalated between the 4th and 5th segments of the sternum and covering the external genital ducts.

This genus more clearly than any other connects the *Rhizopinæ* and the *Pinnotheridæ* together.

35. *Chasmocarcinops gelasimoides,* n. sp.

Carapace nearly as long as broad, its surface abundantly sprinkled with vesiculous granules, its free margins rather sparsely ciliated : all the regions are distinguishable, and the cardiac and posterior lobe of the gastric regions are defined by deep impressions : the antero-lateral borders are sharply defined and granular.   Front very distinctly bilobed, prominent.

Chelipeds in the male very unequal, the larger one being twice as long as the carapace, its chief bulk being contributed by the hand, which, with its large swollen polished palm and long crooked fingers meeting only at tip, recalls that of *Gelasimus.* The smaller cheliped (like the female chelipeds) is not much shorter than the larger one and, like it, has the articulation of the wrist confined to a rather prominent postero-inferior lobe of the hand, and the fingers longer than the palm : the chief difference is that the palm is not enlarged and swollen and that the fingers meet throughout almost all their extent.   In both chelipeds the surfaces of all the segments are smooth, and there are sharpish granules along the borders of the arm and at the not very pronounced inner angle of the comparatively slender wrist.

The legs, like the fingers of the smaller cheliped, are fringed, but not very thickly, with hair.   The 3rd pair are very slightly the longest, being twice the length of the carapace.   The edges of the meropodites are furnished with sharp granules and spinules, these being abundant in the case of the first 3 pairs and rather few on the 4th pair.   In the first 3 pairs also the carpopodites are of good length and subcylindrical, and the dactyli straight and almost styliform; but in the 4th pair the two terminal joints are compressed, the carpopodite being shortened and the dactylus recurved.

A male and a female from off Madras, 12 fathoms.   The carapace of the male is 11 millim. long and 12 millim. broad.

Subfamily PINNOTHERELINÆ.

TETRIAS, Rathbun.

Tetrias, Rathbun, Proc. U. S. Nat. Mus. XXI. 1898, p. 607.

Carapace strongly calcified, broader than long, deep, subquadrangular, dorsally flattish, anteriorly declivous, the regions faintly indicated.

Front between a third and a fourth the greatest breadth of the carapace, its edge only deflexed, not directly united to the epistome. Orbits circular, small : eyestalks short, eyes small.   The antennules fold a little obliquely from the transverse.   Antennæ small, the flagellum in the orbital hiatus.

Epistome well defined : buccal cavern broadish, quadrilateral. External maxillipeds large, their palp about as large as their merus and ischium combined : ischium distinct, small ; merus very large, carpus large and triangular and articulating at the antero-external angle of the merus, propodite large and articulating with the end of the carpus, dactylus large and spathulate and articulating with the inner angle of the propodite : exognath small and a good deal concealed.

. Chelipeds equal, short : the chelipeds in the male equal, and much stouter than the legs.

First 3 pairs of legs coarse, not differing much from each other or from the chelipeds in length, though the second pair are slightly the longest. The fourth (last) pair are very much smaller than the others.

The abdomen of the male is narrow and consists of 7 separate segments.

Tetrias differs very little from *Pinnixa* of which it might, perhaps, be regarded as a subgenus.

Distribution : Indo-Pacific, Andamans to California.

36. *Tetrias Fischeri,* (A. M. E.).

or *Pinnixa (Tetrias) Fischeri* (A. M. E.).

Pinnotheres Fischeri, A. Milne Edwards, Ann. Soc. Entomol. France, VII. 1867, p. 287.

Pinnixa Fischeri, A. Milne Edwards, Nouv. Archiv. du Mus. IX. 1873, p. 319, pl. xviii. fig. 3 : de Man, Archiv fur Naturges. LIII. 1887, i. p. 385, pl. xvii. fig. 2.

Carapace and appendages everywhere covered by a close adherent coat of short hair. The regions of the carapace are fairly well indicated and its dorsal surface is closely and finely granular, except in the middle where also the hair is somewhat deficient. Deflexed edge of the front broadly triangular. Eyes well pigmented. The inner edge of the carpus and the inner and distal edges of the large spathulate dactylus of the external maxillipeds are fringed with a close row of hairs of extraordinary length.

Chelipeds in the male much more massive than the legs, and about $1\frac{1}{2}$ times the length of the carapace : their movements are somewhat restricted. There are some spinules at the inner angle of the wrist, and numerous rows of granules—the lowermost row rather acute--on the outer surface of the palm : the fingers, which are shorter than the palm, are stumpy but sharp-pointed.

The first 3 pairs of legs are coarse and are all about $1\frac{1}{2}$ times the length of the carapace, though the second pair are very slightly the longest. The 4th pair are very short—not two thirds the length of the carapace—and are much slenderer than the others. All the legs have a shaggy posterior border, and all end in small hooked dactyli. The posterior border of the meropodite of the last pair is armed with small coarse spines.

. ˙. The abdomen of the male is narrow and consists of 7 segments: the first two segments are very short, the 3rd 4th and 5th gradually increase in length and slightly decrease in breadth, the 6th is a little shorter than the 5th, and the 7th is long and spathulate and encroaches on the buccal cavern.

In the Indian Museum is a single male specimen, from coral, from the Andamans: its carapace is a little over 5 millim. long and 7 millim. broad.

Subfamily PINNOTERINÆ.

*PINNOTERES, Latreille.

Pinnotheres, Latreille, Hist. Nat. Crust. et Ins. VI. p. 78, and Gen. Crust. et Ins., p. 34: Lamarck, Hist. Nat. An. Sans. Vert. (2nd edit. Vol. V. p. 410): Bosc, Hist. Nat. Crust. I. p. 239: Leach, Malac. Pod. Britt.: Desmarest, Consid. Gen. Crust. p. 116: De Haan, Faun. Japon. Crust., p. 34: Milne Edwards, Hist. Nat. Crust. II. 30, and Ann. Sci. Nat., Zool., (3) XX. 1853, p. 216: Dana, U. S. Expl. Exp. Crust. pt. I. p. 378: Bell, British Stalk-eyed Crust. p. 119: Miers, Challenger Brachyura, p. 275: Ortmann, Zool. Jahrb, Syst., VII. 1894, p. 698: Bürger, Zool. Jahrb. Syst., VIII. 1894-95, p. 362: Adensamer, Ann. Nat. Hofmus., Wien, 1897, p. 105.

Carapace often ill calcified, generally convex with ill-defined edges, in shape transversely oval, or circular, or subquadrangular or sub-hexagonal with rounded angles, the surface generally smooth, the regions seldom defined.

Front narrow, generally deflexed in the female if not in the male. Orbits small, circular, eyestalks short, eyes small. Antennules folding obliquely in small pits. Antennæ small, the minute flagellum standing in the inner angle of the orbit.

Epistome well defined. The buccal cavern is of a curious crescentic shape, being arched and very broad from side to side, but very narrow fore and aft. The external maxillipeds completely close the buccal cavern: they consist chiefly of the merus, which is fused with the ischium to form a single large obliquely-directed joint carrying the flagellum at its inner end: the flagellum is small though its propodite may be spathulate, and the dactylus is often inserted on the inner or flexor border of the propodite: the exognath is for the most part concealed.

The chelipeds and legs are short, the chelipeds being equal and generally, even in the female, stouter than the legs.

The abdomen in the male is narrow, in the female it is generally larger than the sternum: it consists of 7 separate segments.

* *Pinnoteres*, the correct transliteration of the Greek word, was used by Rumph in 1705, so that no apology is necessary for reverting to it.

The Pinnoteræ live as parasites or messmates, generally within the mantles of Lamellibranch Mollusks.

Key to the Indian species of Pinnoteres.

I. The dactylus of the external maxillipeds is articulated far back on the inner or flexor edge of the propodite : the eyes in the female are not entirely visible in an ordinary dorsal view :—

    1. The dactyli of all the legs are about equal :—

        i. Carapace somewhat octagonal in outline, with deepish tomentose pits separating the branchial from the median regions: first three pairs of legs nearly equal in length : dactyli of all the legs of fair length............ *P. Edwardsi.*

        ii. Carapace circular, perfectly smooth : second pair of legs decidedly the longest : dactyli of all the legs very short........................... *P. mactricola.*

    2. Dactylus of the 3rd pair of legs longer than any of the others .... *P. purpureus.*

    3. Dactylus of the 4th pair of legs longer than any of the others .... *P. parvulus.*

II. The dactylus of the external maxillipeds is articulated to the tip of the propodite : the eyes in the female are entirely dorsal .... *P. abyssicola.*

37.   *Pinnoteres Edwardsi,* de Man.

Pinnotheres Edwardsi, de Man, Journ. Linn. Soc., Zool., XXII. 1887-88, p. 103, pl. vi. figs. 6-9 (1889).

The description applies to the female.

The length of the carapace is nearly equal to the greatest breadth. Carapace octagonal in shape, with the angles rounded : its dorsal surface little convex, with tomentose depressions of some size and depth separating the median from the branchial regions. The deflexed part of the front is very distinctly triangular. Eyes very small, but deeply pigmented.

Dactylus of external maxillipeds slender and inconspicuous ; placed far back on the inner edge of the spathulate propodite.

Chelipeds and legs more or less downy, especially on their under surface. Chelipeds nearly as long as the carapace, a little longer and much stouter than the legs, unarmed : dactylus as long as the upper border of the palm.

Legs rather coarse : the first 3 pairs are about equal in length, the 4th pair is a little shorter.

Carapace 15 millim. long and 16 millim. broad.

From an *Ostræa* from Mergui.

38. *Pinnoteres purpureus*, n. sp.

Closely related to *P. palaensis*, Bürger.

The description applies to the female.

Carapace and appendages smooth, polished, nude. Carapace transversely oval, strongly convex, the regions not well defined. De-flexed part of front broadly and indistinctly triangular. Eyes very small, but well pigmented.

Dactylus of external maxillipeds slender and inconspicuous, placed far back on the inner (flexor) edge of the propodite.

Chelipeds and legs slender, the chelipeds being little stouter than the legs and about the same length as the first pair of legs. The movable finger is not much more than half the length of the upper border of the palm.

The third pair of legs are the largest of all, their meropodites and carpopodites being longer than those of the first two pairs and nearly twice as long as those of the 4th pair. The dactyli of the 3rd and 4th pairs are several times the length of those of the first two pairs, and *the dactylus of the 3rd pair exceeds that of the 4th pair*. Though the 4th pair have a long dactylus their total length is not greater than that of either of the first two pairs.

Colour either hyaline with numerous minute specks of bluish-black pigment, or the specks may be sufficiently numerous to make the whole animal nearly black.

From an *Ostræa* from the Andaman Islands.

Carapace 7 millim. long and 9 millim. broad.

39. *Pinnoteres parvulus*, Stimpson, de Man.

Pinnotheres parvulus, Stimpson, Proc. Ac. Nat. Sci. Philad. 1858, p. 108 : de Man,, Journ. Linn. Soc., Zool., XXII. 1887–88, p. 105, and Archiv fur Nat. LIII. 1887, i. p. 383 : Ortmann, Zool. Jahrb., Syst., VII. 1893–94, p. 699 : Bürger, Zool. Jahrb.,, Syst., VIII. 1894-95, pp. 363, 376, pl. ix fig. 18 and x. fig. 17.

A single damaged female appears to differ from *P. purpureus* only in the following particulars :—

(1) though the 4th pair of legs are shorter than the 3rd, they are decidedly longer than the 2nd, and still more decidedly longer than the 1st.

(2) the dactylus of the 4th pair of legs is the longest of all.

40. *Pinnoteres mactricola*, n. sp.

Closely related to *P. cardii*, Bürger.

The description applies to the female.

Carapace perfectly circular smooth and polished, convex. Edge of front nearly straight. Eyes minute, well pigmented.

Dactylus of external maxillipeds slender and inconspicuous, arising far back on the inner (flexor) edge of the propodite.

Chelipeds decidedly stouter than the legs and about as long as the first pair of legs: their inner border is scantily fringed with hair: their dactylus is nearly two-thirds the length of the palm.

Legs slender, fringed with hairs: the second pair are decidedly the longest—a little longer than the carapace: the fourth pair are decidedly the shortest: the first and third pairs are about equal in length: in all four pairs the dactyli are equally short.

From *Mactra violacea*, from the mouth of the R. Hooghly.

Diameter of carapace not quite 6 millim.

In the male the front is a little prominent and the chelipeds are very much stouter.

41. *Pinnoteres abyssicola,* Alcock and Anderson.

Pinnoteres abyssicola, Alcock and Anderson, Ann. Mag. Nat. Hist. (7) III. 1899, p. 14: Alcock, Investigator Deep Sea Brachyura, p. 81.

The description applies to the female.

Carapace subcircular, smooth, convex. Front rather prominent, little deflexed, broadly triangular. Eyes of good size but deficient in pigment, entirely dorsal.

The palp of the external maxillipeds is minute and is much concealed by hairs that fringe the prominent internal angle of the merus: the dactylus is borne at the tip of the propodite.

Chelipeds much stouter than the legs, nude except for a fringe of hairs on the lower border of the immobile finger: they are about as long as the carapace, and the dactylus is not much shorter than the upper border of the palm.

Legs slender, nude: the 2nd and 3rd pairs are slightly longer than the 1st and 4th, being nearly 1½ times the length of the carapace: the dactyli also of the 2nd and 3rd pairs are a little longer than those of the 1st and 4th.

From *Lima indica*, from 430 fathoms off the Travancore coast.

Diameter of carapace 8 millim.

XANTHASIA, White.

Xanthasia, White, Ann. Mag. Nat. Hist. XVIII. 1846, p. 176: Dana, U.S. Expl. Exp., Crust., pt. I. p. 383: Milne Edwards, Ann. Sci. Nat., Zool., (3) XVIII. 1853, p. 221: Bürger, Zool. Jahrb., Syst., VIII. 1894-95, p. 386.

Resembles *Pinnoteres* in structure and habit, but differs in the following particulars:—

The edge of the carapace is well defined and, in all but its fronto-orbital portion, forms an upturned crest, so that the dorsal surface of the carapace is depressed and saucer-like. Other crests are found on the dorsal surface of the carapace and, in the centre, a large mushroom-like tubercle.

Though it is on an inferior plane, the narrow front is prominent and not deflexed.

The buccal cavern and mouth-parts have the same curious form, except that (owing to the encroachment of the epistome in the middle line) the anterior edge of the buccal cavern is bilobed or bow-shaped rather than semicircular, and the dactylus of the external maxillipeds is wanting or is represented by a few hairs.

Distribution : Iudo-Pacific, from the east coast of Africa to Fiji.

42.   *Xanthasia murigera*, White.

Xanthasia murigera, White, Ann. Mag. Nat. Hist. XVIII. 1846, p. 177, pl. ii. fig. 3 : Dana, U. S. Expl. Exp., Crust. pt. I. p. 384, pl. xxiv. figs. 6 *a-b* : Milne Edwards, Ann. Sc. Nat., Zool., (3) XX. 1853, p. 221 : A Milne Edwards, Nouv. Archiv. du Mus. IX. 1873, p 321 : Haswell, Cat. Austral. Crust. p. 113 : Miers, Zool. H. M. S. Alert, pp. 518, 546 : de Man, Journ. Linn. Soc., Zool., XXII. 1887-88, p. 106 : Bürger, Zool. Jahrb., Syst. VIII. 1894-95, p. 386, pl. x. fig. 33 : Adensamer, Ann. KK. Nat. Hofmus. Wien, XII. 1897, p. 109 : Nobili, Ann. Mus. Genov. (2) XX. 1899, p. 264.

The edge of the carapace is formed, in all but its short fronto-orbital portion, by a thin sharp upturned overhanging crest, which ends in a curl on the anterior part of either branchial region.

A large mushroom tubercle, having a rough or reticulate surface and a more or less reniform outline, occupies the middle of the dorsal surface of the carapace, and between this and the front is a pair of parallel longitudinal crests.

The front is somewhat prominent and is dorsally grooved or obscurely bilobed, and on each side of it, beyond the small orbits, is a small wing-like projection.

Chelipeds not, or hardly, stouter than the legs : the dactylus in the male is about two-thirds, in the female not much more than half the length of the palm.

Legs rather coarse : the first three pairs, which are about equal to one another and to the chelipeds in length, are about as long as the carapace, the fourth pair are a little shorter : the dactyli in all are about equally short.

In the female the broad abdomen is traversed longitudinally by a sort of coarse interrupted carina.

In the Indian Museum are 5 specimens from the Andamans and Mergui. The carapace of the largest female is 11·5 millim. long and 15·5 millim. broad.

The *Xanthasia* sp., or *Xanthasia Whitei*, from Mergui, referred ·to 'by de Man in Journ. Linn. Soc., Zool., XXII. 1887-88, p. 106, pl. vii. fig. 1 is represented in the collection by a single small male and is characterized by having the upraised edge of the carapace blunt and rounded, instead of thin and acute, and the median tubercle of the carapace ill defined instead of sharply circumscribed: the posterior margin of the carapace, also, is more prominent and is not quite continuous with the lateral margins. The legs also are somewhat longer.

Family OCYPODIDÆ, Ortmann, emend.

Key to the Indian genera of Ocypodidæ.

I. A hairy-edged pouch leading into the branchial cavity, between the bases of the 2nd and 3rd pair of true legs [*Ocypodinæ*] :—

    1. Antennular flagella rudimentary, completely hidden beneath the front: antennæ small, almost rudimentary: eyes very large, occupying the greater part of the ventral surface of the eyestalks: chelipeds very unequal in both sexes .... OCYPODA.

    2. Antennular flagella small, not hidden beneath the front: antennæ of good size: eyes small, terminal on the long slender eyestalks: in the male only, one cheliped is enormously enlarged the other being very small .... · .... GELASIMUS.

II. No pouch or opening between the bases of any of the legs :—

    1. The antennules fold obliquely or nearly vertically: curious membranous spaces, or "tympana," are present on the meropodites of the legs (*Scopimerinæ*) :—

        i. Tympana very well defined: external maxillipeds very large and with a strong almost hemispherical bulge forwards :—

            a. Merus of external maxillipeds larger than the ischium : the distal end of the 4th abdominal segment of· the male is fringed with bristles and overlaps the 5th segment .... DOTILLA.

            b. Ischium cf external maxillipeds larger than the merus: the 4th abdominal segment of the male is normal, but the 5th is constricted in part or all of its extent and gives the abdomen a wasp-like appearance .... SCOPIMERA.

     ii. Tympana ill defined: external maxillipeds of moderate size, the merus larger than the ischium: the chelipeds of the female, though not so stout as those of the male, are stouter than the legs ......   TYMPANOMERUS.

  2. The antennules fold obliquely or quite transversely: no "tympana" are present on any of the joints of the legs (*Macrophthalminæ*) :—

     i. Merus of the external maxillipeds smaller than the ischium, the flagellum coarse and articulating at the antero-external angle of the merus: front deflexed: eyestalks often very long ......   MACROPHTHALMUS.

     ii. Merus of the external maxillipeds as large as or larger than the ischium, at least the two terminal joints of the flagellum are slender: eyestalks not particularly long :—

        a. Front declivous : carapace slightly convex: the flagellum of the external maxillipeds articulates at the antero-external angle of the merus : (the chelipeds of the female, as in all *Macrophthalminæ*, are shorter and slenderer than the legs) ....   CLISTOSTOMA.

        b. Front square-cut, not in the least deflexed; carapace quite flat dorsally: the flagellum of the external maxillipeds articulates near, but not at, the antero-external angle of the merus : eyes not terminal on the eyestalks ....   TYLODIPLAX.

Subfamily OCYPODINÆ, Dana.

OCYPODA, Fabr.

Ocypoda, Fabricius, Ent. Syst. Suppl. p. 347 : Desmarest, Consid. Gen. Crust. p. 119, and Dict. Sci. Nat. XXVIII. p. 239 : De Haan, Faun. Japon. Crust. p. 29 : Milne Edwards, Hist. Nat. Crust. II. 41, and Ann. Sci. Nat., Zool., (3) XVIII. 1852 p. 141 : Dana, U. S. Expl. Exp. Crust. pt. I. p. 324 : Kingsley, Proc. Ac. Nat. Sci. Philad. 1880, p. 179 : Miers, Ann. Mag. Nat. Hist. (5) X. 1882, p. 376, and Challenger Brachyura, p. 237 : Ortmann, Zool. Jahrb., Syst., X. 1897-98, p. 359 (*Revision der Gattung* Ocypoda).

  Carapace deep, square or subquadrilateral, broader (but not much broader) than long, moderately convex, strongly declivous anteriorly, its dorsal surface closely granular with the regions indistinctly and incompletely defined. Front a narrow deflexed lobe, from a seventh to an eighth the greatest breadth of the carapace.

    J. II. 45

Orbits very capacious, occupying the whole face of the carapace between the front and the antero-lateral angles on either side, usually not very deep: their floor is divided into two fossæ, one for the basal portions of the eyestalk, the other for the eye. The basal joint of the eyestalk is visible .throughout: the eye chiefly occupies the ventral surface of the eyestalk, and is often, but not always, tipped by a horn or style formed by a prolongation of the latter.

The basal antennular joint is visible, but the rudimentary antennular flagellum is quite hidden beneath the front. The antennæ, which lie in the orbital hiatus, are, though properly formed in all their parts, little more than rudiments.

The epistome, though short, is quite distinct, and is sculptured. The buccal cavern (in its widest part) is as broad as long, but diminishes in size a little, anteriorly: it is completely closed by the external maxillipeds, which are somewhat narrow and elongate and end in a coarse flagellum that articulates with the antero-external angle of the merus.

Chelipeds shorter than the legs and, in both sexes, remarkably unequal, the larger one being much more massive than the legs. The palm is short and high—especially in the larger cheliped—and is almost always compressed—especially so in the smaller cheliped: the fingers are stout, usually compressed, and strongly toothed. In most cases there is, on the inner surface of the larger palm, near the fingers, a stridulating organ, which can be scraped against the inner surface of the ischium.

Legs stout, the fourth pair much shorter and somewhat less massive than the first three pair, which are of about equal length: between the basal joints of the 2nd and 3rd pair is an orifice, thickly protected by hairs, leading towards the branchial cavity. The branchial cavity is very capacious, and its lining membrane is thick spongy and vascular.

The abdomen of the male is narrow: in both sexes it consists of seven separate segments.

· *Distribution :* Tropical and subtropical coasts, from the American Atlantic, through the Mediterranean and Red Seas, to the American Pacific.

The Ocypodes live together in large companies, and most of them are in the habit of digging long and tortuous burrows in the moist sand near high-water mark, into which they retire with great rapidity when alarmed. As a rule they do not go far from their burrows, but if they do happen to wander and are cut off, they run to sea with marvellous speed. Though the burrows can be but temporary structures, each individual crab, in all the species that I have observed, keeps rigidly to its own. The efficacy of the stridulating-organ as a musical instrument is beyond dispute, and I have published my own observations on that of *O. macrocera* in the Administration Report of the Marine Survey of India for the year 1891–92 (reprinted in the *Annals and Magazine of Natural History* for 1892). Dr. A. R. Anderson has published a note on the sound produced by *O. ceratophthalma* in this *Journal* for the year 1894.

My own opinion is that these crabs use the stridulating-organ when in their burrows—which undoubtedly are private property—to warn intending intruders of the herd that the burrow is occupied, and thus to prevent the burrow becoming crowded to suffocation-point. This, of course, need not be its exclusive use.

Key to the Indian species of Ocypoda.

I. No stridulating ridge on the inner surface of the palm: eyestalks not prolonged beyond the eyes in the form of a style .... *O. cordimana.*

II. A stridulating ridge on the inner surface of the palm: eyestalks (except sometimes in the young) prolonged beyond the eyes to form a horn or style :—

    1. Length of the stridulating organ much more than half the greatest breadth of the palm: antero-lateral angles of the carapace well pronounced : —

        i. Fingers of both chelipeds pointed :—

            a. Stridulating ridge narrow, consisting entirely of small tubercles: no brushes of hairs on the propodites of any of the legs .... *O. platytarsis.*

            b. The stridulating ridge consists of tubercles gradually passing into striæ: the anterior surface of the propodites of the first two pairs of legs thickly furnished with hairs ... *O. ceratophthalma.*

        ii. Fingers of the smaller cheliped expanded at tip: the stridulating ridge consists entirely of striæ... *O. macrocera.*

    2. Length of the stridulating organ much less than half the greatest breadth of the palm: antero-lateral angles of the carapace rounded off .... *O. rotundata.*

The synonomy of the species of *Ocypoda* has been discussed, at length, by Ortmann (Zool. Jahrb., Syst., X. 1897–98, p. 359), who has had access to a great deal more material than I have. It would be inadvisable, therefore, for me, working on a collection made almost entirely in India, to attempt any independent criticism of the older work; so that, in dealing with the Indian species, I shall generally restrict my citations to the papers of Ortmann and the other authors (Kingsley and Miers) who have made a revision of the genus.

43. *Ocypoda ceratophthalma* (Pallas), Ortm.

Cancer ceratophthalmus, Pallas, Spicilegia Zool. IX. p. 83, pl. v. figs. 7, 8.
Cancer cursor, Herbst, Krabben, I. ii. 74, pl. i. figs. 8, 9.
Ocypoda ceratophthalma, Fabricius, Ent. Syst. Suppl. p. 347: Milne Edwards, Hist. Nat. Crust. II. 48, and Cuvier Règne An. Crust. pl. 17: Kingsley, Proc. Ac.

Nat. Sci. Philad. 1880, p. 179: Miers, Ann. Mag. Nat. Hist. (5) X. 1882, pp. 378, 379: C. W. S. Aurivillius, Zur. Biol. Amphib. Decap., p. 17 (Mitg. K. Ges. Wiss. Upsala, 1893).

ORTMANN, ZOOL. JAHRB., SYST. X. 1897-98, pp. 360, 364 (*ubi synon.*).

Carapace square, its greatest breadth, which is about a tenth more than its greatest length, is at the acuminated antero-lateral angles, which coincide with the outer orbital angles and are right angles, or nearly so.

The borders of the carapace, with the exception of the posterior border, are elegantly beaded or serrulate, and the lateral borders in their anterior third are straight and parallel, or nearly so.

The cardiac region can be distinguished, and the anterior ends of the cervical groove are present on either side of the gastric region.

Upper border of orbit sinuous and a little oblique, so that the outer angle of the orbit is considerably behind the front: the lower border has an obscure notch near its middle, but there is *no gap at its outer angle.* The eyestalk is prolonged beyond the eye into a blunt-pointed style of variable length.

The lateral borders of the buccal cavern, though their general direction is slightly convergent anteriorly, have a distinct outward curve. The merus and ischium of the external maxillipeds have their exposed surface circumscribed by a raised row of granules, which is deficient only at the basal attachment of the ischium.

Chelipeds and legs scabrous, the asperities having in many places a tendency to a rugiform or squamiform arrangement, and almost forming serrations on the borders of some of the joints, and becoming spines or teeth on the lower borders of the arms and hands and at both angles of the wrist—especially at the inner angle where there is always at least one distinct spine.

The stridulating organ of the larger palm is of good length (much more than half the greatest breadth of the palm) and is some little distance from the immobile finger, a thick strip of hair intervening: in its upper half it consists of tubercles gradually passing to striæ, in its lower half it consists of a comb of fine regular and very close-set striæ. It plays against a polished ridge that runs across the upper part of the inner surface of the ischium.

The palms and fingers of both hands—but notably of the smaller hand—are compressed, and the fingers of both hands are *pointed.*

The first three pairs of legs have the merus broadened: they do not differ greatly in length, and the 2nd pair, which are slightly the longest, are about two-and-a-half times the greatest length of the carapace. The fourth (last) pair are a good deal shortened—reaching only a little more than half-way along the propodite of the 3rd pair—and have a much narrower merus.   In all the legs the dactylus is stout and fluted like a bayonet and has more or less of its anterior surface hairy : though somewhat laterally-compressed at base and gradually broadening and becoming dorso-ventrally-compressed towards the tip, it may fairly be called styliform.   The propodites of the first two pairs of legs have conspicuous brushes of hairs along their anterior surface.

In the Indian Museum are 84 specimens from all parts of the coasts of the mainland and islands of India.   Large specimens have the carapace 40 millim. long and about 45 millim. in greatest breadth.

Distribution : Indo-Pacific, from the east coast of Africa to the Sandwich Islands.

In young specimens the surface of the appendages is smoother and the eyestalks are not prolonged beyond the eyes, which are of large size. In half-grown specimens the terminal style of the eyes is still short.

44.   *Ocypoda macrocera*, Edw.

Ocypoda macrocera, Milne Edwards, Hist. Nat. Crust. II. 49 : Kingsley, Proc. Ac. Nat. Sci. Philad. 1880, p. 181 : Miers, Ann. Mag. Nat. Hist. (5) X. 1882, pp. 378, 381 : ORTMANN, ZOOL. JAHRB., SYST., X. 1897–98, pp. 360, 368.

Closely related to *O. ceratophthalma*, from which it is distinguished by the following characters :—

(1) the carapace is rather broader and the orbits are a little more oblique :

(2) the raised marginal row of granules on the external maxillipeds is less pronounced :

(3) the fingers of the smaller cheliped are lamellar up to the tips, which are broad and blunt, not pointed :

(4) the stridulating ridge is less hairy and consists entirely of striæ.

(5) it is a smaller species, large specimens having the carapace 31 millim. long and 37 millim. broad.

In the Indian Museum are 78 specimens from the coasts of the Bay of Bengal : there are none from the west coast or from any of the islands, and the species appears to be confined to the Bay.

The colour, in life, is bright red.   This species lives in large warrens in the sands of almost all parts of the east coast of the penin. sula.   One of its most active enemies is the Brahminy kite (*Haliastur indus*).   One almost certain use of the stridulating-organ is to give warning to intending trespassers, of its own species, that a burrow is already occupied by its rightful owner.

45. *Ocypoda platytarsis*, Edw.

Ocypoda platytarsis, Milne Edwards, Ann. Sci. Nat. Zool. (3) XVIII. 1852, p. 141: Kingsley, Proc. Ac. Nat. Sci. Philad. 1880, p. 180: Miers, Ann. Mag. Nat. Hist. (5) X. 1882, pp. 378, 383: ORTMANN, ZOOL. JAHRB, SYST., X. 1897–98, pp. 359, 363 (*ubi synon.*).

This species may be distinguished from *O. ceratophthalma*, which it closely resembles, by the following characters :—

(1) the carapace is very distinctly broader, its length being about four-fifths of its breadth, and the orbits are hardly at all oblique :

(2) the surface of the ischium of the external maxillipeds is often quite smooth :

(3) the stridulating ridge is not, or hardly at all, hairy and consists entirely of granules or small mamillated tubercles; and though the upper edge of the inner surface of the ischium of the larger cheliped is raised and rough, there is no special process against which the stridulating-ridge of the palm can be scraped :

(4) the dactyli of the legs, though fluted as in the other species, are distinctly compressed dorso-ventrally and broadened :

(5) there are no brushes of hairs along the anterior surface of the propodites of any of the legs.

It is a somewhat larger species, the carapace in full-sized adults being 40 millim. long and 54 millim. broad.

In the Indian Museum there are 42 specimens from both coasts of the peninsula and from Ceylon.

46. *Ocypoda rotundata*, Miers.

Ocypoda rotundata, Miers, Ann. Mag. Nat. Hist. (5) X. 1882, pp. 378, 382: Ortmann, Zool. Jahrb., Syst., X. 1897–98, pp. 360, 364.

This species differs from *O. ceratophthalma* in the following important particulars :—

The carapace is less distinctly quadrilateral, owing to the fact that the antero-lateral borders are arched, instead of forming an angle with the upper border of the orbit. These borders sometimes form an unbroken curve with the upper border of the orbit, but sometimes the junction between the two is marked by a notch. The length of the carapace is about five-sixths its greatest breadth, which, owing to the curvature of the antero-lateral borders, is some distance behind the orbits.

There is a notch in the middle of the lower border of the orbit, and a gap at the outer angle, between the upper and lower borders.

The deflexed tip of the front is swollen.

The spines or serrations at the inner angle of the wrist are more numerous, and at the outer angle are better marked.

The length of the stridulating organ is much less than half the greatest height of the palm : the organ consists of about a dozen distant ridges much concealed in hair, and each ridge is sharply serrated.

The scraper on the ischium is placed near the upper angle of the inner face of that joint and consists of an elongate-elliptical longitudinally-grooved cicatrix-like surface, with a patch of hair above it and a much larger patch below it.

The fingers of the smaller cheliped are almost as much dilated at tip as those of *O. macrocera.*

The dactyli of the legs are dorso-ventrally compressed as in *O. platytarsis.*

There is a thick brush of hairs along the anterior surface of the propodite of the first pair of legs only.

The meropodites of the first three pairs of legs are not so broad as in the three preceding species.

In the Indian Museum are 29 specimens from the coasts of Cutch, Sind, and Baluchistan.

This is the largest Indian Ocypode, the carapace of the adult being 52 millim. long and 62 millim. broad.

47.   *Ocypoda cordimana,* Desm.

Ocypoda cordimana, Desmarest, Consid. Gen. Crust. p. 121: Milne Edwards, Hist. Nat. Crust. II. 45 : Kingsley, Proc. Ac. Nat. Sci. Philad., 1880, p. 185 : de Man, Notes Leyden Mus., III. 1881, p. 248 : Miers, Ann. Mag. Nat. Hist. (5) X. 1882, pp. 379, 387 : ORTMANN, ZOOL. JAHRB , SYST., X. 1897–98, pp. 359, 362 (*ubi synon.*).

Carapace deep, quadrilateral, strongly convex fore and aft, its length about seven-eighths its greatest breadth, which is some little distance behind the orbits, owing to the gentle curve of the antero-lateral borders : its antero-lateral angles coincide with the outer orbital angles, and point acutely forwards.

Orbits deep; their upper border sinuous, but not in the least oblique ; there is usually a notch near the middle of their lower border, and always a deep gap at the outer angle. No terminal style to the eyes.

The lateral borders of the buccal cavern are anteriorly convergent and have no outward curve. The marginal row of granules on the outer surface of the ischium of the external maxillipeds is indistinct or absent.

Though the chelipeds and legs are rough and the roughness is in places squamiform, there is no serration of their edges, except in the case of the lower borders of the arms, the inner edge of the wrists, and the lower border of the hands. The palm of the larger hand, though deep, is not particularly compressed, and it has no stridulating ridge.

The propodites and dactyli of the legs are rather short and stout, the dactyli being fluted and more or less hairy : the edges of the propodites of the first 2 pairs of legs are hairy. The third pair of legs, which are slightly longer than the first 2 pairs, are less than twice the length of the carapace.

In the Indian Museum are 59 specimens, from the Laccadives, the Madras coast, Ceylon, Mergui, Tavoy, the Andamans and Nicobars.

The carapace of the largest specimen is 35 millim. long and 40 millim. broad.

GELASIMUS, Latr.

?. *Gelasimus*, Latreille, Dict. des Sciences Nat. XVIII. p. 286 (1820) : Desmarest, Consid. Gen. Crust. p. 122, and Dict. Sci. Nat. XXVIII. p. 241 : De Haan, Faun. Japon Crust. p. 25 : Milne Edwards, Hist. Nat. Crust. II. 49, and Ann. Sci. Nat., Zool., (3) XVIII. 1852, p. 144 : Dana, U. S. Expl. Exp. Crust. pt. I, pp. 312, 315 : Hess, Archiv f. Naturges. XXXI. 1865, p. 145 : A. Milne Edwards, Nouv. Archiv. du Mus. IX. 1873, p. 271 : Kingsley, Proc. Ac. Nat. Sci. Philad. 1880, pp. 135, 136 : Miers, Challenger Brachyura, p. 241 : de Man, Notes Leyden Mus. XIII. 1891, pp. 20-23 : Ortmann, Zool. Jahrb., Syst. VII. 1893-94, pp. 749-753.

" *Uca*," Leach, Trans. Linn. Soc. XI. 1815, pp. 309, 323 : M. J. Rathbun, Proc. Biol. Soc. Washington, XI. 1897, p. 154 : Ortmann, Zool. Jahrb., Syst , 1897-98, p. 346 (cf. notes by Desmarest and Milne Edwards, *ll. cc. supra*).

In obedience to certain interpretations of the rule of priority, which sacrifice everything to a legal precision that defeats the object of classification, some modern authors propose to apply the name *Uca*, which was originally given to and has for nearly seventy-five years been authoritatively used for a land-crab of the Gecarcinoid family, to the species of the Ocypodoid family which have for the same long period been known to everybody by the name *Gelasimus*.

One of the objects of my poor work being to avoid confusion, I cannot consent to this proposal : and if the rules of nomenclature do not permit me to retain a name that has been deliberately chosen, and used without any ambiguity, by such illustrious predecessors as Latreille, Milne Edwards, and Dana, then I think that the rules should be modified.

The introduction of a rule sanctioning the retention of any name that has been accepted and defined by a monographer of repute, and that has thereafter been in common use for fifty years, would probably satisfy those to whom the written authority of the law is a consideration of first importance.

Carapace deep, subquadrilateral but with the antero-lateral angles produced and acute and the lateral borders more or less convergent posteriorly, occasionally subhexagonal, a good deal broader than long, the regions never very strongly defined. The front is a narrow declivous lobe, the breadth of which, between the eyestalks, is from one-sixteenth to one-sixth the greatest breadth of the carapace.

The orbits are narrowish trenches occupying the whole anterior extent of the carapace between the narrow front and the antero-lateral angles, and are more or less sinuous and oblique: the eyestalks are very long and are formed as in *Ocypoda,* but are much slenderer: the eyes, though chiefly ventral in aspect, are always terminal.

The small antennular flagella, which are not hidden under the front, fold obliquely. The antennæ, which stand free at the inner angle of the orbits, have well developed flagella.

Epistome, though short, quite distinct. The lateral borders of the buccal cavern are convex outwards, sometimes so much so as to give the cavern a subcircular outline. The external maxillipeds have a long ischium and a short and somewhat oblique merus with the coarse flagellum jointed to its antero-external angle: they close the buccal cavern except for a chink anteriorly.

The chelipeds differ greatly in the sexes. In the female they are equal, are shorter and slenderer than the legs, and have broad-tipped spoon-shaped fingers. In the male one of the chelipeds resembles those of the female, but the other is of relatively gigantic proportions, the hand alone being often as big and heavy as all the rest of the animal.

The legs are stout and end in very sharp dactyli, and the meropodites of at least the 2nd and 3rd pairs are foliaceous : these two pairs are a little longer than the other two, being about twice the length of the carapace.

As in *Ocypoda,* the branchial cavity is capacious, and its lining membrane thickened and vascular, with a fleshy lobe, shaped like a gill-plume, projecting into the space between the tips of the last two gill-plumes : also, between the basal joints of the 2nd and 3rd pairs of legs, there is an orifice, thickly protected by hairs, leading towards the branchial cavity.

The abdomen of the male is narrow : in both sexes of all the Indian species it consists of seven separate segments.

Distribution : all the warmer regions of the globe, from the Atlantic coasts of America eastwards (including the Mediterranean basin) to the Pacific coasts of America again.

The species of *Gelasimus* are, like the Ocypodes, gregarious, and live in warrens in the mud-flats of tropical and subtropical estuaries. Their intelligence, like that of the Ocypodes, is of a high order.

In one species, at any rate (*Gelasimus annulipes*), the males, which are greatly in excess of the females, use the big and beautifully-coloured cheliped, not only for fighting with each other, but also for "calling" the females. I have described my own observations on these points in the *Administration Report of the Marine Survey of India for* 1891-92—reprinted, as an extract, in the *Annals and Magazine of Natural History* for 1892.

The fact that the males greatly outnumber, and therefore are more commonly captured than, the females, is sufficient justification for the common practice of using the larger cheliped of the male for the discrimination of the species. It must, however, be remembered that—at least in all the Indian species—this organ changes greatly with advancing age.

I must also confess here that the synonomy of species has defied me.

Key to the Indian species of Gelasimus.

I. The breadth of the front, measured exactly between the bases of the eyestalks, is between a fifth and a sixth the greatest breadth of the carapace :—

    1. Two oblique granular ridges on the inner surface of the palm of the large cheliped of the male, one continuous with the dentary edge of the immobile finger, the other running to the lower edge of the same finger :—

        i. Carapace subquadrilateral, the true lateral borders being moderately convergent posteriorly : an enlarged tooth near the tip of the immobile finger of the large cheliped of the male gives the tip of this finger a notched-truncate appearance   *G. annulipes.*

        ii. Carapace subquadrilateral, the true lateral borders nearly parallel : the tip of the immobile finger of the large cheliped of the male is oblique-truncate but not notched ....   *G. lacteus.*

        iii. Carapace distinctly hexagonal, owing to the great obliquity of the orbits and the strong convergence posteriorly of the true lateral borders : tip of the immobile finger of the large cheliped of the male not truncate or notched ....   *G. triangularis.*

    2. The oblique crest running to the lower edge of the immobile finger of the large cheliped of the male is either absent or is represented by a slight and smooth tumescence ....   *G. inversus.*

II. The breadth of the front, measured as above, is very much less than a sixth the greatest breadth of the carapace :—

    1. No row of granules running inside of and parallel with the lower border of the orbit :—

        i. The inner border of the arm of the larger cheliped of the male ends in a sharp tooth or spine, independent of the terminal lobe-like constriction of the arm :—

            a. Front, measured as above, about a tenth the greatest breadth of the carapace : in the large cheliped of the male the wrist is

smooth, the palm full with the granular
ridges on the inner surface indistinct, and
the fingers are not specially compressed...   *G. tetragonum.*

   b. Front, measured as above, not a fifteenth
the greatest breadth of the carapace : in
the large male cheliped the upper surface
of the wrist is granular and the fingers are
remarkably compressed and blade-like :—

     a. In the large male cheliped the crests
on the inner surface of the palm
are moderately prominent, the
dactylus is quite blade-like and the
cutting-edge of the immobile finger
is not much scallopped ....   *G. Marionis.*

     β. The crests on the inner surface of
the palm are extremely prominent,
the cutting edge of the dactylus is
not quite straight and that of the
immobile finger is scallopped into
two large triangular lobes ....   *G. Marionis,* var.
                                           nitidus.

  ii. The arm of the large male cheliped ends in a
constricted lobe, but there is no sharp upstanding
tooth inside it on the inner border :—

   a. Front, measured as above, about a twelfth
the greatest breadth of the carapace ; the
fingers of the large male cheliped have
tips that suggest tongs, owing to the
presence of an enlarged tooth near the
tip : the meropodites of the last pair of legs
are nearly as foliaceous as those of the
preceding pair ....   *G. acutus.*

   b. Front, measured as above, not a fifteenth
the greatest breadth of the carapace : the
fingers of the large male cheliped end in
simple hooked tips : the meropodites of
the last pair of legs are not much broad-
ened ....   *G. Dussumieri.*

2. On the lower wall of the orbit, inside of and parallel
with the middle third of the lower border of
that cavity, is a raised row of granules ....   *G. Urvillei.*

48. *Gelasimus annulipes,* Latr., Edw.

? *Cancer vocans minor,* Herbst, Krabben, I. ii. 81, pl. i. fig. 10.

Gelasimus annulipes, Milne Edwards, Hist. Nat. Crust. II. 55, pl. xviii. fig.
10–13 ; and Ann. Sci. Nat., Zool., (3) XVIII. 1852, p. 149, pl. iv. fig. 15 : Dana, U. S.
Expl. Exp., Crust., pt. I. p. 317 : Heller, Novara Crust. p. 38 : Hilgendorf, in v. d.
Decken's Reis. Ost-Afr. III. i. p. 85, and MB. Ak. Berl. 1878, p. 803 : Hoffmann,
in Pollen and van Dam, Faun. Madagasc., Crust. p. 18 : Kossmann, Reise roth.

Meer., Crust., p. 53 : Miers, Phil: Trans. Roy. Soc. Vol. 168, 1879, p. 488, and Ann. Mag. Nat. Hist. (5) V. 1880, p. 310, and Zool. H. M. S. Alert, pp. 518, 541, and Challenger Brachyura, p. 244: Richters, in Möbius Meeresf. Maurit., p. 155 : Kingsley, Proc. Ac. Nat. Sci. Philad. 1880, p. 148, pl. x. fig. 22: de Man, Notes Leyden Mus. II. 1880, p. 69, and Journ. Linn. Soc., Zool., XXII. 1887-88, p. 118 pl. viii. fig. 5–7, and Archiv f. Naturges. LIII. 1887, i. p. 353, and Notes Leyden Mus. XIII. 1891, pp. 23, 39, and in Weber's Zool. Ergebn. Niederl. Ost-Ind. II. 1892, p. 307, and Zool. Zahrb., Syst., VIII. 1894–95, p. 577 : Lenz & Richters, Abh. Senck. Nat. Ges. Frankf., XII. 1881, p. 423: F. Müller, Verh. Ges. Basel. VIII. 1886, p. 475 : J. R. Henderson, Trans. Linn. Soc., Zool., (2) V. 1893, p. 388 : Ortmann, Zool. Jahrb. Syst. VII. 1893-94, pp. 752, 758, and Jena. Denk. VIII. 1894, p. 57 : Zehntner, Rev. Suisse de Zool. II. 1894, p. 178.

Gelasimus Carionis, Edw. (*nec* Desm.), Hist. Nat. Crust. II. 53.

Gelasimus porcellanus, White, P. Z. S. 1847, p. 85, and in Adams and White, Samarang Crust., p. 50: Milne Edwards, Ann. Sci. Nat., Zool., (3) XVIII. 1852, p. 151 : Kingsley, Proc. Ac. Nat. Sci. Philad. 1880, p. 155.

Gelasimus perplexus, Milne Edwards, Ann. Sci. Nat., Zool., (3) XVIII. 1852, p. 150, pl. iv. fig. 18: A. Milne Edwards, Nouv. Archiv. du Mus. IX. 1873, p. 274.

? *Gelasimus pulchellus*, Stimpson, Proc. Ac. Nat Sci. Philad. 1858, pp. 99, 100.

Uca annulipes, Ortmann, Zool. Jahrb. Syst. X. 1897–98, pp. 351 and 354: Nobili, Ann. Mus. Genov. (2) XX. 1899, p. 274 : Doflein, SB. Ak. Münch. XXIX. 1899, p 193.

Length of the carapace about three-fifths of the greatest breadth at the acute claw-like antero-lateral angles. The posterior border of the dorsum of the carapace—*i.e.*, the border corresponding with the last segment of the sternum—is a good deal over half the greatest breadth of the carapace, so that the lateral borders of the dorsum of the carapace, which are distinctly defined in almost two-thirds of their extent by a fine raised line, are only moderately convergent. The post-gastric and cardiac regions are the only ones that are defined, and they but faintly.

Front, measured between the bases of the eyestalks, from a fifth to a sixth the greatest breadth of the carapace.

Orbits sinuous and considerably oblique ; their upper border defined by a fine raised line which is very distinctly double in a good part of its extent; their lower border very elegantly and regularly serrated—the teeth increasing in size from within outwards. In the female only there is a short row of granules inside of and parallel with the lower border of the orbit.

In the large cheliped of the adult male the greatest length of the hand (including fingers) is at least three times the length of the carapace : the outer surface of the somewhat rounded arm and of the wrist and hand is smooth to the naked eye, with a few small granules on the inner border of the wrist: the lower border of the palm is obscurely margin- ate : and on the inner surface of the palm are two salient granular crests, one of these is deeply grooved and nearly vertical and becomes continuous with the dentary edge of the immobile finger, the other, which is the more prominent, is oblique and runs to the lower border of the same finger. In the adult male the fingers of the large hand are about twice the length of the upper border of the palm: they are not very broad, and owing to the hook-like curve of the dactylus there is a wide space between them when the tips are apposed: the immobile finger is but slightly curved, and is generally shorter than the dactylus, and owing to the presence of an enlarged tooth near the tip, the tip has a characteristic notched-truncate appearance.

The meropodite in the last pair of legs is not at all foliaceous.

The carapace in the adult male is about 11 millim. long and 19 millim. broad.

In the Indian Museum are 300 specimens from all parts of the coast from Karachi on the west to Mergui on the east.

This species is not, as Miers queries, the same as Stimpson's *G. splendidus,* of which we have numerous specimens from Hongkong.

49. *Gelasimus lacteus* (De Haan).

Ocypode (Gelasimus) lactea, De Haan, Faun. Japon., Crust., p. 54, pl. xv. fig. 5.
Gelasimus lacteus, Milne Edwards, Ann. Sci. Nat., Zool., (3) XVIII. 1852, pl. iv. fig. 16: Stimpson, Proc. Ac. Nat. Sci. Philad. 1858, p. 100: Miers, P. Z. S. 1879, pp. 20, 36: Kingsley, Proc. Ac. Nat Sci. Philad. 1880, p. 149, pl. x. fig. 28: Cano, Boll. Soc. Nat. Napol. III. 1889, p. 234: de Man, Notes Leyden Mus. XIII. 1891, p. 22: Ortmann, Zool. Jahrb., Syst, VII. 1893-94, pp. 752, 759.
Uca lactea, Ortmann, Zool. Jahrb., Syst., X. 1897-98, pp. 351, 355.

Easily distinguished from *G. annulipes,* which is its nearest relative, by the following characters :—

(1) the carapace is much more nearly quadrangular, the posterior border of its dorsum being between three-fifths and two-thirds of its greatest breadth, and its true lateral borders being parallel, while the lateral borders of its dorsum are nearly so :

(2) in the larger cheliped of the male the outer end of the upper border of the arm, and the inner border of the wrist, are distinctly denticulated; the dactylus is not so strongly hooked, and the end of the immobile finger though obliquely truncate has an acuminate tip—never a notched-truncate tip:

(3) the colour, in spirit specimens, has a sort of livid bloom never seen in *G. annulipes.*

In the Indian Museum are 47 specimens from Karachi and 3 from the Andamans.

[*Gelasimus inversus,* Hoffmann.

Gelasimus inversus, Hoffmann, in Pollen and van Dam, Faun. Madagasc. Crust. p. 19, pl. iv. figs. 23-26 (1874): Kingsley, Proc. Ac. Nat. Sci. Philad. 1880, p. 155

de Man, Notes Leyden Mus. XIII. 1891, pp. 21, 44, pl. iv. fig. 12: Ortmann, Zool. Jahrb., Syst., VII. 1893-94, p. 751, and Jena. Denk. VIII. 1894, p. 59.

Gelasimus chlorophthalmus, Hilgendorf (*nec* Edw.), MB. Ak. Berl. 1878, p. 803 (*apud* de Man).

Gelasimus Smithii, Kingsley, Proc. Ac. Nat. Sci. Philad. 1880, p. 144, pl. 9, fig. 14 (*apud* Ortmann).

Uca inversa, Ortmann, Zool. Jahrb., Syst., X. 1897-98, p. 351.

There are in the Indian Museum specimens of this species from Madagascar and the Red Sea, and some from Karachi which differ from the type in the form of the dactylus of the large male cheliped, and are here separated as a variety.]

50. *Gelasimus inversus,* var. *sindensis,* nov.

This variety differs from typical *G. inversus* from Madagascar only in having the tip of the dactylus of the large male cheliped simple (instead of furnished with a second tooth that gives it a notched appearance) and the palm of the hand smoother externally.

The species resembles *G. annulipes,* from which it differs in the following characters :—

(1) the lateral borders of the dorsum of the carapace are defined by a fine line which is raised and distinct in the anterior third only, and is a little more oblique :

(2) the lower border of the orbit is much more sinuous, and is either entire or is quite imperceptibly denticulated at its outer angle :

(3) in the large cheliped of the male the arm is trigonal with sharp edges, the upper edge rising into a distinct lobe or crest and the distal end of the inner edge forming a crest or blunt tooth; the inner edge of the wrist is distinctly denticulated, and the upper border of the palm has several longitudinal rows of granules; of the granular ridges on the inner surface of the palm *the lower one that in* G. annulipes *runs to the lower edge of the immobile finger is absent or, at most, is represented by a smooth and slight swelling;* finally the immobile finger, though as in *G. annulipes* nearly straight and shorter than the dactylus, has a simple not a notch-like tip.

In the Indian Museum are 30 specimens from Karachi. The carapace of the largest specimen is 10 millim. long and 18 millim. broad.

51. *Gelasimus triangularis,* A. M. Edw.

Gelasimus triangularis, A. Milne Edwards, Nouv. Archiv. du Mus. IX. 1873, p. 275: Kingsley, Proc. Ac. Nat. Sci. Philad. 1880, p. 150: de Man, Journ. Linn. Soc., Zool., XXII. 1887-88, p. 119, pl. viii. figs. 8-11, and Notes Leyden Mus. XIII. 1891, p. 22, and in Weber's Zool. Ergebn. Niederl. Ost. Ind. II. 1892, p. 307: and Zool. Jahrb., Syst., VIII. 1894-95, p. 577: J. R. Henderson, Trans. Linn. Soc., Zool., (2) V. 1893, p. 388.

Gelasimus perplexus, Heller (nec Edw.), Novara Crust. p. 38, pl. v. fig. 4.

?? Gelasimus minor, Owen, Zool. H. M. S. "Blossom," Crust., p. 79, pl. xxiv. figs. 2, 2a (1839): Milne Edwards, Ann. Sci. Nat., Zool., XVIII. 1852, p. 151: Kingsley, Proc. Ac. Nat. Sci. Philad. 1880, p. 150.

Uca triangularis, Nobili, Ann. Mus. Genov. (2) XX. 1899, p. 274.

Length of the carapace about four-sevenths of the greatest breadth, which is at the spine-like antero-lateral angles.

Carapace strongly convex, almost hexagonal, the regions not indicated. The posterior border of the dorsum of the carapace is less than half the greatest breadth, hence not only the lateral borders of the dorsum of the carapace, but also the true lateral borders, are strongly convergent posteriorly, the former being defined by a fine raised line in more than two-thirds of their extent.

Front, as in *G. annulipes*, from a fifth to a sixth the greatest breadth of the carapace.

Orbits sinuous, much oblique : the upper border defined by a fine microscopically-beaded line, which is double in great part; the lower microscopically beaded, serrulate at its outer end.

In the large cheliped of the adult male the hand is about 2½ times as long as the carapace ; the outer surface of the arm, wrist, and hand are smooth to the naked eye; all the borders of the arm are sharply defined and finely serrulate, the inner border of the wrist is finely serrulate, and the upper and lower borders of the palm are marginate and granulate, especially the upper border; and the two oblique granular crests on the inner surface of the arm are in strong relief.

In the large hand the dactylus, in the adult, is from 1½ to 1¾ times the length of the upper border of the palm; its tip is simply hooked and overhangs the simple upcurved tip of the immobile finger.

The meropodite of the last pair of legs is not nearly so broad as that of the two preceding pairs.

In the Indian Museum are 70 specimens, all but one being from various parts of the Bay of Bengal littoral. The carapace of a large specimen is 10 millim. long and about 18 millim. broad.

The figures of *G. minor*, Owen, agree very well with this species, and if the two names should prove to refer to the same species this name has the precedence.

52. *Gelasimus tetragonum* (Herbst).*

Cancer marinus, minor, vociferans, Seba, Thesaurus, III. p. 48, pl. xix. fig. 15.

Cancer tetragonon, Herbst, Krabben, I. ii. 257, pl. xx. fig. 110, and III. i. 31.

Gelasimus tetragonum, Rüppell, 24 Krab. roth. Meer., p. 25, pl. v. fig. 5: Milne

* I assume that Herbst used *tetragonon* as a noun substantive in apposition to Cancer ; it may therefore continue in apposition to Gelasimos used as a substantive.

Edwards, Hist. Nat. Crust. II. 52, and Ann. Sci. Nat., Zool., (3) XVIII. 1852, p. 147, pl. iii. fig. 9 : Guérin, Voy. Coquille, II. Zool., Crust. p. 10, pl. i. figs. 2, 3 : A. Milne Edwards, in Maillard's l'ile Réunion, Ann. F., p. 6, and Nouv. Archiv. du Mus. IX. 1873, p. 273 : Heller, Novara Crust., p. 37 : Hilgendorf, in v. d. Decken's Reisen Ost. Afr. Crust. p. 84 : Hoffmann, in Pollen and Van Dam, Faun. Madag. Crust. p. 16 : Kossmann, Reis. roth. Meer. Crust. p. 52 : Kingsley, Proc. Ac. Nat. Sci. Philad. 1880, p. 143, pl. ix. fig. 11 : de Man, Archiv f. Naturges. LIII. 1887, i. p. 353, and Notes Leyden Mus. XIII. 1891, pp. 20, 24, pl. ii. fig. 6 : Ortmann, Zool. Jahrb. Syst. VII. 1893-94, pp. 750, 754 : Whitelegge, Mem. Austral. Mus. III. 1897, p. 138.

Gelasimus Duperreyi, Guérin, Dana U. S. Expl. Exp. Crust. pt. i. p. 317.

Uca tetragona, Ortmann, Zool. Jahrb., Syst. X. 1897-98, p. 348 : Doflein SB. Ak. Münch. XXIX. 1899, p. 193.

Length of the carapace about two-thirds of its greatest breadth at the acute antero-lateral angles. Carapace somewhat pentagonal, markedly convex fore and aft, the regions all recognizable but not strongly defined : though the posterior border of its dorsum is only half its greatest breadth, the true lateral borders are but slightly convergent posteriorly. In the adult male the fine raised line that bounds the dorsal plane on each side is distinct as such only in the neighbourhood of the antero-lateral angles, but in the female it runs much further backwards.

The breadth of the front, measured between the bases of the eyestalks, is about a tenth the greatest breadth of the carapace.

Orbits much oblique, both borders sinuous, the lower border elegantly denticulated throughout.

In the large cheliped of the adult male the upper border of the arm is fairly prominent and the inner border ends in a sharp tooth, quite independent of the constricted-off terminal lobule ; the wrist is quite smooth to the naked eye, and has the inner angle sharp but not spiniform ; and the hand is about 2½ times the greatest length of the carapace.

In the hand of this cheliped the palm is, to the naked eye, frosted with very fine granules, some of which in the neighbourhood of a scar near the base of the immobile finger are visible to the naked eye ; its upper border is not, and its lower border is but obscurely, defined ; and the two oblique crests on its inner surface are mere swellings, often quite faint, and never strongly salient. The fingers are neither broad nor particularly thin : the dactylus, which is about 1⅔ times the length of the upper border of the palm, tapers and is somewhat hooked at tip ; the immobile finger commonly has two teeth a little enlarged, the second one being near the tip and sometimes giving the tip a somewhat notched (but not truncated) appearance.

The merus of the last pair of legs is not at all foliaceous.

In the Indian Museum are 29 specimens from the Andamans : the carapace of a large one is 17 millim. long and 26 millim. broad.

The " Challenger " specimens referred by Miers to this species have a broad front and are identical with specimens from Hongkong that I take to be *G. splendidus.*

53. *Gelasimus Marionis*, Desm.

Gelasimus Marionis, Desmarest, Consid. Gen. Crust., p. 124, pl. xiii. fig. 1, and Dict. Sci. Nat. XXVIII. 1823, p. 243: Milne Edwards, Ann. Sci. Nat., Zool., (3) XVIII. 1852, p. 145, pl. iii. fig. 5 (*nec* Hist. Nat. Crust. II. 53): de Man, Notes Leyden Mus. II. 1880, p. 67: Miers, Ann. Mag. Nat. Hist. (5) V. 1880, p. 308 : Kingsley, Proc. Ac. Nat. Sci. Philad. XXXII. 1880, p. 141, pl. ix. fig. 8.

Gelasimus cultrimanus, White, P. Z. S. 1847, p. 205, Ann. Mag. Nat. Hist. XX: 1847, p. 205, and Samarang Crust. p. 49 (*apud* Miers *loc. cit. supra*).

Gelasimus cultrimanus var. *Marionis*, Ortmann, Zool. Jahrb. Syst. VII. 1893-94, pp. 750, 754.

Length of the carapace about two-thirds of the greatest breadth, which is at the claw-like antero-lateral angles.

Carapace little convex, all its regions very well defined, the posterior border of its dorsum in the adult male is half its greatest breadth and the true lateral borders are moderately convergent posteriorly: the fine raised line that in some other species defines the greater part of the dorsal plane is here, in the adult male, confined to the neighbourhood of the antero-lateral angles.

The breadth of the front between the bases of the eyestalks is not a fifteenth the greatest breadth of the carapace.

Orbits not very oblique nor very sinuous ; the lower border, which is nearly straight, is elegantly crenulate throughout.

In the large cheliped of the adult male the upper border of the arm is prominent and the inner border ends in a sharp tooth, independent of the terminal constricted-off lobule; the upper surface of the wrist is granular, and the inner border of the wrist has a denticle or spinule at its angle; and the hand (fingers included) is about three times the length of the carapace.

This large hand has a curious twist: its palm is compressed and has the upper and lower margins well defined, the outer surface covered with large granules, and the two granular crests on its inner surface fairly prominent: its fingers are broad thin and laminar ; the dactylus, which may be four times as long as the upper border of the palm, is shaped like a knife-blade; and in the immobile finger, which has a groove or line of pits along its outer surface, the dentary edge has a simple S-shaped curve.

The merus of the last pair of legs is not at all foliaceous.

In the Indian Museum are 9 specimens from the Andamans. The carapace of a large specimen is 18·5 millim. long and 26·5 millim. broad.

54. *Gelasimus Marionis* var. *nitidus*, Dana.

Gelasimus vocans, Milne Edwards, Ann. Sci. Nat., Zool., (3) XVIII. 1852, p. 145, pl. iii. fig. 4 (*nec* Hist. Nat. Crust. II. 54): Stimpson, Proc. Ac. Nat. Sci. Philad. 1858, p 99: Heller, Novara Crust. p. 37: Hilgendorf, in v. d. Decken's Reis. Ost-Afr., p. 83: A. Milne Edwards, Nouv. Archiv. du Mus. IX. 1873, p. 272: Hoffmann, in Pollen and Van Dam's Faun. Madagasc. Crust. p. 16: Miers, Phil. Trans. Roy. Soc. Vol. 168, 1879, p. 488, and Ann. Mag. Nat. Hist. (5) V. 1880, p. 308, and Challenger Brachyura, p. 242: Richters, in Mobius, Meeresf. Maurit. p. 155: de Man, Notes Leyden Mus. II. 1880, p 67, and XIII. 1891, p. 23, pl ii. fig. 5, and Archiv f. Naturges. LIII. 1887, i. p. 352, and in Weber's Zool. Ergebn. Niederl. Ost-Ind. II. 1892, p. 305, and Zool. Jahrb. Syst. VIII. 1894-95, p. 572: Haswell, Cat. Austral. Crust. p. 92.

Gelasimus nitidus, Dana, U. S. Expl. Exp. Crust. pt. I. p. 316, pl. xix. figs. 5a-d: Milne Edwards, Ann. Sci. Nat., Zool., (3) XVIII. 1852, p. 147: Thallwitz, Abh. Mus. Dresden, 1890-91, p. 42.

Gelasimus cultrimanus, Kingsley, Proc. Ac. Nat. Sci. Philad. 1880, p. 140, pl. ix. fig. 7: Ortmann, Zool. Jahrb., Syst., VII. 1893-94, pp. 750-753, and Jena Denk. VIII. 1894, p. 56.

Uca cultrimana, Ortmann, Zool. Jahrb., Syst. X. 1897-98, p. 348.

Differs from *G. Marionis* only in the form of the large hand of the adult male : this member, in var. *nitidus,*

(1) is not much over $2\frac{1}{2}$ times the length of the carapace, its dactylus being but little more than twice the length of the upper border of the palm :

(2) it has the two oblique granular ridges on the inner surface of the palm remarkably salient :

(3) it has the dentary edge of the immobile finger thrown into a characteristic W-shaped curve owing to the strong projection of two large triangular lobes, and

(4) it has the dactylus somewhat hooked at tip.

In the Indian Museum are 103 specimens, chiefly from the Andamans and Nicobars, but also from the Coromandel and Malabar coasts. The length of the carapace in large specimens is 14 millim., the breadth 21 millim.

55. *Gelasimus acutus*, Stimpson, de Man.

Gelasimus acutus, Stimpson, Proc. Ac. Nat. Sci. Philad. 1858, p. 99: Tozzetti, Magenta Crust. p. 107: Kingsley, Proc. Ac. Nat. Sci. Philad. 1880, p. 144: de Man, Journ. Linn. Soc., Zool. XXII. 1887-88, p. 113, pl. vii. figs. 8-9, pl. viii. figs. 1-4, and Notes Leyden Mus. XIII. 1891, p. 21, and in Weber's Zool. Ergebn. Niederl. Ost-Ind. II. 1892, p. 306, and Zool. Jahrb. Syst. VIII. 1894-95, p. 573: Ortmann, Zool. Jahrb., Syst. 1893-94, p. 750.

Uca acuta, Doflein, SB. Ak. Münch. XXIX. 1899, p. 193.

Length of the carapace about three-fifths the greatest breadth, which is at the acute wing-like antero-lateral angles.

Carapace strongly convex fore and aft, the regions moderately well defined : its lateral borders are strongly convergent, and still more so are the lateral borders of the dorsal plane, which are defined in more than two-thirds of their extent by a fine raised line : the posterior border of the dorsal plane is contained from $2\frac{1}{2}$ to $2\frac{4}{5}$ times in the geatest breadth.

Front, measured between the eye-stalks, about a twelfth the greatest breadth of the carapace, its moulded and bevelled edges do not together take up half its breadth.

Orbits moderately oblique, both upper and lower borders much sinuous ; the lower border finely, the upper border still more finely and more distantly crenulate.

In the large cheliped of the adult male all three borders of the arm are well defined, the inner and the lower borders being crenulated, but the inner border having no tooth independent of the terminal constricted-off lobule; the upper surface of the wrist and the outer surface of the palm are closely covered with vesiculous granules; and the hand (fingers included) may be $3\frac{1}{2}$ times the length of the carapace.

In this large hand the upper and lower borders of the palm are well defined, and of the two oblique granular crests on the inner surface of the palm the upper one that runs to the dentary edge of the immobile finger is short and indistinct : the fingers are not particularly broad or thin, and however the teeth may be disposed, there is always one near the end of each finger that is enlarged so as to give the ends of the fingers, when apposed, a sort of tongs-like or forceps-like grip : the dactylus is from 2 to nearly $2\frac{2}{3}$ times the length of the upper border of the palm.

The merus of the last pair of legs is distinctly foliaceous.

In the Indian Museum are 92 specimens chiefly from the Sunderbunds and Mergui, but also from Karachi and the Andamans. In a large specimen the carapace is 14 millim. long and 25 broad.

56. *Gelasimus Dussumieri*, Edw.

Gelasimus Dussumieri, Milne Edwards, Ann. Sci. Nat. Zool. (3) XVIII. 1852, p. 148, pl. iv. fig. 12 : A. Milne Edwards, Nouv. Archiv. du Mus. IV. 1868, p. 71, and IX. 1873, p. 274 : Hoffmann in Pollen and van Dam's Faun. Madag. Crust. p. 17, pl. iii. figs. 19-22 : Kingsley, Proc. Ac. Nat. Sci. Philad. 1880, p. 145, pl. x. fig. 16 : de Man, Notes Leyden Mus. II. 1880, p. 68, and XIII. 1891, pp. 20, 26, and Journ. Linn. Soc. Zool. XXII. 1887-88, p. 108, pl. vii. figs. 2-7, and in Weber's Zool. Ergebn. Niederl. Ost-Ind. II. 1892, p. 306, and Zool. Jahrb., Syst., VIII. 1894-95, p. 576 : Lenz and Richters, Abh. Senck. Nat. Ges. Frankf. XII. 1881, p. 423 : Haswell, Cat. Austral. Crust. p. 93 : Miers, Zool. H. M. S. Alert, pp. 518, 541 : Ortmann, Zool. Jahrb. Syst. VII. 1893-94, pp. 750, 755.

Gelasimus longidigitum, Kingsley, Proc. Ac. Nat. Sci. Philad. 1880, p. 144, pl. ix. figs. 10, 13 (*fide* Ortmann *l. c. infra*).

Uca Dussumieri, Ortmann, Zool. Jahrb , Syst , X. 1897-98, p. 348 : Nobili, Ann. Mus. Genov. (2) XX. 1899, p. 273 : Doflein, SB. Ak. Münch. XXIX. 1899, p. 193.

Closely related to *G. acutus,* from which it can be distinguished by the following characters when fully adult males are compared :—

(1) the regions of the carapace are much more strongly defined, and the raised lines that bound the dorsal plane of the carapace on each side are more curved, less rapidly convergent, and less distinct in their posterior part, which gives the carapace a much less posteriorly-contracted look ; and the orbits are less oblique :

(2) the front, measured between the bases of the eyestalks, is about a fifteenth the greatest breadth of the carapace, and its moulded and bevelled edges together take up more than two-thirds of its breadth :

(3) in the large cheliped the arm is longer and more slender, both the oblique granular ridges on the inner surface of the palm are very strongly defined, and the fingers may be fully 3 times the length of the upper border of the palm :

(4) these large fingers are broader and thinner, their tips are somewhat hooked and have no enlarged tooth near them, but near the middle of the immobile finger there is a enlarged tooth or triangular lobe :

(5) the merus of the last pair of legs, though it is compressed and somewhat broadened, is not a short foliaceous joint.

In the Indian Museum are 52 specimens, from Mergui, Andamans and Nicobars, and Bimlipatam.

57. *Gelasimus Urvillei,* Edw.

Gelasimus Urvillei, Milne Edwards, Ann. Sci. Nat., Zool., (3) XVIII. 1852, p. 148, pl. iii. fig. 10 : Kingsley, Proc. Ac. Nat. Sci. Philad. 1880, p. 145, pl. ix. fig. 15 : de Man, Notes Leyden Mus. XIII. 1891, pp. 21, 34 : Ortmann, Zool. Jahrb., Syst., VII. 1893-94, p. 750.

Gelasimus Dussumieri, Hilgendorf (*nec* Edw.), in v. d. Decken's Reis. Ost-Afr. Crust. p. 84, pl. iv. fig. 1.

This species closely resembles *G. acutus* and *G. Dussumieri,* but is distinguished from both by the presence of a raised row of granules behind and parallel with the middle third of the lower border of the orbit—*i.e.,* just inside the orbital cavity.

As in *G. acutus,* the fine raised lines that define the dorsal plane of the carapace laterally are distinct throughout and rapidly convergent, which gives the carapace a look of breadth in front and of unusual narrowness behind ; and, as in *G. acutus,* the meropodites of the last pair of legs are, even in the male, decidedly shortened and foliaceous joints.

On the other hand the front is, as in *G. Dussumieri,* extremely narrow, and its bevelled and moulded edges take up most of its breadth between the eye-stalks. The regions of the carapace, also, are as strongly defined as they are in *G. Dussumieri.*

The large hand of the male resembles that of *G. Dussumieri* in having both the oblique granular ridges on the inner surface of the palm strongly salient, and in having very long fingers with simple hooked tips: the fingers however are not so broad and thin, and the lobe near the middle of the dentary edge of the immobile finger may be present or not.

In the Indian Museum are 10 specimens, from Karachi, Madras, and the Nicobars.

The carapace of the largest specimen is 20 millim. long and 36 millim. broad.

Subfamily SCOPIMERINÆ.

DOTILLA, De Haan, Stimpson.

Doto, De Haan, Faun. Japon. Crust. p. 24 (1835) *nom. præoc.* : Milne Edwards, Hist. Nat. Crust. II. 38, and Ann. Sci. Nat., Zool., (3) XVIII. 1852, p. 152.
Dotilla, Stimpson, Proc. Ac. Nat. Sci. Philad. 1858, p. 98.

Cephalothorax so deep as to be subcubical, as long as broad or a little broader than long. Anteriorly the sidewalls of the carapace have a curious gyrous-sulcate sculpture resembling brain-convolutions: often also a similar kind of sculpture is found on the dorsum of the carapace and on the meropodites of the external maxillipeds.

Front a narrow deflexed lobe much as in *Ocypoda.* The orbits, which occupy all the rest of the anterior border of the carapace, are more or less oblique and shallow—in one species so shallow as to be almost obsolete. Eyestalks rather long and slender, with the eyes at the end.

Antennules, like those of *Ocypoda,* having the basal joint of good size, and the flagellum small and hidden by the front. The antennæ stand at the inner angle of the lower orbital border and have a rather short flagellum.

The epistome would be linear but for a large median triangular lobe that projects between the external maxillipeds.

Buccal cavern enormous, suboval or subcircular in outline: the external maxillipeds, which completely cover it and are also very large, have a strong almost hemispherical bulge ; their merus is much larger than the ischium and carries the flagellum at the antero-external angle: the exognath is extremely slender and inconspicuous.

Chelipeds equal, stouter than the legs : fingers usually slender and a little deflexed, usually without conspicuous teeth.

Legs not much differing in length, which is moderate : their meri (as also those of the chelipeds) have on the upper surface a curious membranous area or "tympanum." Similar "tympana" may also be present on some of the segments of the sternum.

The abdomen in the male consists of 7 separate segments, and though narrow is nowhere linear or compressed : the distal end of the fourth segment is thickly fringed with bristles, and overlaps and partly conceals the fifth tergum. In the female, according to De Haan, the abdomen consists of 5 separate segments.

Distribution : Tropical shores and mud-flats, from East Africa and the Red Sea eastwards to Japan. Found in the same situations as *Gelasimus* and *Ocypoda.*

Key to the Indian species of Dotilla.

I. Carapace broader than long : chelipeds not much longer than the carapace, and not much differing from the legs in point of length : no " tympana " on the sternum :—

    1. Meropodites of legs not dilated : fingers of chelæ slender, without any conspicuous teeth :—

        i. Whole surface of merus of external maxillipeds gyrous-sulcate: fingers not longer than palm.............................. *D. affinis.*

        ii. Only the outer-half of the merus of the external maxillipeds is gyrous-sulcate :—

            a. Fingers slightly longer than the palm............................... *D. Blanfordi.*

            b. Fingers more than twice as long as the palm .... *D. intermedia.*

    2. Meropodites of the legs dilated :—

        i. Fingers of chelæ without any conspicuous tooth : dactyli of the legs, even of the last pair, shorter than the propodites...... *D. brevitarsis.*

        ii. A large tooth on each finger of the chelæ, arranged so that when the tips of the fingers are closed these large teeth meet, and an hour-glass-shaped space is left between the closed fingers : dactyli of the legs longer than the propodites............... *D. clepsydrodactylus.*

II. Carapace at least as long as broad : chelipeds 3 or 4 times as long as the carapace, and much longer than the legs : " tympana " present on the sternum......... *D. myctiroides.*

58. *Dotilla affinis,* n. sp.

Differs from *D. sulcata,* with specimens of which, from the Red Sea, I have compared it, only in the following characters :—

(1) there is no spine on the under surface of the arm, (2) the fingers are not so long as the palm, (3) there is a small tympanum on the dorsal surface of the merus of the last pair of legs, whereas in *D. sulcata* only the tympanum on the ventral surface is present.

The carapace behind the gastric and inside the branchial regions, forms a smooth semicircular facet, but all its anterior and lateral regions have a curiously convoluted sculpture, the convexities of the convolutions being finely granular.

The grooves that define these convolutions form, when viewed as a whole, a sort of five-rayed star, the anterior ray (which runs up between the eyes on to the front) being the shortest, the antero-lateral rays (which run towards the outer angles of the orbit) being a little longer, and the postero-lateral rays (which really are triple) being the longest of all.

The pterygostomian regions and neighbouring part of the side-walls of the carapace, and the meropodites of the external maxillipeds have the same curious convoluted sculpture. The orbits are shallow but are perfectly defined.

The merus of the external maxillipeds is more than twice the size of the ischium.

Chelipeds (measured round their curve) not twice the length of the carapace : no spine on any of their segments : fingers not so long as the palm.

Legs slightly longer than the chelipeds, their meropodites not at all broadened but all having a " tympanum " : except in the case of the last pair of legs—in which the dactylus is remarkably long—the dactyli are rather shorter than the propodites.

No tympana on the sternum.

In the Indian Museum are 4 specimens from Aden and the Baluchistan coast. The carapace of the largest is 5·3 millim. long and 7·3 millim. broad.

59. *Dotilla intermedia,* de Man.

Dotilla intermedia, de Man, Journ. Linn. Soc., Zool., XXII. 1887-88, p. 135, pl. ix. figs. 4-6 (1888).

Carapace sculptured in much the same way as in *D. affinis,* only the grooves are not so deep and distinct, and there is an additional groove running parallel with the posterior margin.

The merus of the external maxillipeds is not twice as large as the ischium, and the sculpturing consists of a single loop parallel with the outer border of the merus; the inner half of that joint being quite smooth.

Fingers more than twice as long as the palm. In the last pair of legs the dactylus is about twice as long as the propodite : in all the other legs the dactyli are very little longer than the propodites.

In other respects this species agrees with *D. affinis.*

In the Indian Museum are 15 specimens from Mergui. The carapace is 4 millim. long and a little over 4 millim. broad.

60. *Dotilla Blanfordi*, n. sp.

The whole of the dorsal surface of the carapace is areolated and grooved (the areolæ being finely granular and the grooves smooth) as follows :—

A very distinct groove runs parallel with either lateral border, and a scarcely less distinct one runs parallel with the posterior border, and in the space bounded by these grooves a six-rayed star of grooves of nearly equal length can be made out. This " star " is formed by a groove running fore and aft down the middle of the carapace and having, on either side of it, a semicircular chord joining the outer angle of the orbit with a point near the postero-lateral angle of the carapace. The intersection of these grooves cuts the post-gastric sub-region into 4 symmetrical tubercles.

The whole side-wall of the carapace is finely granular, and the sub-hepatic and pterygostomian regions have the characteristic convoluted sculpture. The orbits are shallow but are perfect.

The external maxillipeds are finely granular : the merus is twice as big as the ischium, and its sculpture consists of a single loop parallel with the outer border and a single groove parallel with the inner border.

Chelipeds as in *D. affinis*, except that the fingers are a little longer than the palm.

Legs as in *D. affinis*, the meropodites being slender and all having a " tympanum," but in the last pair the dactylus is about twice as long as the propodite, and in the other pairs the dactyli are very slightly longer than the propodites. No sternal tympana.

In the Indian Museum are 4 specimens from the coast of Sind and Baluchistan. The carapace of the type is a little over 5 millim. long and not quite 7 millim. broad. Collected by Mr. W. T. Blanford, F.R.S.

61.   *Dotilla clepsydrodactylus*, n. sp.

Near *D. Wichmanni*, de Man.

The sculpture of the dorsum of the carapace is like that of *D. Blanfordi,* only the grooves are much deeper cut and the groove between the post-gastric region and the postero-lateral angle of the carapace is double: the sculpture of the sidewall of the carapace is like that of *D. Blanfordi.*

In the external maxillipeds the merus is not twice as big as the ischium, and its sculpture consists of a single, simple convolution parallel with the outer border, the inner half of its surface being quite smooth—as is *D. intermedia.*

The orbits are shallow but are quite perfect.

The chelipeds, measured all round their curve, are not twice the length of the carapace and have no spine on the arm. The fingers are much longer than the palm : *in the adult male they are extremely slender, and each has a large tooth arranged so that when the tips of the fingers are closely apposed these two teeth meet and leave an hour-glass-shaped space between the closed fingers.*

Legs a little longer than the chelipeds ; their meropodites are slightly but distinctly dilated and all have a *tympanum :* their dactyli are all longer than their propodites, and in the last pair the dactylus is very long, slender, straight, and fluted. No sternal tympana.

Colours, speckled like the sand in which they live.

In the Indian Museum are eight specimens from False Point on the sea face of the Mahanaddi Delta. The carapace of the largest is 5 millim. long and 6 millim. broad.

62.   *Dotilla brevitarsis*, de Man.

: *Dotilla brevitarsis*, de Man, Journ. Linn. Soc., Zool., XXII. 1887-88, p. 130, pl. ix. figs. 1-3 (1888).

The whole carapace is grooved and areolated (but the sculpture is not very deep) as follows :—

A strong groove runs fore and aft down the middle of the carapace, another runs parallel with the posterior border, and on each side another takes a sinuous course along each lateral border : other short and rather indefinite grooves join the median and lateral grooves.

The sublhepatic and pterygostomian regions have the usual convoluted sculpture. Orbits shallow, but distinct.

The merus of the external maxillipeds is much larger than the ischium : its whole surface is sculptured, the sculpture taking the form of a W-shaped convolution.

J. II. 48

Chelipeds short, without any spine on the arm : palm short, high, and compressed, with sharp edges, traversed by a fine raised line near and parallel with the lower border : fingers thin and compressed, about as long as the palm, the upper edge of the dactylus—like that of the palm—fringed with hair.

Legs a little longer than the chelipeds, the meropodites—especially of the first 3 pairs—*much broadened and compressed,* all having a tympanum. *The dactyli, even of the last pair of legs, are shorter than the propodites.*

No tympana on the sternum.

In the Indian Museum are fragments of 3 specimens from Mergui : de Man states that the breadth of the cephalothorax of the largest specimen is nearly 10 millim.

63. *Dotilla myctiroides,* Edw.

Doto myctiroides, Milne Edwards, Ann. Sci. Nat. Zool. (3) XVIII. 1852, pl. iv, fig. 24.

Dotilla myctiroides, Stimpson, Proc. Ac. Nat. Sci. Philad. 1858, p. 98 : A. O. Walker, Journ. Linn. Soc. Zool. XX. p. 111 : Aurivillius, Zur Biologie amphibischer Dekapoden, p. 5, pl. i. figs. 1–13, pl. iii. fig. 13, (Mitg. Ges. Wiss. Upsala, 1893).

Scopimera myctiroides, Henderson, Trans. Linn. Soc., Zool., (2) V. 1893, p. 390.

Carapace about as long as, or slightly longer than, broad, little sculptured dorsally, though its antero-lateral parts are studded with vesiculous granules. Front grooved : a groove runs parallel with either lateral border, and a faint groove crosses either postero-lateral angle. The side-walls anteriorly have the usual "brain-convolution" sculpture.

Orbits very oblique and very shallow, almost obsolete.

The merus of the external maxillipeds is nearly twice as big as the ischium and is finely granular; a single faint groove, most distinct anteriorly, runs parallel with its outer border.

Chelipeds between three and four times the length of the carapace, all the joints long, slender, and unarmed : fingers longer than the palm, without any conspicuous teeth.

Legs long, but much shorter than the chelipeds : the meropodites strongly dilated, and with a large "tympanum" : the dactylus of the last pair is longer than the propodite, but in the other three pairs it is a little shorter than the propodite.

On either side of each of the last four thoracic sterna is a large tympanum.

In the Indian Museum are 19 specimens from the Andamans and 11 from the Coromandel coast. The carapace is 10 millim. long.

SCOPIMERA, De Haan.

Scopimera, De Haan, Faun. Japon. Crust., p. 24 (1835): Milne Edwards, Ann. Sci. Nat., Zool., (3) XVIII. 1852, p. 153.

Scopimera has the same deep " cubical " carapace and the same general facies as *Dotilla*, but differs in the following characters :—

The carapace is much broader than long and has none of the curious sculpture, resembling brain convolutions, that is found, at any rate on the sidewalls, in *Dotilla* : the external maxillipeds are unsculptured and their merus, though large, is smaller than their ischium : the abdomen of the male has a curious wasp-like form owing to the length and narrowness of its fifth segment, which segment may even become elongate-linear by constriction ; it has no bristles either on the 4th tergum or elsewhere : in the female the abdomen consists of 7 separate segments.

Distribution : Indo-Pacific shores, from Karachi to Japan.

Key to the Indian species of Scopimera.

I. Chelipeds and legs with a reticulate or subsquamiform granulation, the chelipeds in the male about twice the length of the carapace: most of the tympana on the legs are traversed by a longitudinal ridge : fifth abdominal tergum of male long and narrow, but not linear... *S. investigatoris.*

II. Chelipeds and legs finely and uniformly granular, the chelipeds in the male nearer 3 times than twice the length of the carapace : the tympana not subdivided by a ridge : the fifth abdominal tergum of the male is long and linear .... *S. crabricauda.*

According to F. Müller, *S. globosa*, De Haan, is found in Indian waters. The form of the abdomen in this species is similar to that of *S. investigatoris*, but the carapace is smooth, and the tympana of the legs are different.

64.   *Scopimera investigatoris*, n. sp.

Carapace much broader than long, decidedly pentagonal, without distinction of regions, smooth except anteriorly and laterally where there are numerous irregularly-scattered granules : the sidewalls and pterygostomian regions finely granular.

Orbits broad as in *Ocypoda*, shallow, the upper border very oblique, the lower border finely denticulated and very prominent as in *Gelasimus*.

External maxillipeds with some obsolescent granulation. Chelipeds and legs finely granular in a somewhat reticulate or subsquamiform way.

Chelipeds about twice as long as the carapace: tympanum on the inner surface of the arm large, that on the outer surface of the arm small: fingers about as long as the palm, without any enlarged teeth.

First 3 pairs of legs about the same length as the chelipeds, the 4th pair shorter: the merus of all much dilated and with large well-defined tympana, all of which, except only the one on the dorsal surface of the last pair, are longitudinally subdivided by a fine ridge: the dactylus in the first 3 pairs is about the same length as the propodite, but in the last pair is considerably longer.

In the male abdomen the first 2 segments are horizontal-linear, the 3rd and 4th, though distinct, form a " butterfly " plate, the 5th is long and narrow and longitudinally grooved and gradually expands to meet the 6th, which is long and broad, while the 7th is transversely oval.

In the female the abdomen is of the usual shape, but in its broadest part is little more than half the breadth of the sternum.

In the Indian Museum are 11 specimens, from Diamond Island off C. Negrais in Burma. The carapace of the largest male is 4·5 millim. long and 7 millim. broad.

65. *Scopimera crabricauda,* n. sp.

Carapace subpentagonal, the regions indistinctly indicated, the surface of the mid-dorsal region is symmetrically puckered or vesiculous ; the sidewalls and pterygostomian regions granular.

Orbits moderately broad and deep, the upper border oblique, the lower border prominent and finely denticulate.

External maxillipeds smooth: chelipeds and legs "frosted" under the lens.

In the male the chelipeds are more than $2\frac{1}{2}$ times the length of the carapace and are longer and much stouter than the legs : there is a large tympanum on the inner surface of the arm, and a very small one on the outer surface: the dactylus is a little shorter than the palm and has one large tooth. In the female the chelipeds are shorter and not much stouter than the legs : the fingers are shorter than the palm, and the dactylus has no large tooth.

The meropodites of the legs are much dilated: all have tympana but these are not subdivided by any ridge: in the first 3 pairs of legs the dactyli are a little longer, in the fourth pair considerably longer, than the propodites.

In the male abdomen the first 2 segments are linear-horizontal and concealed, the 3rd and 4th form a triangular plate deeply grooved down the middle line, the 5th is long linear and grooved, the 6th and 7th, though separate, together form a racket-head.

In the female the abdomen is of normal shape.

In the Indian Museum are a male and female from Karachi. The carapace of the male is 6·5 millim. long and barely 10 millim. broad.

TYMPANOMERUS, de Man, Rathbun.

Diozippe, de Man, Journ. Linn. Soc., Zool., XXII. 1887–88, p. 137 (1888): *nom. præocc.*

Tympanomerus, Rathbun, Proc. Biol. Soc., Washington, XI. 1897, p. 164.

Carapace deep, quadrilateral, broader than long, the regions not defined. Front narrow, deflexed : the orbits are trenches occupying the whole anterior border of the carapace between the front and the antero-lateral angles.

Eyes, antennules, antennæ and epistome as in *Dotilla*. Buccal cavern large, a little narrowed and rounded anteriorly : the external maxillipeds completely close the buccal cavern, the anterior outer corner of the ischium is marked off as a distinct facet as in *Dotilla* and *Scopimera*, the merus is much larger than the ischium, the palp arises near the antero-external angle of the merus, and the exognath is small and linear.

Chelipeds in both sexes stouter, and in the male longer, than the legs : fingers a little deflexed.

Legs rather compressed, the two middle pairs a little longer than the first and last pair : there are ill-defined tympana on the meropodites.

The abdomen in both sexes consists of separate segments, and in the male is narrow.

Distribution : Japanese and Andaman Seas.

The name *Tympanomerus* is a most unfortunate one, since the " tympana," compared with those of *Dotilla* and *Scopimera*, are ill-defined and inconspicuous.

66. *Tympanomerus orientalis* (de Man).

Diozippe orientalis, de Man, Journ. Linn. Soc. Zool. XXII. 1887–88, p. 138, pl. ix. figs. 8–10.

Carapace square-cut, the length about four-fifths of the greatest breadth, dorsally nearly flat with the lateral borders well defined especially anteriorly, the surface a little lumpy in places : a perfectly straight fine transverse ridge runs close to and parallel with the posterior border.

Front grooved dorsally, hardly a fourth the breadth of the carapace. The outer angle of the lower border of the orbit forms a prominent tooth. The merus of the external maxillipeds is grooved along the outer border.

Chelipeds in the male nearly three times the length of the carapace: wrist elongate, somewhat cuboid, with a strong laterally-compressed lobe or tooth at its inner angle : palm rather high, both borders marginate and a second fine ridge runs close to and parallel with the lower border: fingers a little shorter than the palm, finely denticulate.

·In the female the chelipeds are not twice the length of the carapace, the wrist is not elongate, though the tooth at its inner angle is present, and the fingers are a little longer than the palm.

The meropodites of the legs are slightly dilated, the dactyli are shorter than the propodites, and the carpopodites and propodites of the first two pairs are densely tomentose.

The fifth abdominal tergum of the male, though not particularly elongate, is a little constricted at base.

In the Indian Museum are 6 specimens from Mergui. The carapace of the largest is 4 millim. long and 5 millim. broad.

Subfamily MACROPHTHALMINÆ, Dana.

CLISTOSTOMA, De Haan restr.

Cleistostoma (= *dilata* nec *pusilla*) De Haan, Faun. Japon. Crust. p. 26 : Milne Edwards, Ann. Sci. Nat., Zool., (3) XVIII. 1852, p. 160.

Carapace of no great depth, broader than long, its sides slightly arched, its regions ill-defined.

Front of moderate breadth, more than a fourth the greatest breadth of the carapace, declivous : orbits well defined, of good depth, occupying all the rest of the anterior border of the carapace : eyestalks stout, eyes terminal. The antennules fold obliquely: the antennæ are small and stand in the inner orbital hiatus.

Epistome well defined, very short fore and aft, with a prominent lobe or tooth in the middle line projecting between the external maxillipeds.

Buccal cavern squarish, but with the sides a little arched, completely closed by the external maxillipeds. These are large, and have the inner angle of the ischium strongly produced, the merus as large as or larger than the ischium, and the palp articulating at the antero-external angle of the merus : the carpus is ovate, but the two terminal joints are very short and slender : the exognath is in great part concealed.

Chelipeds in the female shorter and slenderer than any of the legs, in form exactly like those of the female of *Gelasimus.*

Of the legs the first two pairs are the shortest and slenderest, while the middle two pairs are much the largest and have very broad mero-podites.   There are no " tympana."

The abdomen of the female consists of 7 separate segments, and is very broad.

67.   *Clistostoma dotilliforme*, n. sp.

Carapace rather depressed, slightly convex, smooth, with the regions ill-defined ; its lateral borders are slightly arched and are finely serrated anteriorly behind the acute, almost dentiform, antero-lateral angles.   Front between a third and a fourth the greatest breadth of the carapace, concave in the middle line.   Upper border of the orbit sinuous, lower border prominent and finely serrated.

Merus of the external maxillipeds larger than the ischium, sculptured (somewhat as in the *Dotillæ*) with a sort of Y-shaped sulcus starting from the antero-external angle.   The pterygostomian regions also are sculptured with branching or convoluted grooves much as in the *Dotillæ*.

The second and third pair of legs, which are much longer than the other two pair, are a little over $1\frac{1}{2}$ times the length of the carapace and have an almost foliaceous meropodite with the anterior border finely serrulate and the posterior border elegantly spinate : the anterior border of the carpus and propodite of the second and third pair of legs is tomentose.

A single egg-laden female is in the Indian Museum : it was found at Karachi, and its carapace is 7 millim. long and 9 millim. broad.

TYLODIPLAX, de Man.

Tylodiplax, de Man, Zool. Jahrb., Syst. VIII. 1894-95, p. 598 (1895).

Carapace deepish, quite flat dorsally, broader than long and broader behind than in front, the lateral borders being posteriorly divergent and having a distinctly convex curve, the regions more or less defined.

Front between a third and a fourth the greatest breadth of the carapace, not deflexed, grooved longitudinally.   The orbits occupy the rest of the anterior border of the carapace, but as the extent of this border is a good deal less than the greatest breadth of the carapace, and as the front is broad, the orbits have not the same elongate form as they have in most species of *Macrophthalmus*, though otherwise similar. Antennules and antennæ as in *Macrophthalmus*. *Eyes small, not terminal on the eyestalks.*

The epistome would be linear, were it not for a septum-like fold or lobe that projects strongly between the meropodites of the external maxillipeds : owing to this fold the anterior edge of the buccal cavern has a bilobed appearance. The external maxillipeds completely close the buccal cavern : their merus is at least as long as and decidedly broader than their ischium : the flagellum, which is slender, is articulated near, but not at, the antero-external angle of the merus : the exognath is not much concealed, though not completely exposed.

The chelipeds in the adult male are unknown : in the young male they are equal and are shorter and slenderer than the legs, except perhaps the very small 4th pair.

The legs have somewhat the same relations as in *Macrophthalmus*—i.e., the first and last pairs are much the shorter and the two middle pairs are much the longer and stouter.

The abdomen in the female is unknown : in the male it is narrow, and consists of 5 separate joints, the 3rd 4th and 5th segments being fused, but without obliteration of sutures.

It seems to me of very doubtful utility to separate this form from *Paraclistostoma*, de Man, or either of them from *Clistostoma*, De Haan (as restricted by de Man).

68. *Tylodiplax indica*, n. sp.

Two young males from Karachi are in the Indian Museum : their chelipeds are still of the female *Macrophthalmus* type; so that it is impossible to give a complete diagnosis of the species.

Carapace more or less hairy, finely punctate, its length less than two-thirds its greatest breadth which, owing to the strong divergence, from before backwards, of the lateral borders, is posterior; its antero-lateral angle is an obtuse angle. The gastric region is defined by a perfectly circular line.

Front square-cut, laminar, but not projecting beyond the inner angles of the orbits, from which it is separated by a groove : the front is concave in the middle line.

The pigment of the eyes is small in amount, and is placed some distance behind the end of the eyestalks.

The merus of the external maxillipeds is longer and much broader than the ischium, and has its antero-external angle considerably dilated, and its surface somewhat granular.

The chelipeds of the immature male, and the legs, are hairy, much as in *Macrophthalmus depressus*, the hairs on the posterior border of the merus of the 2nd pair of legs and on the dorsal surface of carpus and propodite of the 2nd and 3rd pair of legs being particularly thickset. The length of the longest (second) pair of legs is $2\frac{1}{3}$ times that of the carapace, that of the last pair of legs is very little more than that of the carapace.

Two young males from Karachi: the carapace 6·5 millim. long and 11 millim. broad.

MACROPHTHALMUS, Latreille.

Macrophthalmus, Latreille, in Cuvier Règne An. (ed. 2) Vol. IV. p. 44 (1829): De Haan, Faun. Japon. Crust. p. 26: Milne Edwards, Hist. Nat. Crust. II. 63, and Ann. Sci. Nat., Zool. (3) XVIII. 1852, p. 155: Dana, U. S. Expl. Exp., Crust. pt. I. p. 312: Miers, Challenger Brachyura, p. 248.

Carapace depressed, quadrilateral, broader than (sometimes more than twice as broad as) long: the regions are well defined, the cervical and branchial grooves being characteristically conspicuous both on the dorsum of the carapace, and on the lateral border where they cut out two prominent teeth or lobes.

Front deflexed, narrow, often a narrow lobe as in *Gelasimus*: its free edge never approaches the epistome. The orbits are narrow trenches occupying the whole anterior border of the carapace between the front and the antero-lateral angles: eyestalks usually very long and slender, as in *Gelasimus*. The antennular flagella, which are rather small, fold transversely beneath, but are not concealed by, the front. The antennæ stand at the inner angle of the orbit: the basal joint is short, and the flagellum is of good length.

Epistome very short fore and aft, almost linear, but well delimited from the palate. Buccal cavern somewhat arched anteriorly. The external maxillipeds have a broad foliaceous ischium and merus (the latter about half the length of the former) and a coarse flagellum articulating with the antero-external angle of the merus: though the ischium and merus may not quite meet across the middle of the buccal cavern, the narrow interval that may exist between them is largely filled by the flagella, so that the underlying parts are concealed.

The chelipeds differ greatly in the sexes: in the female they are equal, and are shorter and slenderer than any of the legs except, perhaps, the short and weak last pair: in the *adult* male they are equal or subequal, and are longer and stouter than any of the legs except, perhaps, the particularly large and stout penultimate pair: in both sexes the fingers are curiously deflexed and bent or curved inwards distally.

Of the legs, the first and last pairs are usually singularly short and slender compared with the second and third pairs: the third pair are the longest and stoutest, being nearly or quite as large as the chelipeds,

J. II. 49

and the fourth (last) pair much the shortest and weakest of all. The dactylus in all is broad, stout, and laterally compressed.

The abdomen in both sexes consists of 7 separate segments, and in the male is narrower at base than the breadth of the sternum.

Key to the Indian species of Macrophthalmus.

I. Carapace much broader than long, its sides are distinctly convergent posteriorly and the antero-lateral angles are acute and spiniform: front narrow :—

   1. The eyestalks project *nearly half their length* beyond the antero-lateral angles of the carapace .... *M. Verreauxi.*

   2. The eyestalks project slightly beyond the antero-lateral angles of the carapace : the true first tooth of the lateral border of the carapace belongs to the upper border of the orbit, and the antero-lateral angle of the carapace is formed by the true second tooth....... ...... *M sulcatus.*

   3. The eyestalks do not project beyond the antero-lateral angles of the carapace :—

      i. Some of the borders of some of the leg joints are denticulate or spiny...... .... *M. pectinipes.*

      ii. Legs smooth, except for a small subterminal denticle on the anterior border of the meropodites .... *M. convexus.*

II. Carapace broader than long, its sides are parallel :—

   1. The tooth at the antero-lateral angle of the carapace is truncate and square-cut: front about an eighth the greatest breadth of the carapace : inner surface of the palm of the male smooth........................ .... *M. depressus.*

   2. Front about a fourth the greatest breadth of the carapace : inner surface of the palm of the male armed with a spine........ .... *M. erato.*

III. Carapace broader than long, its sides divergent posteriorly : two nearly parallel, obliquely longitudinal, finely beaded lines on the posterior part of each epibranchial region .... *M. tomentosus.*

Besides the fore-named, the four following species, of which I have not seen specimens, are said to occur in Indian Seas :—

(1) *M. simplicipes,* Guérin, Mag. de Zool. II. 1838, pl. xxiv. fig. 1 : it appears to differ from *M. pectinipes* in having no spines or denticles on the leg-joints.

(2) *M. carinimanus,* Milne Edwards, Hist. Nat. Crust. II. 65, and Ann. Sci. Nat. Zool. (3) XVIII. 1852, p. 156: it appears to differ from *M. convexus* only in having a spine on the inner surface of the palm of the male cheliped.

(3) *M. pacificus,* Dana, U. S. Expl. Exp., Crust. pt. I. p. 314, pl. xix. fig. 4: it appears to differ from *M. erato* only in not having a spine on the inner surface of the palm of the male cheliped.

M. bicarinatus, Heller, Novara Crust. p. 36, pl. iv. fig. 2, which I am unable from the descriptions to distinguish from *M. pacificus.*

69.   *Macrophthalmus Verreauxi*, Edw.

Macrophthalmus Verreauxi, Milne Edwards, Ann. Sci. Nat., Zool., (3) IX. 1848, p. 358, and XVIII. 1852, p. 155, pl. iv. fig. 25: Hess, Archiv f. Nat. XXXI. 1865, i. pp. 142, 171: de Man, Notes Leyden Mus. II. 1880, p. 184: Haswell, Cat. Austral. Crust. p. 89.

Carapace finely granular on the branchial regions, its length about two-thirds its greatest breadth, its sides slightly convergent posteriorly and cut anteriorly into 3 teeth, the first of which is the antero-lateral angle.

Front only very moderately deflexed, its least breadth (between the eyestalks) is about a fifth the greatest breadth of the carapace, very obscurely bilobed.

Orbits oblique, sinuous, their borders microscopically beaded.   *The eyestalks project nearly half their length beyond the antero-lateral angles of the carapace.*

The external maxillipeds, when the flagella are folded, completely occlude the buccal cavern : the suture between the merus and ischium is oblique.

The legs are darkly variegated or incompletely banded, and are unarmed except for a subterminal spine on the anterior border of the meropodites of the first 3 pairs.

The chelipeds in the young male are not as long as, though more massive than, the 2nd and 3rd pairs of legs.

In the Indian Museum are 4 specimens, more or less damaged, from the Andamans and Mergui (" Investigator " collection).   The largest male (which wants the chelipeds) has a carapace 9 millim. long and 14 millim. broad.

70.   *Macrophthalmus pectinipes*, Guérin.

Macrophthalmus pectinipes, Guérin, Voy. Favorite, p. 167, pl. 49 (1839), and Mag. de Zool. II. 1839, Crust. (Cl. VII.) pl. xxiii (1838) : Milne Edwards, Ann. Sci. Nat., Zool., (3) XVIII. 1852, p. 158 : Henderson, Trans. Linn. Soc., Zool., (2) V. 1893, p. 389 : Ortmann, Zool. Jahrb., Syst., X. 1897–98, p. 340.

Carapace studded with large conspicuous pearly granules, its length in the adult male is about six-elevenths of its greatest breadth at the level of the second tooth of the lateral border : the lateral borders are slightly but distinctly convergent posteriorly where they are beaded or denticulate, anteriorly they are cut into three acute teeth the last of which is minute, the first being the outer orbital angle.

The front, measured at its narrowest part between the eyestalks, is barely a sixteenth the greatest breadth of the carapace : its free edge is distinctly bilobed.   Orbits sinuous, a little oblique ; their upper border elegantly denticulate, the lower border unevenly crenulate. Eyestalks slender and curved: the eye does not reach to the end of the orbital trench.

When their flagella are folded the external maxillipeds completely occlude the buccal cavern : the suture between the ischium and merus is hardly at all oblique.

In the adult male the chelipeds are from $2\frac{1}{2}$ to 3 times the length of the carapace and longer than any of the legs except the 3rd (penultimate) pair: except the hand, their joints are not more massive than those of the 2nd and 3rd pair of legs. The arm is trigonal, its inner border being prominent and rising into a crest, on the most convex part of which is a short horny plate, called by de Man the " musical ridge " : this border of the arm, as also the inner border and angle of the wrist and the extreme proximal end of the upper border of the palm, is serrated. The palm is nearly as long as the arm and is perfectly smooth and unsculptured, it has a tuft of hair at its extreme distal end, continuous with a thick fringe of hair along the upper border of the dactylus : the dactylus is about two-thirds the greatest length of the palm and has a molariform tooth at its basal end, but there is no such tooth on the immobile finger: the fingers meet only at the distal inbent end.

In the female and young male the chelipeds are short and slender, a good deal fringed with hair, but unsculptured, and the fingers are longer than the palm.

In both sexes the legs are alike, the 2nd and 3rd pairs being remarkably long and strong and the 1st and 4th (last) pairs being short and comparatively slender. The 3rd pair, which are the longest of all; are from $2\frac{1}{2}$ to nearly 3 times the length of the carapace, the 4th pair are only about $1\frac{1}{2}$ times the length of the carapace. In all but the last pair the meropodites carpopodites and propodites are scabrous, the anterior border of all these joints and the distal end of the posterior border of the meropodites being serrated : in the third pair only the posterior border of the propodite is very strongly serrated.

In the Indian Museum are 7 specimens from Karachi and one from Orissa. In a large male specimen the carapace is 35 millim. long and 62 millim. broad.

The great changes that occur in the chelipeds during the growth of the male indicate that caution is necessary in basing specific distinctions on the form of these organs in this genus.

71. *Macrophthalmus convexus,* Stimpson.

Macrophthalmus convexus, Stimpson, Proc. Ac. Nat. Sci. Philad.. 1858, p. 97:
Miers, Ann. Mag. Nat. Hist. (5) V. 1880, p. 307 : Haswell, Cat. Austral. Crust. p. 89 :

.de Man, Archiv f. Naturges. LIII. 1887, i. p. 354, pl. xv. fig. 4: Ortmann, Zool. Jahrb., Syst., VII. 1893-94, p. 745 and X. 1897-98, pp. 342, 344.

Macrophthalmus inermis, A. Milne Edwards, Ann. Soc. Ent. France, (4) VII. 1867, p. 286, and Nouv. Archiv. du Mus. IX. 1873, p. 277, pl. xii. fig. 5 (*apud* de Man).

Carapace smooth, becoming finely granular near the lateral margins, its length in the male is half, in the female decidedly more than half, its greatest breadth : on either branchial region, behind the branchial groove, are two granular eminences, one behind the other : 3 teeth arranged as in *M. pectinipes* at the anterior end of the posteriorly-convergent lateral borders, the first (outer orbital angle) being the most prominent and much the largest, the third minute.

Front, in its narrowest part between the eyestalks, about one-eleventh the greatest breadth of the carapace, its free edge obscurely bilobed. Orbits considerably oblique, the upper border microscopically beaded, the lower border finely and elegantly serrate. The eyestalks are slender and curved, and the eyes reach to the end of the orbital trench.

The suture between the ischium and merus of the external maxillipeds is decidedly oblique, and there is a distinct gap between these appendages even when their flagellum is folded.

The chelipeds have the same general proportions as in *M. pectinipes :* all the borders of the arm are granular or denticulate, but there is no "musical ridge" on the inner border : a bunch of spinules at the inner angle of the wrist : both borders of the palm, but particularly the lower border, are finely granular, and a fine raised granular line runs along the outer surface of the palm parallel with the lower border : the inner surface of the palm, like that of the fingers, is hairy, but quite smooth and unarmed beneath the hair : there is a small molariform tooth at the base of the dactylus, and a larger one having a forward slant on the immobile finger.

The legs have the same general proportions as in *M. pectinipes,* but they are quite smooth and unarmed, except for a small subterminal spine on the anterior border of the meropodites of the 2nd and 3rd pair.

In the Indian Museum are 5 specimens from the Andamans. The carapace of the largest specimen is 10·5 millim. long and 21·5 millim. broad.

72. *Macrophthalmus sulcatus,* Edw.

Macrophthalmus sulcatus, Milne Edwards, Ann. Sci. Nat. Zool. (3) XVIII. 1852, p. 156 : Ortmann, Zool. Jahrb. Syst. X. 1897-98, pp. 344, 345 (*nec synon.*).

Carapace free of granules in the female, studded with minute granules in the male, its length in the male only about three-eighths, in the female nearly half, its greatest breadth. · On the branchial region, behind the branchial groove, are, in both sexes, three granular eminences, one behind the other, the last being on the posterior border. The lateral borders are convergent : their true *first* tooth, which in other species is at once the antero-lateral angle of the carapace and the outer angle of the orbit, appears in this species to belong to the *upper border* of the orbit, so that the antero-lateral angle of the carapace is formed by the much larger *second* tooth which also is the *apparent* outer orbital angle.

The least breadth of the front, between the eyestalks, is about an eighth the greatest breadth of the carapace : its free edge is very obscurely bilobed.

Orbits sinuous and oblique : the upper border microscopically beaded and furnished near its outer end with a sharp recurved tooth, which is really the outer orbital angle, though the apparent angle is the much larger tooth of the lateral border of the carapace : the lower orbital border is finely denticulated in its inner two-thirds, but is broken and indistinct beyond this. Eyestalks long, slender, curved : the eyes reach not only beyond the true limits of the orbit, but also beyond the antero-lateral angle of the carapace.

The external maxillipeds do not quite meet across the buccal cavern : the suture between the ischium and merus is decidedly oblique.

The legs and chelipeds have the same general proportions as in *M. pectinipes*, but the legs are unarmed.

In the male chelipeds the anterior border of the arm is hairy and strongly denticulated, but there is no " musical ridge : " the inner angle of the wrist and the proximal part of the upper border of the palm are also denticulated. On the outer surface of the palm there is a crest running close to, and parallel with, the lower border ; and on the inner surface of the palm, near the middle line, is a longitudinal row of denticles the first one of which is considerably enlarged : the surface above this ridge, as also the inner surface of the fingers, is densely hairy. The dactylus is not nearly two-thirds the length of the palm : the immobile finger, but not the dactylus, has a strong molariform tooth at its basal end.

In the female the chelipeds are short and weak as usual, and the hand is quite smooth and has the borders — but specially the lower border — thin and sharp.

In the Indian Museum are a male and a female from the Andamans : the carapace of the male is 9 millim. long and 24 millim. broad.

73.　*Macrophthalmus depressus*, Rüpp.

Macrophthalmus depressus, Rüppell, 24 Krabben Roth. Meer. p. 19, pl. iv. fig. 6, pl. vi. fig. 13 : Milne Edwards, Hist. Nat. Crust. II. 66, and Ann. Sci. Nat. Zool. (3)

XVIII. 1852, p. 159 : Heller, SB. Ak. Wien, XLIII. 1861, i. p. 362 : de Man, Notes Leyden Mus. III. 1881, p. 255, and Archiv f. Naturges. LIII. 1887, i. pl. xv. fig. 3, and Journ. Linn. Soc., Zool., XXII. 1887-88, p. 124, and Zool. Jahrb., Syst. VIII. 1894-95, p. 578 : J. R. Henderson, Trans. Linn. Soc., Zool., (2) V. 1893, p. 389 : Ortmann, Zool. Jahrb., Syst. VII. 1893-94, p. 745 (?) and X. 1897-98, pp. 341, 342.

Macrophthalmus affinis, Guérin, Mag. de Zool. II. 1838, pl. xxiv. fig. 2 : Milne Edwards, Ann. Sci. Nat. (3) XVIII. 1852, p. 158 : Haswell, Cat. Austral. Crust. p. 88 (*apud* Ortmann).

Carapace studded with minute granules not always plainly visible to the naked eye, its length in the male about two-thirds of its breadth. The lateral borders are parallel and the antero-lateral angle is rather a square-cut lobe than a tooth. On the epibranchial regions, behind the branchial groove, are two nearly parallel obliquely-longitudinal finely-granular lines, the inner of which is faint.

Front, at its narrowest part, about an eighth the breadth of the carapace, longitudinally grooved, but its free edge is straight and not bilobed.

Orbits little sinuous and little oblique, their upper border microscopically, their lower border finely and evenly denticulate. Eyestalks slender, hardly curved, the eyes reach almost to the end of the orbital trenches.

When the flagella are folded there is not much space between the external maxillipeds : the suture between the ischium and merus of these appendages is hardly oblique.

In the male the chelipeds and legs have much the same general proportions as in *M. pectinipes*, but they are unarmed, except for a small subterminal denticle on the anterior border of the meropodites of the first three pairs of legs : on the other hand the inner surface of the joints of the chelipeds, and the upper surface of the leg-joints (especially of the meropodites) are densely hairy. The dactylus is more than two-thirds the length of the palm, which is smooth and unsculptured : there is a molariform tooth near the basal end of the dactylus, and a similar, but less distinct and more oblique, tooth on the immobile finger.

In the Indian Museum are 2 males from Mergui, besides several specimens from Aden. The carapace of the largest specimen is 14 millim. long and 22 millim. broad.

74. *Macrophthalmus erato*, de Man.

Macrophthalmus erato, de Man, Journ. Linn. Soc., Zool., XXII. 1887-88, p. 125, pl. viii. figs. 12-14, and Zool. Jahrb. Syst., VIII. 1894-95, p. 579.

Carapace quadrilateral, not granular to the naked eye, its length about two-thirds of its breadth, the cervical groove plain, but the branchial groove faint : the second tooth of the lateral border is a little more prominent than the first. *Front about two-ninths the breadth of the carapace,* square cut, longitudinally grooved, but not bilobed. Orbits slightly sinuous, hardly oblique : eyestalks little curved, stoutish, not quite reaching end of orbit. In the male the lower border of the orbit is peculiar : it is finely denticulate at its internal extremity and has a small lobule at its outer angle, and in between these it has the form of a prominent deflexed somewhat triangular lobe. In the female the lower border of the orbit is finely crenulate throughout. The external maxillipeds do not quite meet across the buccal cavern, and the suture between their ischium and merus is a little oblique.

All three borders of the arm are serrated, and the inner angle of the wrist and upper border of the arm are very finely denticulated. There is a strong "musical crest" obliquely parallel with the inner border of the arm and in the middle third of that border. Palm longer than the arm, its inner surface is hairy and carries a spine near the carpal end about midway between the upper and lower borders. The fingers are considerably less than two-thirds the length of the palm : there is a molariform tooth at the base of the dactylus and a larger slanting one on the immobile finger.

The upper surface of the legs, especially in the case of the third pair, is hairy.

In the Indian Museum are 4 specimens from Mergui and Akyab : the carapace of the largest specimen is 10 millim. long and 14 millim. broad.

75. *Macrophthalmus tomentosus,* Eyd. and Soul.

Macrophthalmus tomentosus, Eydoux and Souleyet, Zool. Voy. Bonite, I. p. 243, pl. iii. fig. 8, (1841) : Milne Edwards, Ann. Sci. Nat., Zool., (3) XVIII. 1852, p. 159 : A. Milne Edwards, Nouv. Archiv. du Mus. IX. 1873, p. 279 : de Man, Journ. Linn. Soc., Zool., XXII. 1887-88, p. 122.

Carapace studded with very fine granules : its length is about two-thirds its greatest breadth, *which is behind the middle of the lateral border, the lateral borders being decidedly divergent posteriorly.* On either epibranchial region, behind the branchial groove, are two finely beaded obliquely-longitudinal lines. The first two teeth of the lateral borders are square-cut.

Front, in its narrowest part, about one-eleventh the greatest breadth of the carapace ; though longitudinally grooved it is not bilobed.

Orbits hardly sinuous, not oblique ; their upper border microscopically beaded, their lower border finely crenulate. The eyestalks are hardly curved, and the eyes do not reach to the end of the orbits.

The chelipeds and legs have the same general proportions as in *M. pectinipes,* but are shorter. Chelipeds unarmed and unsculptured, except for some spinules along the inner angle of the wrist and some denticles along the proximal part of the upper border of the palm: in the distal half of the inner border of the arm is a short upstanding horny "musical crest": the borders of the arm and the inner border of the fingers are hairy. The dactylus has a small molariform tooth near the base, and the immobile finger has a much larger one.

The legs are unarmed, except for a small subterminal denticle on the anterior border of the meropodites of the first 3 pairs: the upper surfaces of their joints are more or less hairy.

In the Indian Museum is a single specimen from Mergui: its carapace is 23 millim. long and 34 millim. broad.

Family MICTYRIDÆ, Dana.

MICTYRIS, Latreille.

Mictyris, Latreille, Gen. Crust. et Ins. p, 40 (1806), and in Cuvier Règne Animal, III. p. 21: Desmarest, Consid. Gen. Crust. p. 115, and Dict. Sci. Nat. XXVIII. 1823, p. 235: De Haan, Faun. Japon. Crust. p. 24: Milne Edwards, Hist. Nat. Crust. II. 36, and in Cuvier Règne An., Crust. p. 67, and Ann. Sci. Nat, Zool., (3) XVIII. 1852, p. 154: Miers, Challenger Brachyura, p. 278.

Carapace elongate globose, oval but truncated posteriorly by the short and perfectly straight posterior border, the cervical and cardio-branchial grooves well developed and making the regions very distinct and convex, the posterior border fringed with bristles, as is also the apposed very prominent edge of the first abdominal tergum.

The afferent branchial orifice is a singular valvular recess, formed dorsally by a semicircular notch in the margin of the carapace, and ventrally by a curious cup-shaped dilatation of the base of the epipodite of the external maxillipeds.

Front a narrow deflexed lobe as in *Ocypoda.* Orbits represented by a small post-ocular spine, the eyes, which are borne on shortish stalks, being quite unconcealed.

Antennules as in *Ocypoda,* the basal joint being large and exposed, while the flagellum is rudimentary and concealed beneath the front. Antennæ small but well formed, standing in the usual position.

Epistome short lozenge-shaped. Buccal cavity enormous, somewhat oval in outline. External maxillipeds very large and foliaceous, with a hemispherical bulge causing them to face as much laterally as ventrally: their greater part is formed by the ischium, the inner margin of which is hairy, especially at base: the merus is very much smaller

than the ischium and carries the coarse hairy ·flagellum at its antero-
external angle : the exognath is small, slender, and very inconspicuous.

Chelipeds moderately long and rather slender, stouter and a little
shorter than the legs; their freest motion is in a vertical plane :· the
wrist is a rather elongate trigonal obconical joint.·

Legs somewhat compressed : the first pair are the longest and the
others decrease slightly in length in posterior succession.

The abdomen in both sexes is of a· broad truncate-oval shape,· the
segments from· the 2nd to the 6th gradually increasing in length but
the 7th being narrow : in both sexes the abdomen is fringed with hairs.

Distribution : Indo-Pacific from China and Australia to the
Andamans. ·   .   ·   ·

In habits the species of *Mictyris* resemble the Ocypodes, · Gelasimi
and Dotillæ.

76. *Mictyris longicarpus,* Latreille.

Mictyris longicarpus, Latreille, Gen. Crust. et .Ins. p. 41 (1806): Desmarest,
Consid. Gen. Crust. p. 115, pl. xi. fig. 2, and Dict. Sci. Nat. XXVIII. p. 236: Guérin,
Icon. Règne An. Crust. pl. iv. fig. 4: Milne Edwards, Hist. Nat. Crust. II. 37, and
in Cuvier Règne An. Crust. pl. xviii. fig. 2, and Ann. Sci. Nat., Zool., (3) XVIII.
1852, p. 154: Dana, U. S. Expl. Exp. Crust. pt. I. p. 389: Stimpson, Proc. Ac. Nat.
Sci Philad. 1858, p. 99 : Hess. Archiv f. Nat. XXXI. 1865, p. 142 : Heller, Novara
Crust. p. 40 : A. Milne Edwards, Nouv. Archiv. du Mus. IX. 1873, p. 276: Tozzetti,
Magenta Crust. p. 185, pl. xi. figs. 5, 5a-c : Nauck, Zeits. Wiss. Zool. XXXIV. 1880,
p. 22, pl. i. figs 5-7 (gastric teeth) : Haswell, Cat. Austral. Crust. p. 116 : Miers, Zool.
H. M. S. Alert, pp. 184, 248, and Challenger Brachyura, p. 278 : de Man, Archiv f.
Naturges. LIII. 1887, i. p. 358, and Notes Leyden Mus. XII. 1890, p. 83 : Henderson,
Trans. Linn. Soc., Zool., (2) V. 1893, p. 390 : Aurivillius, Zur Biol. Amphib. Dekap.
p. 38, pl. iii. figs. 10-11 (Mitg. K. Ges. Wiss. Upsala, 1893) : Ortmann, Zool. Jahrb.
Syst. VII. 1893-94, p. 748, and in Semon's Forschungr. Crust. p. 58 (Jena. Denks.
VIII): Stead, Zoologist, (4) II. 1898, p. 307 : Nobili, Ann. Mus. Genov. (2) XX.
1890, p. 272.

Carapace smooth, the regions moderately convex and dividing the
dorsal surface into four lobes : edge of front broadly triangular : *linea
anomurica* very distinct.

Chelipeds a little over 1½ times the length of the carapace : a
strong spine at the inner angle of the ischium (sometimes absent in the
female) :⁻ usually some spinules along the distal part of the lower
border of the arm : wrist with the upper border of the outer surface
marginate, and with a tooth near the middle of the distal border of the
inner surface : palm much shorter than the wrist and not much more
than half the length of the fingers; the upper and lower borders of its
outer surface are marginate, and the middle of its outer surface is
·traversed· by two divergent ridges which are continued along ·the

fingers : fingers slender and tapering, in the male there is an enlarged tooth near the base of the dactylus.

The legs, like the chelipeds, are rough under the lens : the edges of their propodites and dactyli are finely plumed : none of their joints are dilated : the first pair, which are slightly the longest, are about $1\frac{3}{4}$ times the length of the carapace.

In the Indian Museum are 5 specimens from the Andamans and 2 from the Nicobars.

Family HYMENOSOMIDÆ, Ortm.

Key to the Indian Genera or Sub-genera.

I. Front conspicuously tridentate : the external maxillipeds do not quite meet across the buccal cavern and their exognath is not hidden in its proximal portion : chelipeds much more massive than the legs ... ... HYMENICUS.

II. Front broadly triangular, or truncated : the external maxillipeds completely close the buccal cavern and their exognath is completely hidden :—

    1. The interantennular septum is a prominent plate : chelipeds in the male much more massive than the legs ... ... ELAMENA.

    2. The interantennular septum is a mere ridge ; chelipeds in both sexes slender, not stouter than the legs ... ... TRIGONOPLAX.

ELAMENA, Edw.

Elamena, Milne Edwards, Hist. Nat. Crust. II. 33, and Ann. Sci. Nat., Zool., (3) XX. 1853, p. 223 : Dana U. S. Expl. Exp. Crust. pt. I. p. 379 : A Milne Edwards, Nouv. Archiv. du Mus. IX. 1873, p. 321.

Carapace flat dorsally, thin and almost lamellar, triangular or sub-circular, its edges are usually turned up to form a thin circumscribing ridge and are without any teeth. Front broadly triangular, or sometimes truncated. There are no orbits and the eyes, though they may be hidden beneath the front, are exposed and non-retractile : a small post-ocular tooth may be present or not. The antennules fold beneath the front and are not visible from above when folded : the interantennular septum is a prominent plate. Antennal peduncle slender, the flagellum of no great length.

Epistome well defined and remarkably long fore and aft. Buccal cavern square ; the external maxillipeds, which completely close it, have the merus about as large as the ischium, the palp articulating not far from the antero-external angle of the merus, and the exognath slender and concealed.

Chelipeds in the male subequal, massive, especially as to the palm. Legs long and slender.

The abdomen of the male does not quite fill all the space between the last pair of ambulatory legs.

77. *Elamena sindensis*, n. sp.

Carapace broadly piriform, smooth, flat, with no distinction of regions: its edge, which is slightly turned up, is entire and unarmed. Front a prominent broad triangular lamina, somewhat rounded at tip. No post-ocular tooth. Interantennular septum very prominent. Eyes not quite concealed beneath the front.

Male chelipeds about $1\frac{2}{3}$ times as long as the carapace, palm massive and somewhat swollen, fingers stout and pointed and meeting throughout their length. Female chelipeds little longer than carapace, slender, with a slender palm and longish fingers spooned at tip.

Legs slender, the 1st pair not three times as long as the carapace: in all, there is a distinct tooth at the end of the anterior border of both the merus and carpus, and the dactylus is long compressed and falcate with two or three teeth at the end of its posterior border.

In the Indian Museum are 7 specimens from Karachi : the carapace of a male is 5 millim. long and 6 in greatest breadth.

78. *Elamena truncata* (Stimpson ?).

? *Trigonoplax truncata*, Stimpson, Proc. Ac. Nat. Sci. Philad. 1858, p. 109.
Elamena truncata, A. Milne Edwards, Nouv. Archiv. du Mus. IX. 1873, p. 323 : J. R. Henderson, Trans. Linn. Soc., Zool., (2) V. 1893, p. 395.

Carapace orbiculate-ovate, smooth, flat, with no distinction of regions, its edge, which is slightly turned up and entire and unarmed, shows the faintest traces of angulation in 2 or 3 places. No post-ocular tooth ; eyes quite concealed beneath the front. The front, though it projects slightly beyond the carapace is *broadly truncated*, having its free margin cut quite straight. Interantennular septum very prominent. The female chelipeds and the legs are as in the preceding species, the anterior border of the merus and carpus of all the legs ending in a strong tooth.

In the Indian Museum is a female from the Nicobars.

TRIGONOPLAX, Edw.

Trigonoplax, Milne Edwards, Ann. Sci. Nat. Zool. (3) XX. 1853, p. 224.

This is best regarded as a subgenus of *Elamena*, from which it differs only in the following unimportant particulars :—(1) the edge of the carapace is not turned up, (2) the interantennular septum is a mere ridge, (3) the chelipeds in the male, as in the female, are very slender.

79.　*Elamena (Trigonoplax) unguiformis*, De Haan.

Elamene unguiformis, De Haan, Faun. Japon. Crust. p. 75, pl. xxix. fig. 1 and pl. H : J. R. Henderson, Trans. Linn. Soc. Zool., (2) V. 1893, p. 394.

Trigonoplax unguiformis, Milne Edwards, Ann. Sci. Nat. Zool. (3) XX. 1853, p. 224 : Ortmann, Zool. Jahrb., Syst , VII. 1893-94, p. 31.

Carapace smooth, flat, lamellar, broadly pentagonal with the postero-lateral sides about a third as long as any of the others, the regions not defined, the sides entire, unarmed.　Front a broad, horizontal, triangular lamina.　No post-ocular tooth : eyes not concealed by the front, though the eyestalks are.　Interantennular septum a mere ridge.

Epistome as long as broad.　Chelipeds and legs smooth and slender.

Chelipeds not stouter than the legs, about 1½ times as long as the carapace : fingers slender, as long as the slender sub-cylindrical palm, their tips spooned.

The anterior border of the meropodite of all the legs ends in an inconspicuous denticle, the dactylus of all is long, subfalciform, and strongly compressed, and has two or three denticles at the tip of the posterior border.　The 2nd and 3rd pair of legs, which are the longest, are more than three times the length of the carapace.

In the Indian Museum are 5 specimens from the Andamans, The carapace of one is 12 millim. long and 14 in greatest breadth.

HYMENICUS, Dana.

Hymenicus, Dana, Amer. Journ. Sci. (2) XII. 1851, p. 290, and U. S. Expl. Exp. Crust. pt. I. p. 387 : Milne Edwards, Ann. Sci. Nat., Zool., (3) XX. 1853, p. 224.

Differs from *Elamena* only in the following particulars :—
(1) the front is tridentate and the ridge that defines the edge of the carapace dorsally is continued across its base between the eyes : (2) the interantennular septum, as in *Trigonoplax*, is a mere ridge : (3) on either lateral border of the carapace teeth are sometimes present : (4) the external maxillipeds do not quite meet across the buccal cavern and their exognath is not hidden in its proximal portion.

Rhynchoplax of Stimpson (Proc. Ac. Nat. Sci. Philad. 1858, p. 109) is probably synonymous.

Key to the Indian species of Hymenicus.

I.　Median spine of the rostrum of moderate length :
3 teeth on either lateral border of the carapace　　...　H. Wood-Masoni.

II.　Median spine of the rostrum very long : no teeth on
the lateral borders of the carapace　　...　　...　　...　H. inachoides.

80. *Hymenicus Wood-Masoni*, n. sp.

Body and chelipeds tomentose. Carapace dorsally flat or sunken, longer than broad, circular without the rostrum, the regions demarcated by fine grooves.

The front, which is delimited from the rest of the carapace by a fine raised line running across its base between the eyes, is cut into 3 prominent teeth, the middle one of which is somewhat the largest. The antennules fold beneath the front.

A small post-ocular denticle: a large tooth on the lateral border of the carapace above the base of the 1st pair of legs, another, hardly smaller, midway between this and the front, a third, much smaller, midway between this and the post-ocular denticle.

Chelipeds in the adult male more than twice the length of the carapace, very much stouter than the legs, the palm being specially massive. When denuded, the upper border of the arm is dentate and there is a stout spine near the far end of the outer border of this joint: there are several sharp tubercles on the upper surface of the wrist, the outer surface of the palm is reticulate in places, and the fingers which are stout and as long as the palm, have elegantly interlocking teeth.

In the female the chelipeds are considerably shorter and, though stouter than the legs and formed on the male pattern, are not nearly so stout as in the male.

The legs have long, curved dactyli, which are armed with small recurved teeth at the distal end of the posterior border: the 2nd pair, which are a little the longest, are over $2\frac{1}{2}$ times the length of the carapace.

Carapace of male (including rostrum) 7·5 millim. long and 6 broad.

Specimens were collected by the late Professor Wood-Mason at Port Blair in the Andamans, and at Port Canning near Calcutta.

81. *Hymenicus inachoides*, n. sp.

Carapace somewhat tomentose, flat, elongate-triangular, ending in a rostrum of three long teeth of which the middle one is about a third the length of the rest of the carapace, the other two being more than half the length of the middle one. The regions are all well defined by grooves. No spines on the lateral borders of the carapace. Post-ocular denticle hardly distinguishable. The antennules fold beneath the front.

Chelipeds of the adult male somewhat tomentose, not $1\frac{1}{2}$ times the length of the carapace: arm slender, with a tooth near the distal end of the outer border; palm short, high, produced and somewhat swollen below; the fingers a little longer than the palm, stout, and finely toothed.

Legs long and slender, with long dactyli furnished with hook-like teeth at the end of the posterior border : the 2nd pair of legs are nearly three times the length of the carapace.

A single male from Port Canning near Calcutta : its carapace is 8·5 millim. long and 6 millim. in its greatest breadth.

Family GRAPSIDÆ Dana.

Key to the Indian Genera.

I. The antennules fold beneath the front in the ordinary way :—

    1. No oblique hairy ridge on the exposed surface of the external maxillipeds :—

        i. A very wide gap between the external maxillipeds, the exopodites of which appendages are narrow, and the palp of which appendages articulates at or near the antero-external angle of the merus : the abdomen of the male fills all the space between the last pair of ambulatory legs (*Grapsinæ*) :—

            A. Front less than half the greatest breadth of the carapace : merus of the external maxillipeds longer than broad :—

                a. Fingers with broad spooned tips : flagellum of exopodite of external maxillipeds well developed.   GRAPSUS.

                b. Fingers acute, not spooned : flagellum of exopodite of external maxillipeds absent ....   GEOGRAPSUS.

            B. Front more than half the greatest breadth of the carapace : merus of the external maxillipeds broader than long :—

                a. Antennæ completely excluded from the orbit...... ....   METOPOGRAPSUS.

                b. Antennæ in the orbital hiatus...   PACHYGRAPSUS.

        ii. A moderate gap between the external maxillipeds, the exopodites of which appendages are broad, and the palp of which appendages articulates near the middle of the anterior border of the broad merus : the abdomen of the male does not quite fill all the space between the last pair of legs (*Varuninæ*) :—

            A. Exognath of the external maxillipeds not as broad as the ischiognath : terminal joints of legs thin broad and compressed ....   VARUNA.

B. Exognath of the external maxillipeds
as broad as or broader than the ischiognath : dactyli of the legs compressed but not broadened —:

    a. Carapace flat and depressed.... PTYCHOGNATHUS.

    b. Carapace deepish, strongly con-
    vex in both directions......... PYXIDOGNATHUS.

2. An oblique hairy ridge on the exposed surface of
the external maxillipeds (*Sesarminæ*) :—

  i. Carapace little, sometimes not at all, broader
than long, the pterygostomian regions and
sidewalls with a sieve-like reticulation : lower
border of orbit not abnormally prominent :—

    A. Antennæ lodged in the orbital hiatus :—

      a. Carapace nearly square : front
      abruptly and vertically de-
      flexed............ .... SESARMA.

      b. Antero-lateral borders of cara-
      pace arched : front obliquely
      deflexed...... .... SARMATIUM.

    B. The tooth at the inner angle of the
    lower border of the orbit meets the
    front, so as to exclude the antennæ
    from the orbit :—

      a. Carapace dorsally smooth and
      nude .,..... .... METASESARMA.

      b. Carapace dorsally verrucose and
      densely tomentose... .... CLISTOCŒLOMA.

  ii. Carapace much broader than long, the
pterygostomian regions, etc., not reticulated :
lower border of orbit prominent beyond the
front. Front gradually declivous. General
appearance much like *Macrophthalmus*........ METAPLAX.

II. The antennules fold nearly longitudinally in deep notches
in the front visible in a dorsal view (*Plagusiinæ*) :—

  1. Merus of the external maxillipeds of good size and
  as broad as the ischium...... . ...…...... PLAGUSIA.

  2. Merus of the external maxillipeds small and much
  narrower than the ischium............... .... LIOLOPHUS.

Sub-family GRAPSINÆ, Dana (pt.).

GRAPSUS, Lamk., Kingsley.

Grapsus (part) Lamark, Syst. Anim. Sans Vertebr. : Latreille, Hist. Nat. Crust.
et Ins. VI. p. 56, and Gen. Crust. p. 32.

Grapsus, Leach, Trans. Linn. Soc. XI. 1815, pp. 309, 323.

Grapsus (part) Desmarest, Consid. Gen. Crust., p. 129, and Dict. Sci. Nat.
XXVIII. p. 247 : Milne Edwards, Hist. Nat. Crust. II. 83, and Ann. Sci. Nat., Zool.,
(3) XX. 1853, p. 166 : Dana, U. S. Expl. Exp. Crust. pt. I. p. 336.

Grapsus, Kingsley, Proc. Ac. Nat. Sci. Philad. 1880, pp. 188 and 192 : Miers, Challenger Brachyura, p. 254.

Goniopsis, De Haan, Faun. Japon. Crust. p. 33.

Carapace little broader than long, much depressed, the regions fairly well defined, the branchial groove particularly clear, the branchial regions with regular obliquely transverse ridges, the gastric region with a transverse squamiform sculpture. The lateral borders are arched and are armed with a tooth, placed immediately behind the acute outer orbital angle.

Front about half the breadth of the anterior border of the carapace, strongly deflexed : along the line of flexion are 4 tubercles, the outer of which on either side correspond with the supra orbital angles.

Orbits of moderate size, deep, distinctly divided into two fossæ : their lower border is deeply notched near the outer angle : the wide inner orbital hiatus is filled partly by the antennal peduncle and partly by a strong isolated tooth that belongs to the inner of the two fossæ into which the orbit is divided.

The antennules fold nearly transversely in rather narrow fossæ : the interantennulary septum is very broad. The antennal flagellum is short, and lies practically in the orbital cavity : the excretory tubercle of the basal antenna-joint is singularly prominent.

Epistome of good length fore and aft, well defined ; its wings run up towards the orbital hiatus. Buccal cavity square with the antero-lateral corners rounded off. The external maxillipeds are widely distant, leaving between them a rhomboidal gap in which the mandibles are exposed : the ischium and merus are both narrow, the merus being slightly shorter than the ischium, and the palp, which is coarse—especially as to its carpus—articulates at the antero-external angle of the merus.

Chelipeds subequal in both sexes and much shorter than the legs, though, in the male, of a somewhat stouter make : hands and fingers short and stout, the tips of the fingers broad and hollowed *en cuillère.*

Legs broad and compressed, especially as to the merus : the dorsal surface of some of the joints has a sort of reticulate or squamiform sculpture, and the dactyli are thorny.

The abdomen in both sexes consists of 7 segments, and in the male its base is as broad as the sternum between the last pair of legs.

Distribution : rocks and reefs of all the tropical and subtropical seas.

The *Grapsi* of Indian seas are found in considerable number wherever there are rocks. They live out of water and are very cunning and active : if they cannot succeed in dodging their pursuer they fling themselves into the sea and in that way escape capture. Their colour in life is a dark|bottle-green.

82. *Grapsus grapsus* (Linn.).

Seba,'Thesaurus, III, p. 43, pl. xviii. figs. 5, 6.

Cancer grapsus, Linnæus, Syst. Nat. (ed. xii.) p. 1048 : Fabricius, Ent. Syst. II. p. 438 and Suppl. p. 342.

Grapsus maculatus (Catesby, 1743), Milne Edwards, Ann. Sci. Nat. Zool. (3) XX. 1853, p. 167, pl. vi. fig. |1 : Hoffmann, in Pollen and Van Dam, Faun. Madagasc., Crnst., p. 21: Brocchi, Ann. Sci. Nat., Zool., (6) II. 1875, Art. 2, p. 78 (male appendages) : Kingsley, Proc. Ac. Nat. Sci. Philad. 1879, p. 401, and 1880, p. 192 : de Man, Notes Leyden Mus. V. 1883, p. 159 : Miers, Zool. H. M. S. Alert, pp. 518, 544, and Challenger Brachyura, p. 255 : Cano, Boll. Soc. Nat. Napol. III. 1889, p. 236 : R. 1. Pocock, Journ. Linn. Soc., Zool., XX. 1890, p. 512 : de Man, Notes Leyden Mus. XIII. 1891, p. 49 : Koelbel, Ann. Nat. Hofmus. Wien, VII. 1892, p. 114 : Henderson, Trans. Linn. Soc., Zool., (2) V. 1893, p. 391 : A. Milne Edwards and Bouvier, Hirondelle Crust. (Monaco, 1894) p 47: de Man, Zool. Jahrb. Syst. IX. 1895-97, p. 79: Whitelegge, Mem. Austral. Mus. III. 1897, p. 139 : Nobili, Boll. Mus. Torino, 'XII. 1897, p. 3. *Grapsus maculatus* var. *pharaonis,* A. Milne Edwards, Nouv. Archiv. du Mus. IX. 1873, p. 285.

Grapsus pictus, Latreille, Hist. Nat. Crust. et Ins. VI. p. 69, pl. xlvii. fig. 2, and Genera Crust. p. 33 : Lamarck, Hist. Nat. Anim. Sans Vert. V. p. 248 : Dumeril, Dict. Sci. Nat. XIX. p. 322 : Desmarest, Consid. Gen. Crust. p. 130, pl. xvi. fig. 1 : Milne Edwards, in Ouvier Règne An. pl. xxii. fig. 1, and Hist. Nat. Crust. II. 86 : Milne Edwaids and Lucas, Voy. Amer. Merid., Crust. p. 28 : Gay, Hist. Fisica Chili, pt. III. Zool. p. 166 : Dana, U. S. Expl. Exp. Crust. pt. I. p. 336 : ?Desbonne and Schramm, Crust. Guadal. p. 49 : Martens, Archiv f. Nat. XXXVIII. 1872, p. 106 : Miers, Cat. Crust. New Zeal. p. 36, and P. Z. S. 1877, p. 73, and Phil. Trans. 1879, p. 489, and Ann. Mag. Nat. Hist (5) V. 1880, p. 310 : Smith, Trans. Connect. Acad. IV. 1880, p. 256 : Tenison Woods, Proc. Linn. Soc. N. S. W. V. 1880-81, p. 117 : Ozorio, Journ. Sc. Nat. Lisb. XI. 1885-87, p. 227. *Grapsus pictus* var. *ocellatus,* Studer, Abh. Ak. Berlin, 1882, Gazelle Crust. p. 14: *Grapsus pictus* var. *Webbi,* Hilgendorf, SB. Nat. Freunde Ges. 1882, p. 24.

Grapsus ornatus, Milne Edwards, Ann. Sci. Nat. Zool. (3) XX. 1853, p. 168.

Grapsus pharaonis, Milne Edwards, Ann. Sci. Nat. Zool. (3) XX. 1853, p. 168 : Heller, SB. Ak. Wien. XLIII. 1861, i. p. 362 : Hoffmann in Pollen and Van Dam, Faun. Madagasc. Crust. p. 20, pl. v. figs. 32-35 : Richters, in Mobius, Meeresf. Maurit. Crust. p. 156.

Grapsus Webbi, Milne Edwards, Ann. Sci. Nat. (3) XX. 1853, p. 167 : Stimpson, Proc. Ac. Nat. Sci. Philad. 1858, p. 102.

Grapsus altifrons, Stimpson, Ann. Lyc. Nat. Hist. N. Y. VII. 1862, p. 230.

Grapsus grapsus, Ives, Proc. Ac. Nat. Sci. Philad. 1891, p. 190 : Ortmann, Zool. Jahrb. Syst. VII. 1893-94, p 703 : Faxon, Mem. Mus. Comp. Zool., XVIII. 1895, p. 30 : Rathbun, Proc. U. S. Nat. Mus. XXI. 1898, p. 604.

Goniopsis picta, De Haan, Faun. Japon. Crust. p. 33, and Krauss Sudafr. Crust. p. 46.

Carapace somewhat discoidal in shape, owing to the curvature of the sides : its regions well defined : the transverse and oblique ridges are salient, and the surface between the latter is coarsely reticulate.

Front deep and almost vertically deflexed, overhanging the epistome and much concealing the antennules, its free edge crenate.

Length of the epistome one-third or more of its greatest breadth. The tooth at the inner angle of the orbit is blunt.

Chelipeds in the male hardly longer than the carapace, shorter in the female : inner border of ischium and arm strongly spinate, and there are one or two less acute spines at the far end of the outer border of the arm : wrist with fine scattered tubercles on its upper surface, and with its inner angle produced to form a talon-shaped spine : palm nearly as high as long, its outer surface sculptured, its upper border culminating in a tooth : the fingers have very broad rounded tips, and the length of the dactylus in the male is nearly twice the length of the upper border of the palm.

Of the legs the 1st pair are very decidedly the shortest and the 3rd pair the longest, the latter being about twice the length of the carapace : the 4th pair are longer than the first by a dactylus, and shorter than the 2nd by about two-thirds of a dactylus. Only in the last pair of legs does the breadth of the merus approach half the length of the same joint : the far end of the upper border of the merus is spine-like and there are usually 2 or 3 spines at the far end of the lower border.

In the Indian Museum are 18 specimens from the Laccadives, the Andamans, the Coromandel coast, and Ceylon. The carapace of a large specimen is 64 millim. long and 68 millim. broad.

83.   *Grapsus strigosus* (Herbst).

Cancer strigosus, Herbst, Krabben, III. i. p. 55, pl. xlvii. fig. 7. *Grapsus strigosus,* Bosc, Hist. Nat. Crust. I. p. 203 : Latreille, Hist. Nat. Crust. et Ins., VI.   p. 70, *etc.* : Milne Edwards, Hist. Nat. Crust. II. 87 : Gay, Hist. Fis. Chili, III. Zool.   p. 168 : Dana, U. S. Expl. Exp. Crust. pt. I. p. 338 : Milne Edwards, Ann. Sci. Nat., Zool., (3) XX. 1853, p. 169 : Stimpson, Journ. Bost. Soc. Nat. Hist. VI. 1857. p. 466 : Kinahan, Journ. Roy. Soc. Dubl. I. 1858, p. 340 : Stimpson, Proc. Ac. Nat. Sci. Philad. 1858, p. 102 : Hess, Archiv f. Nat. XXXI. 1865, i. pp. 147, 171 : Heller, Novara Crust. p. 47 : A. Milne Edwards, Nouv. Archiv. du Mus. IV. 1868, p. 71 and IX. 1873, p. 286 (*ubi synon.*) : Hilgendorf in v. d. Decken's Reis. Ost–Afr. III. i. p. 87 : Hoffmann, in Pollen & Van Dam, Faun. Madag. Crust. p. 20, pl. v. fig. 31 : Lockington, Proc. Calif. Acad. VII. 1876, p. 151 : Kossmann, Reise roth. Meer., Crust. p. 60 : Miers, P. Z. S. 1877, p. 136, and Ann. Mag. Nat. Hist. (5) II. 1878, p. 410 : Hilgendorf, MB. Ak. Berl. 1878, p. 808 : E. Nauck, Zeits. Wiss., Zool. XXXIV. 1880, p. 32 (*gastric teeth*): Kingsley, Proc. Ac. Nat. Sci. Philad. XXXII. 1880, p. 194 : Haswell, Cat. Austral. Crust. p. 97 : Miers, Zool. H. M. S. Alert, pp. 518, 544, and Challenger Brachyura, p. 256 : Müller, Verh. Nat. Ges. Basel, VIII. p 475 : de Man,

Archiv f. Nat. LIII. 1887, i. p. 365, and Journ. Linn. Soc. Zool. XXII. 1888, p. 148 : Cano, Boll. Soc. Nap. III. 1889, p. 236 : Walker, Journ. Linn. Soc., Zool., XX. 1886–1890, p. 110 : Henderson, Trans. Linn. Soc., Zool., (2) V. 1893, p. 390: de Man, Zool. Jahrb. Syst. IX. 1895–97, p. 80: Ortmann, Zool. Jahrb., Syst., 1893–94, p. 705 : Wedenissow, Bull. Soc. Ent. Ital. 1894, p. 415.

 Grapsus albo-lineatus, Lamarck, Hist. Nat. Anim. Sans Vert. V. p. 249 (*fide* Edw.).

 Gonioposis flavipes, Macleay, Ill. Ann. S. Africa, p. 66, and Krauss, Sudafr. Crust. p. 46 (*apud* Miers).

 Goniopsis strigosa, De Haan, Faun. Jap. Crust. p. 33 : Macleay, *loc. cit.* : Krauss, *loc. cit.*

 Grapsus granulosus, *pelagicus*, aud *Peroni*, Milne Edwards, Ann. Sci. Nat. (3) XX. 1853, p. 169 (*fide* A. M. E.).

 The chief differences between this species and *G. grapsus* are the following :—

 The branchial grooves of the carapace are not so well cut, the transverse and oblique ridges are low and smooth, and the surface between the oblique ridges is quite smooth.

 The front is not so deep and is obliquely deflexed, hardly overhanging the epistome and not concealing the antennules, and its free edge is not so distinctly crenulate. The tooth at the inner angle of the orbit is subacute. The length of the epistome is not nearly a third its greatest breadth.

 In the chelipeds, the tooth at the inner angle of the wrist is nearly straight, not talon-like, the length of the upper border of the palm is nearly two-thirds the length of the dactylus, and the tips of the fingers are not so broad and blunt.

 In the legs the meropodite is broader, its greatest breadth being half its length. Moreover the difference in size between the 1st and 4th pairs of legs is much less marked.

 In the Indian Museum are 76 specimens, from the Baluchistan and Sind coast, the Malabar coast, Ceylon, the Coromandel coast, the Arakan and Tenasserim coast, Mergui, the Andamans, and the Nicobars.

 The carapace of the largest specimen (a female) is 59 millim. long and 63 millim. broad.

GEOGRAPSUS, Stimpson.

 Geograpsus, Stimpson, Proc. Ac. Nat. Sci. Philad. 1858, p. 101 : Kingsley, Proc. Ac. Nat. Sci. Philad. 1880, pp. 188, 195 : Miers, Challenger Brachyura, p. 260.

 Orthograpsus, Kingsley, *l. c.* pp. 188, 194.

 Closely resembles *Grapsus*, but differs in the following important particulars :—

 The carapace is more quadrate, the sides being very little arched, it is also broader and less depressed. The lobe at the inner inferior angle of the orbit is not so completely isolated. The antennal peduncle is not so massive, nor is its "urinary tubercle" conspicuous. The epistome is shorter fore and aft, and is much less well defined.

The chelipeds are altogether of a different type, being vastly more massive than the legs, and in the adult male at least as long as the longest legs: the fingers are pointed. Though the dactyli of the legs are thorny, they are not so closely covered with thorns, nor are the thorns so coarse, as in *Grapsus*. Between the coxæ of the 2nd and 3rd pair of legs is a narrow fossa fringed with hair leading to the branchial cavity.

The two Indian species of the genus are land-crabs and are found in the jungles of the Andaman and Nicobar islands and in the villages of the Laccadive islands. They are extremely vigilant and active.

84. *Geograpsus Grayi* (Edw.).

Grapsus Grayi, Milne Edwards, Ann. Sci. Nat. Zool. (3) XX. 1853, p. 170.

Geograpsus rubidus, Stimpson, Proc. Ac. Nat. Sci. Philad. 1858, p. 103.

Geograpsus Grayi, A. Milne Edwards, Nouv. Archiv. du Mus. IX. 1873, p. 288; Miers, Phil. Trans. 1879, p. 489, and Zool. H. M. S. Alert, pp. 518, 545, and Challenger Brachyura, p. 261: Kingsley, Proc. Ac. Nat. Sci. Philad. 1880, p. 196; Richters, in Mobius, Meeresf. Maurit. p. 156 : Haswell, Cat. Austral. Crust. p. 98; Ortmann, Zool. Jahrb., Syst., VII. 1893-94, p. 707 : de Man, Zool. Jahrb., Syst. IX. 1895-96, p. 80 : Nobili, Ann. Mus. Genov. (2) XX. 1899, p. 266.

Carapace subquadrilateral, a little convex, the lateral borders well defined anteriorly, ill defined and slightly convergent posteriorly : transverse markings fine, curved or oblique on the branchial regions, almost invisible on the gastric region.

The four tubercles along the line of flexion of the front are not salient; the edge of the front in a dorsal view is concave. The notch near the outer end of the lower border of the orbit is small and narrow. The epistome is rather ill defined.

Chelipeds in both sexes a little unequal: squamiform markings are present but, except on the arm, are indistinct, as also are the scattered granules on the upper surface of the palm. The larger cheliped may be a little under or a little over twice the length of the carapace. The inner border of the ischium is denticulate, the inner border of the arm is expanded to form a dentate lobe, and the inner angle of the wrist is spiniform.

The greatest breadth of the meropodites of the legs is less than half their length. The first pair of legs are slightly shorter than the 4th: the 2nd pair are the longest of all, being about twice the length of the carapace. The last 3 joints of all the legs are bristly.

Colours in life yellow-ochre, the greater part of the dorsum of the carapace livid bluish or purplish.

In the Indian Museum are 24 specimens from the Andamans, Nicobars, and Laccadives.

The carapace of a large male is 40 millim. long and 49 broad.

85. *Geograpsus crinipes* (Dana).

Grapsus crinipes, Dana, Proc. Ac. Nat. Sci. Philad. 1851, p. 249, and U. S. Expl. Exp. Crust., pt. I. p. 341, pl. xxi. fig. 6.

Geograpsus crinipes, Stimpson, Proc. Ac. Nat. Sci. Philad. 1858, p. 101 : Heller, Novara Crust. p. 48 : Streets, Bull. U. S. Nat. Mus. VII. 1877, p. 115 : Kingsley, Proc. Ac. Nat. Sci. Philad. 1880, p. 196 : Ortmann, Zool. Jahrb., Syst. VII. 1893-94, p. 706 : de Man, Zool. Jahrb., Syst. IX. 1895-97, p. 83 : Whitelegge, Mem. Austral. Mus. III. 1897, p. 139.

Grapsus rubidus, Hilgendorf, in v. d. Decken's Reisen Ost-Afr. Crust., p. 87, pl. v. : Hoffmann, in Pollen & Van Dam, Faun. Madagasc. Crust. p. 22.

Differs from *G. Grayi* in the following particulars : —

The carapace is quite flat, and the lateral borders, which are thin and well defined throughout their extent, are slightly *divergent* posteriorly : the transverse markings are distinct and *nearly straight.*

The four tubercles along the line of flexion of the front are salient, and the free edge of the front is quite straight. The notch near the outer end of the lower border of the orbit is large, and the lobule external to the notch is denticulate. The epistome is well defined from the palate by a granular or pectinate ridge.

The chelipeds in the male are nearly equal, but in the *female* they are unequal. The squamiform markings on the arm, wrist, and lower portion of the hand are distinct, as also are the vesiculous granules on the upper surface of the palm and dactylus.

The greatest breadth—near the far end—of the meropodites of the last 3 pairs of legs is *more than half* their length.

Colour in life bright red.

In the Indian Museum are 2 males and a female from the Andamans, a male from the Nicobars, and a female from the Laccadives. The carapace of a female is 40 millim. long and 45 broad.

Metopograpsus, Edw.

Metopograpsus, Milne Edwards, Ann. Sci. Nat., Zool., (3) XX. 1853, p. 164 : Kingsley, Proc. Ac. Nat. Sci. Philad. 1880, pp. 188, 190 : Miers, Challenger Brachyura, p. 257.

Carapace quadrate, little broader than long, somewhat depressed, the regions not well defined, the branchial groove distinct, fine oblique grooves are present on the lateral parts of the branchial regions : the antero-lateral, or outer orbital angle, is acute, but there are no teeth on the lateral border behind it.

Front very broad, more than half the extreme width of the carapace, deflexed : along the line of flexion are four depressed lobes, the outer one of which on either side sometimes shows a tendency to split into two.

Orbits of moderate size, occupying the corners of the carapace : the lower border is notched near its outer end : the orbital hiatus is filled by a special lobe which belongs to the inner of the two fossæ into which the orbit is divided and this lobe completely excludes the antennæ from the orbit. The antennules fold nearly transversely in fossæ of good size. The antennæ have a short and slender flagellum : the basal joint of the peduncle is not very massive.

Epistome well defined, but short fore and aft. Buccal cavity square with the anterior corners rounded off. The external maxillipeds leave between them a rhomboidal gap in which the mandibles are exposed : the merus is shorter than the ischium, and carries the coarse palp at or near the antero-external angle.

Chelipeds either subequal or unequal, the larger one much more massive than the legs but shorter than the 2nd and 3rd pairs of these : fingers rather short and stout, with the tip spooned.

Legs broad and compressed, especially as to the merus, which joint—like the arm of the chelipeds—usually has some squamiform markings : the last three joints have bristly edges and the dactylus is thorny.

The abdomen in both sexes consists of 7 separate segments, and in the male its base is as broad as the sternum between the last pair of legs.

An Indo-Pacific genus.

86. *Metopograpsus messor* (Forskal) Edw.

Cancer messor, Forskal, Descrip. Anim. in itin. orient. p. 88. *Grapsus messor,* Milne Edwards, Hist. Nat. Crust. II. 88 : Krauss, Sudafr. Crust. p. 43 : Hoffmann in Pollen & Van Dam, Faun. Madag. Crust. p. 23 : Sluiter, Tijds. Nederl. Ind. XL. 1881, p. 164. *Metopograpsus messor,* Milne Edwards, Ann. Sci. Nat., Zool., (3) XX. 1853, p. 165 : Heller, SB. Ak. Wien, XLIII. 1861, p. 362, and Novara Crust. p. 44 : A. Milne Edwards, Nouv. Archiv. du Mus. IV. 1868, p. 71 : Kossmann, Reise roth. Meer., Crust. p. 57 : Hilgendorf, MB. Ak. Berl. 1878, p. 808 : Miers, Phil. Trans. 1879, p. 489, and Zool. H. M. S. Alert, pp. 184, 245, 518, 545, and Challenger Brachyura, p. 258 : de Man, Notes Leyden Mus. II. 1880, p. 183, and Journ. Linn. Soc. Zool. XXII. 1887-1888, p. 144, pl. ix. fig. 11, and Archiv f. Naturges. LIII. 1888, i. p. 361, pl. xv. fig. 6, and in Weber's Zool. Ergebn. Niederl. Ost. Ind. II. p. 314 :

Richters, in Mobius, Meeresf. Maurit. p. 156: Kingsley, Proc. Ac. Nat. Sci. Philad. 1880, p. 190: Lenz & Richters, Abh. Senck. Nat. Ges. XII. 1881, p. 425: Müller, Verh. Nat. Ges. Basel, VIII. p. 475: Ozorio, Journ. Sci. Nat. Lisb. XI. p. 227: Henderson, Trans. Linn. Soc., Zool., (2) V, 1893, p. 390: Ortmann, Zool. Jahrb., Syst., VII. 1893-94, p. 701: Whitelegge, Mem. Austral. Mus. III. 1897, p. 139: Nobili, Ann. Mus. Genov. (2) XX. 1899, p. 265.

　Grapsus Gaimardi, Savigny, Descr. Egypt. Crust. pl. ii. fig. 3.

　Metopograpsus Eydouxi and *intermedius,* Milne Edwards, Ann. Sci Nat., Zool., (2) XX. 1853, p. 165 (*sec.* Kingsley, *l.c.*).

　Pachygrapsus æthiopicus, Hilgendorf, in v. d. Decken, Reisen Ost-Afr., Crust. p 88, pl. iv. fig. 2 (*fide* Kossmann, *l.c.,* and Hilgendorf, *l.c.*).

Carapace about four-fifths as long as broad, the sides distinctly convergent posteriorly; besides the oblique markings on the lateral parts of the epibranchial regions, there are some fine transverse markings on the post-frontal region.

Front about three-fifths the greatest breadth of the carapace, its free edge beaded, thin and prominent but hardly laminar, and slightly sinuous. Orbits little oblique, their major diameter is a little more than a third the width of the front: the inner angle of the lower border is denticulate.

Chelipeds unequal, the length of the larger one about $1\frac{1}{2}$ times that of the carapace: there are wrinkles or squamiform markings on the upper surface of the arm and wrist and—along with some vesiculous granules—on the upper and lower borders of the hand. The inner border of the ischium is denticulate, the inner border of the arm is spinate and is expanded distally to form a laciniate lobe, and there is a spine, which may be double, at the inner angle of the wrist: the fingers have blunt tips, and the dactylus is not very much longer than the upper border of the palm.

Of the legs the 1st pair is the smallest and the 3rd pair the longest —about twice the length of the carapace: in all, the upper border of the merus ends in a spine and the lobe at the far end of the lower border is spinate: in the last three pairs the greatest breadth of the merus is half its length.

The terminal segment of the male abdomen is simply triangular.

In the Indian Museum are 56 specimens, from Karachi, Bombay, the Orissa coast, the Ganges Delta, the Arakan coast, and the Andamans. The carapace of the largest specimen is $23\frac{1}{2}$ millim. long and 30 millim. broad.

87.　*Metopograpsus maculatus,* Edw.

　Metopograpsus maculatus, Milne Edwards, Ann. Sci. Nat. Zool. (3) XX. 1853, p. 165: de Man, Journ. Linn. Soc., Zool., XXII. 1887-88, p. 145, pl. x. figs. 1-3.

Distinguished from the only other Indian species by the following characters :—

The carapace is much more elongate, its length being seven-eighths of its breadth, its sides are very markedly convergent posteriorly, and there are no transverse markings on the post-frontal region.

The front is nearly three-fourths the greatest breadth of the carapace, and its free edge is decidedly laminar and nearly straight.

The orbits are oblique : their major diameter is less than a third the breadth of the front and the inner angle of their lower border is not denticulate.

The fingers of the chelæ, though their tips are spooned, are not very blunt: the dactylus is much longer than the upper border of the palm.

Except perhaps in the last pair of legs, the meropodites are narrower, their greatest breadth being decidedly less than half their length.

In the male abdomen the terminal segment has a somewhat three-lobed appearance.

In the Indian Museum are two specimens from Mergui. It seems to me very doubtful whether they are distinct from *M. latifrons,* White (Jukes, Voy "Fly," II. 337, pl. ii. fig. 2).

PACHYGRAPSUS, Randall, Stimpson.

Pachygrapsus, Randall, Proc. Ac. Nat. Sci. Philad. 1839, p. 126 : Milne Edwards, Ann. Sci. Nat., Zool., (3) XX. 1853, p. 166 : Stimpson, Proc. Ac. Nat. Sci. Philad. 1858, p. 101 : Kingsley, Proc. Ac. Nat. Sci. Philad. 1880, pp. 188, 198: Miers, Challenger Brachyura, p. 259.

Differs from *Metopograpsus* only in the following particulars :—

(1) the tooth or lobe at the inner angle of the lower border of the orbit is small and does not fill the orbital hiatus, so that the antennæ are not excluded from the orbit ; (2) there may be a tooth or two on the lateral border of the carapace immediately behind the outer orbital angle.

Distribution : West Indies eastwards, through the Mediterranean, to the American Pacific coast.

88. *Pachygrapsus minutus,* A. M. Edw.

Pachygrapsus minutus, A. Milne Edwards, Nouv. Archiv. du Mus. IX. 1873, p. 292, pl. xiv. fig. 2 : Kingsley, Proc. Ac. Nat. Sci. Philad. 1880, p. 201 : de Man, Notes Leyden Mus. V. 1883, p. 158, and Archiv f. Naturges. LIII. 1887. i. p. 368, and Journ. Linn. Soc., Zool., XXII. 1888, p. 148 : Cano, Boll. Soc. Nat. Napol. III. 1889, p. 240.

Carapace a good deal broader than long, its whole dorsal surface marked with fine transverse and oblique lines : the lateral borders are

J. II. 52

strongly convergent posteriorly, and have no spine behind the acute outer orbital angle.

Front about three-fifths the greatest breadth of the carapace, moderately deflexed, its free edge slightly sinuous. Orbits little oblique, their major diameter more than a third the breadth of the front, their lower border not denticulate.

The chelipeds in the male are subequal and vastly more massive than the legs, and are about twice the length of the carapace, and, except for some squamiform markings on the arm, are smooth : the inner border of the ischium and both borders of the arm are crenulate, and the distal end of the inner border of the arm is expanded to form a denticulate lobe : the inner angle of the wrist is dentiform : the fingers are stout and blunt.

Of the legs the two middle pairs are the longest, being not twice the length of the carapace. In all the last three joints are bristly, and the merus has a spine at the far end of the anterior border and two largish spines at the far end of the posterior border.

The terminal joint of the male abdomen is simply triangular.

A small species: the carapace of the single specimen (from Mergui) in the Indian Museum is 6·5 millim. long and 10 millim. broad.

Subfamily VARUNINÆ.

VARUNA, Edw.

Varuna, Milne Edwards, Dict. Hist. Nat. XVI. p. 511 (1830), and Hist. Nat. Crust. II. 94, and Ann. Sci. Nat. Zool., (3) XX. 1853, p. 176 : Kingsley, Proc. Ac. Nat. Sci. Philad. 1880, pp. 188, 205 : Miers, Challenger Brachyura, p. 265.

Trichopus, De Haan, Faun. Japon. Crust. p. 32.

Carapace very little broader than long, depressed, with thin sharp edges, the regions fairly well indicated. Front a little more than half the breadth of the anterior border and a little more than a third the greatest breadth of the carapace, straight, prominent, sublaminar, little deflexed. Antero-lateral borders of the carapace arched, cut into 3 teeth including the outer orbital angle.

Orbits small, of good depth, their lower border broken and incomplete. The antennules fold obliquely and the interantennulary septum is broad. Antennæ of fair size, standing in the orbital hiatus.

Epistome of good length, well defined. Buccal cavern square. The external maxillipeds gape, but not very widely : their exognath is not nearly as broad as the ischium : their merus is shorter, but anteriorly much broader, than the ischium, its antero-external angle being considerably produced, so that the palp articulates near the middle of the anterior border.

Chelipeds equal, but variable in size. In old males they are considerably longer, and vastly more massive, than the legs: in the female they are shorter, and though stouter are not vastly stouter than the legs. The fingers, though sharp pointed, are a little hollow-tipped.

The legs have the three terminal joints compressed, dilated, and plumed, for swimming: the 2 middle pairs are the longest, the last pair is the shortest.

The abdomen in both sexes consists of 7 separate segments: in the male it does not completely cover the sternum between the last pair of legs.

Distributed throughout the Indo-Pacific, ascending estuaries even into freshwater. Commonly found at sea on drift logs.

89.   *Varuna litterata* (Fabr.) Edw.

Cancer litteratus, Fabricius, Ent. Syst. Suppl. p. 342: Herbst, Krabben, III. i. 58, pl. xlviii. fig. 4.

Grapsus litteratus, Bosc, Hist. Nat. Crust. I. p. 203, and Latreille, Hist. Nat. Crust. et Ins. VI. p. 71.

Varuna litterata, Milne Edwards, Dict. d'Hist. Nat. XVI. p. 511.

Trichopus litteratus, De Haan, Faun. Japon. Crust. p. 32: Dana, U. S. Expl. Exp. Crust. pt. I. p. 336, pl. xx. fig. 8.

Varuna litterata, Milne Edwards, Hist. Nat. Crust. II. p. 95, and Ann. Sci. Nat. Zool., (3) XX. 1853, p. 176: Lucas, Hist. Nat. Anim. Artic., Crust., p. 72, pl. iii. fig. 4: Stimpson, Proc. Ac. Nat. Sci. Philad. 1858, p. 103: Heller, Novara Crust. p. 51, A. Milne Edwards, Nouv. Archiv. du Mus. IV. 1868, p. 71, and IX. 1873, p. 295: Brocchi, Ann. Sci. Nat. (6) II. 1875, *(male appendages)*: Miers, Cat. Crust. New Zealand, p. 40, and Ann. Mag. Nat. Hist. (5) V. 1880, p. 310, and Challenger Brachyura, p. 265: Tozzetti, Magenta Crust. p. 122, pl. viii. figs. 2 *a–g*: Hilgendorf, MB. Ak. Berl. 1878, p. 808: Neumann, Crust. Heidelb. Mus., p. 27: Nauck, Zeits. Wiss. Zool. XXXIV. 1880, p. 29 *(gastric teeth)*: Kingsley, Proc. Ac. Nat. Sci. Philad. 1880, p. 205: Sluiter, Tijds. Nederl. Ind. XL. 1881, p. 164: Haswell, Cat. Austral. Crust. p. 103: Filhol, Crust. Nouv. Zel. in Miss. l'ile Campbell, p. 390: de Man, Archiv für Nat. LIII. 1887, i. p. 371, and in Weber's Zool. Ergebn. Niederl. Ost-Ind. II. 1892, p. 315, and Zool. Jahrb., Syst. IX. 1895, p. 112: Henderson, Trans. Zool. Soc. (2) V. 1893, p. 391: Ortmann, Zool. Jahrb. Syst., VII. 1893–94, p. 713: Max Weber, Zool. Jahrb. Syst. X. 1898, p. 157: Nobili, Ann. Mus. Genov. (2) XX. 1899, p. 267.

Carapace curiously pitted and frosted above, the regions well enough defined by grooves, which in places are broad shallow and uneven; the disposition of these grooves in the middle of the carapace makes a letter H. The borders of the carapace are thin and are sharply defined and finely beaded or milled: the antero-lateral borders are arched and are cut into three teeth, including the outer orbital angle: the postero-lateral boundary of the carapace, on each side, is a distinct facet.

The chelipeds vary, according to sex and age, from a little over once (in the female) to a little over twice (in old males) the length of the carapace. The borders of the arm are denticulated, especially the inner border; the inner angle of the wrist forms a large sharp spine with some spinules at its base; the inner surface of the palm is more or less granular, the outer surface has some fine reticulate markings and —running parallel with the lower border, on to the fixed finger— a raised line: the fingers are stout and strongly toothed, the dactylus being longer than the upper border of the palm.

The 2nd and 3rd pair of legs, which are about equal, are over 1½ times the length of the carapace: the 1st pair are a little more than a dactyl-length, the 4th pair a little less than a dactyl-length longer than the carapace. The only armature of the legs, which are typical swimming paddles, is a subterminal spine on the anterior border of the meropodite.

In the Indian Museum are 63 specimens from the seas of India. The carapace of the largest male is 50 millim. long and 56 millim. broad.

PTYCHOGNATHUS, Stimpson.

Ptychognathus, Stimpson, Proc. Ac. Nat. Sci. Philad. 1858, p. 104 : Kingsley, Proc. Ac. Nat. Sci. Philad. 1880, pp. 188, 203 : de Man, Zool. Jahrb., Syst., IX. 1895, p. 90.
Gnathograpsus, A. Milne Edwards, Nouv. Archiv. du Mus. IV. 1868, p. 180.
Cœlochirus, Nauck, Zeits. Wiss. Zool. XXXIV. 1880, pp. 30, 66 (*teste* de Man).

Very closely resembles *Varuna,* from which it differs only in the following particulars :—

(1) the exopodite of the external maxillipeds is of remarkable breadth, being at least as broad as, and usually much broader than, the ischium of those appendages :

(2) the regions of the carapace are not always so well defined.

(3) the dactyli of the legs, though compressed, are not so broad.

Distribution : Islands of the Indo-Pacific, entering fresh water above any tidal influence.

Key to the Indian species of Ptychoynathus.

I. Carapace hardly broader than long : front prominent, straight or hardly sinuous : the antennules fold very obliquely :—

    1. Teeth of the antero-lateral border sharp and salient : regions of the carapace fairly well defined : fingers of the female chelæ nude :—

        i. Inner angle of the wrist dentiform, but not produced : a large shaggy patch of hairs on the inner surface of the hand of the male... *P. dentata.*

       ii. Inner angle of the wrist produced to form a
          long spine : a patch of hair on the outer
          surface of the hand of the male, near the
          finger cleft... *P. onyx.*

  2. Teeth of the antero-lateral border not salient, in-
     conspicuous : regions of the carapace not, or hardly,
     indicated :—

       i. A subterminal patch of bristles on the outer
          surface of the fixed finger of the female...... *P. andamanica.*

       ii. Fingers of female nude .... *P. pusilla.*

II. Carapace decidedly broader than long : front little pro-
    minent and decidedly sinuous: the antennules fold nearly
    transversely.. *P. barbata.*

90. *Ptychognathus dentata*, de Man.

Ptychognathus dentatus, de Man, in Weber's Zool. Ergebn. Niederl. Ost-Ind. II. 1892, p. 318, pl. xviii. fig. 9.

Carapace inappreciably broader than long, flat but not particularly depressed, its regions quite distinct, as also are the cervical and branchial groves and a pair of post-frontal tubercles: on the posterior part of each epibranchial region, obliquely parallel with the postero-lateral borders, is a fine ridge.

Front prominent, laminar, nearly straight, its extent is two-fifths the greatest breadth of the carapace.

Antero-lateral borders of the carapace cut into three sharp salient teeth, of which the first is much the largest, and the third much the smallest.

Upper border of the orbit very sinuous. The antennules fold very obliquely. Anterior border of the buccal cavern not granular, but having a median horizontal tooth.

Exognath oval, with a smooth and strongly convex surface : its greatest breadth in the male is more than twice that of the ischiognath, but in the female is only a little more than that of the ischiognath.

Chelipeds of the male more than 1½ times the length of the carapace, smooth : inner angle of the wrist acute, but not spiniform : palm higher than long, inflated at the postero-inferior angle, and having a tussock of hairs in the middle of its inner surface: dactylus more than twice the length of the upper border of the palm, longer slenderer and less strongly toothed than the fixed finger : both fingers though hollowed at the tip are sharp-pointed. In the female the chelipeds are about as long as the carapace; the inner angle of the wrist is spiniform; the palm is not swollen and is nude, and its outer surface is traversed, near the lower border, by a fine raised line which extends nearly to the tip of the fixed finger.

The 2nd and 3rd pairs of legs are about 1⅘ times, the 1st pair are not quite 1½ times, and the 4th pair are are not 1⅓ times, the length of the carapace : on the anterior border of the merus of the first three pairs is a subterminal spine.

The sidewall of the carapace and the basal joints of the legs have little tomentum.

In the Indian Museum are 2 males and an egg-laden female from "the Bay of Bengal" and 2 young females from Upper Tenasserim.

The carapace of the largest male is 19 millim. long and not quite 20 millim. in its greatest breadth.

91. *Ptychognathus onyx*, n. sp.

Very closely related to P. *spinicarpus*, Ortm., and to P. *Polleni* and *affinis*, de M., if these species are distinct.

This species very nearly resembles *P. dentata*, from which it differs, young males being compared with females of the same size, only in the following particulars :—

(1) the carapace though otherwise similar is much thinner and more depressed and its markings are not quite so distinct :

(2) in the middle of the anterior border of the buccal cavern is a slight prominence, but no distinct tooth :

(3) the exognath (*in the young male*) is, as in the female of *P. dentata*, but little broader than the ischiognath :

(4) in the chelipeds of the young male the inner angle of the wrist is produced to form a long spine ; there is no hair on the inner surface of the palm, but on the outer surface, in the finger-cleft and extending along the fixed finger, there is a tuft of hair ; the outer surface of the palm also, as in the female of *P. dentata*, is traversed, close to the lower border, by a raised line, which runs to the tip of the fixed finger ; finally the fingers are blunter, and the dactylus is only about twice as long as the upper border of the palm.

Practically the chief distinction between this species and *P. dentata* is that in the male of this species the inner angle of the wrist forms a long spine, and the hair is on the outside instead of on the inside of the hand.

In the Indian Museum are two young males probably from Tavoy. The carapace is a little over 12 millim. long and 13 millim. broad.

92. *Ptychognathus andamanica*, n. sp.

Closely related to P. *pusilla*, of which it may be an Andaman variety.

Carapace not much broader than long, quite flat, much depressed, the regions are hardly indicated, even when the carapace is quite dry, but the H-shaped mark in the middle is always plainly visible, the whole surface is closely and finely punctate : there are no post-frontal tubercles, but on the posterior part of either epibranchial region there is a fine line running obliquely-parallel with the postero-lateral, borders.

Front prominent, laminar, slightly sinuous, its extent is two-fifths the greatest breadth of the carapace.

The antero-lateral borders are cut into 3 not very acute or distinct lobes (including the outer orbital angle), of which the first is much the largest, and the last much the smallest.

Upper border of the orbit slightly sinuous : the antennules fold very obliquely. The anterior border of the buccal cavern is granular and a little concave.

The exognath is long and elliptical; its breadth in the female, is nearly twice that of the ischiognath.

The chelipeds in the female (male unknown) are about as long as the carapace, and their outer surface is very finely reticulate-granular : inner angle of wrist pronounced, but not spiniform : palm without hair, but there is a characteristic brush of stiffish hair at the tip of the fixed finger on its outer surface. The fingers have broad tips, especially the fixed finger, which is stouter and more strongly toothed than the dactylus : the dactylus is about twice as long as the upper border of the palm : the outer surface of the palm and fixed finger is traversed, near the lower border, by a fine raised granular line.

The legs have not much tomentum on the basal joints, but the anterior border of the meropodites is rather thickly fringed: the subter-minal denticle on the anterior border of the meropodites is small, blunt, inconspicuous, or obsolescent. The 2nd and 3rd pair of legs, which are the longest, are about $1\frac{1}{2}$ times, the 1st pair are not $1\frac{1}{4}$ times, and the 4th pair are little more than once, the length of the carapace.

In the Indian Museum are two young females from a freshwater stream at the base of Saddle Hill in North Andaman Island. Their colour is dark mottled green. The carapace is a little over 13 millim. long and about 14 millim. broad.

93.   *Ptychognathus pusilla*, Heller.

Ptychognathus pusillus, Heller, Novara Crust. p. 60 : Kingsley, Proc. Ac. Nat. Sci. Philad. 1880. p. 204 : de Man, Notes Leyden Mus. V. 1883, p. 161, and Zool. Jahrb. Syst. IV. 1888–89, p. 440, and in Weber's, Zool. Ergebn Niederl. Ost-Ind. II. p. 325, and Zool. Jahrb. Syst. IX. 1895, p. 99, and X. 1898, pl. xxviii. fig. 22 : Ortmann, Zool. Jahrb., Syst. VII. 1893–94, p. 712.

This species, which was first found in the Nicobar Islands, is not represented in the Museum collection and I have never seen it.

94.   *Ptychognathus barbata* (A. M. Edw.).

Gnathograpsus barbatus, A. Milne Edwards, Nouv. Archiv. du Mus. IX. 1873, p. 316, pl. xvii. fig. 4.

Ptychognathus barbatus, Ortmann, Zool. Jahrb. Syst. VII, 1893-94, p. 712 : de Man, Zool. Jahrb., Syst., IX. 1895, p. 105.

Carapace decidedly broader than long, flat, depressed, the regions indistinct : the two postfrontal · tubercles are fairly distinct, but there is no distinct raised line on the posterior part of the epibranchial regions, running obliquely parallel with the posterior borders, such as is present in all the other Indian species. There is a good deal of tomentum on the sides of the carapace.

· · Front decidedly sinuous, not prominent, its extent is a little more than two-fifths the greatest breadth of the carapace.

· · The antero-lateral borders of the carapace are cut into 3 not very conspicuous teeth (including the outer orbital angle) of which the first is much the largest and the third much the smallest, as usual.

Upper border of the orbit little sinuous : the antennules fold nearly transversely. Anterior border of the buccal cavern finely granular.

The exognath is elliptical, with a slightly convex surface : in the male its greatest breadth is more than that of the ischiognath, in the female it is slightly narrower than in the male.

Chelipeds in the male about 1⅔ times the length of the carapace, the inner angle of the wrist little pronounced ; the hand massive, with a tuft of hair in the finger-cleft and running some little distance along the outer surface of both fingers ; the fingers are rather blunt, the dactylus, which is about twice the length of the upper border of the palm is longer slenderer and less strongly toothed than the fixed finger, against which it closes rather obliquely. In the female the chelipeds are about as long as the carapace and are not very massive, the inner angle of the wrist is dentiform, there is no hair on the hand or fingers, and the outer surface of the hand and fixed finger is traversed near the lower border by a raised line.

The leg-joints are less expanded and less abundantly plumed than in the other Indian species, and there is no subterminal spine on the anterior border of the meropodites. The 2nd and 3rd pairs of legs are about 1⅔ times, the 1st pair about 1½ times, and the last pair a little over once, the length of the carapace.

In the Indian Museum are 3 specimens from Diamond Island off the Pegu coast and from Akyab, (besides numerous specimens from Samoa). The carapace of an apparently adult male is 11 millim. long and 14 millim. broad.

PYXIDOGNATHUS, A. M. Edw.

Pyxidognathus, A. Milne Edwards, Bull. Soc. Philom. Paris (7) III. 1878, p. 109 : de Man, Notes Leyden Mus. V. 1883, p. 160, and Journ. Linn. Soc., Zool., XXII. 1888, p. 148.

Hypsilograpsus, de Man, Notes Leyden Mus. I. 1879, p. 72 (*ipso teste*).

This genus is closely related to *Varuna* and *Ptychognathus*. It differs from *Varuna* in the same particulars that *Ptychognathus* does, that is to say, the exognath of the external maxillipeds is much broader than the ischiognath, and the dactyli of the legs though compressed are not dilated. It further differs, both from *Varuna* and *Ptychognathus* in the following characters :—

(1) the carapace is decidedly transverse, is deep, and is dorsally strongly convex in both directions : it is also anteriorly declivous with the front deflexed, and its antero-lateral borders are hardly arched :

(2) the antennules fold transversely :

(3) the lower border of the orbit is complete, except of course at the orbital hiatus :

(4) the carpopodites and propodites of the legs are not particularly broad.

Distribution : Indo-Pacific in fresh or brackish water.

Key to the Indian species of Pyxidognathus.

I. A single spine on the posterior border of the meropodites of the legs . ... *P. fluviatilis.*

II. More than one spine on the posterior border of the meropodites of the legs .... *P. deianira.*

95. *Pyxidognathus deianira*, de Man.

Pyxidognathus deianira, de Man, Journ. Linn. Soc., Zool., XXII. 1888, p. 148, pl. x. figs. 4-6.

Carapace about $\frac{3}{4}$ as long as broad, convex, smooth, without distinction of regions excepting a faintish H-shaped mark in the middle. Free edge of front sinuous or four-lobed, as in the next species.

Antero-lateral borders of the carapace cut into three prominent acute teeth (including the outer orbital angle), the first of which is the largest, and the last of which is spine-like.

Upper border of orbit slightly sinuous, lower border finely denti-culate.

Exognath of the external maxillipeds, in the male, very much broader than the ischiognath, and having a smooth convex surface.

Chelipeds in the young male about $1\frac{1}{2}$ times the length of the cara-pace : inner border of ischium, arm, and wrist denticulate ; inner angle of

J. II. 53

wrist spiniform ; the upper border of the palm is granulate, a finely beaded raised line traverses the lower part of the outer surface of the palm and fixed finger, and there is a very short series of granules near the middle of the inner surface of the palm : the palm is nearly as high as long, and the dactylus is much longer than the upper border of the palm and closes against the fixed finger by the tip only.

The 2nd pair of legs, which are the longest, are not much short of twice the length of the carapace ; the 4th pair, which are the shortest, are but little longer than the carapace. In all the legs, the meropodite has some fine rugosities on its upper surface, a spine near the far end of the anterior border, and some spines on the posterior border—these being most numerous in the case of the 4th pair of legs : and in all, the edges of the 3 terminal joints are hairy but not plumose, nor are these joints broadened or compressed.

In the Indian Museum are two very small male specimens from Mergui.

96. *Pyxidognathus fluviatilis*, n. sp.

Carapace transverse, markedly convex, finely punctate, the regions indicated only by an H-shaped mark in its centre.

Front between two-fifths and a third the greatest breadth of the carapace, deflexed, sinuous or four-lobed, the two middle lobes broad, the outer lobes (= inner orbital angles) subacute.

Antero-lateral borders of the carapace slightly arched, cut into three prominent acute teeth (including the outer orbital angle) of which the first is the largest and least acute, and the third is spine-like.

Orbits of good depth, the upper border slightly sinuous, the lower border defined by a granular ridge running close behind the prominent denticulated ridge that bounds the infra-orbital region of the carapace.

Anterior border of buccal cavern prominent, finely crenulate. Exognath in the female broader than the ischiognath, and having a smooth convex surface.

Chelipeds in the female about as long as the carapace, more massive than the legs : inner angle of wrist acuminate : a raised line runs along the outer surface of the palm and fixed finger, close to the lower border : fingers rather sharp though spooned at tip, dactylus hardly twice the length of the upper border of the palm, longer and rather less strongly toothed than the fixed finger.

All the leg-joints are plumed, and all the dactyli are long compressed and recurved. In all the legs there is a very strong spine in the distal half of the posterior border of the meropodite, and in the first 3 pairs there is a smaller subterminal spine on the anterior border of the same joint. The 2nd and 3rd pairs of legs are about $1\frac{3}{4}$ times, the 1st pair are not quite $1\frac{1}{2}$ times, and the 4th pair are about $1\frac{1}{4}$ times the length of the carapace.

Colour mottled dark green. A single female was found clinging to the floats of a fisherman's net in the R. Ichamutty above Bongong in the Jessore District : its carapace is 15 millim. long and 19 millim. broad.

The legs are obviously adapted for swimming, and the recurved dactyli and spiny meropodites appear to be adaptations to a swift current.

The chief difference between this species and *P. deianira*—the female of the former being compared with the male of the latter—is that in this species the three terminal joints of the legs are more compressed and the posterior border of the meropodites is armed with a single spine.

Sub-family SESARMINÆ, Dana.

SESARMA, Say.

Sesarma, Say, Journ. Acad. Nat. Sci. Philad. I. 1817, p. 76 : Milne Edwards, Hist. Nat. Crust. II. 71, and Ann. Sci. Nat. Zool. (3) XX. 1853, p. 181 : A. Milne Edwards, Nouv. Archiv. du Mus. IX. 1873, p. 301 : Kingsley, Proc. Ac. Nat. Sci. Philad. 1880, p. 213 : Miers, Challenger Brachyura, p. 269 : de Man, Zool. Jahrb., Syst., II. 1886–87, p. 641 and IX. 1895–97, p. 128 : Bürger, Zool. Jahrb., Syst., VII, 1893–94, p 613.

Pachysoma, De Haan, Faun. Japon. Crust., p. 33.

Holometopus, Milne Edwards, Ann. Sci. Nat. Zool. (3) XX. 1853, p. 187.

Carapace squarish or actually square (the sides being straight and usually nearly parallel), usually deep (though occasionally shallow and much depressed), seldom very convex : the gastric region is almost always very well delimited, and is commonly divided into 5 subregions, and in most cases the 4 antero-lateral subregions project as 4 prominent post-frontal tubercles.

The side-walls of the carapace have everywhere a characteristic fine-meshed reticulate texture as regular as that of a sieve. This appearance is due to a multitude of small uniform granules arranged in pairs in close-set parallel rows : between each pair of granules is a little row of bristles, one of which in each row is long and points diagonally forwards.

The front occupies half, or more, of the anterior border of the carapace, and is obliquely or vertically deflexed.

The orbits, which occupy the rest of the anterior border of the carapace, are oval and of good depth : below their outer angle is a deepish gap leading into a system of grooves which open into a notch at the antero-lateral angle of the buccal cavern. At the inner angle of the orbit is the usual tooth, belonging to the inner of the two fossæ into which (as in all the crabs of this subfamily) the orbit is so plainly divided. The eyes are of no great length.

The antennules fold nearly transversely into rather narrow fossæ : the inter-antennular septum is very broad.

The antero-external angle of the 2nd joint of the antennal peduncle is a good deal produced : the antennal flagellum, which is slender and rather short, lies in the orbital hiatus.

Epistome well defined, prominent, rather short fore and aft. Buccal cavern square. The external maxillipeds leave between them a large rhomboidal gap, which is a good deal filled up by a hairy fringe : they are obliquely traversed, from a point behind the antero-external angle of the ischium to the antero-internal angle of the merus, by a conspicuous line or crest of hairs : the palp, which is rather coarse, is attached to the rounded summit of the obliquely-directed merus.

Chelipeds massive—not always so in the female—usually subequal, of no great length : palm high and short, the fingers though subacute, are hollowed at the tip.

The legs do not usually differ very markedly in length, though the third pair are the longest and the first and last (4th) pairs the shortest : the meropodites are thin, and are usually, but not always, broad.

The abdomen in both sexes consists of 7 separate segments : in the male it occupies the whole breadth of the sternum between the bases of the last pair of legs. In both sexes the second segment, as well as the exposed portion of the first, are narrow fore and aft. In the female the last segment is small and narrow from side to side, and is more or less impacted in the broad 6th segment : in the male also the last segment is much narrower than the one that precedes it.

Distribution : all tropical and subtropical seas : not found in the Mediterranean.

I am not inclined to adopt the subgenera proposed by Dr. de Man, although I must admit that his system is convenient in practice, for identifying species.

I may also mention here that specific distinctions based merely on the sculpture of the dactylus of the male chelæ are inadmissible, as the sculpturing frequently differs in the two fingers of the same individual.

Key to the Indian species of Sesarma.

I. Carapace deepish, its length decidedly less than its breadth between the antero-lateral angles, its sides nearly parallel—never markedly divergent posteriorly :—

    1. The inner border of the arm bears, near its far end, a large acute tooth : on the upper surface of the palm of the male are at least two characteristic *oblique* comb-like ridges : the upper surface of the movable finger of the male is milled :—

        i. Posterior border of the meropodites of the legs entire :—

            a. No tooth on the lateral border of the carapace behind the orbital angle :—

                α. Front more than half the extent of the anterior border of the carapace............................ *S. quadratum.*

                β. Front exactly half the extent of the anterior border of the carapace............................ *S. pictum.*

            b. A tooth on the lateral border of the carapace, behind the orbital angle .... *S. bidens.*

        ii. Distal end of the posterior border of the meropodites of the legs acutely serrate (no tooth behind the outer orbital angle)... .... *S. Andersoni.*

    2. The inner border of the arm does not end in a large spine or acute lobe, though it may be a little dilated distally : there are no oblique pectinated ridges on the upper surface of the palm, and the upper surface of the movable finger of the male though it may be granular is not milled :—

        i. A tooth at the inner angle of the wrist (a tooth on the lateral border of the carapace behind the orbital angle) :—

            a. The breadth of the carapace between the antero-lateral angles is equal to, or more than, the breadth between the epibranchial teeth............................ *S. Edwardsi.*

            b. The breadth of the carapace between the antero-lateral angles is decidedly less than the breadth between the epibranchial teeth, the sides of the carapace being markedly sinuous............ .... *S. Meinerti.*

        ii. No spine at the inner angle of the wrist :—

            a. Carapace and appendages not uniformly tomentose : two acute teeth—the second of which is hardly visible—on the lateral border, behind the acute orbital angle.. *S. intermedium.*

   b. Carapace and appendages covered with
   a short but very dense fur, amid which
   are prominent tubercle-like tufts of
   hair: lateral borders cut into three
   blunt lobes (including the orbital angle)
   of equal size.................................... *S. lanatum.*

II. Carapace nearly square, its length being little less than its
  breadth between the antero-lateral angles : the inner border
  of the arm ends in an acute serrated lobe : a very finely
  pectinated ridge traverses the upper surface of the palm,
  fore and aft, close to the upper border : (a tooth on the
  lateral border of the carapace behind the orbital angle) :—

   1. Carapace deep, its sides nearly parallel : a transverse
    granular ridge on the inner surface of the palm :
    dactyli of the legs of good length :—

     i. Upper border of movable finger of male with
      an elegantly milled crest of 40 to 60 fine
      lamellæ... *S. tæniolatum.*

     ii. Upper border of movable finger of male with
      a coarsely crenulate crest .... *S. tetragonum.*

   2. Carapace shallow and depressed, its sides divergent
    posteriorly : no transverse granular crest on the inner
    surface of the palm : dactyli of the legs short : (a
    milled crest of about 25 very fine lamellæ on the
    upper border of the movable finger of the male)........ *S. Brockii.*

III. Carapace somewhat elongate (its length being decidedly
  more than its breadth at the antero-lateral angles), shallow
  and depressed :—

   1. No tooth on the lateral border of the carapace behind
    the orbital angle : legs with remarkably broad mero-
    podite and remarkably short propodite : upper border
    of movable finger of male with an elegantly milled
    crest of about 40 fine lamellæ .... *S. latifemur.*

   2. Two teeth on the lateral border behind the orbital
    angle : movable finger without any milling :—

     i. Post-frontal tubercles of the gastric region
      serrated : legs with meropodites of good
      breadth and dactyli of good length............... *S politum.*

     ii. Post-frontal tubercles smooth : legs with rather
      narrow meropodites and short dactyli............ *S. oceanicum.*

IV. The length of the carapace is just equal to its breadth at
  the antero-lateral angles : legs long and slender, with
  elongate dactyli :—

   1. Carapace shallow, depressed, perfectly square, its
    sides quite parallel : two little teeth on the lateral
    border behind the orbital angle..... ... . *S. Finni.*

   2. Carapace deepish, its sides strongly divergent pos-
    teriorly where its breadth is much greater than its
    length : two teeth (not including the orbital angle)

on the lateral border, the posterior one being very small :—

    i. Third pair of legs not three times the length of the carapace.. *S. longipes.*

    ii. Third pair of legs more than three-and-a-half times the length of the carapace................... *S. kraussi.*

97. *Sesarma quadratum* (Fabr.).

Cancer quadratus, Fabricius, Ent. Syst. Suppl. p. 341.

Ocypoda quadrata, Bosc, Hist. Nat. Crust. I. p. 198.

Ocypoda plicata, Latreille, Hist. Nat. Crust. &c. VI. p. 47.

Sesarma quadrata, Milne Edwards, Hist. Nat. Crust. II. 75, and Ann. Sci. Nat., Zool , (3) XX. 1853, p. 183.

Sesarma quadratum, A. Milne Edwards, Nouv. Archiv. du Mus. IX. 1873, p. 302 : Miers, Phil. Trans. Vol. 168, 1879, p. 490.

Sesarma quadrata, Richters, in Mobius' Meeresf. Maurit. p. 157 : Kingsley, Proc. Ac. Nat. Sci. Philad. 1880, p. 217 : Lenz and Richters, Abh. Senck. Nat. Ges. XII. 188, p. 425 : de Man, Zool. Jahrb. Syst. II. 1887, p. 655, pl. xvii. fig. 2 and p. 683, and IV. 1889, p. 434, and IX. 1895–97, pp. 181, 182, and Notes Leyden Mus. XII. 1890, p. 99, and in Weber's Zool. Ergebn. Niederl. Ost-Ind. II. p. 328 : Thallwitz, Abh. Mus. Dresden, 1890–91, No. 3, p. 37 : Henderson, Trans. Linn. Soc. Zool., (2) V. 1893, p. 392 : Ortmann, Zool. Jahrb. Syst. VII. 1893–94, p. 724.

Grapsus (Pachysoma) affinis, De Haan, Faun. Jap. p. 66, pl. xviii. fig 5.

Sesarma affinis, Krauss, Sudafr. Crust. p. 45 : Milne Edwards, Ann. Sci. Nat. Zool. (3) XX., 1853, p. 183 : Heller, Novara Crust. p. 62 : de Man, Notes Leyden Mus. II. 1880, p. 22 : Miers, Ann. Nag. Nat. Hist. (5) V. 1880, p 312 : Kingsley, *l.c. supra,* p. 213 : Ortmann, *l.c. supra,* p. 724.

Sesarma ungulata : Milne Edwards, Ann. Sci. Nat., Zool. (3) XX. 1853, p. 184 : Kingsley, *l.c. supra,* p. 218.

Sesarma aspera, Heller, Novara Crust. p. 63, pl. vi. fig. 1 : Kingsley, *l.c. supra,* p. 214 : Müller, Verh. Nat. Ges. Basel, 1886, p. 476 : de Man, Zool. Jahrb. II. 1887, p. 656 and Journ. Linn. Soc. Zool. XXII. 1887–88, p. 169.

Sesarma melissa, de Man, Zool. Jahrb. Syst., II. 1887, p. 656, and Journ Linn Soc., Zool., XXII. 1888, p. 170, pl. xii. figs. 5–7, and Zool. Jahrb. Syst., IV. 1889, p. 434.

Carapace hardly convex, decidedly broader than long, its length being about four-fifths its breadth between the antero-lateral angles, deep ; the 4 post-frontal lobes prominent equal and a little rugose trans. versely, the rugæ being sparsely tufted with hair ; the cardiac and intes. tinal regions very much less distinct than the gastric : some oblique striations on the epibranchial regions.

Front decidedly more than half the greatest breadth of the carapace, not very deep, its free margin usually but slightly sinuous. Lateral borders of carapace nearly parallel, a little divergent ante. riorly, without any tooth behind the acute orbital angle.

The chelipeds differ in the sexes, being about $1\frac{3}{4}$ times the length of the carapace in the male and much more massive than the legs, but in the female hardly 1⅓ times the length of the carapace and not more massive than the legs. In both sexes the outer surface of the arm wrist and palm are granular, the granules on the arm and wrist having a squamiform arrangement, the inner border of the arm bears a subterminal spine of large size, the upper border of the arm ending in a much smaller spine, the inner angle of the wrist is not dentiform, and the inner surface of the palm is more or less granular. In the male the palm is a little swollen below and has, on its upper surface, some short oblique crests, of which two are most elegantly pectinated : in the female the palm is not swollen and the crests are simply granular. The dactylus is less than twice the length of the upper border of the hand (palm) and its dorsal surface is elegantly milled with from 11 to 19 blunt, rather coarse, transverse lamellæ : in the female this milling is incomplete and very indistinct. In neither sex is there any great gap between the closed fingers.

The meropodites of the legs are foliaceous, their greatest breadth in the 2nd and 3rd pairs being more than half their length, their anterior border ends at an acute subterminal spine, and their dorsal surface has some fine transverse squamiform sculpture. The anterior border of the last three joints of the legs, and part of the posterior border of the last two, is fringed with tufts of bristles. The 3rd pair of legs, which are slightly the longest, are about twice the length of the carapace, and their dactylus is about three-fourths the length of their propodite.

In the Indian Museum are 42 specimens from both coasts of the Peninsula, Ceylon, the Andamans and the Nicobars.

In a male of good size the carapace is 16 millim. long and 20 millim. broad.

98. *Sesarma pictum,* De Haan.

Grapsus (Pachysoma) pictus, De Haan, Faun. Japon. Crust. p. 61, pl. xvi. fig. 6.

Sesarma picta, Krauss, Sudafr. Crust. p. 45 : Milne Edwards, Ann. Sci. Nat., Zool., (3) XX. 1853, p. 184 : Stimpson, Proc. Ac. Nat. Sci. Philad. 1858, p. 106 : de Man, Notes Leyden Mus. II. 1880, p. 22, and Zool. Jahrb., Syst., II. 1887, p. 657, and IX. 1895–97, pp. 181, 182, and Journ. Linn. Soc., Zool., XXII. 1888, p. 171.: Bürger, Zool. Jahrb., Syst., VII. 1893–94, p. 626 : Ortmann, Zool. Jahrb., Syst., VII. 1893–94, p. 725.

Agrees with *S. quadratum* in everything but the following particulars :—

(1) the carapace is not so broad, its length being about five-sixths of its breadth between the antero-lateral angles :

(2) the front is not so broad, its extent being only half the breadth of the carapace :

(3) the meropodites of the legs are not so broadly foliaceous, their greatest breadth, in the middle two pairs, being less than half their length.

The Indian Museum possesses a single specimen from Mergui.

99.   *Sesarma bidens* (De Haan).

Grapsus (Pachysoma) bidens, De Haan, Faun. Japon. Crust. p. 60, pl. xvi. fig. 4, and pl. xi. fig. 4.

Sesarma bidens, Dana, U. S. Expl. Exp. Crust. pt. I. p. 353 : Milne Edwards, Ann. Sci. Nat., Zool., (3) XX. 1853, p. 185 : Stimpson, Proc. Ac. Nat. Sci. Philad. 1858, p. 105 : Heller, Novara Crust. p. 64 : Hilgendorf, in v. d. Decken's Reisen Ost-Afr., Crust., p. 91, pl. iii. fig. 3a : Hoffmann, in Pollen & Van Dam, Faun. Madag. Crust. p. 24 : Miers, Ann. Mag. Nat. Hist. (5) V. 1880, p. 313, and Zool. H. M. S. Alert, pp. 184, 246 : Kingsley, Proc. Ac. Nat. Sci. Philad. 1880, p. 214 : de Man, Notes Leyden Mus. II. 1880, p. 28, and Zool. Jahrb., Syst., II. 1887, p. 658, and in Weber's Zool. Ergebn. Niederl. Ost-Ind. II. p. 330 : Lenz & Richters, Abh. Senck. Nat. Ges. XII. 1881, p. 425 : Bürger, Zool. Jahrb., Syst., VII. 1893–94, p. 628 : Ortmann, *ibid.* p. 726 : Nobili, Ann. Mus. Genova (2) XX. 1899, p. 269.

Sesarma Dussumieri, Milne Edwards, *l. c. supra* : Tozzetti "Magenta" Crust. p. 145, pl. ix. figs. 3 *a-f* : Kingsley, Proc. Ac. Nat. Sci. Philad. 1880, p. 215 : de Man, Zool. Jahrb. Syst. II. 1887, p. 659, and IX. 1895–97, p. 208, and Journ. Linn. Soc., Zool., XXII. 1888, p. 177, pl. xii. figs. 8–12 : Ortmann, Zool. Jahrb., Syst., VII. 1893–94, p. 726.

Sesarma lividum, A. Milne Edwards, Nouv. Archiv. du Mus., V. 1869, Bull. p. 25, and IX. 1873, p. 303, pl. xvi. fig. 2 : Brocchi, Ann. Sci. Nat., Zool., (6) II. 1875, Art. 2, p. 83 (*male appendages*) : Kingsley, *tom. cit. supra*, p. 216 : de Man, Archiv. f. Naturges. LIII. 1887, i. p. 381, pl. xvii. fig. 1, and Zool. Jahrb. Syst. II. 1887, p. 659, and Journ. Linn. Soc., Zool., XXII. 1888, p. 180.

Sesarma Haswelli, de Man, Zool. Jahrb., Syst., II. 1887, p. 658, and Journ. Linn. Soc., Zool., XXII. 1888, p. 175.

This species very closely resembles *S. quadratum*, from which it differs in the following characters :—

(1) there is a small sharp tooth on the lateral border of the carapace, immediately behind the outer orbital angle :

(2) the carapace is slightly less transverse (though decidedly broader than long) :

(3) the transverse ridges on the upper surface of the dactylus of the male chelæ are coarser and shorter and more tubercle-like.

In the Indian Museum are 52 specimens from the coasts of the Bay of Bengal, Andamans, Nicobars and Ceylon.

100. *Sesarma Edwardsi*, de Man.

Sesarma Edwardsi, de Man, Zool. Jahrb., Syst , II. 1887, p. 649, and Journ. Linn.
Soc., Zool., XXII. 1888, p. 185, pl. xiii. figs. 1–4 : Ortmann, Zool. Jahrb., Syst., VII.
1893-94, p. 721.

Differs from *S. quadratum* in the following particulars :—

(1) the carapace is squarer and less transverse, and the four post-frontal lobes of the gastric region are more prominent; the front also is slightly, but distinctly, broader:

(3) there is a sharp tooth on the lateral border of the carapace immediately behind the antero-lateral angle :

(3) the upper border of the arm does not end in a spine, and though there may be a slight subterminal dilatation of the crenulated inner border of the arm there is *no large spine :*

(4) there is a *sharp tooth or spine just below the inner angle of the wrist :*

(5) the upper surface of the wrist and outer surface of the palm are covered—usually very closely covered—with vesiculous tubercles; and there are smaller and sharper tubercles on the upper surface of the dactylus and the lower surface of the fixed finger of the chelæ :

(6) there are no oblique pectinated crests on the palm :

(7) *the male abdomen is singularly broad.*

In the Indian Museum are 126 specimens, most of which came from the Burma coast from Arakan to Tavoy, the rest from the Gangetic delta, the Andamans and Ceylon.

In the variety separated by de Man as *crassimana* the abdomen is not quite so broad as it is in the typical form, and the palm of the male is larger and more swollen.

101. *Sesarma intermedium* (De Haan).

Grapsus (Pachysoma) intermedius, De Haan, Faun. Japon. Crust. p. 61, pl. xvi.
fig. 5.

Sesarma intermedia, Milne Edwards, Ann. Sci. Nat., Zool., (3) XX. 1853, p. 186 :
Stimpson, Proc. Ac. Nat. Sci. Philad. 1858, p. 105: Heller, Novara Crust. p. 64 :
Kingsley, Proc. Ac. Nat. Sci. Philad. 1880, p. 216 : Miers, Ann. Mag. Nat. Hist.
(5) V. 1880, p. 314: de Man, Notes Leyden Mus. II. 1880, p. 25, and Zool. Jahrb.,
Syst., II, 1887, p. 649, and Journ. Linn. Soc., Zool., XXII. 1888, p. 182 : Ortmann,
Zool. Jahrb. Syst. VII. 1893–94, p. 721.

Differs from *S. quadratum* in the following particulars :—

(1) the carapace is more quadrate and less transverse, the post-frontal lobes are less prominent and much smoother, and the front is broader :

(2) there is a tooth—and sometimes also a second rudimentary tooth—on the lateral border immediately behind the orbital angle :

(3) there is no large subterminal spine on the inner border of the arm, nor does the upper border end in a spine:

(4) in the corner of the upper surface of the palm there are in the male some oblique granular lines, but no pectinated crests; and on the inner surface of the palm there is a conspicuous transverse granular crest:

(5) the upper surface of the dactylus of the male chelæ is granular in its proximal half, but is not milled with transverse lamellæ.

From *S. Edwardsi* it is distinguished by numerous characters, but the absence of a spine at the inner angle of the wrist is sufficiently characteristic.

In the Indian Museum are 5 specimens from Mergui.

102. *Sesarma Meinerti*, de Man.

Sesarma Meinerti, de Man, Zool. Jahrb., Syst., II. 1887, pp. 648, 668, and IX. 1895-97, p. 166: Bürger, Zool. Jahrb. Syst. VII. 1893-94, p. 617, and Ortmann, *ibid.* p. 720.

Sesarma tetragona, Edw. (*nec* Fabr.), Milne Edwards, Hist. Nat. Crust. II. 73, and Ann. Sci. Nat., Zool., (3) XX. 1853, p. 184: A. Milne Edwards, Nouv. Archiv. du Mus. IV. 1868, p. 71, and IX. 1873, p. 304, pl. xvi. fig. 4: Hilgendorf, in v. d. Decken's Reisen Ost-Afr., Crust. p. 90: Hoffmann, in Pollen & Van Dam, Faun. Madag. Crust. p. 23: Hilgendorf, MB. Ak. Berl. 1878, p. 809: Kingsley, Proc. Ac. Nat. Sci. Philad. 1880, p. 218.

Carapace convex, especially fore and aft, a little broader than long, deep: the 4 post-frontal lobes prominent, unequal—the outer ones being much narrower than the middle pair; the cardiac and intestinal regions are quite distinct, and the usual oblique striations are found on the epibranchial regions: the whole dorsal surface of the carapace is rather profusely covered with tufts of hair.

Front decidedly more than half the greatest breadth of the carapace, which is just behind the orbital angles, not very deep, its free edge sinuous. Lateral borders of the carapace somewhat sinuous, armed with a large tooth behind the orbital angle: there may even be a trace of a second epibranchial tooth.

Chelipeds subequal, almost equally massive in both sexes, about twice as long as the carapace. The outer surface of the arm and wrist is finely rugose, that of the palm is only pitted: neither the upper nor the inner border of the arm end in a tooth: inner angle of wrist pronounced but not dentiform: no pectinated crests of any kind on the palm: the fingers are a good deal arched and meet only at tip, the upper surface of the dactylus in the male has a row of inconspicuous denticles: on the inner surface of the palm there is an oblique granular crest.

The meropodites of the legs are foliaceous, but their breadth is not twice their length; but otherwise the legs are as in *S. quadratum.*

The abdomen of the male is decidedly narrow.

In the Indian Museum are 26 specimens from the Andamans and one from Madras. The carapace of a large one is 33 millim. long and 38 millim. broad : in the female the carapace is not so broad.

103.   *Sesarma Andersoni,* de Man.

Sesarma Andersoni, de Man, Zool. Jahrb., Syst., II. 1887, p. 657, and Journ. Linn. Soc. Zool. XXII. 1888, p. 172, pl. xii. figs. 1–4.

Carapace moderately deep, hardly convex, considerably broader than long, the four post-frontal lobes of the gastric region only moderately prominent, nearly equal, pitted ; the cardiac and intestinal regions faintly indicated ; the oblique striations of the epibranchial regions very sharp and distinct, one of them almost projects beyond the lateral border as a tooth behind the orbital angle.

Front more than half the greatest breadth of the carapace, not very deep, its free margin a little convex but nearly straight. The lateral borders of the carapace are slightly convergent posteriorly : except for the afore-mentioned projection of the first branchial ridge there is no tooth behind the orbital angle.

Chelipeds much larger in the male than in the female, but the difference is not so marked as in *S. quadratum.* The inner border of the arm ends in a very acute denticulated lobe : the palm is traversed on the outer surface, near the lower border, by a fine raised line, and on the upper surface in the male are numerous short parallel oblique striæ one of which at least is most elegantly pectinate : in the female these crests are less numerous and less distinct : the upper surface of the dactylus of the male is milled, the lamellæ increasing in size and coarseness from behind forwards.

At the distal end of the posterior border of the meropodites of the legs are three or four strong spines, decreasing in size from behind forwards, but there is no subterminal spine on the anterior border : in other respects, except that the dactyli are slightly shorter, the legs are very similar to those of *S. quadratum.* The male abdomen is broad.

In the Indian Museum are 8 specimens from Mergui : the carapace of the largest is 7 millim. long and 9 millim. broad.

104.   *Sesarma lanatum,* n. sp.

Carapace deepish, dorsally flat, everywhere covered, as also are the appendages, with a dense fur amid which are freely scattered little dense adherent tufts of hair resembling tubercles. When this covering is removed the surface of the carapace is smooth and polished, with the gastric region and its four post-frontal tubercles distinct.

The length of the carapace is considerably less than its breadth between the antero-lateral angles.

Front a little more than half the breadth of the carapace, obliquely deflexed, its free margin nearly straight. The lateral borders of the carapace are nearly parallel and anteriorly are cut into three blunt lobes of nearly equal size—including the outer orbital angle.

The chelipeds when denuded have a smooth surface and sharp borders: they are similar in the two sexes, except that they are much more massive in the male. There is a blunt angular projection at the far end of the inner border of the arm, the inner angle of the wrist is pronounced but not dentiform, and the upper border of the palm is traversed fore and aft by a fine sharp crest: in the male the palm is at least as high as long: the upper border of the dactylus is faintly crenulate in its proximal two-thirds.

The meropodites of the legs are foliaceous, but their breadth is less than half their length: their borders are entire. The dactyli of the legs are claw-like, their length being about three-fourths that of the propodites.

The abdomen of the male is narrow.

In the Indian Museum are 4 specimens from Bombay and Karachi: the carapace of the largest is $8\frac{1}{2}$ millim. long and 10 millim. broad.

105.   *Sesarma tæniolatum*, White.

Sesarma tæniolatum, White, List Crust. Brit. Mus. p. 38 (1847): Miers, P. Z. S. 1877, p. 137, and Ann. Mag. Nat. Hist. (5) V. 1880, p. 313: de Man, Notes Leyden Mus. II. 1880, p. 26: Kingsley, Proc. Ac. Nat. Sci. Philad. 1880, p. 218: de Man, Zool. Jahrb., Syst., II. 1887, pp. 647, 666, and IX. 1895-97, p. 166, and Journ. Linn. Soc., Zool., XXII 1888, p. 181, and in Weber's Zool. Ergebn. Niederl. Ost-Ind. II. p. 330: Bürger, Zool. Jahrb., Syst., VII. 1893-94, p. 615, and Ortmann, *ibid.* p. 720.

Sesarma Mederi, Milne Edwards, Ann. Sci. Nat. Zool. (3) XX. 1853, p. 185: Tozzetti, "Magenta" Crust. p. 136, pl. ix. figs. 1 *a–i*.

Carapace deep, nearly flat dorsally, square, its length being slightly less than its breadth between the antero-lateral angles. All the regions are quite well defined, and the 4 post-frontal tubercles—the middle two of which are not very much broader than the outer ones—are very prominent. The whole dorsum of the carapace is covered with tufts of hair, which are largest and longest anteriorly. There are some oblique striæ on the sides of the epibranchial regions.

Front half, or a little more than half, the breadth of the carapace, not very deep, its free margin strongly sinuous. Lateral borders of the carapace nearly parallel, armed with one acute tooth behind the acute outer orbital angle.

The chelipeds are similar in the two sexes, except that they are a good deal more massive and more sharply sculptured in the male. They are not quite twice the length of the carapace : the outer surface of the arm and wrist are granular-rugose, the outer surface of the palm is granular, and there is a transverse granular ridge on the inner surface of the palm : the upper border of the arm is crest-like and ends in a sharp tooth, and the distal end of the inner border forms an acute angular serrate lobe : the inner angle of the wrist is dentiform : close to and nearly parallel with the upper border of the palm runs a fine and very finely and evenly pectinate crest : along the upper border of the dactylus runs a very elegantly milled crest of from 40 to 60 fine teeth. In the male the palm is at least as high as long, the fingers meet only at tip, and the dactylus is about twice the length of the upper border of the palm.

The meropodites of the legs are foliaceous, but their greatest breadth is not quite half their length : there is a sharp subterminal spine on their anterior border only. The dactyli of the legs are two-thirds, or more, the length of the propodites. The 3rd pair of legs, which are the longest, are a little more than twice the length of the carapace.

In the Indian Museum are 9 specimens, from Mergui, the Andamans, and Penang. The carapace of a large specimen is nearly 38 millim. long and nearly 40 broad.

106. *Sesarma tetragonum* (Fabr.).

Cancer tetragonus, Fabricius, Ent. Syst., Suppl. p. 341.
Cancer fascicularis, Herbst, Krabben etc. III. i. 49, pl. xlvii. fig. 5.
Sesarma tetragona, de Man, Zool. Jahrb., Syst., II. 1887, p. 646 : Henderson, Trans. Linn. Soc., Zool., (2) V. 1893, p. 392.

This · species closely resembles *S. tæniolatum,* from which it differs in the following characters :—

(1) the carapace is slightly broader :

(2) the subterminal lobe of the inner border of the arm is smaller, while the tooth at the inner angle of the wrist is more pronounced :

(3) the fine striated crest along the upper border of the palm is shorter :

(4) the crest of the upper surface of the movable finger of the chelæ is coarsely crenulate.

In the Indian Museum are 8 specimens from Ceylon, Madras, the Mahanaddi Delta, and the Ganges Delta. The carapace of a large one is 40 millim. long and 43 millim. broad.

107. *Sesarma Brockii*, de Man.

Sesarma Brockii, de Man, Zool. Jahrb., Syst., 1887, p. 651, and IX. 1895–97, p. 171, and Archiv f. Naturges. LIII. 1887, i. p. 373, pl. xvi. fig. 3 : Thallwitz, Abhand. Zool. Mus. Dresden, 1890–91, No. 3, p. 39 : Ortmann, Zool. Jahrb., Syst., VII. 1893–94, p. 721.

Resembles *S. tæniolatum*, but differs in the following characters :—

(1) the carapace is *shallow* and much depressed, its length is just equal to its breadth between the antero-lateral angles, its dorsal surface is not so hairy, and its sculpture though similar is not so deeply cut : the front is not so sinuous.

(2) the lateral borders of the carapace are *slightly divergent* posteriorly, and there are *two* teeth—the posterior of which is, however, extremely small—behind the outer orbital angle :

(3) no subterminal spine on the upper border of the arm : no transverse granular crest on the inner surface of the palm :

(4) the milled crest along the upper border of the dactylus of the chelæ is lower and has only about 25 teeth :

(5) the legs are longer, their meropodites are narrower and their dactyli—except in the case of the 1st pair of legs—are *barely half the length of their propodites.*

In the Indian Museum there is a young male from the Andamans. In this specimen the chelipeds are not massive and are very little longer than the carapace.

108. *Sesarma latifemur*, n. sp.

Closely related to *S. elongatum*, A. M. Edw.

This species belongs to the same natural group as *S. tæniolatum*, from which it differs only in the following characters :—

(1) the carapace is *shallow* and much depressed, and its length is decidedly *more* than its breadth between the antero-lateral angles, its dorsal surface is not quite so hairy and its post-frontal lobes are deeper cut :

(2) the lateral borders of the carapace are decidedly *divergent* posteriorly and have no tooth behind the orbital angle :

(3) the male chelipeds are little longer than the carapace : the crest-like upper border of the arm does not end in a spine : the inner

angle of the wrist, though well pronounced, is not spiniform : the transverse beaded ridge on the inner surface of the palm is very short :

(4) the dactylus of the chelæ is not nearly twice the length of the upper border of the palm, and the milled crest on its upper surface consists of not more than 40 teeth :

(5) the meropodites of the legs are remarkably foliaceous, their greatest breadth, in the case of the 2nd and 3rd pairs, being more than half their length : all the leg joints are thinner and flatter :

(6) the dactyli of the legs are remarkably short, their length, in the case of the 2nd and 3rd pairs, being *less than half* the length of their propodites.

In the Indian Museum is a single male from the Andamans : its carapace is nearly 35 millim. long, and a little over 30 millim. broad across the antero-lateral angles.

109. *Sesarma politum*, de Man.

Sesarma polita, de Man, Zool. Jahrb. Syst. II. 1887, p. 654 : Journ. Linn. Soc., Zool., XXII. 1888, p. 189, pl. xiii. figs. 7–9.

Carapace shallow and much depressed, a good deal longer than broad, all the regions well defined : the four post-frontal lobes of the gastric subregions are deep-cut and very prominent, their anterior overhanging edges are serrated and their surface bears some transversely arranged sharpish tubercles : the two middle lobes are decidedly larger than the outer ones.   There are no oblique striæ on the epibranchial regions.

Front more than half the breadth of the carapace, its free margin markedly sinuous.   The lateral borders of the carapace are nearly parallel though slightly sinuous : there are two well cut teeth behind the outer orbital angle.

Chelipeds equal, and not so very much longer than the carapace : the outer surface of the arm wrist and hand are closely beset with small tubercles, which in places have a squamiform look, and the inner surface of the palm is granular but has no transverse ridge : the inner and outer borders of the arm, the inner border of the wrist, and the upper border of the palm and movable finger are conspicuously serrulate, and there is also a noticeable dilatation near the far end of the inner border of the arm.   There are no pectinated crests of any sort on the palm, and the fingers—both surfaces of which are smooth and polished—have no large gap between them when closed.

The legs are shortish, the 3rd pair being hardly $1\frac{2}{3}$ times the length of the carapace, and rather slender. The meropodites are nearly three times as long as broad, they have a subterminal spine on the anterior border and in the case of the 1st pair their posterior border is distinctly serrulate. The dactyli are rather short, their length, in the third pair, being less than two-thirds the length of the propodite: they are remarkably tomentose.

In the Indian Museum there is a single specimen from Mergui: its carapace is 38 millim. long and 35 millim. broad.

110. *Sesarma oceanicum*, de Man.

Sesarma oceanica, de Man, Zool. Jahrb., Syst., IV. 1889, p. 429, pl. x. fig. 9, and Notes Leyden Mus. XIII. 1891, p. 52.

Carapace shallow, depressed, its length greater than its breadth between the antero-lateral angles; all the regions are fairly well defined and the 4 post-frontal lobes of the gastric subregions are prominent, the middle pair being more than twice as broad as the two outer ones: the surface of the carapace is granular anteriorly and punctate posteriorly, and near the sides are numerous short oblique striæ.

Front half the breadth of the carapace, deepish, its free margin a a little sinuous: orbits not at all oblique: the lateral borders of the carapace have a slight, but distinct, convex curve, and there are two teeth—the posterior of which is extremely small—behind the outer orbital angle.

Chelipeds equal, not much longer than the carapace: the outer surface of the arm and wrist are rugose and both surfaces of the palm are studded with sharpish granules: there is a small angular lobe near the far end of the inner border of the arm, and the inner angle of the wrist is dentiform: the palm is not quite as high as long, close to and nearly parallel with its upper border is a fine and finely granular ridge: the dactylus is about half as long again as the upper border of the palm, and there are some sharpish granules along its upper surface.

The legs are slender: their meropodites are more than three times as long as broad and are not foliaceous, they have a subterminal spine on the anterior border only: their dactyli are shortish, those of the 3rd pair being less than two-thirds the length of their propodites, and are densely plumed: the 3rd pair of legs are about $2\frac{1}{3}$ times the length of the carapace.

In the Indian Museum is a single specimen from the Nicobars: its carapace is 20 millim. long, and 16·5 millim. across the antero-lateral angles.

111. *Sesarma Finni,* n. sp.

Near *S. maculata,* de Man.

Carapace shallow, depressed, flat, perfectly square, its length being equal to its breadth at the antero-lateral angles and its sides being parallel: the regions are indicated, but not emphasized, and the 4 post-frontal lobes are sharply prominent, the middle pair being much broader than the outer ones.

Front half the breadth of the carapace, deepish, its free edge nearly straight: two little teeth on the lateral border of the carapace, behind the outer orbital angle.

In the chelipeds of the female the outer surface of the arm wrist and hand are granular; the upper border of the arm ends acutely, and the inner border ends in a spine; the inner angle of the wrist is pronounced, but is not dentiform; and the upper surface of the palm is traversed, fore and aft, close to the upper border, by a fine and finely milled ridge.

Legs long and slender, the 3rd pair being more than $2\frac{1}{2}$ times the length of the carapace: their meropodites are not foliaceous, being about three times as long as broad, and they have a subterminal spinule on the anterior border only: their dactyli are long and slender, those of the 3rd pair being more than three-fourths the length of the propodite: the propodites and dactyli of all the legs are fringed with short stiff sharp bristles.

The species is represented by a small female from the Andamans: its carapace is not quite 11 millim in either diameter.

112. *Sesarma longipes,* Krauss.

Sesarma longipes, Krauss, Sudafr. Crust. p. 44, pl. iii. fig. 2: Milne Edwards, Ann. Sci. Nat., Zool., (3) XX. 1853, p. 199: Kingsley, Proc. Ac. Nat. Sci. Philad. 1880, p. 216: de Man, Zool. Jahrb., Syst., II. 1887, p. 651.

The length of the carapace is equal to its breadth at the antero-lateral angles, but as the lateral borders of the carapace diverge considerably, from before backwards, the *greatest* breadth of the carapace (at the level of the 2nd pair of legs) is considerably more than the length.

Carapace deepish, very slightly convex; its regions are not very well defined, but the median longitudinal groove of the gastric region is deep, and the 4 post-frontal lobes are sharply prominent, the middle pair being much broader than the outer ones.

Front half the extent of the anterior border of the carapace, the free margin slightly sinuous: the divergent lateral borders of the carapace have a tooth of good size behind the outer orbital angles.

Chelipeds in the female not half as long again as the carapace : the outer surface of the arm and wrist are rugulose, and both surfaces of the palm are studded with sharpish granules : the upper border of the arm ends acutely, but there is no spine at the end of the inner border; the inner angle of the wrist is pronounced, almost dentiform ; there are no granular or pectinated crests of any kind on the palm : the fingers are little bent and leave no large gap between them when closed, there are some sharpish granules along the upper border of the dactylus, and along the lower border of the fixed finger.

The legs are remarkably uneven in length, the third pair being more than $2\frac{1}{2}$ times the length of the carapace ; the meropodites are not exactly foliaceous, their greatest breadth being hardly two-fifths of their length, and they have a subterminal spine on the anterior border only ; the dactyli are remarkably long, those of the third pair being as long as their propodites.

In the Indian Museum are 2 females from the Andamans : the carapace of the larger one is 18 millim. long and 20 millim in its greatest breadth posteriorly.

113. *Sesarma Kraussi*, de Man.

Sesarma Kraussi, de Man, Zool. Jahrb., Syst., II. 1887, p. 652, and Journ. Linn. Soc., Zool., XXII. 1888, p. 193, pl. xiv. figs. 1–3.

Differs from *S. longipes*, which it closely resembles, in the following characters :—

(1) the four post-frontal lobes are not so prominent, the outer ones, indeed, being very inconspicuous :

(2) the free edge of the front is more sinuous, owing to the depth of the median notch :

(3) there are two distinct teeth on the lateral border of the carapace, behind the outer orbital angle :

(4) the outer surface of the wrist and both surfaces of the palm are nearly smooth, and there is a row of sharp granules along the *outer* surface of the fixed finger: the upper border of the arm does not end acutely :

(5) the legs are even longer and slenderer, the 3rd pair being more than $3\frac{1}{2}$ times the length of the carapace: the meropodites of the legs are at least 3 times as long as broad.

In the Indian Museum is a single male from the Nicobars : its carapace is 9 millim. long and 11 millim. in greatest breadth.

Heller ("Novara" Crust. pp. 64, 65) includes the following species in the Indian fauna :—

S. Eydouxi, Milne Edwards, Ann. Sci. Nat., Zool., (8) XX. p. 184 (Madras).

S. indica, Milne Edwards, *tom. cit.* p. 186 (Ceylon, Nicobars).

S. gracilipes, Milne Edwards, *tom. cit.* p. 182 (Nicobars).

SARMATIUM, Dana.

Sarmatium, Dana, Silliman's Amer. Journ. Sci. (2) XII. 1851, p. 288, and Proc. Ac. Nat. Sci. Philad. 1851, p. 251, and U. S. Expl. Exp. Crust. pt. I. p. 357 : Kingsley, Proc. Ac. Nat. Sci. Philad. 1880, p. 212 : de Man, Zool. Jahrb., Syst , II, 1887, p. 659.

Metagrapsus, Milne Edwards, Ann. Sci. Nat., Zool., (3) XX. 1853, p. 188.

This genus, which I almost agree with Dr. de Man in regarding as only a subgenus of *Sesarma,* differs from *Sesarma* in the following particulars :—

(1) the front, instead of being abruptly and vertically deflexed, is gradually declivous and obliquely deflexed :

(2) the antero-lateral borders of the carapace are usually a little arched, instead of being in the same straight line with the postero-lateral borders :

(3) the abdomen of the male does not completely coincide with the breadth of the sternum at the level of the 5th pair of legs ; and in the female the terminal segment is not deeply impacted in the penultimate segment.

Distribution : West Indies, West coast of Africa, Indo-Pacific.

114. *Sarmatium crassum,* Dana.

Sarmatium crassum, Dana, Proc. Ac. Nat. Sci. Philad. 1851, p. 251 ; U. S. Expl. Exp. Crust. pt. I. p. 358, pl. xxiii. figs. 1 *a–d* : Milne Edwards, Ann. Sci. Nat., Zool., (3) XX. 1853, p. 189 : Kingsley, Proc. Ac. Nat. Sci. Philad. 1880, p. 212 : de Man, Zool. Jahrb. Syst. II. 1887, p. 660.

Carapace deep, broader than long, broader behind than in front, smooth, with very faint indications of regions and no oblique striæ on the epibranchial regions : of the post-frontal lobes the two middle ones alone are distinct, and they are not prominent, they occupy almost all the space between the orbits.

The front is half the extent of the anterior border of the carapace, its free edge is very little concave in the middle line. The antero-lateral borders of the carapace are distinctly arched and are cut into 2 broad blunt lobes (one of which is the orbital angle) followed by a small tooth.

Chelipeds "of male short, hand above transversely four to five-plicate, externally nearly smooth, moveable finger with four short rudiments of spines, carpus mostly smooth, a few seriate granules above " In the female the transverse plications of the upper surface of the hand are very indistinct and the dactylus is smooth.

Legs not much compressed: the meropodites are not broadened, there is a spinule at the distal end of their anterior border: the dactyli are slender but are shorter than the propodites.

In the Indian Museum is a young female from the Nicobars: its carapace is 8 millim. long and 9 broad.

Henderson (Trans. Linn. Soc., Zool., (2) V. 1893, p. 393) describes a variety of *Sarmatium indicum* (Milne Edwards, Nouv. Archiv. du Mus. IV. 1868, p. 174, pl. xxvi. figs. 1–5) from Cochin.

METASESARMA, Edw.

Metasesarma, Milne Edwards, Ann. Sci. Nat. Zool. (3) XX. 1853, p. 188 : Kingsley, Proc. Acad. Nat. Sci. Philad. 1880, p. 211 : de Man, Zool. Jahrb., Syst. IX. 1895–97, p. 128.

The most marked difference between this genus and *Sesarma*, which it closely resembles, is that the tooth at the inner angle of the orbit meets the thickened angle of the front, so as to completely exclude the antennæ from the orbit.

The regions of the carapace are not defined, and the post-frontal tubercles are inconspicuous : the front is vertically deflexed as in *Sesarma*, but is deeper and overhangs the epistome : the reticulate appearance of the pterygostomian and neighbouring regions is finer, closer, and more confused: the orbits are more open below : the antennæ are much smaller : the legs are not so broad and compressed.

The *Metasesarmata* are land and fresh-water crabs of the Indo-Pacific region.

115. *Metasesarma Rousseauxii*, Edw.

Metasesarma Rousseauxii, Milne Edwards, Ann. Sci. Nat., Zool. (3) XX. 1853, p. 188, and Archiv. du Mus. VII. 1855, p. 158, pl. x. figs. 1 *a-c* : Kingsley, Proc. Ac. Nat. Sci. Philad. 1880, p. 211: de Man, Zool. Jahrb., Syst., IV. 1889, p. 439, and IX. 1895–97, p. 138, and X. 1898, pl. xxix. fig. 28, and in Weber's Zool. Ergebn. Niederl. Ost-Ind. II. p. 350: Henderson, Trans. Linn. Soc., Zool., (2) V. 1893, p. 392: Ortmann, Zool. Jahrb., Syst., VII. 1893–94, p. 717.

Sesarma Aubryi, de Man (*nec* A. M. Edw.), Journ. Linn. Soc., Zool., XXII. 1888, p. 168.

Carapace deepish, a little broader than long, smooth to the naked eye, slightly convex fore and aft : a short semilunar groove separates the gastric from the cardiac region, and there is a median longitudinal post-frontal groove of some depth : the middle pair of post-frontal tubercles are distinct, though not prominent, but the outer ones are hardly distinguishable.

Front a little more than half the breadth of the carapace, vertical, deep, somewhat spathulate, the free edge convex and very slightly sinuous.

Sides of the carapace slightly curved and convergent posteriorly, no tooth behind the outer orbital angle.

The chelipeds are longer and more massive in the male, but are otherwise similar in both sexes : in the male they are less than $1\frac{1}{2}$ times the length of the carapace. To the naked eye they are smooth, except for a patch of vesiculous granules in the middle of the inner surface of the palm. The inner angle of the wrist is sharply pronounced, and the upper border of the palm and of the base of the dactylus have a few small blunt serrulations. The palm is as high as long, the dactylus is about $1\frac{1}{2}$ times the length of the upper border of the palm, the fingers, though a little hollowed at tip, are subacute and have no gap between them when closed.

Legs rather slender, smooth and unarmed to the naked eye : the meropodites are not broadened : the dactyli are as long as their pro-podites and like them are fringed with dark spine-like bristles. The 3rd pair of legs, which are the longest, are less than twice the length of the carapace.

In the Indian Museum are 61 specimens from the Andamans and Nicobars, Mergui, Ganges Delta, Madras, and Minnikoy (Laccadives). Many of the specimens were taken on land, hiding under timber, in which situation their curious mottled coloration must be protective. The largest specimen has a carapace 14 millim. long by nearly 17 broad.

CLISTOCŒLOMA, A. M. Edw.

Clistocœloma, A. Milne Edwards, Nouv. Archiv. du Mus. IX. 1873, p. 310 : Kingsley, Proc. Ac. Nat. Sci. Philad. 1880, p. 219.

Differs from *Sesarma* only in the following characters :—

(1) the tooth at the inner angle of the lower border of the orbit meets the front, as in *Metasesarma,* so as to completely exclude the antennæ from the orbit :

(2) the reticulation of the sidewalls of the carapace resembles that of *Sesarma,* but, on denudation, the lines of granules are found to be absent, so that the meshwork is made up of hairs entirely :

(3) the merus of the external maxillipeds is shorter.

From *Metasesarma* this genus is distinguished by the lobulation of the dorsum of the carapace and the dentate lateral borders.

If *Metasesarma* is to be classed as a subgenus of *Sesarma* as it has been, and with undoubted reason, by Dr. de Man, the same course might be taken with *Clistocœloma.*

116.   *Clistocœloma balansæ*, Edw.

Clistocœloma balansæ, A. Milne Edwards, Nouv. Archiv. du Mus. IX. 1873, p. 311, pl. xvii. fig. 1.

The whole body and the appendages, except the tips of the dactyli of the legs, are everywhere covered with a dark dense adherent fur, amid which, on the dorsal aspect, are numerous clumps of tomentum that look like tubercles : the legs, in addition, have a shaggy fringe of coarse hair.

Carapace square, as long as broad, somewhat depressed : when denuded it is smooth and polished, with all the regions well defined and boldly and symmetrically lobulated, and the post-frontal lobes prominent, the outer ones being again subdivided into two tubercles.

Front much more than half the breadth of the carapace, nearly vertically deflexed, deepish, its free margin sinuous and turned up to form a trenchant horizontal edge.

The lateral borders of the carapace are cut, anteriorly, into three lobes including the outer orbital angle.

Chelipeds subequal, nearly similar in size in both sexes, not more massive than the legs, shorter even than the 1st pair of legs, which are little longer than the carapace. When denuded they are smooth, except that the upper surface of the wrist is a little lumpy : the inner border of the arm is a little convex distally, but does not expand into an undoubted lobe : the palm is higher than long, but is by no means swollen or massive, and in the male only its upper surface is traversed, obliquely fore and aft, as close as possible to the upper border, by a fine microscopically-pectinate crest : the fingers are subacute, though slightly hollowed at tip, and have no wide gap between them when closed, and the fixed finger is shorter and deeper than the dactylus, the dactylus is nearly twice as long as the upper border of the palm, and in the male its upper border is milled with about 14 or 15 lamellæ.

Legs markedly unequal : the third pair, which are the longest, are not quite twice as long as the carapace. In all, the meropodites are thin and broad, and the dactyli are not two-thirds as long as their propodites.

In the Indian Museum are a male and two females from the Nicobars. The carapace of the largest is 19 millim. in either diameter.

117.   *Clistocœloma merguiense*, de Man.

Clistocœloma merguiensis, de Man, Journ. Linn. Soc., Zool., XXII. 1888, p. 195, pl. xiii. fig. 10, and Notes Leyden Mus. XII. 1890, p. 92 : and Zool. Jahrb., Syst., IX. 1895-97, p. 339, and X. 1898, pl. xxxi. fig. 40.

This species differs from *C. balansæ* in the following particulars :—

(1) the carapace is decidedly broader than long, its lobulations are not nearly so bold and convex, and the outer post-frontal lobules may be entire :

(2) the free edge of the front is not turned up to form a trenchant horizontal crest, although it is well defined :

(3) the chelipeds of the male are far more massive than any of the legs ; the inner border of the arm is dilated distally ; the palm is a good deal swollen, the pectinate crest that traverses its upper surface is longer, and its inner surface is more granular; the fingers are more widely separated when closed, and the lamellar tubercles along the upper border of the dactylus are more numerous :

(4) it is a smaller species.

In the Indian Museum are 10 specimens from the Nicobars: the carapace of the largest egg-laden female is 10 millim. long and 12 broad.

Metaplax, Edw.

Metaplax, Milne Edwards, Ann. Sci. Nat. Zool. (3) XVIII. 1852, p. 161.

Rhaconotus, Gerstaecker, Archiv f. Naturges. XXII. 1856, i. p. 140, and Kingsley, Proc. Ac. Nat. Sci. Philad., 1880, p. 213.

Metaplax, de Man, Journ. Linn. Soc., Zool., XXII. 1888, pp. 153-155.

Carapace quadrilateral, somewhat depressed, a good deal broader than long, the regions well or fairly defined and the cervical and branchial grooves distinct.

Front declivous, its breadth about a third or a fourth that of the carapace, the convexity of its free edge impinges on the epistome to help in forming the broad interantennulary septum.

Lateral borders of the carapace straight, or a little arched anteriorly, nearly parallel, cut into 4 or 5 teeth of which the last one or two are very inconspicuous. The posterior part of the sidewalls of the carapace with some hairs curving towards the incurrent branchial opening.

Orbits of good depth : their outer wall incomplete, their lower border crenulate : the eyes do not fill the orbits and the eyestalks are not prolonged.

The antennules fold nearly transversely : the septum between them is broad. The antennæ lie in the orbital hiatus, their basal joint is extremely short, their flagellum is of fair length.

Epistome short, but well defined and prominent : buccal cavern squarish : the external maxillipeds leave between them a large rhomboidal gap, in which the mandibles are exposed : a broad oblique groove, bounded internally by a line of hairs, runs from a point behind the antero-external angle of the ischium to the anterior edge of the merus : the merus is truncated, and the foliaceous propodite articulates near its antero-external angle.

The chelipeds differ very markedly in the sexes : in the female they are shorter and slenderer than the legs, but in the male they are longer and much more massive than the legs. In the male there is always a short oblique horny crest, either on or close to and parallel with, the inner border of the arm, as in many species of *Macrophthalmus* : it probably, as Dr. de Man suggests, is scraped against the lower border of the orbit to produce a musical sound.

Legs slender, the first and last pairs much shorter than the 2nd and 3rd pairs—the 3rd pair the longest.

The abdomen in the male does not quite cover the sternum between the bases of the last pair of legs : it may have all 7 segments distinct, or, rarely, the 3rd 4th and 5th segments may be fused together : in the female all 7 segments are separate and the 7th is small and deeply impacted in the 6th, as in *Sesarma.*

Distribution : Estuaries and mudflats of the Oriental littoral.

The species of *Metaplax* have many points of resemblance with the Ocypodoid genus *Macrophthalmus*, and this is all the more likely to lead to confusion as the two genera share the same habitat and have the same manner of life ; but there is no doubt of the true position of *Metaplax* among the *Sesarminæ.*

Key to the Indian species of Metaplax.

I. Anterior border of carpi and propodites of legs spiny : chelipeds in the male 3 times the length of the carapace.......   *M. crenulata.*

II. Anterior border of carpopodites and propodites of legs smooth : male chelipeds less than 3 times the length of the carapace :—

    1. Dactylus of chelæ of male without any prominent lobe on its dentary edge : chelipeds of male equal :—

        i. 3rd 4th and 5th abdominal segments fused together in the male...............................   *M. indica.*

        ii. All the abdominal segments separate :—

            a. Length of the carapace about three-fourths the breadth : orbital portion of lower border of orbit with 4 or 5 teeth...   *M. dentipes.*

            b. Length of the carapace less than three-fourths the breadth : orbital portion of lower border of orbit with 9 or 10 teeth...   *M. distincta.*

Dactylus of chelæ of male with a prominent lobe
projecting on the dentary edge : chelipeds of male
markedly unequal :—

    i. Palm of larger cheliped of male longer than high. *M. elegans.*

   ii. Palm of larger cheliped of male higher than long. *M. intermedia.*

118. *Metaplax indica*, Edw.

Metaplax indicus, Milne Edwards, Ann. Sci. Nat., Zool., (3) XVIII. 1852, p. 161,
and Archiv. du Mus. VII. 1855, p. 165, pl. xi. figs. 2–2c.

Carapace about two-thirds as long as broad, deepish, a little convex,
its surface smooth, the regions and the cervical and epibranchial grooves
faint.

Front about a third the greatest breadth of the carapace.   Lateral
borders of the carapace nearly straight, cut into 4 teeth, of which the
first 2 are large, the 3rd very small, and the 4th very inconspicuous.

Lower border of the orbit of the male continued to the level of the
first notch in the lateral border of the carapace, unevenly crenulate.

Chelipeds of the male equal, more than $2\frac{1}{4}$ times the length of the
carapace, smooth and unarmed, to the naked eye: arm long and slender,
projecting far beyond the carapace, its musical crest is almost on the
inner border, close to its proximal end : palm nearly twice as long as
high, increasing in height from its proximal to its distal end : fingers
slender, acute, not noticeably channelled and only moderately incurved,
neither of them have any large lobes on their dentary edge, the dactylus
is hardly shorter than the upper border of the palm, and though it is
deflexed is not hooked.

Legs quite unarmed, the carpopodites and propodites of the two
middle pairs remarkably tomentose : the third pair of legs are a little
more than twice as long as the carapace.

The 3rd 4th and 5th abdominal segments of the male are fused
together—though the sutures are not obliterated on either side, but
only in the middle—to form a single piece.

In the female the chelipeds are very slender, quite smooth, a
little longer than the carapace, and the lower border of the orbit is
finely and evenly serrulate.

In the Indian Museum are a male and a female from Karachi : the
carapace of the male is 10 millim. long and 14·5 millim. broad.

119. *Metaplax distincta*, Edw.

Metaplax distinctus, Milne Edwards, Ann. Sci. Nat., Zool., (3) XVIII. 1852,
p. 162, pl. iv. fig. 27 : de Man, Journ. Linn. Soc., Zool., XXII. 1888, p. 158, pl. x.
figs. 7–9 : Henderson, Trans. Linn. Soc., Zool., (2) V. 1893, p. 391.

Differs from *M. indica* in the following characters :—

(1) the carapace is more than two-thirds—nearly three-quarters—as long as broad :

(2) the lower border of the orbit of the male is prolonged to the level of the second notch in the lateral border of the carapace, and its orbital portion is cut into 9 or 10 little, blunt, obscurely-bilobulate teeth, which decrease very regularly in size from within outwards :

(3) the chelipeds of the male are hardly $2\frac{1}{2}$ times the length of the carapace; the arm has denticulate borders—the inner border being a little dilated distally—and is not elongate and slender, its musical crest runs obliquely away from the inner border and is nearer to the middle of that border: the palm is only about half again as long as high: the fingers are obliquely-truncated and channelled at tip, the fixed finger has a lobe (though not a very large one) on its dentary edge, the dactylus is hardly shorter than the upper border of the palm and has a strong hook-like curve :

(4) the anterior border of the meropodites of the legs is armed, in the first and last pairs with a single subterminal spine, in the middle two pairs with several spines; the tomentum on the carpopodites and propodites of the middle two pairs of legs is not so thick :

(5) the abdomen of the male consists of 7 separate segments.

In the Indian Museum are 8 specimens from Madras, Coconada, Mergui, and the Nicobars : the carapace of the largest male is 15 millim. long and 21 broad.

120. *Metaplax dentipes* (Heller).

Helice dentipes, Heller, Novara Crust. p. 62, pl. v. fig. 5: Kingsley, Proc. Ac. Nat. Sci. Philad. 1880, p. 220.

Metaplax dentipes, de Man, Journ. Linn. Soc., Zool., XXII, 1888, p. 162, pl. xi. figs. 1-3.

This is little more than a large variety of *M. distincta*, from which it differs in the following particulars :—

(1) the carapace is less transverse, its length being slightly more than three-fourths of its breadth :

(2) the lower border of the orbit of the male is divided, in its orbital portion, into 4 or 5 blunt, broad, compressed teeth decreasing in size from within outwards, and each tooth has a little cusp at its outer end :

(3) in the chelipeds, the inner border of the arm is more dilated distally; the lobe on the dentary edge of the fixed finger is not so convex, and the dactylus is as long as the upper border of the palm

and is not so strongly hooked; the dactylus also sometimes has an enlarged tooth—not a distinct lobe—near the middle of its dentary border:

(4) the anterior border of the meropodites of the legs is very often quite free from spines, but sometimes there are inconspicuous spinules where spines exist in *M. distincta.*

In the Indian Museum are 23 specimens from the banks of the Hooghly, the mud-flats of Arakan and Tenasserim, and Mergui. The carapace of a large male is a little over 21 millim. long and 29 millim. broad.

121. *Metaplax elegans,* de Man.

Metaplax elegans, de Man, Journ. Linn. Soc., Zool., XXII. 1888, p. 164, pl. xi. figs. 4–6, and Zool. Jahrb., Syst., VIII. 1894–95, p. 596.

Metaplax crassipes, de Man, in Weber's Zool. Ergebn. Niederl. Ost-Ind. II. 1892, p. 325, pl. xix. fig. 12 (*ipso teste*).

Resembles *M. indica* in the form of the carapace, but can be recognized by the following characters :—

(1) the groves of the carapace are fainter:

(2) the lower border of the orbit instead of being irregularly cut into dentiform lobules is very finely and regularly pectinate :

(3) the chelipeds in the male are not $2\frac{1}{2}$ times as long as the carapace and are distinctly unequal, the hand of one being decidedly larger than its fellow : the arm is not elongate, its edges are granular, and its musical crest, which is very fine, stands at the middle of the inner border, running obliquely parallel with that border: the larger palm is only a little longer than high and its inner surface is granular, its fingers are obliquely truncate and strongly channelled, and both of them have a lobe near the middle of their dentary border, the dactylus also is strongly curved, at any rate in the larger hand :

(4) in the first pair of legs the meropodites have a single subterminal spine on the anterior border, in the 2nd pair there are from three to six spines, and in the 3rd and 4th pairs from seven to ten : moreover at the extreme distal end of the *posterior* border of the meropodites of the two middle pairs of legs there may be two or three spinules :

(5) the abdomen of the male is broader, and has all 7 segments separate.

In the Indian Museum are 32 specimens from the Godavari Delta and from Mergui : the carapace of the largest male is 10 5 millim. long and 16 broad.

122.   *Metaplax intermedia*, de Man.

Metaplax intermedius, de Man, Journ. Linn. Soc., Zool., XXII. 1888, p. 166, pl. xi. figs. 7-9.

Differs from *M. indica* in the following characters:—

(1) In the male the lower border of the orbit is continued a little beyond the first notch in the antero-lateral border of the carapace, and at its inner end it is cut into a series of 5 or 6 little even teeth that decrease in size from within outwards, and then it gradually becomes minutely and regularly pectinate:

(2) the chelipeds of the male are markedly unequal, the difference in size being in the hand: their length is about $2\frac{1}{2}$ times that of the carapace: the arm is of no great length and is somewhat broadened across the middle, its edges are granular, and its musical crest lies in the middle of the inner border, close to and nearly parallel with that border: the palm has granular edges and is much compressed at its antero-inferior corner; in the larger cheliped *the hand is at least as high as long:* the fingers are obliquely truncated and strongly channelled; in the larger hand the dactylus is hooked and has a lobe on its cutting edge near the proximal end, while the fixed finger is broad, is thin and compressed at its basal end, and presents on its cutting edge a notch (corresponding with the lobe on the dactylus) followed by a high lobe that descends obliquely to the tip of the finger:

(3) near the far end of the anterior border of the meropodites of the legs is a spine:

(4) the abdomen of the male has all 7 segments distinct, and is rather broadly triangular.

In the Indian Museum are 11 specimens from the Godavari Delta, the Gangetic Delta and Mergui. The carapace of the largest male is $9\frac{1}{2}$ millim. long and 15 broad.

123.   *Metaplax crenulata*, Gerstaecker.

Rhaconotus crenulatus, Gerstaecker, Arch. f. Naturges. XXII. i. 1856, p. 142, pl. v. fig. 5: Kingsley, Proc. Ac. Nat. Sci. Philad. 1880, p. 213.

Metaplax crenulatus, de Man, Journ. Linn. Soc., Zool., XXII. 1888, p. 156, and Zool. Jahrb., Syst., IV. 1889, p. 439.

Carapace about three-fourths as long as broad, convex, with the regions well defined and the cervical and epibranchial furrows deep and coarse, its surface pitted.

Front about a fourth the greatest breadth of the carapace. Lateral borders of the carapace cut into five teeth, the edges of which are serrated ; the anterior part of the lateral borders is distinctly arched.

The lower border of the orbit, in the male, extends beyond the first notch of the lateral border of the carapace, its inner end is sharp entire and sinuous, but all the rest of its extent is elegantly beaded.

Chelipeds of the male three times the length of the carapace, the borders of the wrist and hand, and the inner border of the wrist, sharply granular or serrulate: arm long and slender, somewhat dilated at its proximal end, the musical crest close to the proximal end and almost on the inner border: the palm gradually increases in height from behind forwards, its greatest height is about half its length, along the middle of its inner surface is a row of granules ending in a granular patch: fingers slender, acute. incurved, not channelled, the extreme length of the dactylus is only about three-fourths that of the upper border of the palm: there are no prominent lobes on the dentary edges of the fingers.

Both borders of the meropodites of the legs, as well as the anterior border of the carpopodites and propodites, are spinulate. The third pair of legs are nearly as long as the male chelipeds.

In the abdomen of the male, which is narrow, all 7 segments are distinct, the penultimate segment being square.

In the female the chelipeds are very slender and are about $1\frac{1}{2}$ times the length of the carapace, and the lower border of the orbit is elegantly pectinate.

In the Indian Museum are 11 specimens from the Sunderbunds and Mergui. The carapace of the largest male is 30 millim. long and 40 broad.

Sub-family PLAGUSIINÆ, Dana.

PLAGUSIA, Latreille.

Plagusia (part), Latreille, Gen. Crust. et Ins. p. 33 (1806) : Desmarest, Consid. Gen. Crust. p 126 (part): De Haan, Faun. Japon. Crust. p. 31: Milne Edwards (part), Hist. Nat. Crust. II. 90, and Ann. Sci. Nat. Zool., (3) XX. 1853, p. 178: Miers, Ann. Mag. Nat Hist. (5) I. 1878, p. 148, and Challenger Brachyura, p. 271: Kingsley, Proc. Ac. Nat. Sci. Philad., 1880, pp. 189, 223.

Philyra, De Haan, *l.c. supra.*

Carapace subcircular, depressed, the antero-lateral borders toothed. The interorbital space is broad, being nearly a third the greatest breadth of the carapace ; but there is no true front, so that the antennular fossæ, into which the antennules fold nearly vertically, are visible in a dorsal view as deep clefts in the anterior border of the carapace. The interantennular septum is broad. Orbits deep : the antennæ stand in the wide orbital hiatus, their flagellum is short.

Epistome short: buccal cavern squarish, its anterior border is crenate and projects strongly in a horizontal direction. The external maxillipeds do not meet across the buccal cavern, but the space between them, which is not very broad, is occluded by bristles : their merus is as broad as the ischium and carries the palp at its summit : *their exognath has no flagellum.*

Chelipeds and legs dorsally rugose. Chelipeds subequal : in the male they are more massive than the legs, and longer than those of the first and last pairs, in the female they are shorter and slenderer than any of the legs : the fingers are stout and have rounded hollowed-out tips.

Legs very stout, with broad massive meri and short stout serrated dactyli.

The abdomen of the male is triangular and rather broad : it covers all the sternum between the last pair of legs, and it may have all 7 segments distinct or the 3rd 4th and 5th fused. In the female the abdomen is broad and consists of 7 segments, but the 3rd 4th and 5th do not move independently of one another.

Distribution : all warm seas, and extending into the Mediterranean.

In habit the *Plagusiæ* to a certain extent resemble the *Grapsi,* dodging about rocks that are awash at high tide, and hiding in crannies when pursued. They also resemble *Varuna* in being able to make themselves at home on drift timber in the open sea. This will account for the very wide range of some of the species.

The presence of two species in the Mediterranean implies nothing, of itself, for they may very probably have been carried there by ships. On the "Investigator" one could always see a *Plagusia* adhering to the ship's side near the water-line.

124. *Plagusia depressa* var. *squamosa* (Hbst.).

? Cancer depressus, Herbst (*nec* Fabr.), Krabben &c. I. ii. 117, pl. iii. figs. 35 a-b.

Cancer squamosus, Herbst, I. ii. 260, pl. **xx.** fig. 113 (*v.* Hilgendorf, SB. Ges. Nat. Freunde, 1882, p. 24).

Plagusia squamosa, Latreille, Gen. Crust. p. 34, and Nouv. Dict. Hist. Nat. XXVI. p. 533, and (?) Encycl. Méthod. X. 1825, p. 145 : Lamarck, Hist. Nat. Anim. Sans Verh. p. 246 : Milne Edwards, Hist. Nat. Crust. II. 94 : Krauss, Sudafr. Crust. p. 42 : Milne Edwards, Ann. Sci. Nat., Zool., (3) XX. 1853, p. 178 : Heller, SB. Akad. Wien, XLIII. 1861, p. 363, and Novara Crust. p. 51 : A. Milne Edwards, Nouv. Archiv. du Mus. IX. 1873, p. 298 : Richters, in Möbius, Meeresf. Maurit. p. 157 : Hilgendorf, SB. Ges. Nat. Freunde, Berlin, 1882, p. 24.

Plagusia tuberculata, Lamarck, *l. c.* p. 247 : Latreille, Encycl. Méthod. X. p. 146 : Milne Edwards, *l c.* p. 94 : Miers, Ann. Mag. Nat. Hist. (5) I. 1878, p. 148 : Haswell, Cat. Austral. Crust. p. 110 : Müller, Verh. Ges. Basal, VIII. 1886, p. 476 : de Man, Notes Leyden Mus. V. 1888, p. 168, and Zool. Jahrb., Syst., IX. 1895-97,

p. 358 : Ortmann, Zool. Jahrb., Syst., VII. 1893–94, p. 730 : M. J. Rathbun, P. U. S. Nat. Mus. XXI. 1898, p. 605.

Plagusia immaculata, Lamarck, *l. c.* p. 247 : Miers, *l. c.*, p. 150, and Challenger Brachyura, p. 273, pl. xxii. fig. 1 : Haswell, *l. c.* : de Man, Archiv für Naturges. LIII. 1887, i. p. 371 : Cano, Boll. Soc. Nat. Napol. III. 1889, p. 246 : Henderson, Trans. Linn. Soc., Zool., (2) V. 1893, p. 391 : Ortmann, *l. c.* : Nobili, Ann. Mus. Genov. (2) XX. 1899, p. 271.

Plagusia depressa, Latreille (*nec* Fabr.), Encycl. Méth. X. 145 : Milne Edwards, Hist. Nat. Crust. II. 93, and Ann. Sci. Nat., Zool., (3) XX. 1853, p. 179 : Heller, Novara Crust. p. 51.

Plagusia orientalis, Stimpson, Proc. Ac. Nat. Sci. Philad. 1858, p. 103, and Ann. Lyc. Nat. Hist. New York, VII. 1860, p. 231.

All the regions of the carapace are distinct, and the surface is covered with flat pearly or squamiform tubercles which are fringed anteriorly with little close-set bristles of uniform length.

The tubercles vary: sometimes they are prominent, sometimes depressed, and sometimes they are almost obsolete on the most convex portions of the carapace. The little fringes of bristles also vary: sometimes they fill all the space between the tubercles, somtimes they can only be made out with a lens, sometimes they are absent.

The antero-lateral border of the carapace is armed with four teeth (including the orbital angle) which decrease in size from before backwards. The epistome is prominent beyond the anterior border of the carapace and is usually cut into seven lobes.

The chelipeds of the adult male are massive and are about half again as long as the carapace, but in the female they are slender and only about as long as the carapace. The inner angle of the wrist is coarsely dentiform : the tubercles on the upper surface of the palm and dactylus are arranged in high relief in longitudinal rows, those on the outer surface of the palm—especially at the upper part of it—have a tendency to fall into transverse rows.

On the posterior edge of the dorsal surface of the basipodites of the legs is a subacute tooth or blunt lobe with entire edges, this tooth being most conspicuous in the 2nd and 3rd pair of legs: on the anterior border of the meropodites there is a single strong spine, subterminal in position : the upper surface of the carpopodites propodites and dactyli is traversed longitudinally by a dense strip of long bristles. The 3rd pair of legs, which are the longest, are not quite twice the length of the carapace.

In the Indian Museum are 31 specimens from the Bay of Bengal and Arabian Sea: many of them were taken from drift timber in the open sea. Old specimens are commonly encrusted with barnacles and acorn-shells. The largest specimen in the collection has a carapace 54 millim. long and 56 broad.

LIOLOPHUS, Miers.

Leiolophus, Miers, Cat. Crust. New Zealand, p. 46 (1876), and Ann. Mag. Nat. Hist. (5) I. 1878, p. 153.

Acanthopus, De Haan, Faun. Japon. Crust. p. 29 : Dana, U.S. Expl. Exp. Crust. pt. I. p. 372 : Milne Edwards, Ann. Sci. Nat., Zool., (3) XX. 1853, p. 180 (*nom. præocc.*).

As in *Plagusia*, the antennæ fold nearly vertically in deep slits—visible in a dorsal view—cut in the anterior border of the carapace, the slits dividing the interorbital space into three deep lobes; and the exognath of the external maxillipeds has no flagellum.

The difference from *Plagusia* is as follows :—

The carapace is extremely flat and depressed—being quite disk-like—and is longer than broad : the interantennular septum is of no great breadth : the epistome is almost linear : the merus of the external maxillipeds is very small, being much narrower than the ischium, and is disposed obliquely in repose : the chelipeds and legs, though in places spiny, are not rugose : the legs are much slenderer, and though the meropodites are broad they are very thin : the copulatory organ of the male ends in a claw : finally, the exognath of the external maxillipeds is extremely short and slender.

As in the Indian species of *Plagusia*, the abdomen of the male consists of 5 segments, the 3rd 4th and 5th being fused. The abdomen of the female is similar in this respect to that of the male,

Distribution : as *Plagusia*, but not in the Mediterranean.

125. *Liolophus planissimus* (Hbst.).

Cancer planipes, Seba, Thesaurus III. p. 49, pl. xix. fig. 21 (1758).

Cancer planissimus, Herbst, Krabben &c. III. iv. 3, pl. lix. fig 3 (1804).

Plagusia clavimana, Latreille, Gen. Crust. p. 34 : Lamarck, Hist. Nat. Anim. Sans Vert., Crust., p. 247 : Desmarest, Dict. Sci. Nat. XXVIII. p. 246 : Latreille, Encycl. Méthod. X. p. 146 : Desmarest, Consid. Gen. Crust. p 127, pl. xiv. fig. 2 : Milne Edwards, Hist. Nat. Crust. II. 92, and in Cuvier, Règne Animal, Crust. pl. xxiii. fig. 3 : Hess, Archiv f. Nat. XXXI. 1865, i. p. 154 : Desbonne et Schramm, Crust. Guadaloupe, p. 50 : Richters in Möbius' Meeresf. Maurit. p. 157.

Plagusia serripes, Lamarck, *loc. cit.* : Latreille, Encycl. Méthod *loc. cit.*

Acanthopus planissimus, De Haan, Faun. Japon. Crust. p. 30 : Dana, U. S. Expl. Exp. Crust. pt I. p. 372 : Milne Edwards, Ann Sci. Nat , Zool., (3) XX. p 180 : Heller, SB. Ak. Wien, XLIII. 1861, p. 364 : Stimpson, Ann. Lyc. Nat. Hist. New York, VII. 1862, p. 232 : Heller, Novara Crust. p. 51 : A. Milne Edwards, Nouv. Archiv. du Mus. IX. 1873, p. 299 : Brocchi, Ann. Sci. Nat., Zool., (6) II. 1875, Art. 2 (*male appendages*) : Nauck, Zeits. Wiss. Zool. XXXIV. 1880, p. 31 (*gastric teeth*).

Acanthopus clavimanus, Krauss, Sudafr. Crust., p. 42.

Acanthopus Gibbesi, Milne Edwards, Ann. Sci. Nat. Zool. *loc. cit.*

Leiolophus planissimus, Miers, Cat. Crust. N. Z. p. 46, and Ann. Mag. Nat. Hist. (5) I. 1878, p. 153, and P. Z. S. 1879, p. 38, and Zool. H. M. S. Alert, pp. 518, 545 :

Filhol, Crust. N. Z., Miss. l'ile Campbell, p. 394: Haswell, Cat. Austral. Crust. p. 112 : Müller, Verh. Ges. Nat. Basel, 1886, p. 476 : de Man, Arch. f. Nat. LIII. 1887, i. p. 372, and Notes Leyden Mus. XV. 1893, p. 287, and Zool. Jahrb., Syst., IX. 1895–97, p. 358 : Pocock, Journ. Linn. Soc., Zool., XX. 1890, p. 513 : Henderson, Trans. Linn. Soc. Zool. (2) V. 1893, p. 391 : Ortmann, Zool. Jahrb., Syst., VII. 1893–94, p. 731 : Whitelegge, Mem. Austral. Mus. III. 1897, p. 139 : M. J. Rathbun, Ann. Inst. Jamaica, I. 1897, p. 36.

Carapace thin, disk-like, covered with little short bristles which, however, leave certain symmetrical raised linear patches bare : the meropodites of the legs are clad in the same way, and have two long bare stripes.

The front, the antennular and supra-orbital angles, and the epistome are all acutely spinous : the antero-lateral border of the carapace is armed with 4 acute spines : the middle of the upper border of the orbit is more or less serrate. The eyes are large and reniform.

The chelipeds vary according to age and sex, but the arm and wrist are always armed with spines ; the palm is smooth, nude, oval, and somewhat compressed ; and the fingers are short, blunt, and hollowed at tip. In the adult male the palms, or one of them, are remarkably deep.

The anterior border of the meropodites of all the legs is armed along its whole length with remarkably large and even spines, the posterior border ends in a spine : in the case of the first two meropodites there is a second row of spinules parallel with the anterior border, but this is very indistinct in the meropodites of the 3rd pair, and quite absent in those of the 4th.

The colour in life is dark green, the nude streaks being bright green. In the Indian Museum are 36 specimens from the Andamans, Ceylon, and Laccadives : the carapace of the largest is 23 millim. long and 21 broad.

Family GEOCARCINIDÆ, Dana.

Key to the Indian Genera.

I. Fronto-orbital border more than half the maximum breadth of the carapace : interantennular septum broad : epistome well defined and prominent : dactyli of legs with 4 rows of spines :—

    1. Buccal cavern not elongate : exognath of external maxillipeds without a flagellum : opposed edges of the basal joints of the 2nd and 3rd pairs of legs heavily fringed with hair much as in *Ocypoda* :—

        i. Antero-lateral borders of carapace dentate..... GRAPSODES.

        ii. Antero-lateral borders of carapace entire......... EPIGRAPSUS.

2. Buccal cavern elongate : exognath of external maxillipeds with a flagellum : no hairy fringe on the basal joints of the 2nd and 3rd pairs of legs..................   CARDIOSOMA.

II. Fronto-orbital border less than half the greatest breadth of the carapace : interantennular septum narrow : epistome ill-defined and sunken : dactyli of legs with 6 rows of spines : exognath of external maxillipeds without a flagellum...   PELOCARCINUS.

GRAPSODES, Heller.

Grapsodes, Heller, Novara Crust. p. 58 : Kingsley, Proc. Ac. Nat. Sci. Philad. 1880, pp. 188, 197.

Carapace depressed, little broader than long, declivous anteriorly, the regions faintly indicated, the dorsal surface without ridges or wrinkles, the lateral borders well arched and irregularly dentate.

Front about half the width of the anterior border, or about a third the greatest breadth of the carapace, strongly deflexed, its free edge nearly straight.

Orbits small, shallow, the lower border is wanting except for the tooth at the inner angle. The antennules fold nearly transversely in fossæ which are widely open externally : interantennular septum very broad. Antennal flagella slender and very short, standing in the orbital hiatus.

Epistome of moderate length fore and aft. External maxillipeds having a rhomboidal gap between them, in which the mandibles are visible : the merus is narrower than, but about the same length as, the ischium, and is a little oblique : the palp, which though coarse is small, articulates at the antero-external angle of the merus.

Chelipeds in both sexes subequal : in the male they are very much more massive than the legs and longer than the first and last pairs : in the female they are relatively shorter and much less massive than in the male. The tips of the fingers are acute.

Legs stout, their joints are not particularly broad or compressed but have their edges armed with stout bristles : the dactyli are long, acute, and thorny. The 2nd and 3rd pair of legs are the longest, and between their bases is a recess fringed with hairs resembling that found in *Ocypoda* and *Gelasimus*, and probably indicating terrestrial or amphibious habits.

The abdomen in both sexes consists of seven segments, and in the male its base covers all the breadth of the sternum between the last pair of legs.

Distribution : Islands of East Indian Archipelago.

This genus is really identical with *Epigrapsus* (= *Nectograpsus*), from which it only differs in having the regions of the carapace even more indistinct, the lateral borders of the carapace entire, the male chelipeds remarkably unequal, and the dactylus of the legs alone hirsute.

126.   *Grapsodes notatus*, Heller.

Grapsodes notatus, Heller, Novara Crust. p. 58, pl. v. fig. 2 : Miers, P. Z. S. 1877, p. 136 : J. S. Kingsley, Proc. Ac. Nat. Sci. Philad. 1880, p. 197 : de Man, Notes Leyden Mus. V. 1883, p. 160.

Carapace five-sixths as long as broad, the regions defined, though faintly, the surface smooth except sometimes for some granules near the lateral borders. The antero-lateral borders are cut into three shallow teeth or lobes behind which are some inconspicuous crenulations. On the line of flexion of the front are two eminences separated by a notch. Epistome and pterygostomian regions tomentose.

The chelipeds differ considerably in the sexes, though always smooth. In the adult male they are nearly twice the length of the carapace, the inner angle of the wrist is pronounced but not spiniform, the palm (which is as high as long) has a strong bulge at the infero-posterior angle, the dactylus (which is twice as long as the upper border of the palm) is much longer than the immobile finger and closes very obliquely, and there are two molariform teeth, one near the base of the dactylus, the other nearer the tip of the immobile finger.

In the female the chelipeds are hardly 1½ times the length of the carapace, the inner angle of the wrist is dentiform or spiniform, the palm is not enlarged or inflated, and the fingers are of nearly equal length, meet in the greater part of their extent, and are finely denticulated except near the tips.

The second pair of legs, which are the longest, are about twice the length of the carapace, the third pair are a little shorter than the second, and the first and last pair are about 1¼ times the length of the carapace.

In the Indian Museum are 8 specimens from the Nicobars. The carapace of the largest male is 25 millim. long and 30 millim. broad, but a female is somewhat larger than this.

That this species is probably terrestrial is evidenced by the vaulted branchial cavities, and also by the folding of the membrane that lines them, which is practically the same as that of *Ocypoda, Cardiosoma,* and *Pelocarcinus.*

EPIGRAPSUS, Heller.

Epigrapsus, Heller, Verh. zool.-bot. Ges. Wien, XII. 1862, p. 522: Kingsley, Proc. Ac. Nat. Sci. Philad. 1880, pp. 188, 192: Miers "Challenger" Brachyura, p. 265.

Nectograpsus, Heller, Novara Crust. p. 56.

This genus is really identical with *Grapsodes*, from which it differs in no single point of importance.

The trivial characters that separate it from *Grapsodes* are the following :—

The regions of the carapace are hardly distinguishable, and the lateral margins are entire: the chelipeds in the male are markedly unequal, one of them being longer and vastly more massive than the legs, the other being hardly larger than those of the female (which resemble those of *Grapsodes*): though the legs resemble those of *Grapsodes* in proportions and in the singular length of the dactyli, they differ in having only the terminal joint hirsute.

Distribution : Islands of the East Indian Archipelago and Polynesia.

127. *Epigrapsus politus*, Heller.

Epigrapsus politus, Heller, Verh. zool.-bot. Ges. Wien, XII. 1862, p. 522: Kingsley, Proc. Ac. Nat. Sci. Philad. 1880, p. 192: Miers, Challenger Brachyura, p. 266: Ortmann, Zool. Jahrb., Syst., VII. 1893-94, p. 703: de Man, Zool. Jahrb. Syst. IX. 1895-97, p. 79.

Nectograpsus politus, Heller, Novara Crust. p. 57, pl. v. fig. 3.

Carapace about seven-eighths as long as broad, perfectly smooth, the outer orbital angle not pronounced and the lateral margins entire in the adult. The line of flexion of the front is a little concave in the middle. Epistome and pterygostomian regions tomentose.

Chelipeds smooth, equal in the female, markedly unequal in the male. In the male the larger cheliped is more than twice the length of the carapace, the inner angle of the wrist is not pronounced, the palm is about as high as long and has a strong bulge at its postero-inferior angle, the dactylus is much longer than the immobile finger, and the dactylus has 2 or 3 small molariform teeth while the immobile finger has a single one.

In the female the chelipeds are little longer than the carapace, have the inner angle of the wrist pronounced, the palm not enlarged or inflated, and the fingers finely and inconspicuously toothed and nearly equal in length.

The smaller cheliped of the male is but little larger than those of the female.

Of the legs the 2nd pair are the longest, being twice the length of the carapace, and the 3rd pair are slightly shorter: the 1st pair are nearly $1\frac{1}{2}$ times, the 4th pair about $1\frac{1}{3}$ times the length of the carapace.

In the Indian Museum are 4 specimens, from the Andamans and Nicobars: the carapace of the largest male is 14 millim. long and 16 millim. broad.

CARDIOSOMA, Latreille.

Cardisoma, Latreille, Encycl. Méthod. X. p. 685 (1825) : De Haan, Faun. Japon. Crust. p. 27 : Milne Edwards, Hist. Nat. Crust. II. 22, and Ann. Sci. Nat. Zool. (3) XX. 1853, p. 203 : Smith, Trans. Connect. Acad. Sci. II. 1870, p. 142 : Miers, Challenger Brachyura, p. 219 : Ortmann, Zool. Jahrb., Syst., VII. 1893-94, p. 732.

? Discoplax, A. Milne Edwards, Ann. Soc. Entom. France, (4) VII. 1867, p, 248, and Nouv. Archiv. du Mus. IX. 1873, p. 293.

Carapace deep, convex fore and aft, transversely oval, with the lateral borders tumid and strongly arched owing to the vault-like expansion of the gill-chambers, the pterygostomian regions densely tomentose.

The fronto-orbital border is much more than half, and the deflexed and nearly straight front is about a fourth, the greatest breadth of the carapace. Orbits deep, with the outer angle defined by a denticle, and with the tooth at the inner angle well developed but distant from the front : the eyes are very loose in the orbits.

The antennules fold obliquely beneath the front, by which they are a good deal concealed : the inter-antennular septum is very broad. The antennæ lie in the orbital hiatus, which their broad basal joint nearly fills : their flagellum is very short.

Epistome short, prominent and well defined : buccal cavern elongate squarish, the external maxillipeds do not close it but leave between them a rhomboidal gap in which the mandibles are exposed. In the external maxillipeds the merus is a· longish joint and carries the palp, which is large and not at all concealed, at its antero-external angle : the exognath, which carries a flagellum, is exposed in much the greater part of its extent. The exognaths of the other maxillipeds are heavily fringed with coarse hair.

The chelipeds, which are much more massive than the legs, may either be equal or markedly unequal, differing little in the sexes : they alter considerably with age—one or both—the arm and fingers becoming elongated, and the whole hand increasing in size until it becomes longer than the carapace is broad and more than half as high as the carapace is long.

The legs are stout: some of their joints are fringed with bristles, and their long strong dactyli are square in section and have a series of spines along all four edges.

The abdomen in both sexes consists of 7 separate segments, and in the male its base covers the whole width of the sternum between the last pair of legs.

The branchiæ are eight in number on either side: the gill chambers are vaulted and remarkably capacious, and they are lined by a thick vascular membrane folded to form a sort of pocket, and as in several other crabs—such as *Gelasimus* and *Ocypoda*—that spend most of their time out of water, a sort of "choroid process" of this membrane, shaped like a gill-plume, projects laterally over the pleura of the penultimate pair of legs.

The species of this genus live on land. They are very common in the jungles of the Andamans where they may be found in the day time crouching under roots, fallen logs, &c., sometimes in burrows near the shore.

Distribution: West Indies and neighbouring coasts of America, Cape Verde Is. and West Coast of Africa, Indo-Pacific from Madagascar to Chili.

Key to the Indian species of Cardiosoma.

I. Carapace very strongly convex fore and aft, the regions indistinct : breadth of the orbit not much more than half its length : merus of the legs with bristles only at its distal end.... ....   *C. carnifex.*

II. Carapace very moderately convex fore and aft, the regions distinct : breadth of the orbit about two-thirds its length : merus of the legs with bristles along its whole length..... ....   *C. hirtipes.*

128.   *Cardiosoma carnifex*, (Hbst.).

Cancer carnifex and *hydromus*, Herbst, Krabben etc. II. v. 163, 164, pl. xli. figs. 1, 2 (1794).

Cardisoma carnifex, Latreille, Encycl. Méthod. X. p. 685 : Milne Edwards, Hist. Nat. Crust. II. 23 : Guérin, Icon. Règne An., Crust. pl. v. fig. 2 : Dana, U. S. Expl. Exp. Crust. pt. I. p. 377 : Milne Edwards, Ann. Sci. Nat. Zool. (3) XX. 1853, p. 204 : Heller, Novara Crust. p. 35 : A. Milne- Edwards, Nouv. Archiv. du Mus. IV. 1868, p. 71, and IX. 1873, p. 264 : Hoffmann, in Pollen and van Dam, Faun. Madagasc., Crust. p. 12 : Broçchi, Ann. Sci. Nat., Zool., (6) II. 1875, Art. 2, p. 85, pl. xvii. figs. 117, 118 (*male appendages*) : Miers, P. Z. S. 1877, p. 137, and Phil. Trans. 1879, p. 490, and Challenger Brachyura, p. 220 : Hilgendorf, MB. Ak. Berl. 1878, p. 801 : de Man, Notes Leyden Mus. II. 1880, p. 31, and in Weber's Zool. Ergebn. Niederl. Ost.-Ind. II. p. 285 : Richters, in Möbius, Meeresf. Maurit. p. 157 : Lenz and Richters, Abh. Senck. Nat. Ges. XII. 1881, p. 422 : Taschenberg, Zeitschr. f.

Naturwiss. LVI. 1883, p. 171 : Henderson, Trans. Linn. Soc., Zool., (2) V. 1893, p. 380 : Nobili, Ann. Mus. Genov. (2) XX. 1899, p. 271.

Cardisoma Guanhumi var. *carnifex*, Ortmann, Zool. Jahrb., Syst., VII. 1893-94, p. 735.

Cardisoma obesum, Dana, Proc. Ac. Nat. Sci. Philad. V. 1851, p. 252, and U. S. Expl. Exp. Crust. pt. I. p. 375, pl. xxiv. fig. 1 : Milne Edwards, Ann. Sci. Nat., *t.c.* p. 205 : Stimpson, Proc. Ac. Nat. Sci. Philad. 1858, p. 100 : Streets, Bull. U. S. Nat. Mus. VII. 1877, p. 114 : ? de Man, Notes Leyden Mus. II. 1880, p. 35.

Cardisoma Urvillei, Milne Edwards, Ann. Sci. Nat. *t.c.* p. 204 : de Man, Notes Leyden Mus. *t.c.*, p. 33.

Carapace strongly convex fore and aft, especially in the young, the regions are indicated by inequalities of level, but the posterior limit of the gastric region and the cardiaco-intestinal region are defined by grooves : the posterior areola of the gastric region is always tumid.

The antero-lateral border of the carapace is defined by a fine raised line, becoming indistinct with age, which is not continuous with the small tooth at the outer orbital angle, but starts at a little denticle of its own.

The sides of the front (inner boundaries of the orbit) are very oblique : the sinuous upper border of the orbit runs very slightly backward to the base of the outer orbital tooth : the greatest width (height) of the orbit is little more than half the length of that cavity. The basal antenna-joint is large, touching the front.

The breadth of the buccal cavern, measured across the middle of the external maxillipeds, is equal to its length in the middle line.

In both sexes the chelipeds are unequal : they are smooth, except for a few small tubercles or wrinkles or denticles or granules along the edges of some of the joints : the inner angle of the wrist is dentiform, the palm is higher than long, especially in the larger hand, the stout fingers meet only at tip, especially in the larger hand.

The size of the larger cheliped varies with age. In adults of moderate size it is about twice the length of the carapace, the ischium hardly projects beyond the carapace, and the length of the dactylus is about equal to the height of the palm. In old specimens, especially in the male sex, it is about $2\frac{3}{4}$ times the length of the carapace, the ischium projects far beyond the carapace, and the length of the dactylus is $1\frac{1}{2}$ times the height of the palm.

In the legs there are stiff bristles, not very thickly set, at the distal end of the merus, on the anterior border and surface of the carpus and on both borders of the propodite.

The 7th segment of the male abdomen is half or less than half the length of the 6th, measured in the middle line.

In the Indian Museum there are 13 specimens from the Andamans and the Coromandel coast (besides specimens from Tahiti and Madagascar).

Ortmann considers that this form is only a variety of the West Indian *C. Guanhumi*, with which he regards the West African *C. armatum* as synonymous. So far as I can judge from single specimens of these two supposed species, I should think that this view is correct.

129.   *Cardiosoma hirtipes*, Dana.

Cardisoma hirtipes, Dana, Proc. Ac. Nat. Sci. Philad. 1851, p. 253, and U. S. Expl. Exp. Crust. pt. I. p. 376, pl. xxiv. figs. 2, *a-d* : Milne Edwards, Ann. Sci. Nat., Zool., (3) XX. 1853, p. 205 : Hess, Archiv f. Naturges. XXXI. 1865, i. p. 140 : Heller, Novara Crust. p. 35 : Miers, Cat. Crust. New Zealand, p. 53 : de Man, Notes Leyden Mus. II. 1880, p. 34, and Archiv f. Naturges. LIII. 1887, i. p. 349, pl. xiv. fig. 3 : E. Nauck, Zeits. Wiss. Zool. XXXIV. 1880, p. 26 (*gastric teeth*) : Filhol, Crust. N. Z. in Miss. l'ile Campbell, p. 460 : Ortmann, Zool. Jahrb., Syst., VII. 1893–94, p. 737 : Whitelegge, Mem. Austral. Mus. III. 1897, p. 138 : Nobili, Ann. Mus. Genov. (2) XX. 1899, p. 271.

Discoplax longipes, A. Milne Edwards, Ann. Soc. Entomol. France, (4)' VII. 1867, p. 284, and Nouv. Archiv. du Mus. IX. 1873, p. 294, pl. xv. (*sec* Ortmann, *l.c.*).

This species is easily distinguished from *C. carnifex* by the following characters :—

(1) the carapace is much less convex, the regions are much more distinctly defined, and the gastric region is distinctly subdivided, by grooves, into 3 areolæ : moreover there are some fine oblique striæ on the sides of the epibranchial regions :

(2) the sides of the front, or inner boundaries of the orbit, are not nearly so oblique : the upper border of the orbit is less sinuous and runs slightly forwards to the outer orbital angle : the greatest width of the orbit is nearly two-thirds the length of that cavity. The basal antenna joint does not touch the front :

(3) the breadth of the buccal cavern, measured across the middle of the meri of the external maxillipeds, falls considerably short of the length measured in the middle line :

(4) the chelipeds may be unequal but are far more commonly equal, even in old specimens in which the palms and fingers have grown long and the palm become enlarged :

(5) the bristles on the legs are more thickly set, and they occur along the whole of the anterior border of the merus :

(6) the 7th segment of the male abdomen is more than half the length of the 6th, measured in the middle line.

In the Indian Museum are 18 specimens from the Nicobars and Andamans (besides 4 from the " South Seas " and Madagascar).

In life the carapace is dark violet and the chelæ bright cinnabar red.

PELOCARCINUS, Edw.

Gecarcoidea, Milne Edwards, Hist. Nat. Crust. II. 25 (1837).

Pelocarcinus, Milne Edwards, Ann. Sci. Nat. Zool. (3) XX. 1853, p. 203, and Archiv. du Mus. VII. 1854–55, p. 183 : A. Milne Edwards, Nouv. Archiv. du Mus. (3) II. 1890, p. 171 (*et synon.*).

Hylæocarcinus, Wood-Mason, J.A.S.B. XLII. 1873, pt. 2, p. 258, and Ann. Mag. Nat. Hist. (4) XIV. 1874, p. 189.

Limnocarcinus, de Man, Notes Leyden Mus. I. 1879, p. 65.

Gecarcoidea, Ortmann, Zool. Jahrb., Syst., VII. 1893–94, pp. 732, 738.

Carapace transversely oval, somewhat depressed, with the lateral borders tumid and strongly arched owing to the vault-like expansion of the gill-chambers : the gastric region particularly well defined.

The extent of the fronto-orbital border is less than half the greatest breadth of the carapace, that of the strongly deflexed and nearly straight front is from a sixth to a seventh the greatest breadth of the carapace.

Orbits deep, broadly oval, demarcated dorsally by a sharpish slightly raised border, their outer angle not defined, a wide gap in their lower border : at the inner angle there is a strong tooth which may or may not, even in the same species from the same jungle, meet the front : if it does so, the antennæ, which are much reduced in size, are excluded from the orbit.

The antennules fold obliquely beneath the front, and the inter-antennular septum is not very broad.

Epistome sunken, hairy posteriorly so as to appear ill defined from the palate. Buccal cavern rounded anteriorly, not nearly closed by the external maxillipeds, which leave between them a wide rhomboidal gap in which the mandibles are exposed

The external maxillipeds are rather short : their merus lies obliquely, and its anterior edge is excavated for the insertion of the palp, which is short and coarse and is completely exposed : their exognath is very short and almost entirely concealed and is without a flagellum. The exognaths of the other maxillipeds are heavily fringed with hair.

Chelipeds much more massive than the legs, usually equal in both sexes, though larger and longer in the male than in the female.

Legs stout : in all, the anterior border of the carpus and all the borders of the propodite and dactylus are spiny, there being six rows of spines on the dactylus.

The abdomen in both sexes consists of 7 separate segments, and in the male its base covers all the breadth of the sternum between the last pair of legs.

The gill-chamber and its lining membrane, and the number of branchiæ, are as in *Cardiosoma.*

The *Pelocarcini* are land-crabs. The single Indian species is very common in the jungles of the Andamans, where, especially on the smaller islets, it grows to a large size.

Distribution: Brazil, Andamans and Nicobars, Celebes, Philippines, New Guinea, Loyalty Is.

Ortmann (*l.c.*) throws doubt on the locality Brazil, but, as it appears to me, without sufficient reason, seeing that the elder Milne Edwards states definitely that the type of the species was found in that country by a collector of the Paris Museum. *Pelocarcinus* is by no means the only form of animal life that has this very curious and suggestive distribution, which we also find, among Mammals in the Tapirs, among Birds, as Mr. Finn informs me, in the Piculets of the genus *Picumnus*, among Reptiles in the *Ilysiidæ*, and among fishes in the freshwater eels of the genus *Symbranchus.*

130. *Pelocarcinus Humei* (Wood-Mason).

Hylæocarcinus Humei, Wood-Mason, Journ. As. Soc. Bengal, Vol. XLII. 1873, pt. 2, p. 260, pls. xv, xvi, and Ann. Mag. Nat. Hist. (4) XIV. 1874, p. 190.

Carapace transversely oval, becoming broader with age, its lateral borders tumid and ill defined. The gastic region is particularly well delimited and is divided into three subregions—two antero-lateral and one postero-median—the anterior two of which are separated from one another by a deep groove: the cardiac-intestinal region is fairly well defined.

In adults the carapace is smooth, except for some oblique striæ on the lateral borders, which become squamiform markings on the pterygostomian regions, these regions being devoid of tomentum.

Front nearly vertically deflexed, somewhat spatulate but with the free edge straight. The tooth at the inner angle of the orbit does not usually touch the front, but sometimes it does and excludes the small antennæ from the orbit.

The chelipeds in the adult male are usually equal and are about $2\frac{1}{2}$ times the length of the carapace : the arm projects a long way beyond the carapace, and its upper and inner borders are rugose or irregularly tuberculate; the inner angle of the wrist is truncated; the palm is enlarged, its length is about $1\frac{1}{2}$ times its height and about as long as the dactylus ; the fingers, though they only meet at tip, are not widely separated.

In the adult female the chelipeds are about 1½ times the length of the carapace : the arm projects but little, the hand is not much enlarged, and the fingers almost meet throughout their length.

In many young females the inner edge of the wrist is serrated and there are also a few denticles along the upper border of the palm.

The second pair of legs, which are the longest, are hardly twice the length of the carapace.

Colours in life : carapace violet with some dirty yellow markings : chelipeds and legs yellowish with a livid reddish tinge.

In the Indian Museum are specimens from the Nicobars and from numerous islands of the Andaman group. The largest one has a carapace 82 millim. long and 110 broad.

Family PALICIDÆ, Rathbun.

Palicus, Philippi.

Cymopolia, Roux, Crust. Médit. pl. xxi. 1828 : Milne Edwards, Hist. Nat. Crust. II. 158 : Miers, Challenger Brachyura, p. 333 (*nom præocc.*).

Palicus, Mary J. Rathbun, Proc. Biol. Soc., Washington, XI, 1897, pp. 93, 165 ["Philippi, Zweiter Jahresber. d. Vereins f. Naturk. in Cassel, 11, 1838."].

Carapace depressed, broader than long, covered with granules and with symmetrical tubercles or rugosities that have a tendency to fall into transverse series.

Front about a third the greatest breadth of the carapace, little or not at all deflexed, usually lobed or toothed. Lateral borders of the carapace hardly curved, serrated anteriorly.

Orbits deep, the upper border is cut into several teeth by deep clefts, and there are usually two clefts in the lower border.

The antennules fold nearly transversely beneath the front : the interantennular septum is a narrow plate. The antennæ commonly have the basal joint, which stands in the orbital hiatus, enlarged : the flagellum is well developed.

Epistome sunken, not defined. Buccal cavern square. The external maxillipeds do not close the buccal cavern anteriorly : their merus is very small and is much narrower than the ischium : the ischium has its antero-internal angle and the merus its antero-external angle much produced : the palp articulates near the middle of the concave summit of the obliquely-placed merus.

Chelipeds short and usually slender in the female : in the adult male one of them may be enlarged—rarely both.

The two middle pairs of legs are much the largest: the first pair, except that they are much shorter and slenderer, resemble the middle pairs, but the fourth pair are weak, sometimes filiform, and are elevated above the third pair as in *Dorippe*, etc.

The abdomen in both sexes consists of 7 separate segments, the basal segments being very narrow fore and aft and the 1st linear.

In the female the genital openings are on the 2nd segment of the sternum close to the suture between it and the first.

Distribution : Atlantic coasts of Central America and of the United States, Cape Verde and Mediterranean, Indo-Pacific from Scychelles to California.

The Indian species of *Palicus* live among coral shingle at a depth of from 10 to 40 fathoms, where their mottled coloration and granular rugose carapace afford a good concealment.

Key to the Indian species of Palicus.

I. Posterior border of the propodites and dactyli of the first
 3 pairs of legs entire :—
   1. Front cut into two lobes :—
      i. Lobes of front broad: propodites and
       dactyli of the two middle pairs of legs
       sub-foliaceous.................................... *P. Jukesii.*
      ii. Lobes of front subacute: propodites and
       dactyli of the two middle pairs of legs
       compressed but not broadened .... *P. Whitei.*
   2. Front cut into four lobes, the middle two sub-
     acute, the outer ones broad .. . . *P. Wood-Masoni.*
II. Posterior border of the propodites and dactyli of the
 first 3 pairs of legs elegantly serrate :—
   1. Front cut into four blunt teeth : propodites and
     dactyli of the two middle pairs of legs broadly
     foliaceous .... *P. serripes.*
   2. Front cut into four acute teeth : propodites and
     dactyli of the two middle pairs of legs compressed
     but not foliaceous .... *P. investigatoris.*

131. *Palicus Jukesii* (White).

Cymopolia Jukesii, White, in Jukes' Voy. H. M. S. "Fly," p. 338, pl. ii. fig. 1: Miers Zool. H. M. S. "Erebus" and "Terror," Crust. p. 3, pl. iii. figs. 4-4c, and Challenger Brachyura, p. 335: Haswell, Cat. Austral. Crust. p. 138: Henderson, Trans. Linn. Soc., Zool., (2) V. 1893, p. 405.

Carapace with the regions well defined, and with the surface thrown into four transverse wrinkles, the two middle ones of which are the most convex and best defined : the whole surface is also closely covered with vesiculous and crystalline granules, which are largest on the convexities.

Front divided into two broad rounded lobes : antero-lateral border of the carapace cut into three teeth including the orbital angle : posterior border of the carapace raised, but not cut into well-spaced lobules.

Upper border of the orbit with two deep notches between the inner and outer orbital angles, both of these angles having a concave margin : lower border with two deep notches. There is a leaf-like lobule on the granular eye-stalk, another at the outer angle of the basal antenna-joint, and another in the gap between the antenna and the outer angle of the buccal cavern. The exposed surface of the ischium of the external maxillipeds is obliquely traversed by two ridges which meet at the produced antero-internal angle of the joint.

The chelipeds of the *adult* male are granular and downy and are usually markedly unequal. The larger one is stout, is more than $1\frac{1}{2}$ times the length of the carapace and has a swollen (subcylindrical) club-shaped palm of which the length is not twice the greatest height : the fingers are short and stumpy, the dactylus being little more than a third the length of the palm, and meet only at tip : the smaller cheliped of the male is short and slender, sometimes however it is almost as large as its fellow.

In the female the chelipeds are equal, are hardly longer than the carapace and hardly stouter than the last pair of legs : they have a palm which is as slender and nearly as long as the ischium, and incurved fingers which nearly meet throughout their length.

In the first 3 pairs of legs the merus is stout and broad with a granular dorsal surface and coarsely and unevenly serrulate edges, the anterior edge ending in a crest-like tooth ; the carpus is dorsally carinate, and its anterior border has the form of a two-lobed carina ; and the propodite and dactylus are subfoliaceous owing to the depth of the thin sharp carinæ of their edges—these carinæ being plumed. The 4th pair of legs are short weak and granular as far as the dactylus, which is much shorter than the propodite.

The 1st pair of legs are a little longer, the 4th pair a little shorter, than the carapace : the 2nd and 3rd pairs are about $1\frac{3}{4}$ times the length of the carapace.

In both sexes all the abdominal terga, except the last, are transversely carinate, the carinæ of the 2nd and 3rd terga being most conspicuous. Also on either side of the sternum there are two crests, one behind the base of the last pair of legs, the other almost in a line with the 3rd abdominal carina.

In the Indian Museum are 32 specimens, from the Andamans (up to 36 fath.), the Maldives (15–30 fath.), and Ceylon (34 fath.). The carapace of the largest female is 13 millim. long and 15 broad.

132.   *Palicus Whitei* (Miers).

Cymopolia Whitei, Miers, Zool. H. M. S. " Alert," pp. 518, 551, pl. xlix. fig. C.

At once distinguished from *P. Jukesii,* which it closely resembles, by the sharper and more prominent lobes of the front, and by the slenderer form of the first 3 pairs of legs, in which the edges of the meri are not serrated, the anterior borders of the carpi are not cristiform, and the propodites and dactyli are not in any way subfoliaceous, their edges not being produced to form high thin carinæ.

Other differences, to be noted on closer inspection, are the following :—

The transverse arrangement of the rugæ of the carapace is not marked : the faint transverse carinæ of the 5th and 6th abdominal terga are absent.

In the Indian Museum are 2 adult females and a non-adult female, from the Andamans.

133.   *Palicus Wood-Masoni,* n. sp.

Carapace with the regions distinct and areolated in high relief : except posteriorly, the areolæ have no tendency to arrange themselves transversely : the convexities of the areolæ, but not the interspaces, bear clumps of crystalline granules.

Front cut into 4 teeth, the middle pair narrower, slightly more prominent, and on a rather lower plane than the others : lateral border of the carapace cut into three teeth, including the very large and acute, orbital angle : posterior border raised and irregularly lobulate.

In the upper border of the orbit there are three deep notches, in the lower border a notch and a fissure.

There is only one cheliped in the single specimen known : it is short, not stouter than the legs, and has some blunt denticles on the far end of the arm, on the wrist, and on the upper surface of the hand.

In the first 3 pairs of legs the meri are stout and have a granular, dorsal surface and coarsely serrulate edges, the anterior edge ending in a coarse spine; the carpi are dorsally carinate, and their anterior edge has the form of a two-lobed crest; while the propodites and dactyli are, elongate and compressed with thin, but not cristiform, plumed edges., The filamentous 4th pair are granular up to the dactylus, which is not much shorter than the propodite.

The 1st pair of legs are about 1½ times, the 2nd and 3rd pairs are about 2¼ times, the length of the carapace, while the 4th pair are about as long as the carapace.

In the male (female unknown) the first 5 abdominal terga are transversely carinated, but the 4th and 5th carinæ are faint. The sternum is also carinated on either side of the abdomen, as in *P. Jukesii,* but the crests are much lower.

In the Indian Museum is a single male specimen from the Andamans: its carapace is 9 millim. long and 11 broad.

134. *Palicus serripes,* Alcock & Anderson.

Cymopolia serripes, Alcock & Anderson, Journ. As. Soc. Bengal, Vol. LXIII. pt. 2, 1894, p. 208: Illustrations of the Zoology of the Investigator, Crust. pl. xxiv. fig. 7.

Carapace with the regions well defined and cut up into a multitude of symmetrical convex areolæ, its whole surface is covered with crystalline granules which are enlarged on the convexities of some of the areolæ.

Front cut into 4 teeth, the middle two of which, though deflexed and on a lower plane, are much sharper and more prominent than the others: lateral borders of the carapace posteriorly divergent, cut into five ragged teeth, inclusive of the orbital angle: posterior border cut into from eight to ten well spaced even tooth-like lobes.

Upper border of orbit with 3 deep notches, lower border with a notch and a fissure: eyestalks sharply granular. Ischium of the external maxillipeds longitudinally grooved.

The chelipeds of the female (male unknown) though shorter than the carapace are stouter than the first pair of legs: they may be subequal or unequal: the arm, wrist, and the upper surface of the palm are sharply granular, the palm is rather full and is not elongate, being about half again as long as high and less than half again as long as the fingers.

The 1st pair of legs are about as long as the carapace: their merus is sharply granular and its anterior border ends in a spine: their propodite and dactylus are thin and compressed but not broadened, and their posterior border is evenly serrated.

The 2nd and 3rd pair of legs are a little over 1½ times the length of the carapace: their merus is very stout and broad, with a granular dorsal surface and sharply though irregularly serrated edges: their carpus has the anterior border cristiform and irregularly serrate, and the posterior border subcristiform up to a terminal spine: their propodite and dactylus are short and broadly foliaceous, with the posterior border elegantly and evenly serrated and the anterior border fringed with long hair.

The 4th pair are filiform, not nearly as long as the carapace, and are granular up to the dactylus which is slightly longer than the propodite.

In the female the first 3 abdominal terga are transversely carinate : the carina of the first tergum, which alone is prominent, ends off in a sort of scroll, which flanks the postero-lateral angles of the carapace.

In the Indian Museum are 9 specimens, all adult females, from off the Madras coast in the neighbourhood of Palk Strait and from off Ceylon 34 fathoms. The carapace of the largest is 9·5 millim. long and 11 broad.

135. *Palicus investigatoris,* n. sp.

This species is closely related to *P. serripes,* but differs in the following characters :—

The areolæ of the carapace are capped, not by clusters of granules, but by sharp little tubercles between which the surface is smooth : except on the lateral regions of the carapace there is only one such tubercle to each areola :

(1) the four teeth of the front are all equally acute : the five teeth of the lateral borders of the carapace, though irregular in size, are all very sharp and clean cut : the teeth of the posterior border are smaller and sharper :

(2) there is no fissure towards the inner end of the lower border of the orbit :

(3) there are denticles or sharp tubercles, instead of granules, on the arm, wrist, and upper surface of the hand :

(4) the legs only differ in the case of the 2nd and 3rd pairs in which none of the joints are so broad : the serration of the edges of the merus is different, the terminal spine of the anterior border being greatly enlarged ; the anterior border of the carpus has a spine at each end, but is not otherwise serrated ; and the dactylus and propodite, though thin and compressed, and otherwise quite like those of *P. serripes,* are not broadened, being much less foliaceous.

In the Indian Museum is a single non-adult male from off the Andamans : its carapace is nearly 7 millim. long and 8 millim. broad.

Family PTENOPLACIDÆ.

PTENOPLAX, Alcock & Anderson.

Archæoplax, Alcock and Anderson, Journ. As. Soc. Bengal, Vol. LXIII. pt. 2, 1894, p. 180.

Ptenoplax, Alcock and Anderson, Illustrations of the Zoology of the Investigator, Crust. pl. xv. 1895 : Alcock, Investigator Brachyura, p. 78.

As the generic diagnosis has already been published in this Journal (*loc. cit.* Archæoplax) the above references are sufficient.

136.   *Ptenoplax notopus,* Alcock & Anderson.

Archæoplax notopus, Alcock and Anderson, Journ. As. Soc. Bengal, LXIII. pt. 2, 1894, p. 181, pl. ix. fig. 3.

Ptenoplax notopus, Alcock and Anderson, Ill. Zool. Investigator, Crust., pl. xv. fig. 2 : Alcock, Investigator Brachyura, p. 79.

NOTICE.

The map illustrating this article will be
published in a subsequent No. of this *Journal*.

Editor.

JOURNAL

OF THE

ASIATIC SOCIETY OF BENGAL.

---•●●•---

Vol. LXIX. Part II.—NATURAL SCIENCE.

No. IV.—1900.

XVII.—*The relationship of the water-supply, water-logging, and the distribution of* Anopheles Mosquitos *respectively, to the prevalence of Malaria north of Calcutta.*—*By* LEONARD ROGERS, M.D., M.R.C.P., I.M.S., *Professor of Pathology, Medical College.*

[*With a Map.*]

[Received and read 4th July, 1900.]

The tract of country in which the present inquiry was carried out extends along the East bank of the Hooghly river from Calcutta to Naihati, a distance of 25 miles. The area is fairly typical of Lower Bengal, and has for a long time been looked on as water-logged and very malarious. In 1889 Dr. Gregg, then Sanitary Commissioner of Bengal, after a careful inspection, came to the conclusion that the unhealthiness was due to certain drainage channels having been silted up, and a scheme for re-excavating some of them was prepared, but has not yet been carried out. Owing to an unusual prevalence of fever in 1899 a further inquiry into the health of the tract was ordered, and was carried out by me in February last.

The plan of the enquiry was as follows. As the essential point to be determined was the proportion of the inhabitants of the various parts of the area who were suffering from malaria, a large number of persons were examined for enlargement of the spleen; its size being noted as either just felt, two fingers breadth below the ribs, four fingers breadth below, or extending beyond the navel. The spleen-count as a test for the degree of malaria in a tract of country was used by Major Dyson in the Punjab in a similar inquiry, and by others, and is perhaps the most reliable and easily carried out method, especially in the season of the year when fever is at a minimum. Secondly, the level of the ground-water was taken in as many wells as possible, and inquiries were made as to the height to which the ground-water rose during the rainy season, so as to enable the degree of water-logging to be estimated. Thirdly, the drinking-water supply was carefully noted. Fourthly, the number of fever cases treated at various dispensaries month by month was compared with the monthly rainfall over a series of years, and worked out in charts, in order to ascertain the influence of seasons and rainfall on the fever rate. Lastly, some observations have been made on the distribution and monthly variations of the distribution of the *Anopheles* Mosquitos, which have furnished some rather surprising results.

In carrying out the spleen-count the whole area was divided up into thirteen Municipalities, and as far as possible 100 persons, about half of whom were children, were examined in each Ward of each Municipality, over 5,000 persons having been examined in all. As children suffer from enlarged spleen more commonly than adults, just as Koch has recently shown that the malarial organism is also found in a larger percentage of children, the figures have been corrected so as to represent the spleen rate of 50 children and 50 adults in each Ward, so that the figures of the different areas should be strictly comparable. Visits were made from house to house so as to get a fair sample of the actual inhabitants of the Wards, and every precaution was taken to obtain accurate results, every single person being examined by myself, a month being taken over the inquiry.

The results are embodied in the accompanying map, in which the different municipal areas are shaded in accordance with the percentage of persons who were found to have enlarged spleens, the darker areas representing the highest percentages and *vice versâ*. The dotted lines within the municipal areas enclose the Wards or areas separately examined, and the large figures within them indicate the spleen percentage, while the figures enclosed in a circle are those of the distance of the ground water-level below the surface in feet and inches, the upper figures being the distance in the dry cold weather taken in the month of February, while the lower ones indicate the distance during the height of the previous rainy season. The small figures in brackets refer to the number of the Wards given in the left-hand margin of the map, and correspond with those in the tables given further on.

THE GENERAL RESULTS OF THE SPLEEN-COUNT.

The following table shows the percentage of people who were found to be suffering from enlarged spleen in each Municipality. They are arranged in order from above downwards as they are situated on the map from north to south, while the westernly ones, which lie on the east bank of the Hooghly, are placed on the left, and the easterly ones, which are at a little distance from the river, are placed in the right-hand column, so that the table roughly represents their position on the map.

TABLE I.

Municipality.	Spleen percentages.	Municipality.	Spleen percentages.
Naihati	19·9	(Gobardanga)	(55·5)
Bhatpara	20·0		
Garulia	33·8		
North Barrackpore ...	36·5	(Busirhat)	(52·8)
Titagarh	37·8	Baraset	52·9
South Barrackpore (West)	25·2	South Barrackpore (East)	56·0
Kamarhati (West) ...	18·8	Kamarhati (East) ...	34·8
Baranagar	17·8	North Dum Dum ...	68·1
Chitpore-Cossipore ...	11·2	South Dum Dum ...	32·3
		Maniktolla	13·2
Average	24·5	Average	41·0

A glance at the above table or at the shaded map will show that the places situated on the bank of the Hooghly river have a much lower spleen percentage than those further to the east, even when the latter are but two miles from the river as in the case of the last five in the right-hand column of the table, with the exception of part of North Dum Dum. This having been ascertained, the question arose whether the lower rate on the banks of the Hooghly was to be regarded as the normal rate, and the higher figures of the inner tract as being due to water-logging or other abnormal conditions, or whether the latter must be taken as the usual state of affairs in this part of Lower Bengal, and the banks of the Hooghly as being exceptionally healthy. In order to solve this problem it was necessary to visit other places still further to the east, and Gobardanga and Basirhat, which are situated on the next flowing river to the east of the Hooghly, namely, the Ichahamati, were selected as the most suitable for the purpose. The former is some 20 miles to the east of Naihati, while the latter is 26 miles to the east of Baraset. The former is nearly surrounded by a bend of the river on two sides and by marsh land on the other sides, so that cannot be considered to be well situated from the health point of view, but Basirhat, on the other hand, would appear to be likely to be as healthy as any place in this portion of the Gangetic delta. Nevertheless, both show a spleen-rate of over 50 per cent., which, agreeing as it does with Dr. Gregg's statements about this tract of country, may be taken as approximately the normal figure for this part of Lower Bengal. It would, therefore, appear that the east bank of the Hooghly is exceptionally healthy, although some of the Municipalities in the low-lying tract a little to the east of the river show very high spleen-rates, more especially North Dum Dum and the portion of South Barrackpore to the east of the railway, whose figures are 68 and 56 per cent. respectively. It may also be at once mentioned that last year, namely, 1899, was an exceptionally feverish one on account of the excess and uneven distribution of the rainfall.

On looking more closely at the figures it will be observed that there is one marked exception to the rule above pointed out, for Maniktolla, although situated away from the river-bank and on extremely low-lying and water-logged land, has, nevertheless, the second lowest spleen-rate; an exception which has proved to be the key to a very important factor in the causation of the variations of the spleen-rate in the tract under consideration. The only ground on which the low spleen-rate of Maniktolla can be accounted for is the enjoyment by this advanced Municipality of a good filtered water supply. It is also worthy of note that Chitpore-Cossipore, which has the lowest rate of all, namely, 11·2, has the double advantage of a filtered water supply and a situation on the east bank of the Hooghly. That these are the true reasons of its marked immunity is shown by the fact that the average rate of the two western Wards is only 7·4, while that of the two easternly Wards, situated from one to two miles from the river, is 14·7, that is almost the same as that of Maniktolla. The density of the population of Chitpore, and consequent smaller number of tanks, etc., may also be a slight factor in its healthiness, but the details to be given immediately with regard to the spleen-rates of different parts of Maniktolla and other places show that this is not a factor of any great importance, but on the other hand they will prove conclusively the intimate relationship between a filtered water-supply and a low spleen rate, but as this point is one of the utmost practical importance it will be necessary to go somewhat into detail with regard to the spleen-rates of different Wards of the same Municipalities, more especially of those parts of which are being supplied with good water by mills situated within their boundaries. At the same time the data with regard to the ground water-levels will be given, so that the question of water-logging can also be discussed.

WARD VARIATIONS IN THE SPLEEN-RATES. 1. MANIKTOLLA.—This Municipality, as will be evident from the accompanying map, is situated between the Circular and the New Cut Canals, and this area is so flat that there is only a fall of some eight feet from west to east in a distance of two miles. Its drainage is dependant on channels by the sides of the four main roads, and is carried under the New Cut Canal by means of siphons into the Great Salt Water Lake, but these have to be closed at high tides to prevent the salt water running up into the drains, and they do not work very efficiently at present. The portion of the main drains in the western and more densely populated portion of the Municipality are brick-lined, but the eastern portions are of earth only. The water-level was taken in several wells, and in February it was found to average 5 feet from the surface of the ground, while evidence was obtained that it rises to within from one to two feet during the rainy season. A more typically water-logged place it would be difficult to find. For purposes of comparison it was divided up into western and eastern portions, and the spleen-rates were found to be 12·4 for the former and 14 for the latter, although it might have been expected that the less densely inhabited and more water-logged eastern portion would have had a decidedly higher rate. It was in the west part of this Municipality also in which the larvæ of the malarial-bearing mosquito was found in from half to two-thirds of the tanks as well as in some other pools, as will be detailed further on, so that none of the known causes of malaria were absent, in spite of which this Municipality, together with that of Chitpore and Cossipore, were the two which showed considerably the lowest spleen-rates of all the thirteen, and these two are the only ones which have a full filtered water supply.

The following table shows the above figures in a convenient form.

TABLE II.
MANIKTOLLA.

Area.	Gonnd Water-Level.		Water-supply.	Corrected Spleen percentages.		
	Feb., 1900.	Rains, 1899.		Adult Males.	Children.	General Total.
West part (1)	5 ft.	1 to 2 ft.	Filtered.	13	11·8	12·5
East (2)	5 ft.	2 ft. 6 in.	do.	13	15·0	14·0

CHITPORE-COSSIPORE.—This Municipality is situated immediately to the north of Calcutta, and extends eastwards as far as the Eastern Bengal Railway and northwards to the southern border of Baranagar. It is divided into four Wards, namely, Chitpore and Cossipore West extending from the river to the Grand Trunk Road, the spleen percentages of which are 4·8 and 9·9 respectively; and Chitpore and Cossipore East, extending from the Grand Trunk Road to the Railway, and consequently distant from one to two miles from the river bank, the spleen-rates of which are 13 and 16·75 per cent. respectively. The whole area is supplied fully with filtered water, while those people who do not drink this (and they are certainly a decided minority) will take chiefly river water in the western Wards, and tank water in the eastern ones. The water-level in three wells varried only between 4 and 5 feet from the surface in February, and in the rainy season it had been within from 1 to 2 feet of it; so that here again there was considerable water-logging but the minimum amount of fever, while although the western portion is more densely populated, the eastern part presents numerous tanks, and is generally favourable to the development of malaria, yet, apparently owing to the filtered water-supply, the spleen-rate is very low.

TABLE III.

CHITPORE-COSSIPORE.

Area.	Ground Water-Level.		Water-supply.	Corrected Spleen percentages.		
	Feb., 1900.	Rains, 1899.		Adult Males.	Children.	General Total.
Chitpore, West (3)	4 ft. 3 in.	1 ft.	Filtered.	2 05	7·7	4·85
Cossipore, West (4)	4 ft. 9 in.	2 ft. 9 in.	do.	10·6	9·3	9·95
Chitpore, East (5)	5 ft. 1 in.	2 ft.	do.	16·0	10 0	13·00
Cossipore, East (6)	......		do.	18·3	15·2	16·75

SOUTH DUM DUM.—To the east of the railway, which bounds the Chitpore-Cossipore Municipality, lies South Dum Dum, the most thickly inhabited portions of which are situated on the Jessore, Belgatia and Dum Dum roads, and it is divided into three Wards, which may roughly be taken as respectively including the parts adjoining these three roads. The inhabitants of Ward II who were examined mostly resided near the easternmost portion of the Dum Dum road, and the spleen rate was 37·9. Those of Ward I. mostly lived around that portion of the Jessore road which joins the eastern ends of the Belgachia and Dum Dum roads, and its spleen-rate was 45·3. Lastly, most of those examined in Ward III. lived around the western end of the Belgachia road just to the east of the railway, and consequently close to the Western Ward of Cossipore, and the spleen-rate among them was only 13·7, by far the lowest rate of any place to the east of the railway. Here again the probable explanation of this exception is that many of the inhabitants of this Ward obtain filtered water from the Cossipore Municipality as I ascertained both by inquiry and by seeing them carrying the water myself, while the portion of the other Wards which were examined were too far from Cossipore for the people to resort there for water to any extent. The conditions favourable to malaria are very similar in each Ward, for the Bajulla Khal flows right through Wards II. and III. as a broad swampy track with little or no current except during the rainy season, while the tide flows up it from the Salt Water Lakes at high water, there being no sluice gate where it passes through the bund, while part of the houses of the Municipality are surrounded by rice fields. The water-level in a well in Ward I. was 8 ft. 9 in. below the surface in February, while during the previous rainy season it had risen within 9 inches of the ground, when the water could be dipped out by hand without the use of any rope, so that there is no doubt about this Municipality being very water-logged.

TABLE IV.
SOUTH DUM DUM.

Ward.	Ground Water-Level.		Water-supply.	Corrected Spleen percentages.		
	Feb., 1900.	Rains, 1899.		Adult Males.	Children.	General Total.
I. (7)	8 ft. 9 in.	1 ft.	Tank.	41·8	48·8	45·3
II. (8)	...	...	do.	35·4	40·4	37·9
III. (9)	...	...	Partly filtered.	11·8	15·6	13·7

BARANAGAR.—This Municipality lies between the Hooghly and the Eastern Bengal Railway, extending northwards for nearly two miles above Cossipore. It is divided into four Wards, the first three of which are between the river and the Grand Trunk Road, and the fourth lies to the east of the former, being mostly between the Grand Trunk Road and the railway, and consequently is dependant for its water-supply on tanks, while the first three get theirs mainly from the river, although Ward I., which is the most southernly bordering on Cossipore, obtains a certain amount of filtered water from that Municipality. Ward I. has the lowest spleen-rate, it being only 11·6, Wards II. and III. have intermediate rates of 14·3 and 18·1 respectively, while Ward IV. has the highest rate, namely, 26 ;' differences which can only be explained by the varying water-supply, for although the last Ward also has a larger area under rice cultivation, that portion of it, whose inhabitants were examined, did not present materially different conditions from the other three Wards. Nor will the differences in the ground water-level, which are given in the Table below, account for those of the spleen-rates.

TABLE V.
BARANAGAR.

Ward.	Ground Water-Level.		Water-supply.	Corrected Spleen percentages.		
	Feb., 1900.	Rains, 1899.		Adult Males.	Children.	General Total.
I. (10)	7 ft. 7 in.	4 ft.	River and Tank & some filtered.	15·0	8·3	11·6
II. (11)	4 ft. 1 in.	2 ft.	River and Tank.	12·3	16·3	14·3
III. (12)	7 ft. 6 in.	3 ft.	do. do.	22·9	13·4	18·1
IV. (13)	...	...	Tank only.	14·6	37·3	26·0

KAMARHATI.—This Municipality lies immediately to the north of Baranagar, and consists of two Wards, namely, No. I. between the river and the Grand Trunk Road,. and No. II. from the latter up to the Eastern Bengal Railway, and including Belguria. The spleen-rate of the river Ward was found to be 18·8, while that of the inland Ward was 34·8, a notable difference, while the first Ward mainly relies on the river for its water-supply, and the latter is dependant on tanks; for although there are a few wells in all the municipalities, mostly belonging to private individuals, yet they appear from my inquiries to be little if at all used by the people for drinking purposes, especially if filtered water is available, while many intelligent natives informed me that those who drank filtered water suffered much less than those who drank that from any other source, including well water. The ground water was 7 feet below the surface in February in the riverine Ward, and had been within 1 ft. 8 in. of it in the rainy season of 1899, while it was 1 foot further down in both seasons in the case of the eastern Ward, so that from this point of view the latter should have been slightly the more healthy of the two, instead of entirely the reverse obtaining.

TABLE VI.

KAMARHATI.

Ward.	Ground Water-Level.		Water-supply.	Corrected Spleen percentages.		
	Feb., 1900.	Rains, 1899.		Adult Males.	Children.	General Total.
I. (14) West	7 ft.	1 ft. 8 in.	River and Tank.	17·3	20·4	18·8
II. (15) East	8 ft.	2 ft. 8 in.	Tank only.	32·5	36·6	34·8

NORTH DUM DUM.—This Municipality is situated to the east of Kamarhati, and extends from the railway to Nowi Nadi, a distance of some four miles, and it consists for the most part of rice fields surrounding several villages. It contains two Wards, the westernly of which includes the large village of Nimta, while the easternly one includes Gouripur and Kadihati, which are situated on the Nowi Nadi, a sluggish stream which carries the surface drainage away to the south-east into the Kocho bhil. The water-level in a well in the western Ward was 7 ft. 3 in. below the surface in February, and had risen to within 2 ft. 3 in. in the rainy season of 1899, so that this part is certainly water-logged. The spleen-rate in the western Ward was no less than 76·6 per cent., while among 58 boys of the Nimta High School, who mostly belonged to well-to-do families, it was 67. In the eastern Ward the percentage worked out at 59·6, which is also very high, the average of .the two Wards being 68·1; an extremely high figure. The water-supply is solely from tanks and a very few wells, while the villages are surrounded by flooded rice fields during the rainy season; both a bad water-supply and water-logging being present and factors in causing the marked unhealthiness of this area.

TABLE VII.

NORTH DUM DUM.

Ward.	Ground Water-Level.		Water-supply.	Corrected Spleen percentages.		
	Feb., 1900.	Rains, 1899.		Adult Males.	Children.	General Total.
Western (16)	7 ft. 3 in.	2 ft. 3 in.	Tank	73·3	80·6	76·6
Eastern (17)	...	...	do.	56·6	62·5	59·6

SOUTH BARRACKPORE.—This Municipality is a very large and scattered one, mainly consisting of a riverine portion situated between the Hooghly and the Grand Trunk Road, the following four Wards of which (beginning from the south) were examined, namely; Agarpara, with a spleen-rate of 30·8 and a ground water-level of 7 ft. in February and 1 ft. 8 in. below the surface in the rains of 1899 : Punihati, with a spleen-rate of 31·25 : Sukchar, with a spleen-rate of 12·1 and a ground water-level of 8 ft. in February, and 2 ft. below the surface in the rains of 1899 : and Khardaha, situated just to the south of the khal of the same name, with a spleen-rate of 26·75 and a ground water-level of 6 ft. 6 in. down in February. All these depend mainly for their water-supply on the river, while the exceptionally low rate of Sukchar appears to be due to the unusual number of good pukka houses, many of which are two stories high, the inhabitants of which must have been much better to do than the majority of those in most of the other Wards, while tanks are also fewer than usual in this Ward.

This Municipality also includes a large area of rice land with scattered villages to the east of the Grand Trunk Road, and extending across the Eastern Bengal Railway. Two portions of this were examined, namely, one to the east of Punihati and Sukchar, consisting mainly of the village Sodepore on either side of the Eastern Bengal Railway, but mostly to the east of it, and another village called Natagore to the east of the former. The spleen-rate of this area was 60·4, that of Sodepore having been 61·7, and that of Natagore 64·4. The ground water-levels in February were 10 ft. 6 in. and 9 ft. respectively, and in the rainy season of 1899, 2 ft. and 4 ft. below the surface, measurements which, it will be observed, are very similar to those of the riverine portions of this Municipality, the slight difference being in favour of the inland portions, although their spleen-rates are very much higher than those of the parts on the banks of the

J. II. 61

Hooghly, so that the water-levels do not help in explaining the difference. On the other hand, the dwellers near the river will mostly drink river water, while those who live more inland are entirely dependant on tank water. The other part of the South Barrackpore Municipality which was examined lies to the east of the railway opposite North Barrackpore, and consists of the villages of Chandanpukuria and Nona. The spleen-rate was found to be 51·6, while the ground water-level was 10 ft. 4 in. below the surface in February, but had risen to within 5 ft. during the rains of 1899, figures which are much more favourable than those of Maniktolla and Chitpore-Cossipore, which have the lowest spleen-rates. This, the most north-easterly Ward of the South Barrackpore Municipality, is also dependant on tank water for its drinking supply. The much lower spleen-rate, then, of the parts near the river, as compared with those at a distance of two miles or more from it, is again borne out by this Municipality, the figures of which are given in the table below.

TABLE VIII.

SOUTH BARRACKPORE.

| Area. | Ground Water-Level. | | Water-supply. | | Corrected Spleen percentages. | | |
	Feb., 1900.	Rains, 1899.			Adult Males.	Children.	General Total.
Agarpara (18)	7 ft.	1 ft.	River and Tank		33·3	28·5	30·8
Punihati (19)	...	...	do.	do.	20·0	42·5	31·2
Sukchar (20)	8 ft	2 ft.	do.	do.	19·3	6·9	12·1
Khardaha (21)	6 ft. 6 in.	...	do.	do.	32·5	21·0	26·7
Sodepore (22)	9 ft.	4 ft.	Tank only.		50·0	70·9	60·4
Nona (23)	10 ft. 4 in.	5 ft.	do.		36·5	66·8	51·6

TITTAGHAR.—This is a small Municipality which lies on the east bank of the Hooghly between South and North Barrackpore, and is bounded on the south by the Khardaha Khal, and on the north by the Tittaghar Khal, and on the east by the Grand Trunk Road. It is divided into four Wards numbered I. to IV. from north to south. Two Mills in Wards II. and III. supply a limited amount of filtered water more especially to the inhabitants of Ward II., but Wards I. and IV. on either side of the other two drink nearly entirely river and tank water. Here, then, was a very good opportunity of putting to a crucial test the question as to whether filtered water drinkers suffer less from enlargement of the spleen than do those who drink other kinds, so notes were made regarding nearly all of the people examined in this Municipality as to what water they usually drunk, whether filtered, Hooghly, or tank.

The results are as follows, beginning from the south as before. Ward IV., which is a narrow strip situated on the north bank of the Khardaha Khal, up to which the tidal water flows as far as a sluice gate in a bridge under the Grand Trunk Road, and which contains a series of bustees, had a spleen-rate of 48 per cent., that is a high one for a riparian area. The water-level was 10 ft. 1 in. below the surface in February, but had risen to within 1 ft. 3 in. in the rains of 1899. Only 16 per cent. of those who were examined stated that they drank filtered water. In Ward III. 32 per cent. of those examined had drunk filtered water, and the spleen-rate was 30 per cent. The water-level had been 1 ft. 6 in. below the surface in the rains of 1899, and was 10 ft. 3 in. down in February, so that in this respect the conditions were just the same as in Ward IV., so this factor will not explain the considerable difference between the health of these two Wards; the water-supply only being different. Still more marked, however, was the difference between the spleen-rates of the two northern Wards, that of Ward III., which is opposite the Mills, being 19 per cent., and that of Ward IV. immediately further north, was 54·3; in spite of the ground water-level of the latter having been 6 ft. from the surface at the height of the rains of 1899, and 18 ft. 4 in. down in February last; an exceptionally low rate. On the other hand, the number of the people examined in Ward III. who had drunk filtered water was no less than 82·5 per cent., while only 19·6 of those of the Ward I. stated that they drunk filtered water, and owing to their greater distance from the supply they were probably less regular in obtaining it than were the inhabitants of Ward II. at whose doors it was placed. These figures are sufficiently striking, especially as they confirm the data obtained in several other municipalities, to be given immediately, and they are also in entire agreement with the following results of the differences in the spleen-rate among the drinkers of the different kinds of water in this Municipality. Thus among 140 filtered water drinkers, 37, or 26·4 per cent., had enlarged spleens; while among 179 river water drinkers 74, or 41·3 per cent., were similarly affected; but of 55 tank water drinkers no less than 33, or 67·2 per cent., had enlargement of this organ. Further, if we take the degree of enlargement among the different classes as detailed in Table IX, below, we find that of those who had enlarged spleens the degree of enlargement was very slight in 62 per cent. of the filtered water drinkers, in 43·2 per cent. of the river water drinkers, but only in 27 of those who drank tank water, it being considerable or very enlarged in the remainder. Not only, then, is the percentage of enlarged organs much greater in those who drank unfiltered water (the percentage of mixed river and tank water drinkers being 47·4), but the degree of enlargement of the organ was also much more marked in the latter classes as compared with the filtered water drinkers. (See Table X.).

TABLE IX.

TITTAGARH.

Ward.	Ground Water-Level.		Water-supply.	Corrected Spleen percentages.		
	Feb., 1900.	Rains, 1899.		Adult Males.	Children.	General Total.
IV. (24)	10 ft. 1 in.	1 ft. 3 in.	River and Tank.	36·0	60·0	48·0
III. (25)	10 ft. 6 in.	1 ft. 6 in.	do. but ⅓ of them drank filtered water.	29·2	30·8	30·0
II. (26)	...	...	do. but 82°/₀ drank filtered water.	19·0	18·9	19·0
I. (27)	18 ft. 4 in.	6 ft.	River and tank water.	51·3	57·4	54·3

TABLE X.

SPLEEN ENLARGEMENT AND WATER-SUPPLY IN TITTAGARH.

	Filtered water.	River water.	Tank water.	Total.
Spleen not enlarged	103·0	105·0	18·0	226·0
Spleen slightly enlarged.	23·0 (62°/₀)	32·0 (43°/₀)	10·0 (27°/₀)	65·0
Spleen considerably enlarged.	9·0 (24°/₀)	26·0 (36°/₀)	15·0 (40·5°/₀)	50·0
Spleen markedly enlarged.	5·0 (13°/₀)	16·0 (21°/₀)	12·0 (32·7°/₀)	33·0
Total examined ...	140	179	55	37·4
Percentage with enlarged spleens.	26·4	41·3	67·2	39·5°/₀

NORTH BARRACKPORE.—This is a small Municipality on the east bank of the Hooghly extending from the Tittaghar Khal on the south to the Ichapur Khal on the north, and bounded on the east by the Grand Trunk Road. It consists of three circles. Firstly, Monirampur, situated in the bend of the river to the west of Barrackpore Cantonment, the spleen-rate of which is 24 per cent., while the ground water is low, there having been no water in a well 8 ft. 8 inches deep in February. The water-supply is mainly derived from the river. Secondly, Nawab-gung, also placed on the bank of the river to the north of the last

named, its spleen-rate being 28·6 per cent., while the water-level was
9 ft. below the surface of the ground in February, and had risen to 5 ft.
from the ground in the rains of 1899. The water-supply is mainly derived
from the river. Thirdly, Ichapur, which is situated to the north-east
of the last circle, and the main portion of whose inhabitants reside at a
distance of about one mile from the river, and near the Grand Trunk
Road, and consequently are mainly dependant on tanks for their water-
supply. The spleen-rate of this circle was 56 per cent., although as the
ground water-level was 10 ft. from the surface in February and had not
risen above 4 ft. in the rains of 1899 there was no difference in this respect
from the other two circles which could possible account for the greatly
higher spleen-rate of Ichapur, whose water-supply from tanks instead
of from the river appears to be the only possible explanation of the
facts recorded.

TABLE XI.

NORTH BARRACKPORE.

Area.	Ground Water-Level.		Water-supply.	Corrected Spleen percentages		
	Feb., 1900.	Rains, 1899.		Adult Males.	Children.	General Total.
Monirampur (28)	Below 9 ft.	...	Mainly river.	24·5	23·5	24·0
Nawabgung (29)	9 ft.	5 ft.	do.	37·2	30·1	28 6
Ichapur (30)	10 ft.	4 ft.	Tank.	52·0	66·0	56·0

GARULIA.—This small Municipality is situated between the Hooghly
river and the Grand Trunk Road immediately to the north of the
Ichapur Khal, and its northern half has been supplied with filtered
water from the Dunbar Cotton Mill for the past two years, but the
inhabitants of the southern portion for the most part still drink river
and tank water. As there was a very general opinion among the people
living near the Mill that they had suffered much less from fever
since the filtered water had been introduced, I determined to examine
100 persons who resided near the Mill, and the great majority
of whom (about 80 per cent.) were found on inquiry to have been
drinking the filtered water ; and another 100 a little further to the
south, but all within one mile of the former, and who stated that they
drank river or tank water. Among the former class the spleen-rate
was found to be 21·1 per cent., while among the river and tank water
drinkers it was 46·5 per cent., although the latter included 28 men who
had arrived from the North-West Provinces only in November last,
that is after the fever season is nearly over, and whose spleen-rate was
only 10·7 per cent. If these men are excluded from the calculation, the
spleen-rate of the permanent residents of this southern portion of the

Municipality] rises to] 55·5 per cent., or just over two-and-a-half times as great as among the filtered water drinkers. A well in the northern part showed a water-level 9 ft. 6 inches below the surface of the ground in February, while it had risen to within 1 ft. 6 in. during the rains of 1899, so that [there must have been a considerable degree of water-logging at that time, in spite of which the spleen-rate is low. These facts appear to admit of no other explanation than that the filtered water-supply was the cause of this low rate near the Mill as compared with a precisely similar area in other respects close by which had not the advantage of the stand-pipe water.

TABLE XII.

GARULIA.

Area.	Ground Water-Level.		Water-supply.	Corrected Spleen percentages.		
	Feb., 1900.	Rains, 1899.		Adult Males.	Children.	General Total.
Northern part (31) (near Mill).	9 ft. 6 in.	1 ft. 9 in.	Filtered water.	17·7	24·6	21·1
Southern part (32)	...	...	River and tank water.	50·0	61·1	55·5

BHATPARA.—This Municipality consists of a narrow strip between the river Hooghly and the Eastern Bengal Railway and to the north of Garulia, and it is divided into three Wards, the northern two, Wards I. and II., of which, more particularly, and to a somewhat less extent the southern one, Ward III., obtain some filtered water from the Mills situated within this area. The spleen-rates of all are low, that of the southern one being slightly higher than the other two, although there is not much difference in their water levels, which are slightly in favour of Ward III. The figures are given in Table XIII. below.

TABLE XIII.

BHATPARA.

Ward.	Ground Water-Level.		Water-supply.	Corrected Spleen percentages.		
	Feb., 1900.	Rains, 1899.		Adult Males.	Children.	General Total.
III. (33)	10 ft. 3 in.	4 ft. 4 in.	Mainly river.	19·9	27·8	23·6
II. (34)	9 ft. 8 in.	5 ft.	River and filtered.	22·0	12·0	17·0
I. (35)	7 ft.	2 ft. 6 in.	do.	27·1	11·5	19·3

NAIHATI.—This Municipality is situated between the Hooghly and the Eastern Bengal Railway extending from Naihati itself northwards for five miles as far as the Bhagar Khal, and although narrow to the south it gradually widens out to the north, so that while the lower three Wards are mainly inhabited near the banks of the river, the majority of the people in the two northern Wards live at some little distance from the river at Halishahar and Kanchrapara. Moreover, the Gauripur Jute Mills supply some filtered water to WardsII., so it is worthy of note that this Ward again has the lowest spleen-rate, namely, 10·8, which is little more than half that of the Wards I. and III. on either side of it, which are dependant on the river for their supply. Further, Wards IV, and V. have the highest rates of all, being mainly dependant on tank water for their supplies, so that, and that in spite of their ground water-level being lower than that of Ward III.; so that the only way in which these variations can be explained is by the differences in the water-supplies of the various Wards, which are also in accordance with the results obtained in every previously considered instance.

TABLE XIV.

NAIHATI.

Ward.	Ground Water-Level.		Water-supply.	Corrected Spleen percentages.		
	Feb., 1900.	Rains, 1899.		Adult Males.	Children.	General Total.
I. (36)	...	...	Mainly River.	16·6	22·5	19·5
II. (37)	...	...	Partly filtered.	6·6	15·0	10·8
III. (38)	7 ft. 9 in. 16 ft. 6 in.	... 8 ft.	Mainly River.	7·1	31·0	19·0
IV. (39)	10 ft. 3 in.	4 ft.	Mainly Tank.	15·3	29·4	22·3
V. (40)	10 ft. 3 in.	4 ft.	Tank.	13·3	42·5	27·9

BARASET.—This Municipality is situated on the Soonthee Nudi some eight miles east of the Hooghly river, and its surface drainage flows away to the south-east into the Kocho Bhil. The Soonthee was formerly a large river, but now it resembles an elongated swamp with little or no current except during the rainy season, while its bed is encroached upon in numerous places by series of tanks which in places leave but a few yards between them for the stream, and fishing weirs, etc., also obstruct its course. The Municipality is divided into five Wards, Nos. I. and II. including the town, while Nos. IV. and V. are to the east on the Soonthee Khal, and No. III. to the south. In all the spleen-rates are high, and the ground water-levels do not vary much, but are high in the rains, showing obstructed drainage and water-logging. The water-supply is from tanks, although one or two tube wells have been put down

TABLE XV.

BARASET.

Ward.	Ground Water-Level. Feb, 1900.	Rains, 1899.	Water-supply.	Corrected Spleen percentages. Adult Males.	Children.	General Total.
I. & III. (41)	12 ft. 4 in.		Tank only	52·3	50·4	51·5
III. (42)	16 ft. 3 in.	4 ft.	do.	38·1	80·3	59·2
IV. (43)	8 ft. 2 in.	0 ft. 3 in.	do.	44·2	70·8	57·5
V. (44)	12 ft. 4 in.	4 ft.	do.	38·6	51·2	44·9

WATER-LOGGING AND THE RAILWAY.—It have already been pointed out in discussing the Ward variations of the spleen-rate that there is no definite relationship between the amount of malaria and the ground water-levels. Thus Maniktolla and Chitpore-Cossipore are the most water-logged parts of the whole area, yet they have the lowest spleen-rates on account of their filtered water-supply. The fact that the bank of the Hooghly river is slightly higher than the country further to the east, so that the surface water flows away from the river, and eventually finds its way back through khals to the river or runs off to the south into the Great Salt Water Lakes, might at first sight seem to indicate that the ground water-level would be lower near the river than it is further to the east. Measurements in the wells, however, do not bear this out, for there is very little difference in this respect in the water-level measured in wells on either side of the railway, while the differences noted were rather more frequently in favour of the eastern portions than the contrary. The differences in the spleen-rate, then, of the eastern and western parts cannot be explained on any theory of water-logging, while an examination of the whole area Ward by Ward shows no definite relationship between the spleen-rates and the height of the ground water-level, as a study of the Tables and Map will show.

The Eastern Bengal Railway, which runs from north to south through this area, and, together with the Grand Trunk Road, roughly divides it into western and eastern portions, has frequently been held responsible for the unhealthiness of the country, for it lies across the line of surface drainage. As, however, the drainage flows to the east if it were materially obstructed the western part should be the more unhealthy, instead of which precisely the opposite holds good. Moreover, in the few places in which wells were found on either side of the railway, although at some distance from it, there was no constant or marked difference between either the level of the ground-water in the dry season, or the height to which it rose, during the rains on either side of the road and railway. There is, then, no evidence that the health of this tract has been influenced by the railway or the Grand Trunk Road, and the spleen variations cannot possibly be attributed to their action.

DISTRIBUTION OF THE *Anopheles* MOSQUITOS.

It must now be taken as proved that malaria can be communicated to man by the bites of the *Anopheles* genus of mosquitos, which have previously bitten another case of malaria, and in whose body the plasmodium has undergone developemental changes. It still, however, remains to be proved that this is the only or even the most common way in which the disease is obtained, and it is worthy of note that Laveran, who was the first modern exponent of the mosquito theory, is still of the opinion that it will not explain all that is known of the etiology of the disease. Still enough is known to make it highly advisable to consider the question of the possibility of destroying the particular breed of mosquito which plays a part in distributing malaria. This should not be impossible in limited areas, at any rate, if Major Ross's statement as to their breeding-grounds is correct, namely, that they mainly breed in small pools which are not inhabited by fish, and yet are not so small that they will dry up in a few days, and consequently that such suitable pools are few and far between. In order to test this statement I searched for the larvæ in several Municipalities, but regret to say that I have not been able to confirm Ross's statements. On the contrary, I found the *Anopheles* larvæ in numerous tanks as well as in the small pools which Ross describes, and that too in spite of the former as well as some of the latter abounding in fish. This having been ascertained, a small portion of Maniktolla, measuring about one-sixteenth of a square mile, and containing some thirty tanks, was further examined. During the dry months of from February to May, which are the minimum fever months of the year, I found the *Anopheles* larvæ in from one-third to two-thirds of these tanks, often in enormous numbers, one of them for example, having been estimated to have - contained several million larvæ on one day on which it was examined in May. In the earlier months especially they were also found in several small pools, but the numbers there were nothing as compared to those in the tanks, which are certainly the common breeding-ground of the *Anopheles* in the dry season at any rate. Three pools in a low-lying area are of interest, for in one, some two yards square, and a second which was five yards in diameter, fairly numerous *Anopheles* larvæ were found in spite of the presence of small fish in both, so that it is not surprising that they can also survive in tanks which are swarming with fish. Further, I failed to find any cases of fever near the infected tanks in the hot weather. As there must be several hundred tanks in the five square miles of Manik-tolla alone, the chances of being able to destroy these larvæ appears to be very remote. Further observations are being made on the seasonal distribution of these larvæ and the amount of fever, but it may be mentioned that they nearly disappeared from the tanks after the first burst of the rains, and remained absent during a break which followed, although fever now began to be prevalent, so that up to the present the number of the *Anopheles* has been in inverse proportion to the amount of fever. Possibly the tank forms are different from those of the rainy season in the small pools, but I have not yet been able to settle this point. The differences will be only microscopical, so that this would not lessen the practical difficulties in lessening malaria in Bengal by destroying the mosquitos, the only possible way of partially affecting which would appear to lie in the time-honoured method of extensive drainage in order to lessen the number of suitable breeding-grounds for the mosquitos.

The great difficulty of destroying the *Anopheles* larvæ in Bengal enhances the importance of the influence of a filtered water-supply in reducing so materially the amount of fever, which has been shown to be the case in portions of this tract of country, while the much greater liability of the drinkers of tank water to malaria suggests that the disease may commonly be obtained by drinking infected water, as has for centuries been considered to be the case. Such a mode of infection may be easily reconciled with the mosquito theory if we allow that these insects, in addition to directly inoculating the disease, may also take the parasite back to water, perhaps by means of the black spores described by Ross, in which they may survive for a limited time only, so that the infection has frequently to be renewed by the mosquitos. This is a point which can only be settled by investigation, which I hope shortly to be able to undertake.

Lastly, an examination of charts showing the monthly rainfall and fever-rate in this tract of country revealed the fact there is no constant relationship between either the amount or monthly distribution of the rainfall of different years and the amount of malarial fever. A more detailed examination, however, showed that there is a relationship between the daily distribution of the rain and the fever; those years in which the rainfall is very irregularly distributed with frequent and prolonged breaks, being those in which malarial fevers are most prevalent. This point is also being more closely studied, in conjunction with the observations on the variations in the distribution of the *Anopheles*.

CONCLUSIONS.

The general result of the inquiry has been to show that there is a marked difference between the health of the riverine and more inland portions of the area examined, the former being much more healthy than the latter. The comparatively healthy area extends from the river to the Grand Trunk Road, and in some parts to the Railway, a distance varying from one to two miles. No marked or constant differences in the ground water-level of the healthy and unhealthy parts has been found, which could possibly account for the differences in the spleen-rate noted, so that no theory of water-logging will explain them.

The most striking exception to the rule that the areas at a distance from the river bank have a high spleen-rate is that of Maniktolla, and the eastern portions of Chitpore-Cossipore between the Grand Trunk Road and the railway, both of which, together with the rest of the latter Municipality, have the lowest spleen-rates of all. Moreover they are also the most water-logged portions of the whole area, their ground water-levels both in the dry and in the rainy seasons being the highest met with, so that there must be some other factor to account for their marked immunity from malaria. This is certainly not the absence of the malaria-bearing mosquito, for it was in the first-named place that they were found to be more wide-spread during the minimum fever season than has hitherto been reported from any part of India. The only possible factor remaining is the water-supply, and it is noteworthy that these two water-logged Municipalities are the only ones which have a full filtered water-supply from the same source as Calcutta itself. That this good water-supply is the true explanation of their relative immunity from malaria is borne out by the very low spleen-rate of certain Wards of other Municipalities which have a partial filtered water-supply from various Mills, together with the low rates of the Wards of Baranagar and South Dum Dum, which border on Chitpore-Cossipore, from whose stand-pipes some of their inhabitants were obtaining filtered water, the details of which have already been given. Finally, the figures given in Table X. shows the spleen-rates among river water drinkers to be nearly double, and that of tank water drinkers to be nearly treble that of filtered water drinkers, strongly corroborate the evidence as to the benefit to be derived from filtered water, and affords a key to the whole distribution of the varrying spleen-rates, as can be seen from a study of the accompanying map. Thus, Chitpore-Cossipore West, which has the double advantage of a filtered water-supply and close proximity to the river, so that those who do not drink filtered water will for the most part take river water, has the lowest rate of all, namely, 7·4. The eastern part of the same Municipality, which has tank instead of river water as an alternative to the filtered supply, on the other hand, has a spleen-rate of 14·8, which is almost the same figure as that of Maniktolla, with a similar water-supply. Further, those Wards which are situated immediately on the river bank, but do not possess a filtered water-supply, and consequently get their supply mainly from the river, and to a less extent from tanks, have a rate intermediate between those with filtered water and the inland ones which are dependant entirely on tank water. In short, all the Ward variations in the spleen-rate of the whole area can be explained on the ground of their varying water-supplies in a manner which no other explanation will approach in completeness, so that it is impossible to come to any other conclusion than that the above is the true explanation of the facts recorded. Whether the *Anopheles* mosquitos play a part by taking the malarial parasite back to the tanks from their human hosts or not must be left to be determined by future experiments, but that a good water-supply is an important prophylactic measure in the lessening the prevalence of malaria must I think be admitted, and can be safely acted on.

XVIII.—*I. Further Researches on Mercurous Nitrite and its Derivatives. II. On Mercurous Iodide and a new Method of its Preparation.—By* P. C. RAY, D.Sc.

[Received 16th July ; Read 1st August, 1900.]

CONTENTS.

(1)

PREPARATION OF MERCUROUS NITRITE ON A LARGE SCALE.

As the investigations I am about to describe involve the use of comparatively large quantities of mercurous nitrite at a time I shall begin with describing a method of preparing the salt, which has been found to be more economical and far less troublesome than the usual method.

A tall beaker is taken containing nitric acid (sp. gr. 1041) diluted in the proportion of 1 : 4 with water; mercury is now poured in, care being taken not to fully cover the bottom but to leave an annular or horse-shoe-shaped space. The crystals which are formed on the convex surface of the mercury are continually pushed aside into the empty space by the evolution of gases during the initial stages of the reaction. On standing overnight, however, a crust of the nitrite is formed on the surface of the metal, which acts as a protective layer, thereby hindering further action.

All that is now necessary to do is to incline the beaker gently, when the deposit of the salt slips off into the empty space as explained before, leaving a fresh surface of mercury exposed. This may be repeated 4 or 5 times in the course of the day. Instead of inclining the beaker, the layer of crystals may be carefully scraped off the surface with a glass rod. The process may be allowed to go on for a week, resulting in the continuous growth and accumulation of the salt; the reaction may be started simultaneously in about half a dozen beakers arranged in a row, so as to secure a copious supply.

In the previous papers* it was recommended that each time a layer of crystals is formed, the mother liquor together with the mercury should be decanted off into another beaker. This is a wasteful method, for as soon as the super-incumbent liquid is removed, torrents of red fumes appear on the surface of the mercury. These red fumes are caused by the combination of nitric oxide with the oxygen of the air. It is the nitric oxide that evidently gives rise to the formation of th nitrite, and its loss has to be guarded against.

(2)

PREPARATION OF CHEMICALLY PURE MERCUROUS NITRITE.

The mercurous nitrite prepared as above will answer well enough for ordinary purposes. It generally contains, however, impurities, chiefly in two shapes. First, the crystalline mass encloses minute globules of mercury which cannot be entirely detached. Secondly, as the salt has to be dried on the porous tile, a portion of the heavy mother liquor consisting of mercurous *nitrate* dries up along with it. When it is desirable to obtain the salt in a state of absolute purity, it is mixed with a sufficiently large quantity of water and heated to boiling point for some

---

* For literature on the subject, see Journal, Asiatic Society, 1896, Pt. ii, p. 1, and Transactions, Chem. Soc. for 1897, p. 338.

time. Treated in this way about 18 p.c. of the nitrite undergoes dissociation as already pointed out; thus:

$$Hg_2 (NO_2)_2 = Hg + Hg (NO_2)_2.$$

While by far the larger proportion of it dissolves *as such ;* the saturated solution while still hot is rapidly passed through a "ribbed" filter paper, and the filtrate briskly stirred with a rod. In this way a fine, mealy, crystalline deposit is obtained, which is dried on a porous tile and preserved *inside a dessicator.* The presence of even a trace of atmospheric moisture brings about slow decomposition evolving nitrous fumes. As a test case it may be mentioned that 0·54 g. of the pure salt was placed on the scale pan for three hours, and it lost 5 mgs. during that time.

(3)

Interaction of Mercurous Nitrite and Ethyl Iodide.

Preliminary.

About four years ago while describing mercurous nitrite and its general properties, which were found to bear a remarkable analogy to those of silver nitrite, the author expressed a hope that this new compound would yield nitro ethane by interaction with ethyl iodide.

Since then it has been qualitatively shown that the reaction gives simultaneously both nitro ethane and its isomer, ethyl nitrite.[*] The present investigation embodies a fuller and more systematic study of this reaction.

Experimental.

The general method first described by V. Meyer and O. Stüber has been in the main followed.[†] I shall therefore confine myself to such details only as have a direct bearing on the subject in hand.

Exp. I. 120 g. of mercurous nitrite and 69 g. of ethyl iodide were digested together over a water bath in a round-bottomed flask to which was attached a tubulated funnel and a reflex condenser. The digestion was continued so long as ethyl nitrite was evolved. It is necessary to note here that as soon as ethyl iodide is poured on silver nitrite, an energetic action at once sets in, but when mercurous nitrite is added to the alkyl iodide there is scarcely any' perceptible evolution of heat, and the reaction only begins after digestion has proceeded for some time. The open

---

[*] Proc. Chem. Soc. 1896, p. 218.

[†] Ber. Deut. Chem. Ges., V, pp. 399, 514.

end of the condenser was connected with two tall cylinders (See Fig.)

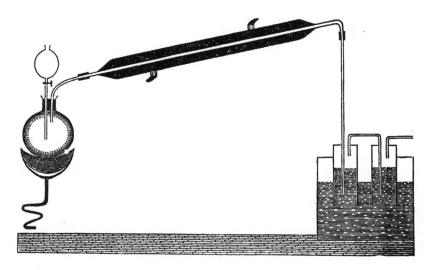

Fig. shewing the formation of Ethyl Nitrite and its absorption by
Alcohol.

containing a measured volume of alcohol for the absorption of ethyl
nitrite. These cylinders were again kept immersed in ice-cooled water
of an average temperature of 10° C., that of the Laboratory varying
from 23° to 25° C. In this manner a concentrated solution of what is
called in the Pharmacopœia *Spiritus Ætheris Nitrosi*, was obtained, the
strength of which was determined by Allen's method.* The yield of
ethyl nitrite was found to be 5·1 g.

Exp. II. 94 g. of mercurous nitrite and 35 gms. of ethyl iodide
were digested as above. The yield of ethyl nitrite was 3·1 g.

Exp. III. In this case 138 g. of mercurous nitrite were digested
with 54 g. of ethyl iodide, yielding 5·2 g. of ethyl nitrite.

It will thus be seen that the yield of ethyl nitrite is only a fraction
of what is demanded by theory. This is partly due to the reaction being
completed only during subsequent digestion on the oil bath, but chiefly
to the fact that when once a certain amount of mercurous iodide has
been superficially formed, a large proportion of the nitrite aggregates
into hard lumps into which the ethyl iodide can only slowly and with
difficulty penetrate.

* Pharmaceutical Journal, 3rd Series, Vol. XV, p. 673.

This is also the case with silver nitrite, though not in so marked a degree.*

Exp. IV. This was a control experiment in which silver nitrite was digested with ethyl iodide: 90 g. of the silver salt were treated with 88 g. of the alkyl haloid. The yield of ethyl nitrite was 4·2 g.

NITRO ETHANE.

After the evolution of ethyl nitrite had ceased, the contents of the flask were subjected to distillation, first over a water bath, and afterwards over an oil bath. The distillates were caught separately.

It was invariably found that during distillation over an oil bath, the receiver was filled with nitrous fumes, a part of which was absorbed by the distillate, imparting to it a bluish tinge. As there was not the slightest trace of yellow colour either in the flask or in the condenser itself, it was suspected that *nitric oxide* was evolved by the slow and gradual decomposition of a portion of mercurous nitrite, which combined with the oxygen of the air in the receiver. The suspicion was confirmed. The presence of the nitrous fumes is highly objectionable, as the crude nitro ethane so obstinately holds them in solution that they cannot be got rid of during fractionation. Distillation in a slow current of carbon dioxide was therefore resorted to for excluding air. In the control experiment with silver nitrite (See *ante*, Exp. IV.) nitrous fumes, though in a far lesser degree, were also noticed in the receiver.†

After the distillation was over, the compact mass of mercurous iodide and nitrite were removed from the flask, well powdered in a mortar, and once more treated with the fraction below 100°, when a further quantity of crude nitro ethane was obtained.

A fair idea of the yield of nitro ethane may be had from the details of one among several experiments. 190 g. of mercurous nitrite and 95 g. of ethyl iodide yielded a distillate of 7 g. between 100°–108°, of 3¼ g. between 108°–110°, and of 4 g. between 111°–114°.

The fraction which came off between 113°·5·–114° (uncorrected) was practically pure nitro ethane. It was treated with an alcoholic solution of caustic soda as recommended by Nef. (Annalen: 280, p. 267). The

---

* " Es gelang uns unter keinen Umständen das ganze Jodäthyl in die Reaction zu verwickeln, sondern stets war das bei der Rectification zuerst übergehende Produkt stark jodhaltig. Wir haben auf jede weise versucht, das Jodäthyl vollständig auszubeuten, doch immer vergeblich." (*Loc. cit.*, p. 402.)

† The behaviour of Mercurous Nitrite in this respect also resembles that of Silver Nitrite. *Cf.* Divers and Schimidzu "Action of heat upon Silver Nitrite, air being excluded"—(Trans. Chem. Soc. Journ., 47, 634), where it is shown that nitric oxide is one of the products of decomposition.

precipitate of sodium nitro-ethane was washed with absolute alcohol. In this manner 2·5 g. of nitro-ethane yielded 2·1 g. of the sodium compound.

It responded to all the characteristic reactions : its aqueous solution turned blood-red with ferric chloride and green with copper sulphate. A concentrated solution of it gave with corrosive sublimate solution a white, mealy, crystalline precipitate. The sodium salt which is extremely hygroscopic exploded with a loud detonation when heated in a narrow test tube. An estimation of sodium is given below.

$$0·2416 \text{ g. gave } 0·1795 \text{ g. of } Na_2 So_4.$$

	Calc. for	Found.
	$Na\ C_2\ H_4\ NO_2$	
Na	23, 60	24·06.

From the above investigations it would appear that by the action of mercurous nitrite on ethyl iodide about equal quantities of nitro ethane and its isomer ethyl nitrite are formed. The yield is, however, somewhat poorer than with silver nitrite, owing to the formation of very compact, hard lumps of mercurous iodide, which interferes with the reaction being completed.

(4)
INTERACTION OF MERCUROUS AND MERCURIC NITRITES WITH THE NITRITES OF SILVER AND SODIUM.

1.
Mercuric Nitrite and Sodium Nitrite.

To the neutral solution of mercurous and mercuric nitrites (the products of dissociation of mercurous nitrite : Journ. Chem. Soc. Trans. 1897, p. 340) is added sodium chloride to remove mercurous mercury. The filtrate which now contains mercuric nitrite and a small quantity of sodium chloride if it was added in excess, as well as *sodium nitrite,* is allowed to evaporate spontaneously. In course of time an orange, crystalline deposit is formed, and this is followed by the appearance of shining iodine-like dark scales; and last of all we obtain a yield of rhombohedral crystals of *sodium nitrate.* Sometimes the orange-red and black compounds are not obtained, but instead we get only sodium nitrate, sometimes again the three compounds are obtained in regular succession, though one kind may predominate over the others.

During the last three years I have repeated the experiments several times, but I have not been able so to control them as to ensure the formation of one variety only to the exclusion of the others.

If there be no sodium chloride present in the mercuric solution, the red and black deposits are not formed, for, as will be seen below, they are oxychlorides of mercury and the presence of sodium chloride is a *sine quâ non* for their formation, and the only yield is one of sodium nitrate. But this last compound has sometimes been observed as the sole product without the formation of the former, even in presence of an excess of sodium chloride. The reaction goes on very slowly, and it takes a month and upwards to complete it.

Analysis and general properties of the red and iodine-like lustrous scales :—

(1) 0·3505 g. gave 0·3485 HgS.
(2) 0· 465 „   „   0· 455 „
(3) 0·5086 „   „   0·2034 AgCl.
(4) 0·6166 „   „   21·1 c.c. moist O at 31° C. and 752 mm.
pressure.

	Theory for $HgCl_2 . 2HgO$	Theory for $HgCl_2 . 2HgO . \frac{1}{2}H_2O$	Found i	ii
Hg	85·35	84·27	84·45	84·34
Cl	10·10	9·97	9·90	
O	4·55	4·50	4·17	
H_2O		1·26		
	100·00	100·00		

The analyses recorded above are of distinct preparations, and they conform to the formula $HgCl_2 . 2HgO . \frac{1}{2}H_2O$.

These salts do not lose in weight when kept in a dessicator over strong sulphuric acid or placed in a steam chamber at 100° C. When heated in a bulb-tube a deposit of moisture is invariably noticed, and a sublimate of mercurous and mercuric chloride obtained, with a residue of orange-yellow oxide. Treated with caustic soda solution, the dark variety changes to orange-yellow.

Millon, and more recently Thummel (Archiv. Pharm.§ [3], 27, 589–605) have exhaustively studied the oxychlorides of mercury, and have described several of them. These were obtained, however, by adding together solutions of mercuric chloride and hydrogen potassium carbonate under varying conditions. Volhard got shining dark crystals by the action of sodium acetate upon corrosive sublimate solution (Annalen : 255, p. 252) ; whilst Haack obtained a reddish-brown crystalline deposit by treating mercuric chloride with phosphate of sodium (*ibid.* 262, 189), all of the formula $HgCl_2 . 2HgO$. The red and black shining compounds, the subject of the present paper, agree in general

§ The Original Memoir is not available here. I am quoting from Watt's Dict. of Chem., New Ed. See also Abs. Chem. Soc. Journ., Vol. LVI, 1050.

properties with those obtained by the above chemists; but the *hydrated* modifications I do not find mentioned anywhere.

I.

Mercurous Nitrite and Silver Nitrite.

A. Concentrated solutions of mercurous nitrite and silver nitrite.

As both mercurous nitrite and silver nitrite are very sparingly soluble in cold water, the solutions used were always kept at about $100°$ C.

Method of experiment :—To the hot or boiling solution of mercurous nitrite containing necessarily mercuric nitrite was added the solution of silver nitrite. No effervescence due to the evolution of gases was noticed, and the liquid which at once became cloudy on account of the separation of metallic mercury, was allowed to stand over night. Next day a perfectly clear solution was obtained, with a deposit of mercury and silver in successive layers at the bottom of the vessel,—the lower one of dirty grey mercury, and the upper one of an arborescent and filamentous growth of shining minute crystals of silver. These metals were estimated in the usual way. The strength of the filtrate was determined by finding out the weights of the *ous* and *ic* mercury as well as that of silver in solution. Control analyses were also simultaneously made to ascertain the original strengths of the mercury and silver nitrite solutions under exactly similar conditions of temperature. For details see Table of Analyses.

In order to estimate the total amount of nitrogen and the transformation, if any, of the nitrite into nitrate, or any other compound of nitrogen, the following method of analysis as exemplified in Exp. I was adopted.

50 c.c. of mercurous and mercuric nitrite solution were boiled for a few minutes with an excess of caustic soda; 25 c.c. of silver nitrite solution were also similarly treated. The filtrates from the mercury and silver precipitates, containing nitrogen in the shape of nitrite of sodium, were now added together and made up to a given volume. After the interaction of mercurous and silver nitrite solution, an aliquot portion of it was boiled with the alkali, and the filtrate set aside as above. The nitrogen in both the cases was estimated by the Crum-Frankland process, as also by the *Urea* method as worked out by Percy Frankland. As a further check a few c.c.'s were in certain instances evaporated to dryness in a porcelain boat and the nitrogen determined by Dumas' method. It is remarkable that the sum total of nitrogen as found by all these different methods was exactly the same, proving that *not only was there no loss of nitrogen during the reaction but that it remained all*

along in the shape of the nitrites of the respective metals, in other words, there was no change in the radical NO_2.

B. Dilute solutions of silver and mercurous nitrite, (*vide* Exp. 5 and 6 in the Table of Results of Analyses).

It is worthy of note that under such conditions of dilution no silver was precipitated.

C. Mercurous Nitrite and Sodium Nitrite.

In this case also the total amount of nitrogen remained constant, and in the shape of nitrites, the only difference being that the mercurous nitrite was completely transformed into mercuric nitrite with precipitation of mercury. In Exp. 7 a 6°/₀ solution of sodium nitrite was used. Sodium nitrite was, however, found to have scarcely any action on very dilute solutions of mercurous nitrite.

DISCUSSION OF RESULTS.

It is not easy to enter into the mechanism of the reaction of mercurous and silver nitrites, when it is remembered that there is no change in the radical NO_2. Mercurous nitrite, it is true, has already been shown to undergo partial dissociation according to the equation,

$$Hg_2 (NO_2)_2 = Hg (NO_2)_2 + Hg.$$

when in solution ; but the reaction, we are at present studying, can scarcely be brought under the same category. At the same time, it must be admitted that, if we were to regard for a moment a molecule of silver nitrite playing the role of a molecule of mercurous nitrite, all the equations under A. could be established on a common basis.

For instance in Exp. 1, 3 $Hg NO_2 + Ag NO_2$ may be regarded as equivalent to 4 $Hg NO_2$, *i.e.*, 2 $Hg_2 (NO_2)_2$ which may be expected to dissociate as follows :—

2 $Hg_2 (NO_2)_2 = 2 Hg (NO_2)_2 + 2 Hg$ [or $Hg + Ag$]. In Exp. 3, 7 $Hg NO_2 + 3 Ag NO_2$ would similarly be equivalent to 10 $Hg NO_2$ *i.e.*, 5 $Hg_2 (NO_2)_2$ which would dissociate thus : 5 $Hg_2 (NO_2)_2 = 5 Hg (NO_2)_2 + 5 Hg$ [or 2 $Hg + 3 Ag$], with this difference, that in place of 2 Hg we get $Hg + Ag$, and in that of 5 Hg we get 2 $Hg + 3 Ag$. In Exp. 5 and 6 bracketted together under B., where dilute solutions of both the nitrites were used, and where there was no precipitate of metallic silver, the nitrite of silver apparently seems to take no part. The same remarks would also apply to Exp. 7 and 8 (*vide* Table of Analyses), where sodium nitrite also appear to act *catalytically*, an expression conveniently used to cover ignorance. The true explanation of the reaction has yet to be found out, and with this view, it is intended to take up another series of experiments under various degrees of dilution of the nitrites.

Table of Results of Analyses.

No of Exp.	C.C. of Mercury Sol. used.	Wt. of ous Mercury in Sol. in grams.	Wt. of, ic, Mercury in Sol.	C.C. of $AgNO_2$ Sol. used.	Wt. of Ag in Sol.	Wt. of ous Mercury after addition of $AgNO_2$ Sol.	Wt. of, ic, Mercury in ditto.	Wt. of "free," i.e., pptd. Mercury.	Wt. of "free" Silver.	Reaction as Expressed in Formula.
(1) A	50	1·074	0·863	50	0·304	0·564	1·205	0·168	0·104	$6 Hg'NO_2 + 5 Hg''(NO_2)_2 + 3 Ag$ $NO_2 = 3 Hg'NO_2 + 7Hg''(NO_2)_2 + 2 AgNO_2 + Hg + Ag$. or $3 HgNO_2 + Ag NO_2 = 2 Hg(NO_2)_2 + Hg + Ag$.
(2)	50	0·685	0·676	25	0·233	0·257	0·973	0·130	0·117	$5 Hg'NO_2 + 5 Hg''(NO_2)_2 + 2 Ag NO_2 = 2Hg'NO_2 + 7 Hg''(NO_2)_2 + Ag + Ag$. or $3 Hg(NO_2)_2 + 2 Hg$ $(N_2)_2 + Hg$.
(3)	25	0·173	0·188	20	0·063	0·053	0·274	0·034	0·027	$10 Hg'NO_2 + 11 Hg''(NO_2)_2 + 7$ $Ag NO_2 = 3 Hg'NO_2 + 16Hg''$ $(NO_2)_2 + 4 AgNO_2 + 2Hg + 3Ag$. or $7 HgNO_2 + 3 AgNO_2 = 5 Hg(NO_2)_2 + 2 Hg + 3 Ag$.
(4)	100	0·564	0·569	50	0·0523	0·488	0·614	0·031	0·0105	$18 Hg'NO_2 + 18 Hg''(NO_2)_2 + 5$ $AgNO_2 = 15 Hg''(NO_2)_2 + 20 Hg''(NO_2)_2 + 4 AgNO_2 + Hg + Ag$. or $3 Hg'NO_2 + Ag NO_2 = 2 Hg(NO_2)_2 + Hg + Ag$.
(5) B	100	0·3454	0·382	50	0·076	0·2284	0·4464	0·057	...	$6Hg'NO_2 + 7 Hg''(NO_2)_2 + AgNO_2 = 4 Hg''NO_2 + 8 Hg''(NO_2)_2 + AgNO_2 + Hg$. or $2 HgNO_2$, i.e., $Hg(NO_2)_2 = Hg(NO_2)_2 + Hg$.
(6)	100	0·392	0·486	50	0·042	0·345	0·507	0·026	...	$15 Hg'NO_2 + 19 Hg''(NO_2)_2 + 9Ag NO_2 = 13 Hg''NO_2 + 20Hg''(NO_2)_2 + AgNO_2 + Hg$. or $2 HgNO_2$, i.e., $Hg(NO_2)_2 = Hg (NO_2)_2 + Hg$.
(7) C	25	0·172	0·242	Na No₂ in c c, 20	Wt. of NaNo₂ in Sol. 1·20	...	0·333	0·081	...	$2 Hg'NO_2 + 3 Hg''(NO_2)_2 + Na$ $NO_2 = 4 Hg''(NO_2)_2 + Na + Hg$. or $2 Hg NO_2$, i.e., $Hg(NO_2)_2 = Hg(N NO_2 + Hg$.
(8)	25	0·700	0·620	20	...	...	0·980	0·34	...	$2 Hg'NO_2 + 2 Hg''(NO_2)_2 + Na$ $NO_2 = 3 Hg''(NO_2)_2 + Na + Na NO_2 + Hg$. or $2 HgNO_2$ i.e., $Hg(NO_2)_2 = Hg(NO_2)_2 + Hg$.

On Mercurous Iodide—*A new Method of its Preparation.*

The yellow residue in the flask (see previous paper) consisting presumably of a mixture of mercurous iodide and the unacted-upon mercurous nitrite was well powdered and introduced into a combustion tube, plugged with asbestos and heated in a tube-heater (Röhren-Oefen.). The powder occupied nearly one-third the length of the tube. When the temperature rose to about 135°C, nitrous fumes began to be disengaged, and an oily liquid collected at the mouth of the tube. This liquid is nitro-ethane, a portion of which obstinately remains absorbed in the hard mass of the mixture referred to above.

On heating more than two hours from 155° to 163°, for the most part stationary at the latter temperature, a thin deposit of lustrous lemon-yellow scales was obtained. The yield however was very poor. Next day the heat was raised to 192°, and the temperature maintained nearly constant for three hours : a sublimate of a compact mass of yellow and orange-yellow crystals was the result.

In another experiment the sublimation was carried on between 190°–210° C. stationary for the most part at 210°; in this case orange-yellow crystals were obtained. In several experiments, however, conducted within the above range of temperatures, the sublimate which was deposited nearest the source of heat was of a dark brown tint; next to it was a deposit of orange-yellow and yellow tablets respectively; and, last of all, near the mouth of the tube was a ring of scarlet crystals of *mercuric* iodide. Sometimes it so happened that by far the larger proportion of the sublimate was of scarlet mercuric iodide ; but whether this was due to the decomposition or dissociation of mercurous iodide formed at first $(Hg_2I_2 = HgI_2 + Hg)$ or not is not clear. More than a dozen experiments were carried on, and the experiences accruing therefrom are recorded above.

General properties :—In Yvon's[*] experiment in which Mercurous Iodide was prepared by the direct union of the elements, only the yellow and orange-yellow crystals are described. According to this chemist sublimation begins at 190° C. My own experience confirms his in the main, though I have noticed that a small quantity of mercurous iodide almost always sublimes between 163° and 170° C.

The dark brown variety when powdered and kept in contact with dilute nitric acid turns dirty yellow, and the orange-yellow under similar conditions orange-red, without undergoing change in the composition ; but boiled for some time with the dilute acid, both these varieties are gradually transformed into *mercuric* iodide, and from the

* Compt. Rend. 76, p. 1607.

hot mother liquor also bright scarlet spangles of mercuric iodide crystallize out on cooling.

Result of Analysis :—Estimation of iodine. The methods described in the standard works of Fresenius, Crookes, &c., for the estimation of iodine in mercurous iodide appeared to be tedious and troublesome in view of the numerous determinations involved. Reduction was effected with zinc and sulphuric acid under certain modifications. The compound is finely powdered and transferred to a flask. Pure granulated zinc and dilute sulphuric acid are then added together with a few pieces of scrap platinum, when evolution of hydrogen at once sets up; the flask is kept actively rotated. After a few minutes the zinc becomes amalgamated and further action ceases. A drop of platinic chloride is now added, reduction proceeds, and a pink colour pervades the liquid, and the flask is shaken as before. When the reduction is complete the solution should be perfectly clear and colourless, and there should not be any trace of a powdery black residue at the bottom of the flask. The iodine is estimated in the usual way, and the additional halogen introduced with the standard $Pt.Cl_4$ drop, corresponding in the present instances to 0·005 AgCl, allowed for. Sometimes the zinc is treated beforehand with the dilute acid and a few drops of $PtCl_4$, and the *platinized* zinc, washed free from chlorine, is added in successive instalments to bring about reduction, which is finished in 20 min. to half an hour.

As control experiments, iodine in resublimed mercuric iodide was estimated according to the above method. Thus:

(1)	0·2232 g.	gave	0 2324	AgI
(2)	0·0604 ,,	,,	0 0625	,,

Calc. for HgI_2 Found

	I	II
I: 55·95	56·27	56·00

Result of analyses of the dark brown, orange-yellow and yellow modifications of mercurous iodide.

(1)	0·4282 g.	gave	0·3043 AgI	whence	I	= 38·40
(2)	0·2204 ,,	,,	0·1577 ,,	,,	,,	= 38·67
(3)	0·2558 ,,	,,	0 1834 ,,	,,	,,	= 38·74
(4)	0 2038 ,,	,,	0·1446 ,,	,,	,,	= 38·34
(5)	0·0824 ,,	,,	0·0587 ,,	,,	,,	= 38·47

Theory for HgI requires I=38·84 %, the Mercury amounted to 61·24 %, that demanded by theory being 61·16 %.

It will much facilitate operation if between 0·2 to 0·15 g. be taken for purposes of analyses. In the case of the dark brown variety of the haloid it is advisable to examine it carefully with a magnifying glass as inside its thick crust minute globules of mercury are often found enclosed.

From the foregoing inquiry it is evident that when the residue in the flask after the interaction of mercurous nitrite and ethyl iodide is heated in a tube between 190°–210°, mercurous iodide sublimes off. The compact mass of crystalline tablets thus obtained varies in all gradations of tint from lemon-yellow and orange-yellow to orange-brown and even dark brown.

Chemical Laboratory, Presidency College.

XIX.—*Description of a new Himalayan genus of* Orobanchaceæ.— *By* J. S. GAMBLE, M.A., F.R.S., *and* D. PRAIN.

[Received 7th August; Read 5th October, 1900.]

GLEADOVIA Gamble & Prain.

Calyx tubulosus, parum inflatus limbo æqualiter 5-lobo. *Corollæ* tubus parum incurvus, labium posticum incurvo-erectum concavum mi-nopere emarginatum, anticum brevius suberectum lobis 3 subæqualibus erectis. *Stamina* inclusa filamentis apice in connectivum conicum dila-tatis, antherarum loculi æquales adnati basi divergentes et mucronato-aristati. *Ovarii* placentæ 4, per paria approximatæ, medioque con-fluentes; stigma dilatatum late æqualiter 2-lobum.—*Herba* parasitica carnosa rhizomate incrassato, squamis ovatis suffulta. *Flores* densius paniculati, pedicellati, 2-bracteolati. *Color* pallide purpurea.—*Species* singula, Himalaica.

The interesting plant for which we propose the above generic des-cription was discovered in Jaunsar in 1898 by the officers of the Imperial Forest School, Dehra Dun; we dedicate it to Mr. F. Gleadow, who was the first actually to find it.

Our plant has all the facies of a *Christisonia*, but cannot be referred to that genus because both anther-cells are perfect, because the corolla is very markedly 2-labiate in place of being sub-equally 5-lobed, and because the two stigmatic lobes are equally large.

The nearest ally of our plant seems to be the American genus *Conopholis* Wallr., with which it agrees as regards corolla and, except that they are not exserted, as regards stamens, but from which it differs in having an equally 5-lobed calyx and a 2-lobed stigma.   From

Boschniackia C. A. Mey., it differs somewhat as regards corolla and very greatly as regards stamens. From *Xylanche* Beck, (*Boschniackia himalaica* H. f. & T.) it further differs in having 2 carpels, not 3. From all the genera mentioned it differs markedly as regards inflorescence, which in those is spicate, in our plant paniculate.

GLEADOVIA RUBORUM *Gamble & Prain.*—A fleshy herb about 6 in. high of which only about one half epigaeal; *root-stock* very thick especially where attached to the host; *scales* ovate, the lower rounded, the upper acute sometimes 2-fid. *Flowers* paniculate; bract solitary, ·7 in. long, sheathing, rounded, pedicel stout ·35 in. long, bracteoles 2, ·7—1 in. long, spathulate, acute, concave. *Calyx* light-red, tubular, somewhat inflated, regularly 5-lobed, 1—1·2 in. long, lobes pale. *Corolla* red with darker veins, tube as long as calyx, slightly curved, distinctly two-lipped; upper lip of 2 connate lobes, rounded, slightly dentate, lower of 3 narrow, spathulate, subequal, acutely dentate lobes. *Stamens* 4, geniculate at point of insertion, anthers elongate, spurred, connective produced in a 2-fid cone, hairy above. *Ovary* 1-celled, ovate-cylindric; style long, incurved at apex; stigma of 2 broad semi-orbicular lobes depressed in the centre; placentæ 2 pairs, free below and above, confluent in the middle, diffuse; ovules very many. *Seeds* very many, minute.

N. W. HIMALAYA :—Bodyar Jaunsar, 8–9,000 ft. ; on the northern slopes in very shady woods of Fir and Deodar on roots of wild Raspberry (*Rubus niveus*); very scarce, *Gleadow ! Gamble ! Duthie ! Duthie's Collectors !*

INDEX.

Names of New Genera and Species have an asterisk (*) prefixed.

INDEX SLIP.

ZOOLOGY.   **N.**

MASSON, W. P.—Note on four Mammals from the neighbourhood of
Darjeeling. Journ. As. Soc. Bengal, LXIX, Pt. ii, 1900, pp.
89-90.

[Mammalia. Carnivora. Ursidæ.]

Ursus malayanus (Tree bear).
Ursus tibetanus (Ground bear).
Occurrence of, in the Darjeeling District.

[Mammalia. Rodentia. Hystricidæ.]

Atherura macrura (Asiatic brush-tailed porcupine).
Occurrence of, in the Darjeeling District.

[Mammalia. Ungulata. Bovidæ.]

Nemorhædus bubalina (goat-antelope or serow).
Cemas goral (goral).
Occurrence of, in the Darjeeling and Sikkim Districts.
Darjeeling and Sikkim Districts, occurrence of tree and ground
bears, Asiatic brush-tailed porcupine, goat-antelope and goral.

MATHEMATICS.   ·   **A.**

DUTT, PROMOTHONATH.—On a new method of treating the properties
of the circle and analogous matters. Journ. As. Soc. Bengal,
LXIX, Pt. ii, 1900, pp. 91-97.
The circle.
New method of treating the properties of.

PHYSICS.   **C.**

DUTT, PROMOTHONATH.—Experimental measurement of the velocity
of sound from observations in a railway train. Journ. As. Soc.
Bengal, LXIX, Pt. ii, 1900, pp. 96-97.
Velocity of sound.
Measurement of, from observations in a railway train.

CHEMISTRY.

MUKERJI, P.—Note on a method of detecting free Phosphorus.
Journ. As. Soc. Bengal, LXIX, Pt. ii, 1900, pp. 97–101.

Phosphorus (free).

Method of detecting.

ZOOLOGY. N.

WALTON, H. J.—Note on the occurrence of Rhodospiza obsoleta
(Licht.) in the Tochi Valley. Journ. As. Soc. Bengal, LXIX,
Pt. ii, 1900, pp. 101–102.

[Aves. Passeres. Fringillidæ.]

Rhodospiza obsoleta (Desert Rose-Finch).

Occurrence of, in the Tochi Valley.

Tochi Valley, occurrence of the Desert Rose-Finch.

ZOOLOGY. N.

BINGHAM, C. T. and THOMPSON, H. N.—On the Birds collected and
observed in the Southern Shan States of Upper Burma. Journ.
As. Soc. Bengal, LXIX, Pt. ii, 1900, pp. 102–142.

Birds, a collection of, from Southern Shan States.

Southern Shan States, a collection of birds from.

ZOOLOGY. N.

FINN, F.—On the form of Cormorant inhabiting the Crozette Islands.
Journ. As. Soc. Bengal, LXIX, Pt. ii, 1900, p. 143.

[Aves. Steganopodes. Phalacrocoracidæ.]

Phalacrocorax verrucosus ? or

Phalacrocorax melanogenys (Blyth).

Occurrence of, in the Crozette Islands.

Crozette Islands, cormorant inhabiting the.

ZOOLOGY. N.

FINN, F. and TURNER, H. H.—On two rare Indian Pheasants.
Journ. As. Soc. Bengal, LXIX, Pt. ii, 1900, pp. 144–146.

[Aves. Gallinæ. Phasianidæ.]

Phasianus humiæ.

Gennæus davisoni ?

Occurrence of, in the Chin Hills.

Chin Hills, occurrence of two rare Indian Pheasants.

ZOOLOGY.   N.

Finn, F.—Notes on the structure and function of the tracheal bulb
in male Anatidæ. Journ. As. Soc. Bengal, LXIX, Pt. ii, 1900,
pp. 147-149.

[Aves.   Anseres.   Anatidæ]

Nettopus coromandelianus.

Absence of tracheal bulb in the male ...   ...   p. 147
Aix galericulata, Cairina moschata × Anas boschas, and
Casarca rutila.

Presence of tracheal bulb in the male ...   . p. 148
Tracheal bulb.

Function of, in drakes   ...   .   p. 148
Difference in size of, influences the voice of drakes   p. 149

ZOOLOGY.   N.

Nicéville, L. de.—Note on Calinaga, an aberrant genus of Asiatic
Butterflies. Journ. As. Soc. Bengal, LXIX, Pt. ii, 1900, pp.
150-155.

[Hexapoda.   Rhopalocera.   Nymphalidæ.]

Calinaga.

Various positions in the Butterfly order.
Description of foreleg.
Geographical range of.
All species probably mimetic.

ZOOLOGY.   N.

Walton, H. J.—Notes on birds collected in Kumaon. Journ. As.
Soc. Bengal, LXIX, Pt. ii, 1900, pp. 155-168.
Birds, a collection of, from Kumaon.
Kumaon, a collection of birds from.

BOTANY.   M.

Prain, D.—Some new plants from Eastern India. Journ. As. Soc.
Bengal, LXIX, Pt. ii, 1900, pp. 168-174 [in part (see also
separate slip King and Prain).]

Tiliaceæ, new, from Eastern India.   Labiatæ, new, from Eastern India.
  Grewia (Eugrewia) nagensium   *Gomphostemma (Pogosiphon)*
    (p. 168).   *inopinatum* (p. 172).

n. spp.

Eastern India, new plants from

BOTANY. M.

KING, SIR GEORGE and PRAIN, D.—Some new plants from Eastern India. Journ. As. Soc. Bengal, LXIX, Pt. ii. 1900, pp. 168–174, [in part (see also separate slip, Prain)].

n. spp.

Eastern India, new plants from

BOTANY. M.

PRAIN, D.—A list of the Asiatic species of Ormosia. Journ. As. Soc. Bengal, LXIX, Pt. ii, 1900, pp. 175–186.

Leguminosæ, new, from China.

Ormosia Henryi, n. sp. (p. 180) ; *O. yunnanensis*, n. sp. (p. 183).

Leguminosæ, new, from Kachin Hills.

Ormosia inopinata, n. sp. (p. 181) ; *O. inopinata*, var. *typica* (nov.) (p. 181) ; *O. inopinata*, var. *dubia* (nov.) (p. 181) ; *O. laxa*, n. sp. (p. 182).

Ormosia polita, n. n. for *O. nitida*, Prain nec Vogel (p. 184).

China and Kachin Hills, new plants from.

ZOOLOGY.   N.

NICÉVILLE, L. DE.—The Food-plants of the Butterflies of the Kanara District of the Bombay Presidency, with a revision of the species of Butterflies there occuring. Journ. As. Soc Bengal, LXIX, Pt. ii, 1900, pp. 187-277.

[Hexapoda. Rhopalocera.]

Food-plants, list of Butterflies in the Kanara District   pp. 193-215
Butterflies, list of in the Kanara District   ...   pp. 215-277
Kanara District, list of Butterflies and their food-plants.

ZOOLOGY.   N.

NICÉVILLE, L. DE.—Butterflies from the Western Himalayas and Kashmir, with the food-plants of their larvæ. Journ. As. Soc. Bengal, LXIX, Pt. ii 1900, pp. 214-215, 277-278.

[Hexapoda. Rhopalocera.]

Food-plants of larvæ.
Of Butterflies of Western Himalayas and Kashmir.
Western Himalayas and Kashmir, butterflies and larvæ, food-plants.

ZOOLOGY.   N.

NICÉVILLE, L. DE.—Note on the avian genus Harpactes, Swainson. Journ. As. Soc. Bengal, LXIX, Pt. ii, 1900, p. 278.

[Aves. Trogones. Trogonidæ.]

Harpactes, Swainson—
Superseded by Pyrotrogon, Bonaparte (vide authors quoted).
Pyrotrogon, Bonaparte—
Supersedes Harpactes, Swainson (vide authors quoted).

BOTANY.   M.

KING, SIR GEORGE.—Materials for a Flora of the Malayan Peninsula.
No. 11. Journ. As. Soc. Bengal, LXIX, Pt. ii, 1900, pp. 1–18 and
44–87.

Melastomaceæ, new, from Malayan Peninsula:—

Melastoma malabathricum var. *perakensis* (nov.) (p. 7), *Oxyspora
stellulata* (p. 9), *O. acutangula* (p. 9), *O. Curtisii* (p. 9), *Allomorphia* exigua var. *minor* (nov.) (p. 11), *A. Wrayi* (p. 11), *A. alata*
(p. 12), *Ochthocharis decumbens* (p. 15), *Anerincleistus macranthus*
(p. 15), *A. Scortechinii* (p. 16), *A. floribundus* (p. 17), *A. sublepidotus* (p. 17), *A. glomeratus* (p. 18), *Phyllagathis tuberculata*
(p. 44), *P. Scortechinii* (p. 45), *P. hispida* (p. 46), *Dissochæta
anomala* (p. 55), *D. Scortechinii* (p. 55), *Anplectrum lepidoto-
setosum* (p. 56), *A. anomalum* (p. 58), *Medinilla scandens* (p. 60),
M. heteranthera (p. 61) and var. nov. *latifolia* (p. 61), *M. venusta*
(p. 61), *M. Scortechinii* (p. 62), *M. Clarkei* (p. 63), *M. perakensis*
(p. 64), *Pterandra Griffithii* (p. 70), *Memecylon epiphyticum*
(p. 74), *M. fruticosum* (p. 74), *M. Kunstleri* (p. 76), *M. Hullettii*
(p. 76), *M.* heteropleurum var. *olivacea* (nov.) (p. 78),
M. cinereum (p. 82), *M. andamanicum* (p. 85), n. spp.

[See also Stapf and King, on separate slip.]

Malayan Peninsula, new species of Melastomaceæ.

BOTANY.   M.

STAPF, O. and KING, SIR GEORGE.—Materials for a Flora of the Malayan
Peninsula. No. 11. Journ. As. Soc. Bengal, LXIX, Pt. ii, 1900, pp.
18–44.

Melastomaceæ, new, from Malayan Peninsula:—

Sonerila epilobioides (p. 22), *S. calaminthifolia* (p. 23), *S. hyssopi-
folia* (p. 23), *S.* erecta, var. *flexuosa* (nov.) p. 24 and var. *discolor*
(nov.) p. 24, *S.* tenuifolia, var. *hirsuta* (nov.) (p. 25),
S. flaccida (p. 25), *S. andamanensis* (p. 26), *S. populifolia*
(p. 26), *S. pallida* (p. 27), *S. rudis* (p. 27), *S. mollis* (p. 28),
S. albiflora (p. 28), *S. lasiantha* (p. 29), *S. suffruticosa* (p. 29),
S. elliptica (p. 30), *S. succulenta* (p. 30), *S. repens* (p. 30),
S. muscicola (p. 31), *S. saxosa* (p. 31), *S. congesta* (p. 32),
S. Cyclaminella (p. 33) and var. *canescens* (nov.) p. 33, *S* integrifolia,
var. *acuminatissima* (nov.) p. 35, *S. bracteata* (p. 35), *S. capitata*
(p. 35), *S. caesia* (p. 36), *S. Nidularia* (p. 37), *S. brachyantha*
(p. 37), *S. microcarpa* (p. 38), *S. costulata* (p. 39), *S. macrophylla*
(p. 39) and var. *laxipilosa* (nov.) (p. 40), *S. globriflora* (p. 42),
S. elatostemoides (p. 42), *S. bicolor* (p. 43), *S. Calycula* (p. 43),
n. spp.

[See also King, separate slip.]

Malayan Peninsula, new species of Melastomaceæ.

ALCOCK, A.—Materials for a Carcinological Fauna of India. No. 6.
The Brachyura Catometopa, or Grapsoidea. Journ. As. Soc.
Bengal, LXIX, Pt. ii, 1900, pp. 279–456.

[Arthropoda. Crustacea. Catometopa.]

Catometopa.

(Restr.), defined ... ... p. 281
Divided into nine families, which are defined ... pp. 283–286
New family of, *Ptenoplacidæ* ... ... p. 285
New subfamilies of, *Pseudorhombilinæ* (p. 286); *Prionoplacinæ*
(p. 286); *Pinnoterinæ* (p. 293); *Pinnotherelinæ* (p. 294); *Xenoph-
thalminæ* (p. 294); *Varuninæ* (p. 296).

Gonoplacidæ, new, from Indian Seas.

Libystes Edwardsi (p. 306) Persian Gulf, Andamans; *L. Alphonsi*
(p. 306) Andamans; *Psopheticus insignis* (p. 310) Gulf of Martaban, 60 and 67 fms.; *Litochira angustifrons* (p. 315) Bombay
and Karachi; *L. Beaumontii* (p. 315) Andamans, off Ceylon,
34 fms.; *Notonyx vitreus* (p. 319) Andaman Sea, 53 fms.; *Ceratoplax hispida* (p. 321) Palk Straits; *Typhlocarcinus rubidus*
(p. 323) Bay of Bengal, 20–65 fms.; *Lambdophallus sexpes* (p. 330)
Bay of Bengal, 65 fms.; n. spp.
Typhlocarcinodes (p. 326); *Lambdophallus* (p. 329) n. genera.

Pinnoteridæ, new, from Indian waters.

Chasmocarcinops (p. 334) n. g. *C. gelasimoides* (p. 334) off
Madras, 12 fms.; *Pinnoteres purpureus* (p. 339) Andamans;
P. mactricola (p. 339) mouth of R. Hooghly, n. spp.

Ocypodidæ, new, from Indian Seas.

Gelasimus inversus, var. *sindensis* (var. nov.) (p. 356) Karachi;
Dotilla affinis (p. 365) Aden and Baluchistan Coast; *D. Blanfordi*
(p. 366) Coast of Sind and Baluchistan; *D. clepsydrodactylus*
(p. 367) Mahanaddi Delta; *Scopimera investigatoris* (p. 369) off
Cape Negrais, Burma; *S. crabricauda* (p. 370) Karachi; *Olisto-
stoma dotilliforme* (p. 373) Karachi; *Tylodiplax indica* (p. 374)
Karachi.

Hymenosomidæ, new, from Indian Seas.

Elamena sindensis (p. 386) Karachi; *Hymenicus Wood-Masoni*
(p. 388) Andamans and near Calcutta; *H. inachoides* (p. 388)
near Calcutta.

Grapsidæ, new, from Indian waters.

Ptychognathus onyx (p. 404) Tavoy; *P. andamanica* (p. 404)
N. Andaman I.; *Pyxidognathus fluviatilis* (p. 408) R. Ichamutty,
Jessore District; *Sesarma lanatum* (p. 418) Bombay and Karachi;
S. latifemur (p. 421) Andamans; *S. Finni* (p. 424) Andamans.

Palicidæ, new, from Indian Seas.

Palicus Wood-Masoni (p. 453) Andamans; *P. investigatoris* (p. 455)
Andamans.

JOURNAL

OF THE

ASIATIC SOCIETY OF BENGAL,

Vol. LXIX. Part II, No. 1.—1900.

EDITED BY

LIONEL DE NICÉVILLE, ESQ., F.E.S., C.M.Z.S., &c.

HON. NATURAL HISTORY SECRETARY.

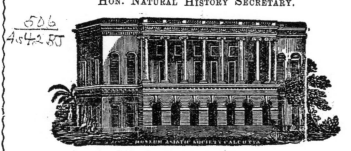

"The bounds of its investigation will be the geographical limits of Asia: and within these limits its inquiries will be extended to whatever is performed by man or produced by nature."—SIR WILLIAM JONES.

** *Communications should be sent under cover to the Secretaries, Asiat. Soc., to whom all orders for the work are to be addressed in India; or care of Messrs. Luzac & Co., 46, Great Russell Street, London, W. C., or Mr. Otto Harrassowitz, Leipzig, Germany.*

CALCUTTA:

PRINTED AT THE BAPTIST MISSION PRESS,

AND PUBLISHED BY THE

ASIATIC SOCIETY, 57, PARK STREET.

1900.

Price (exclusive of postage) to Members, Re. 1-8.—To Non-Members, Rs. 2.
Price in England 8 Shillings.
Issued July 9th, 1900.

JOURNAL
OF THE
ASIATIC SOCIETY OF BENGAL,
Vol. LXIX, Part II, No. 3.—1900.

EDITED BY

THE NATURAL HISTORY SECRETARY.

CALCUTTA:

PRINTED AT THE BAPTIST MISSION PRESS,

AND PUBLISHED BY THE

ASIATIC SOCIETY, 57, PARK STREET.

1900.

Issued July 30th, 1900.

NOTE ON THE PUBLICATIONS

ASIATIC SOCIETY.

—•◦—

The *Proceedings* of the Asiatic Society are issued ten times a year as soon as possible after the General Meetings which are held on the first Wednesday in every month in the year except September and October; they contain an account of the meeting with some of the shorter and less important papers read at it, while only titles or short resumés of the longer papers, which are subsequently published in the *Journal*, are given.

The *Journal* consists of three entirely distinct and separate volumes : Part I, containing papers relating to Philology, Antiquities, etc.; Part II, containing papers relating to Physical Science; and Part III devoted to Anthropology, Ethnology, etc.

Each Part is issued in four or five numbers, and the whole forms three complete volumes corresponding to the year of publication.

The *Journal* of the Asiatic Society was commenced in the year 1832, previous to which the papers read before the Society were published in a quarto periodical, entitled Asiatic Researches, of which twenty volumes were issued between the years 1788 and 1839.

The *Journal* was published regularly, one volume corresponding to each year from 1832 to 1864; in that year the division into two parts above-mentioned was made and since that date two volumes have been issued regularly every year. From 1894 an additional volume, Part III, has been issued.

The *Proceedings* up to the year 1864, were bound up with the *Journal*, but since that date have been separately issued every year.

The following is a list of the Asiatic Society's publications relating to Physical Science, still in print, which can be obtained at the Society's ouse, No. 57, Park Street, Calcutta, or from the Society's Agents in ondon, Messrs. Luzac & Co., 46, Great Russell Street, W. C.; and from r. Otto Harrassowitz, Leipzig, Germany.

SIATIC RESEARCHES. Vols. VII, Vols. XI and XVII, and
              Vols. XIX and XX @ 10/ each   Rs. 50 0
ROCEEDINGS of the Asiatic Society from 1865 to 1869 (incl.) @ /6/ per No.; and from 1870 to date @ /8/ per No.
OURNAL of the Asiatic Society for 1843 (12), 1844 (12), 1845 (12), 1846 (5), 1847 (12), 1848 (12), 1850 (7), 1851 (7), 1857

JOURNAL

OF THE

ASIATIC SOCIETY OF BENGAL,

Vol. LXIX. Part II, No. 2.—1900.

EDITED BY

LIONEL DE NICÉVILLE, ESQ., F.E.S, C.M.Z.S., &c.

HON. NATURAL HISTORY SECRETARY.

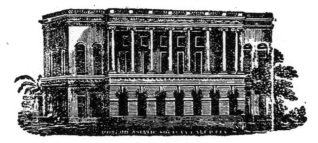

"The bounds of its investigation will be the geographical limits of Asia: and within these limits its inquiries will be extended to whatever is performed by man or produced by nature."—SIR WILLIAM JONES.

*** *Communications should be sent under cover to the Secretaries, Asiat. Soc., to whom all orders for the work are to be addressed in India; or care of Messrs. Luzac & Co., 46, Great Russell Street, London, W. C., or Mr. Otto Harrassowitz, Leipzig, Germany.*

CALCUTTA:

PRINTED AT THE BAPTIST MISSION PRESS,

AND PUBLISHED BY THE

ASIATIC SOCIETY, 57, PARK STREET.

1900.

JOURNAL
of the
ASIATIC SOCIETY OF BENGAL.
Vol. LXIX, Part II, No. 3.—1900.

CALCUTTA:
Printed by the Baptist Mission Press,
and published by the
Asiatic Society, 57, Park Street.
1900.

Issued October 30th, 1900.

NOTE ON THE PUBLICATIONS

ASIATIC SOCIETY.

—◦◦—

The *Proceedings* of the Asiatic Society are issued ten times a year as soon as possible after the General Meetings which are held on the first Wednesday in every month in the year except September and October; they contain an account of the meeting with some of the shorter and less important papers read at it, while only titles or short resumés of the longer papers, which are subsequently published in the *Journal*, are given.

The *Journal* consists of three entirely distinct and separate volumes: Part I, containing papers relating to Philology, Antiquities, etc.; Part II, containing papers relating to Physical Science; and Part III devoted to Anthropology, Ethnology, etc.

Each Part is issued in four or five numbers, and the whole forms three complete volumes corresponding to the year of publication.

The *Journal* of the Asiatic Society was commenced in the year 1832, previous to which the papers read before the Society were published in a quarto periodical, entitled Asiatic Researches, of which twenty volumes were issued between the years 1788 and 1839.

The *Journal* was published regularly, one volume corresponding to each year from 1832 to 1864; in that year the division into two parts above-mentioned was made and since that date two volumes have been issued regularly every year. From 1894 an additional volume, Part III, has been issued.

The *Proceedings* up to the year 1864, were bound up with the *Journal*, but since that date have been separately issued every year.

The following is a list of the Asiatic Society's publications relating to Physical Science, still in print, which can be obtained at the Society's House, No. 57, Park Street, Calcutta, or from the Society's Agents in London, Messrs. Luzac & Co., 46, Great Russell Street, W. C.; and from Mr. Otto Harrassowitz, Leipzig, Germany.

ASIATIC RESEARCHES. Vols. VII, Vols. XI and XVII, and
Vols. XIX and XX @ 10/ each   Rs.   50   0
PROCEEDINGS of the Asiatic Society from 1865 to 1869 (incl.) @
/6/ per No.; and from 1870 to date @ /8/ per No.
JOURNAL of the Asiatic Society for 1843 (12), 1844 (12), 1845
(12), 1846 (5), 1847 (12), 1848 (12), 1850 (7), 1851 (7), 1857

JOURNAL

OF THE

ASIATIC SOCIETY OF BENGAL,

Vol. LXIX. Part II, No. 3.—1900.

EDITED BY

LIONEL DE NICÉVILLE, ESQ., F.E.S , C.M.Z.S., &c.

HON. NATURAL HISTORY SECRETARY.

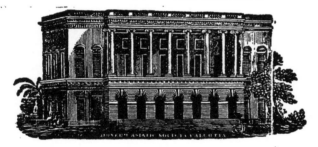

"The bounds of its investigation will be the geographical limits of Asia: and within these limits its inquiries will be extended to whatever is performed by man or produced by nature."—SIR WILLIAM JONES.

** *Communications should be sent under cover to the Secretaries, Asiat. Soc., to whom all orders for the work are to be ·addressed in India; or care of Messrs. Luzac & Co., 46, Great Russell Street, London, W. C., or Mr. Otto Harrassowitz, Leipzig, Germany.*

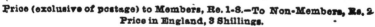

CALCUTTA:

PRINTED AT THE BAPTIST MISSION PRESS,

AND PUBLISHED BY THE

ASIATIC SOCIETY, 57, PARK STREET.

1900.

Price (exclusive of postage) to Members, Re. 1-8.—To Non-Members, Rs. 2
Price in England, 3 Shillings.

Issued November 22nd, 1900.

JOURNAL

OF THE

ASIATIC SOCIETY OF BENGAL,

Vol. LXIX, Part II, No. 3.—1900.

EDITED BY

THE HONORARY SECRETARIES.

CALCUTTA

PRINTED AT THE BAPTIST MISSION PRESS,

AND PUBLISHED BY THE

ASIATIC SOCIETY, 57, PARK STREET.

1900

NOTE ON THE PUBLICATIONS

ASIATIC SOCIETY.

The *Proceedings* of the Asiatic Society are issued ten times a year as soon as possible after the General Meetings which are held on the first Wednesday in every month in the year except September and October; they contain an account of the meeting with some of the shorter and less important papers read at it, while only titles or short resumés of the longer papers, which are subsequently published in the *Journal*, are given.

The *Journal* consists of three entirely distinct and separate volumes : Part I, containing papers relating to Philology, Antiquities, etc.; Part II, containing papers relating to Physical Science; and Part III devoted to Anthropology, Ethnology, etc.

Each Part is issued in four or five numbers, and the whole forms three complete volumes corresponding to the year of publication.

The *Journal* of the Asiatic Society was commenced in the year 1832, previous to which the papers read before the Society were published in a quarto periodical, entitled Asiatic Researches, of which twenty volumes were issued between the years 1788 and 1839.

The *Journal* was published regularly, one volume corresponding to each year from 1832 to 1864; in that year the division into two parts above-mentioned was made and since that date two volumes have been issued regularly every year. From 1894 an additional volume, Part III, has been issued.

The *Proceedings* up to the year 1864, were bound up with the *Journal*, but since that date have been separately issued every year.

The following is a list of the Asiatic Society's publications relating to Physical Science, still in print, which can be obtained at the Society's House, No. 57, Park Street, Calcutta, or from the Society's Agents in London, Messrs. Luzac & Co., 46, Great Russell Street, W. C.; and from Mr. Otto Harrassowitz, Leipzig, Germany.

ASIATIC RESEARCHES. Vols. VII, Vols. XI and **XVII**, and
          Vols. XIX and **XX** @ 10/ each   Rs. 50   0

PROCEEDINGS of the Asiatic Society from 1865 to 1869 (incl.) @ /6/ per No.; and from 1870 to date @ /8/ per No.

JOURNAL of the Asiatic Society for 1843 (12), 1844 (12), 1845 (12), 1846 (5), 1847 (12), 1848 (12), 1850 (7), 1851 (7), 1857

JOURNAL

OF THE

ASIATIC SOCIETY OF BENGAL,

Vol. LXIX. Part II, No. 4.—1900.

EDITED BY

LIONEL DE NICÉVILLE, ESQ., F.E.S., C.M.Z.S., &c.

HON. NATURAL HISTORY SECRETARY.

" The bounds of its investigation will be the geographical limits of Asia : and within these limits its inquiries will be extended to whatever is performed by man or produced by nature."—SIR WILLIAM JONES.

⁎ *Communications should be sent under cover to the Secretaries, Asiat. Soc , to whom all orders for the work are to be addressed in India; or care of Messrs. Luzac & Co., 46, Great Russell Street, London, W. C., or Mr. Otto Harrassowitz, Leipzig, Germany.*

CALCUTTA:

PRINTED AT THE BAPTIST MISSION PRESS,

AND PUBLISHED BY THE

ASIATIC SOCIETY, 57, PARK STREET.

1901.

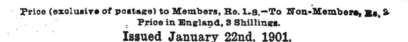

Price (exclusive of postage) to Members, Re. 1-8.—To Non-Members, Rs. 2.
Price in England, 3 Shillings.

Issued January 22nd, 1901.

NOTE ON THE PUBLICATIONS

OF THE

ASIATIC SOCIETY.

—•●•—

The *Proceedings* of the Asiatic Society are issued ten times a year as soon as possible after the General Meetings which are held on the first Wednesday in every month in the year except September and October; they contain an account of the meeting with some of the shorter and less important papers read at it, while only titles or short resumés of the longer papers, which are subsequently published in the *Journal*, are given.

The *Journal* consists of three entirely distinct and separate volumes : Part I, containing papers relating to Philology, Antiquities, etc.; Part II, containing papers relating to Physical Science; and Part III devoted to Anthropology, Ethnology, etc.

Each Part is issued in four or five numbers, and the whole forms three complete volumes corresponding to the year of publication.

The *Journal* of the Asiatic Society was commenced in the year 1832, previous to which the papers read before the Society were published in a quarto periodical, entitled Asiatic Researches, of which twenty volumes were issued between the years 1788 and 1839.

The *Journal* was published regularly, one volume corresponding to each year from 1832 to 1864; in that year the division into two parts above-mentioned was made, and since that date two volumes have been issued regularly every year. From 1894 an additional volume, Part III, has been issued.

The *Proceedings* up to the year 1864, were bound up with the *Journal*, but since that date have been separately issued every year.

The following is a list of the Asiatic Society's publications relating to Physical Science, still in print, which can be obtained at the Society's House, No. 57, Park Street, Calcutta, or from the Society's Agents in London, Messrs. Luzac & Co., 46, Great Russell Street, W. C.; and from Mr. Otto Harrassowitz, Leipzig, Germany.

ASIATIC RESEARCHES. Vols. VII, Vols. XI and XVII, and
Vols. XIX and XX @ 10/ each   Rs. 50 0
PROCEEDINGS of the Asiatic Society from 1865 to 1869 (incl.) @
/6/ per No.; and from 1870 to date @ /8/ per No.
JOURNAL of the Asiatic Society for 1843 (12), 1844 (12), 1845
(2), 1846 (5), 1847 (12), 1848 (12), 1850 (7), 1851 (7), 1857 .

Lightning Source UK Ltd.
Milton Keynes UK
UKHW020815241218
334505UK00011B/940/P

9 781332 302918